Polysaccharide Based Films for Food Packaging: Fundamentals, Properties and Applications

Tawheed Amin • H. R. Naik •
Syed Zameer Hussain • Sajad Mohd Wani
Editors

Polysaccharide Based Films for Food Packaging: Fundamentals, Properties and Applications

Springer

Editors
Tawheed Amin
Sher-e-Kashmir University of Agricultural Sciences and Technology of Kashmir
Shalimar, Srinagar,
Jammu and Kashmir, India

H. R. Naik
Islamic University of Science and Technology
Awantipora, Jammu and Kashmir, India

Syed Zameer Hussain
Sher-e-Kashmir University of Agricultural Sciences and Technology of Kashmir
Shalimar, Srinagar,
Jammu and Kashmir, India

Sajad Mohd Wani
Sher-e-Kashmir University of Agricultural Sciences and Technology of Kashmir
Shalimar, Srinagar, Jammu and Kashmir, India

ISBN 978-981-99-4897-0 ISBN 978-981-99-4898-7 (eBook)
https://doi.org/10.1007/978-981-99-4898-7

This Springer imprint is published by the registered company Springer Nature Singapore Pte Ltd.
The registered company address is: 152 Beach Road, #21-01/04 Gateway East, Singapore 189721, Singapore

Preface

This book presents the fundamentals of various types of polysaccharide-based films for food packaging besides discussing their properties and applications. The newest generation of the films being designed to allow incorporation and controlled release of specific additives are also included in this book. Covering the latest novelties in research conducted in the field of packaging, it considers state-of-the-art innovations in films. This book brings together the vast wealth of scientific knowledge in a systematically organized volume. Food packaging has revolutionized the world and back to the stage, when the packaging was in its infancy, it was not a throwaway layer but has largely become the beautiful exterior to a valuable product. However, with the passage of time, it was felt that apart from being an exterior of the product, the packaging has to perform other functions in that it should be robust in addition to being able to keep food products fresh for longer periods of time. Owing to their low price and favorable properties, petroleum-based polymers and plastics have increasingly found their place in the food packaging marketplace. However, these polymers are biodegradable and are obtained from non-renewable sources. Besides, the environmentalists around the globe are prompting to minimize the use of these plastic-based polymers since they end up in filling waters and landfills thus resulting in health hazard, not only to the environment but also to humans.

Keeping into consideration the importance of packaging, it has become an inseparable part of the food industry and thus it has to be there with the food products. It thus becomes quite imperative to find the alternatives and many such alternatives have been tried, most of which have proven successful. Polysaccharide-, lipid-, and protein-based films could be developed which are safe, biodegradable, and edible. However, in this book, only polysaccharide-based films will be discussed.

Conclusively, the book covers almost all types of polysaccharide-based films, compiled in a systematic manner which can provide a reader both fundamentals and future prospects. The chapters in this book have been written by renowned scientists from across the globe. The book will give a thorough understanding to the researchers working in various R&D sectors since it covers the present and future research prospects.

Jammu and Kashmir, India — Tawheed Amin
Jammu and Kashmir, India — H. R. Naik
Jammu and Kashmir, India — Syed Zameer Hussain
Jammu and Kashmir, India — Sajad Mohd Wani

Contents

About the Editors

Tawheed Amin is Assistant Professor-cum-Junior Scientist (Food Science and Technology) in SKUAST-Kashmir. He did his PhD Food Technology from SKUAST-Kashmir, Jammu and Kashmir. He did his M.Tech Food Technology from Amity University, Noida, Uttar Pradesh, and B.Tech Food Technology from Department of Food Technology, IUST-Kashmir. He is recipient of two (02) Gold Medals for being topper in B.Tech Food Technology and Ph.D Food Technology and has also received First Rank Holder Certificate in M.Tech Food Technology. He has attended several national and international conferences. He has to his credit several national and international papers and popular articles. He has also to his credit five (05) books published by publishers of national and international repute.

H. R. Naik is presently working as Director of Planning & Monitoring, SKUAST-Kashmir. He has an experience of over 30 years in the field of. He has also served as the Head, Division of Food Science and Technology, SKUAST-Kashmir. He has been the Student's Welfare Officer and Placement Officer at SKUAST-Kashmir. He has done his PhD Food Technology from Punjab Agricultural University, Ludhiana, MSc Food Technology from CFTRI, Mysore, and BSc Agriculture from GNDU, Amritsar. He is the recipient of Gold Medal for being the topper in BSc Agriculture and has received Merit Fellowship in MSc and PhD programs.

Syed Zameer Hussain is Professor of Food Science & Technology, SKUAST-Kashmir. He is also Principal Investigator of All India Coordinated Research Project on Post-Harvest Engineering and Technology (AICRP on PHET), Srinagar Centre. He has several publications in peer-reviewed national and international journals. He has been nominated as a finalist for best research paper award by American Association of Cereal Chemistry (AACC) in 2014. He has developed many technologies and a value chain for walnuts. Most of his technologies have been commercialized and has filed six patents as well. He has successfully guided more than 15 postgraduate students as major advisor and is handling many externally funded projects.

Sajad Mohd Wani is Associate Professor-cum-Senior Scientist in SKUAST-Kashmir, Jammu and Kashmir, India. He has served as Assistant Professor at the University of Kashmir for more than a decade. He completed his BSc in Agriculture and MSc in Post-Harvest-Technology from SKUAST-Kashmir and his PhD in Food Technology from the University of Kashmir. He has qualified CSIR-NET in Life Sciences and ICAR-ASRB Net in Horticulture and Food Science and Technology. He has acted as Principal Investigator of four projects and Co-Principal Investigator of ten major research projects. He has been awarded an International Travel Grant by SERB-DST to present a paper in EFFoST held in Greece in 2015. He received the best paper award from AFST(I) in 2016.

Chapter 1
Polysaccharide-Based Films: Need, Status, and Regulatory Aspects

Ashmita Singh, Monika Thakur, Sobiya Manzoor, and Tawheed Amin

Abbreviations

AM	Antimicrobial
CFs	Composite films
CMC	Carboxymethyl cellulose
EFs	Edible films
FDA	US Food and Drug Administration
NFC	Nanofibrillated cellulose
PBFs	Polysaccharide-based films
PBS	Polybutylene succinate
PCL	Polycaprolactone
PE	Polyethylene
PEO	Polyethylene oxide
PET	Polyethylene terephthalate
PGA	Polyglycolic acid
PHA	Polyhydroxyalkanoates
PHB	Polyhydroxybutyrate
PLA	Polylactic acid
PLGA	Polylactic-co-glycolide
PP	Polypropylene
WVP	Water vapour permeability

A. Singh · M. Thakur (✉)
Amity Institute of Food Technology, Amity University Uttar Pradesh, Noida, Uttar Pradesh, India
e-mail: mthakur1@amity.edu

S. Manzoor · T. Amin
Sher-e-Kashmir University of Agricultural Sciences and Technology of Kashmir, Shalimar, Srinagar, Jammu and Kashmir, India

T. Amin et al. (eds.), *Polysaccharide Based Films for Food Packaging: Fundamentals, Properties and Applications*,
https://doi.org/10.1007/978-981-99-4898-7_1

1.1 Introduction

People are leaning more towards natural sources rather than synthetic ones as a result of rising environmental concern. Biopolymeric substances are a potential source of edible films or coatings for packaging materials used in foods as well as preservation because they are biodegradable, edible, and non-toxic in nature (Cazón et al. 2017). Biodegradable substances are those that can be decomposed and returned to nature through the action of microbes. Biopolymers originate from a range of sources, like animals, plants, or marine microbes, all of which create different kinds of biopolymers. Polysaccharides were most abundant biopolymer among biopolymers (Nešić et al. 2020). Starch and its derivatives, cassava, rice starch, cellulose derivatives, chitosan, agar, and a variety of other polysaccharide-derived film-forming compounds were among them. These polysaccharides have a broad spectrum of utilisation in the food sector, such as a binding agent, thickeners in soups, pastry fillings, sauces, as well as candies, and they also play a role in non-food industries (Phan et al. 2005). Biodegradable packaging materials that are desirable include those that are renewable in nature, have excellent mechanical or barrier qualities, and also might be used to incorporate a broad range of additives like antifungal agents, antibacterial agents, antioxidants, colours, as well as other nutrients (Rhim et al. 2013). Polysaccharide has been employed as a coating material in recent years. These natural polymers aid in the prevention of product loss, the extension of material shelf stability, or the preservation of product quality. These systems are typically developed with barrier properties and microbial factors in consideration (Lacroix and Le Tie 2005). Polysaccharides have film-forming capabilities, making them an excellent substitute to conventional food packaging materials. They are stable in nature and of food-grade quality and substantially reduce waste streams from food packing, which lowers total operational expenses (Wang et al. 2007). Polysaccharide-based films provide a flexible or biodegradable coating that helps to prevent foodborne diseases caused by interactions between food ingredients and packaging material (Junter et al. 2016). Polysaccharides are multipurpose polymeric substance with a broader range of application. They have hydrophilic nature and have well-defined structures. Various constraints, such as a lack of mechanical strength, can restrict their functions; however, these limitations could be resolved by implementing minor changes to them and employing them more effectively (Shukla and Maiti 2019). Polysaccharide-based films are the need of the hour and are being discussed as emerging technology in the food industry. Thakur and Modi (2020) also reviewed emerging technologies in the food business, emphasising that polysaccharide-based films have enormous functional potential, but their safety aspects have yet to be investigated. In this chapter, we will discuss about polysaccharide-based films, how they function, and when they will be needed in the future, as well as other safety concerns.

1.2 Rationale for Using Polysaccharide-Based Films

Recently, the demand towards the utilisation of biopolymers for the development of coatings or films has been increased because of their biodegradable nature to safeguard the environment from hazardous waste. Polysaccharide- or protein-based substances were prevalent biopolymers utilised in food packaging sector. Polysaccharide-based films (PBFs) have excellent gas barrier capabilities and can be employed in AM packaging systems as well as in active packaging materials (Kuorwel et al. 2011). Polysaccharides are biomaterials that were broadly employed in the creation of biodegradable films due to their abundance in nature, renewable nature, as well as ability to form films (Feki et al. 2019). Most polysaccharides are readily available and affordable in nature because microbes, plants, or animals can use them as storing energy. They are much more stable as compared to nucleic acids or proteins, although usually they were permanently denatured when heated and have a range of forms and properties (Zheng et al. 2015). PBFs have proven to be a promising polymer material for membranes, polymer films, drug encapsulation, hydrogels, and several applications. The mechanical capability of these films was essentially determined by the bonds that link the compounds, their suitability, and also the molecular structure that forms (Rogovina and Vikhoreva 2006). Due to various properties such as biomolecule immobilisation, adhesion, and controlled release, polysaccharides can form a variety of matrices, like membranes such as coatings or films. PBFs possess a number of applications in the pharmaceutical sector for wound dressing as well as in healing processes (Ribeiro et al. 2020). Conventional packaging of food offers structural support and also protection for the food, but it also results to an aggregation of synthetic polymers that have an environmental impact. The primary requirement for material selection is that there needs to be the little interaction among the prepared food and the packaging material. Biopolymer-based packaging components should preserve food quality, enhance shelf stability, and then also preserve the beneficial effects of processing (Nesic and Seslija 2017).

PBFs are being considered as a possible substitute to synthetic films, and they grow rapidly owing to their renewable nature or ability to reduce environmental plastic waste. For edible coatings or films, PBFs are more cost-effective alternatives than proteins or wax. Pharmaceutical coatings, edible films, and other packing material could therefore benefit from these biopolymers (Prokopovich and Perni 2009).

1.3 Definitions: Polysaccharide-Based Films, Composite Films, and Edible Films

Polysaccharide-based films: Polysaccharides are biopolymers that are composed of monomeric units connected by O-glycosidic bonds. These might have a neutral charge, a positive charge, or a negative charge. Animals and plants use them as a source of energy, which also have a structural function in insect skeletons, plant cell walls, as well as other animals' skeletons. The monosaccharide compositions, chain patterns, linkages, or level of polymerisation all contribute to polysaccharide variation. Polysaccharides are commercially employed in the food business as a gelling or thickening agent, as well as an encapsulation and crystallisation inhibitor (Kumar and Neeraj 2019). These are complex carbohydrates that are found in animals and plants, as well as pectin and gums, that are the most ubiquitous macromolecules in the biosphere. These polysaccharides have been used to make films and membranes, and they have numerous utilisation in the medical, food, and also pharmaceutical sectors (Ferreira et al. 2016). When compared to synthetic plastic films, PBFs are hydrophilic and have a high-water vapour permeability. The molecular weight, degree of branching, conformations, hydrophobicity, as well as electrical charge on the polysaccharide can all influence PBFs. These characteristics directly affect the appearance and physicochemical qualities of coatings (Galgano et al. 2015). PBFs are capable of functioning as a possible barrier against moisture, volatile organic compounds, bacteria, oils, gases, and other pollutants, increasing the shelf life of the material while having strong gas permeability as well as low moisture barrier properties (Véronique Coma 2013). Starch, chitosan, carrageenan, cellulose ethers, alginate, pectins, and other natural polymers could be used to make PBFs. Linear structure of certain polysaccharides cellulose, chitosan, or amylose makes their films flexible, durable, or transparent, as well as resistant to fats and oils. Since they are hydrophilic, they also have poor water vapour barriers (Vieira et al. 2011). These biopolymers have several characteristics, including as antibacterial activity, biocompatibility, and biodegradability, which make it possible to be used in bio-packaging, drug delivery systems, regenerative medicine, and tissue engineering. Derivatives of cellulose and starch, polylactic acid (PLA) and *polycaprolactone* (PCL), polyhydroxybutyrate (PHB), and polybutylene succinate (PBS) are the most often used biopolymers for packaging purposes. These films are an intriguing alternative for elimination of pollutants (Zafar et al. 2016).

Composite films: Composite films (CFs) have heterogeneous structure and can be created by laminating two or more edible films or by forming emulsions from proteins, polysaccharides, or lipids. They aid in enhancing the permeability or mechanical characteristics of the film; however, the functional characteristics of composite films also rely on the interactions of the other components that interact with the film (Umaraw and Verma 2017). In order to achieve barrier and mechanical capabilities, CFs utilise the functional characteristics of each compound of the film. Cellulose ethers, chitosan, carrageenan, starch, and alginates are the most common materials used to make composite films. These substances were typically blended with palmitic or stearic acids, acetylated monoglycerides, beeswax, and also lecithins to create composite films. It has been demonstrated that combining multiple biopolymers to form a composite film improves mechanical capabilities, increases water adsorption, and provides high barrier properties. The use of a combination of biopolymers will help to compensate for the shortcomings of every component and expand the range of applications for these CFs over synthetic films (Cazón et al. 2018). According to some researches, the fabrication of CFs may achieve UV-light barrier capabilities and material strength demonstrated reinforcement impacts resulting in a film with strong elongation values also have good resilience to rupture. All these characteristics of composite films could be beneficial in the food sector by avoiding the oxidation of proteins, lipids, vitamins, as well as the breakdown of antioxidants in foods and also maintaining the quality of food items (Cazón et al. 2020). CFs developed from good polymer matrix, such as chitosan, cellulose derivatives, and others, enable to entrap the essential oils in them, resulting in considerable modifications in the films and contributing to their stability (Sánchez-González et al. 2011). To provide structural support and promote film extensibility, the creation of CFs necessitates the use of two or more key elements. Waxes and fatty acids are good water vapour barriers; however, they lack the power to create self-supporting films. As a result, CFs including both hydrocolloid and lipid constituents may have good water barrier qualities as well as high mechanical integrity (Shaw et al. 2002). The concept of CFs based on biopolymer paves the path for pollution and recycling issues to be addressed. These films outperformed single component films in terms of mechanical or barrier characteristic (Escamilla-García et al. 2013).

Edible films: Biodegradable materials mostly comprised of polysaccharides, lipids, and proteins are known as edible films (EFs). These films have been used in the packaging of seafood, candies, meat products, fruits and vegetables, and other items that require antibacterial, barrier, and antioxidant capabilities. Polysaccharides like starch, pullulan alginate, and cellulose are commonly employed in the development of EFs. However, these polysaccharides do not have a high film-forming capacity, although film performance can be improved by adding additives such as sorbitol, glycerol, and polyethylene glycol (Du et al. 2016). Edible biopolymers as well as food-grade additives were employed to create the EFs as well as coatings. They improve food quality by minimising particle aggregation, as well as protecting food from moisture migration, growth of microorganisms on the surface, or nutrient degradation. EFs are thin layer coatings that are applied on the surface of foods, particularly fruits. Their natural state is hydrophilic. Edibility and biodegradability are the two primary objectives of this form of packaging. EFs improve the mechanical qualities, product's barrier, and shelf life. These films have the ability to cross-link with micronutrients, which improves the packing material's functionality (Nesic et al. 2020). EFs contribute to the development of an interesting and useful tool for monitoring the quality of diverse food products. Various applications for EFs. as wrapping food products and meats, protecting dried fruits and fishes, and controlling transmission of moisture in products like pies or pizzas etc. These films were made by dissolving, dispersing, and emulsifying the film-forming polymers in food-grade fluids like water, ethanol, and organic acids (Guilbert et al. 1995). EFs might be employed as a packaging material for food commodities, as well as for wrapping and pouches. These films protect the food by functioning as flavour and oxygen barrier. The capability of the edible coating applied to freshly picked fruits or vegetables to protect their flavour is beneficial. When compared to uncoated citrus fruits, it has been claimed that edible coatings on citrus fruits help to boost flavour after storage (Miller and Krochta 1997). Numerous polysaccharides such as galactomannans, starch, chitosan, and carrageenans have the tendency to act as edible films and help to decrease the utilisation of synthetic packaging by being placed over or in between food materials in the form of thin layers to protect from aroma, lipids, moisture loss, and gases (O_2 or CO_2) (Mikkonen et al. 2007). EFs aid to create a modified environment in each coated fruit piece, with optimal refrigeration temperature. Relative humidity is very important extrinsic factor in improving the shelf stability of fresh cut fruits. Owing to their colloidal qualities and capacity to produce strong gels, alginate and gellan are utilised as gelling and texturising agents in the food sector that could be used as edible coatings. These coatings function as transporting flavours, nutrients, food additive carriers, colourants, anti-browning, and antimicrobial agents (Rojas-Graü et al. 2008).

1.4 How Films Work?

Films are self-supporting packaging systems that could be utilised to wrap food items. They could be used as thin films on the food's surface or in between the various compounds of food commodities. These films aid in delivering good sensory qualities, excellent mechanical and barrier characteristics, non-toxic and non-polluting nature, and so forth. Protein-based films give structural support, whereas lipid-based films act as a moisture barrier (Debeaufort and Voilley 2009). Barrier characteristics, mechanical properties, and antimicrobial characteristics are all essential in edible films or coatings. Water vapour permeability (WVP), solute permeability, and gas permeability are examples of barrier qualities. Gas permeabilities primarily deal with CO_2 and O_2 as the rate of respiration, and oxidation is regarded in the case of packaged foods. We need to know about water vapour permeability as well as oxygen permeability to comprehend the shelf stability of food. Permanent permeability (P), such as oxygen and water vapour, through a biofilm at a specific temperature is a product of constant diffusion coefficient (D) and solubility (S), resulting in ($P = DS$), assuming no interaction among permanent and the packaging film (Wu et al. 2002).

The elevated permeability of PBFs is owing to the hydrophilic character of the film, which serves as a barrier to restrict the moisture transfer among the food item as well as the surroundings. Water quality is an essential property of biofilms, as it is used to determine the film's water resistance and integrity. PBFs also have gas barrier qualities, biodegradability, and mechanical properties, which ensure the product's structural stability, regulate the rate of release, and offer physical protection (Yoshida et al. 2014).

1.5 Chemistry of Film Forming Materials

Understanding the chemistry of film forming components, like additives or biopolymers, is critical for learning more about their utilisation. Their solubilities in ethanol and also in water are required in choosing the most appropriate solvent for wet casting or mixing active agents. Several characteristics are required, such as water solubility, hydrophilic, hygroscopicity, and lipophilic balance; surface tension, biopolymer hydrodynamic radius, and viscosity of film-forming solutions; and so forth. To comprehend the chemical suitability of plasticisers and additives with other biopolymers, further chemical characteristics of the additives and plasticisers should be determined. This knowledge will aid in the design of processes and the scaling up of commercial production (Han 2013).

Polysaccharides, proteins, lipids, and polyesters are among the four key components that are responsible for biofilm formation as shown in Fig. 1.1.

Polysaccharides as well as proteins have high optical and mechanical characteristics, although they are moisture susceptible, resulting in poor water vapour characteristics; lipids, on the other side, have moisture barrier capabilities; however, they are typically opaque and very inflexible in nature (Gontard and Guilbert 1994).

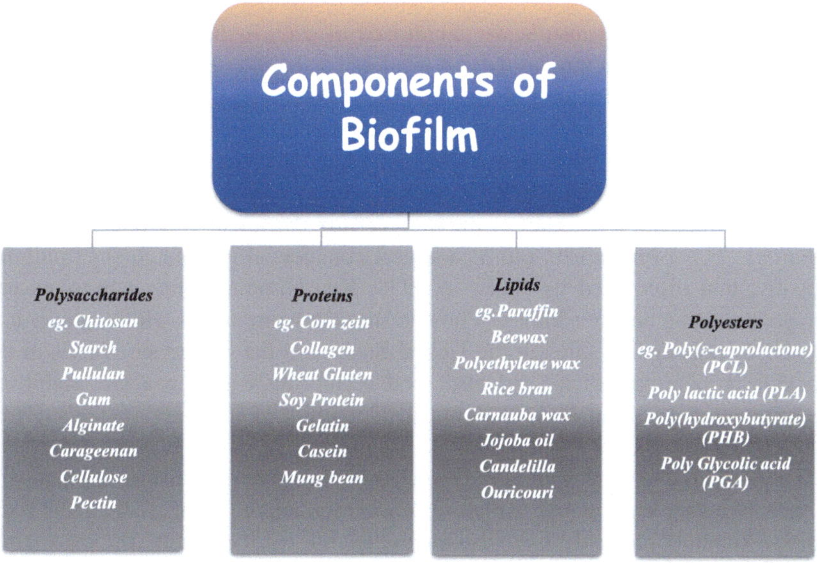

Fig. 1.1 Compounds involved in the biofilm formulation

Cellulose is a polysaccharide comprised of repeated units of D-Glucose that is connected by β-(1–4) glycosidic linkages as a homopolymer. Nanofibrillated cellulose (NFC) is a cellulose derivative utilised as a matrix for the stabilisation and conjugation of enzymes and proteins. These protein conjugates generate biofilms by immobilising proteins on surfaces (Arola et al. 2012).

Carboxymethyl cellulose (CMC) is a linear anionic polysaccharide. It has various applications in foods, paper, pharmaceuticals, as well as textiles. It has high viscosity and therefore is non-toxic. The water binding as well as moisture sorption capabilities of CMC are due to its several carboxylic or hydroxyl groups. CMC hydrogels have a high moisture content, are biodegradable by nature, and have a polymeric structure having high molecular weight. They are employed as a filler in biocomposite films, and they aid in improving the barrier or mechanical capabilities of films based on starch (Tongdeesoontorn et al. 2011).

Chitosan is a linear polysaccharide composed of glucosamine units connected together by β-(1–4) links formed by alkaline chitin deacetylation. Chitosan has a wide variety of abilities, including the capability to form films, biodegradability, and AM activities (Castro et al. 2017). Chitosan is a polymer that is semi-crystalline. Chitosan's biomaterial capability is mostly due to its cationic nature and high charge density (Francis Suh and Matthew 2000). Chitosan is widely used in bio-membranes, biomedicine, and the food industry because of its non-toxic and biocompatible characteristics. Chitosan's water solubility is limited to pH5, limiting its use as a biofilm. Its derivatives have been employed to make extremely water-repellent films with pH ranges of 1–14 using phase separation technique with a broad array of uses, such as biotechnological, fibres, biomedical, and functional membranes, food or nutrition, and environmental or agricultural applications (Song et al. 2010).

Pullulan is a maltotriose trimer comprised of α-(1 → 6)-linked α-(1 → 4)-d-triglucosides. This structure has been considered as an intermediate among dextran and amylose structure because of the presence of α-(1 → 6) and α-(1 → 4)-units in a single molecule; this compound is water soluble and highly flexible (Trinetta and Cutter 2016). Pullulan has received a significant interest owing to its unusual features. It is a non-ionic exopolysaccharide produced under controlled conditions from the fermentation medium of the fungus *Aureobasidium pullulans (Pullularia pullulans*). Hayashibara Co., Ltd. is a largest commercial manufacturer of pullulan globally, first discovered by Bauer in 1938. It is non-mutagenic, non-toxic, and non-carcinogenic and is edible dry powder having white to off-white colour. It is "generally regarded as safe" in the United States and has a number of uses in the biomedical, pharmaceutical, and also food sectors as a gelling agent, stabiliser, thickening agent, as well as prolonging shelf stability of food commodities (Farris et al. 2014).

Starch is an abundant natural polysaccharide that is found in several plants and animals as a source of energy. Amylose (linear) and amylopectin are the most basic types of starch (Branched). It has been employed in biodegradable film packaging, food, paper industries, and cosmetics. However, starch is cold-water insoluble;

retrogradation, gelatinisation, and hydrophilicity treatment for starch is difficult (Bajer et al. 2020).

Gellan, a microbial polysaccharide synthesised by the bacteria *Sphingomonas elodea*, has a broad array of utilisation in the food sector, including texturing as well as gelling. It gels when cations like K^+, Ca^{++}, Mg^{++}, and Na^+ are present, and the gelling abilities are greatly influenced by the cation type, ionic strength, and gum content. Gellan has demonstrated that it has strong mechanical qualities as well as film clarity, and it has a lot of capability as a film-forming agent (Yang and Paulson 2000).

Polylactic acid (PLA) is a biodegradable polymer produced from plants. Antibacterial PLA films are created via solvent casting technology as well as standard film extrusion methods, making it a great alternative for generating commercial, biodegradable packaging material (Byun et al. 2010). Injection moulding, thermoforming, and film casting are some techniques for producing PLA films (Erdohan et al. 2013). Biocompatibility, high strength, grease and oil resistance, strong aroma barrier, crease retention, and thermoplastic fabricability are just a few of the benefits of PLA. Many constituents have been co-polymerised with PLA to enhance the durability and flexibility of the material, for example, PHB, PEO, and PCL etc. Plasticisers aid to reduce the modulus of elasticity in PLA and ensure that the matrix is distributed evenly (Ljungberg and Wesslén 2005).

Polyhydroxybutyrate (PHB) is a part of the polyhydroxybutyrate (PHA) family, which is noted for its biocompatibility and biodegradability. The monomeric unit of PHB, 3-hydroxy butyrate, is a metabolic product in living organisms. Mechanical characteristics, moisture barrier, crystalline nature, and melting point are just a few of the features that make it a biocompatible alternative to PP, PE, and PET. PLGA (polylactic-co-glycolide) is a compostable, crystalline, barrier, and high compatibility linear copolymer of PLA and PGA with FDA approval. Active packaging has made use of PLGA. PHB is a well-known biopolymer in the food sector, as well as in biomedicine in order to develop drug delivery systems (Mlalila et al. 2018).

Waxes are a type of nonpolar lipid that is water insoluble and also does not form a monolayer on the interface, but is soluble in nonpolar solvents like benzene, chloroform, and hexane. They exhibit high hydrophobicity, a nonpolar component, and also a little submerged hydrophilic part that prevents water from adhering to the molecule and aids in its dispersion, thus explains their outstanding lipid barrier qualities. Waxes are commonly used as a fruit and vegetable covering substance. Because of their moisture-barrier qualities, stearyl alcohol, lauric acid, palmitic acid, and stearic acid are commonly employed in edible films (Callegarin et al. 1997).

1.6 Mechanism of Film Formation

Extrusion methods, co-extrusion method, roll coating method, drum coating method, casting method, pan coating method, and several more lamination techniques can all be used to create biofilms, depending on the raw material. Intermolecular forces like

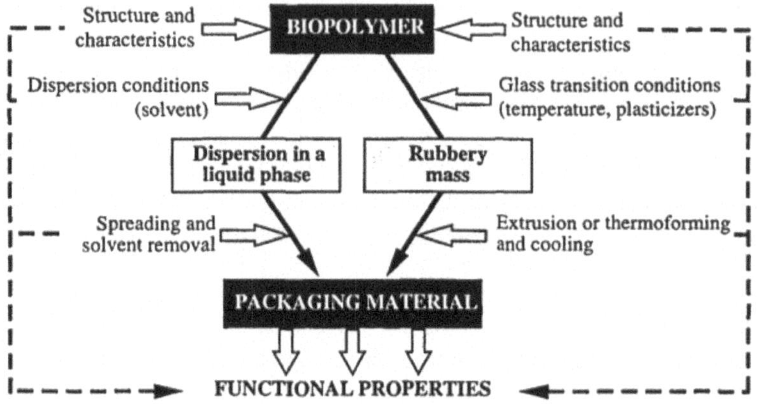

Fig. 1.2 Mechanism for the development of bio-packaging material (Guilbert et al. 1997)

disulphide bonds, cross-links, ionic bonds, electrostatic bonds, and hydrophilic bonds have a significant role in film formation. Food-grade chemicals and solvents, pH modifications, enzymatic alterations, salt addition, heating, and drying should all be used to create edible films. The circumstances of the manufacturing process should be monitored since they affect kinetics and reaction processes (Souza et al. 2010). The mechanism for the production of bio packaging material is shown in Fig. 1.2. The extrusion method and the casting method were the two basic methods for biofilm production. The casting method is the most widely described method. The biopolymer is dissolved and then combined with additives or plasticiser to make the film-forming solution, which is then casted onto the plates and then dried by solvent removal. Extrusion, on the other hand, is dependent on the thermoplastic behaviour of proteins at low moisture levels. Extrusion is used to make the films, which are subsequently heat-pressed at temperatures above 80 °C. Although the extrusion process alters the film's characteristics, it aids in increasing its commercial viability (Gómez-Guillén et al. 2009). Extrusion is preferred because it provides excellent films that are advantageous for packaging; the thermocompression method is a simple process for making high-capacity films without solubilisation; as well as the solvent-casting method, that is widely used for biopolymers, includes casting, solubilisation, and drying steps (Rhim et al. 2006).

Different environmental factors influence the biofilm formation. Surface characteristics and organic material deposition are the two most substantial factors that have impact on the development of biofilms. Several factors, such as topographical, chemical, and physiological properties, affect the initial distribution or adsorption on the surface material and subsequently cause modifications to the surface before the development of biofilm, which results in cell retention, attachment, adhesion, as well as proliferation as shown in Fig. 1.3 (Whitehead and Verran 2015).

Extrusion is the suitable process for forming biofilms on a commercial scale, although casting is considerably more efficient for creating mix films at the pilot scale (Wang et al. 2016). In the absence of drugs, the circular film is generated by

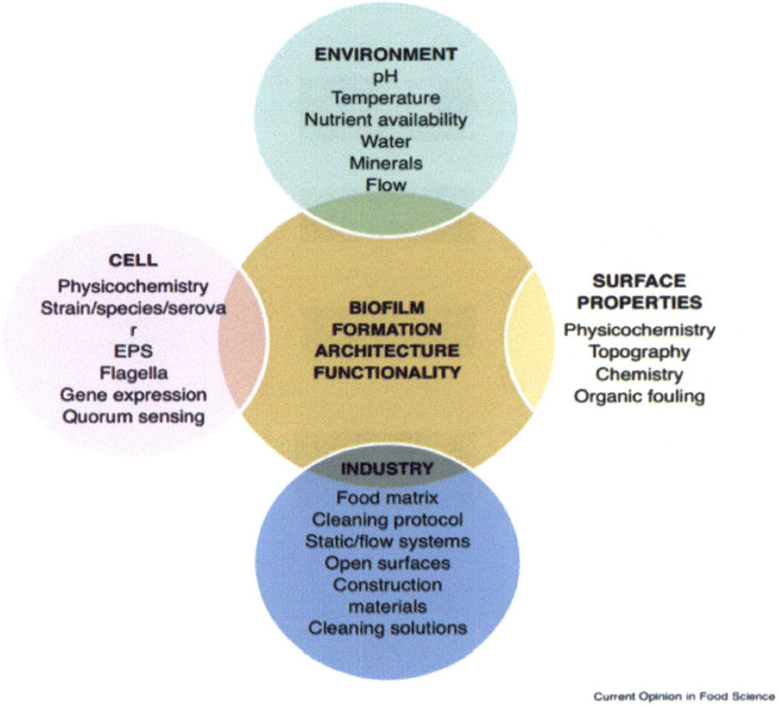

Fig. 1.3 Factors responsible for biofilm formation and functionality (Whitehead and Verran 2015)

casting accompanied by solvent evaporation throughout PBF creation, but in the presence of drugs, the film formation is hindered. The film thickness could be regulated by altering the polysaccharide levels, as well as the film has a regular unevenness. PBFs are spontaneously dissolved in medium, resulting in the emission of chemicals or eventual disintegration (Murata et al. 2010).

The casting technique basically has Wet process, since it involves the dispersion or solubilisation of polysaccharides in a solvent media, while extrusion is referred to as a dry technique because it involves compression or heat pressing moulding for film creation. Because of their brittle qualities, certain biopolymers necessitate plasticisers, which aid in enhancing the polymers' machinability and flowability. Figure 1.4 shows the flowchart for formation of biofilm (Guerrero et al. 2010).

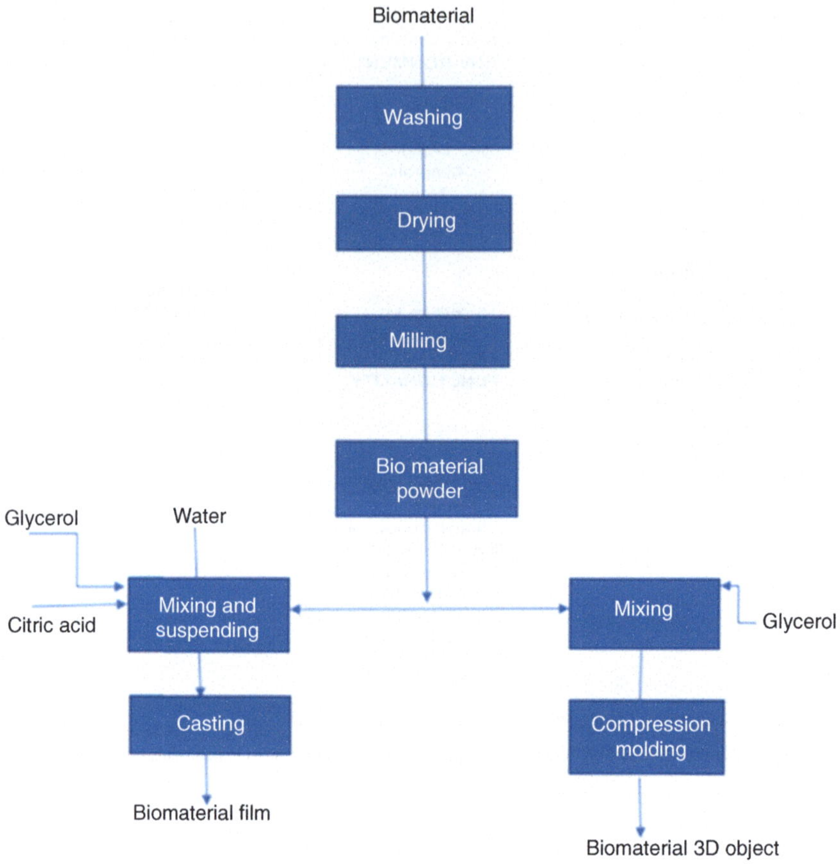

Fig. 1.4 Flow chart for biofilm formation (Gustafsson et al. 2019)

1.7 Status: National and International

Packaging is essential in protecting and preserving food as it passes through the supply network until it reaches the customer. Packaging aids in the food preservation and the enhancement of its shelf stability. Plastics are projected to make up around 37% of packaging materials. Despite their many advantages, plastics are among the most unsustainable available materials, as they never degrade and pollute the environment, increasing greenhouse gas emissions at a cost of $23 billion in natural resources. The food and beverage industry accounts for over 31% of total plastic trash (van Crevel 2016).

A few decades ago, the world was focused on techniques that were effective in terms of output as well as quality, but it is no longer the case. The world now needs a cost-effective and long-term solution. Our current goal is to develop bio-based packaging that meets all of the requirements of a packaging material while also helping to safeguard the environment from toxic plastics. Regarding the technological feasibility and cost, bio-based packaging materials have shown to offer a lot of potential. These materials appear to be a viable solution to the millions of tonnes of plastic garbage that have accumulated in our environment (Tariq 2013).

In today's world, economic factors alone are not enough to create a new product; environmental factors must also be considered. Bio-based films are widely used not only to identify areas for development but also to optimise processing parameters and generate sustainable materials. These materials offer a lot of potential for reducing landfill waste and generating valuable soil enhancers that are beneficial to consumers, companies, and the administration (Leceta et al. 2014).

1.8 Functions and Advantages of Polysaccharide-Based Films and Their Limitations

PBFs have the capability to replace conventional petrochemical plastics in food packaging because of their biodegradability. These biofilms could also be used for fertiliser packing and mulching film (Abdul Khalil et al. 2019). Since the cost of traditional adsorbents has increased, PBFs have been shown to be the most efficient in treating wastewater; they aid in the elimination of contaminants like metal ions and dyes (Crini 2005). Other functional components, such as antibacterial and antioxidants agents, could be carried by PBFs. They have the capability to protect food product from discolouration and also have a potential to display oil and fat resilience (Dutta et al. 2009). They were biodegradable, non-hygroscopic, non-toxic, non-mutagenic, and non-carcinogenic in nature, and they do have good moisture resistance and mechanical qualities, as well as outstanding functional properties such as gelling, foaming, and emulsification (Han et al. 2020). Bio-based packaging materials were used in the food as well as in associated industries as an eco-friendly as well as a safer food packaging option. By adding active chemicals in films, this packaging helps to reduce oil diffusion, dehydration, and surface browning as well as enhance the shelf stability of meat items, confectionary, fruits, and vegetables (Vijayendra and Shamala 2014). Biofilms are being used to alter and regulate the surface conditions of food products. These films aid in the microbiological stability of foods, as well as surface retention to prevent the diffusion of food additives. They also aid in the microbial quality of particular foods (Guilbert et al. 1997).

The efficiency, efficacy, clean, and sustainable behaviour of biomaterials are the major characteristics of bio-based films, which help to advance renewable technologies in the packaging industry. Water and gas permeability; UV protection; acid, alkali, and grease resistance; printability; antifogging capability; sealing ability; and cost are among the features taken into account when creating bio-based substances (Pandit et al. 2018).

These biodegradable polymers can also be used in cosmetics as bio-based sheet masks for face that are composed of useful and easily compostable materials. For face masks, a flexible hydrophilic biopolymer is necessary, and PHA/polysaccharide is another alternative (Beatrice et al. 2018).

Due to their H-bonded thick polymer matrix, PBFs operate as an efficient gas barrier. However, the hydrophilic character of the polysaccharide restricts their water-resistant capabilities and limits their use in the food sector. The usage of biopolymers is still limited, owing to their higher production costs than typical synthetic polymers. To address this issue, numerous industrial or agricultural wastes were employed as a substrate for microbial cultivation to reduce costs, although downstream purification still requires investment (Freitas et al. 2014). Since several polysaccharides do not perform as well as synthetic polymers in terms of vapour barrier, several changes are made, such as the addition of nanoparticles to biopolymers to improve strength as well as barrier properties. However, the problem remains unsolved because nanoparticles could cause environmental damage and are toxic in nature; however, several nanoparticles are sustainable, such as nanocellulose, which would be considered harmless and compostable.

1.9 Conclusion and Future Trends

There is a need to become green these days due to the progressive development in environmental pollution by the accumulation of manufactured polymers. The utilisation of biopolymers as well as the avoidance of synthetic packaging materials has become necessary in the past few years; therefore, scientists are concentrating on materials that are naturally driven and sustainable. PBFs are potential features that aid in enhancing food quality, improving the shelf stability of food commodities, and assuring customer safety. Biopolymers' efficiency and utility are entirely dependent on the materials' inherent characteristics, which include biopolymer (carbohydrates, lipids, proteins) additives, and plasticisers. PBFs have the ability to act as a barrier against bacterial contamination, moisture, or gases, making food preservation easier. However, due of their hydrophilicity, which prevents their usage in the food packaging with high water activity, commercialisation of these biodegradable films remains a challenge. To address these constraints, further study is needed, as well as consideration of other variables such as the commercial feasibility of edible films, safety studies and toxicity evaluations, and customer perceptions. As a result, we may infer that there is a need for more vast study in the field of biopolymers, so that PBFs may soon replace ordinary synthetic plastics, causing it to move closer to environmentally acceptable packaging materials.

References

Abdul Khalil HPS, Chong EWN, Owolabi FAT, Asniza M, Tye YY, Rizal S, Nurul Fazita MR, Mohamad Haafiz MK, Nurmiati Z, Paridah MT (2019) Enhancement of basic properties of polysaccharide-based composites with organic and inorganic fillers: a review. J Appl Polym Sci 136(12):47251. https://doi.org/10.1002/app.47251

Arola S, Tammelin T, Setälä H, Tullila A, Linder MB (2012) Immobilization-stabilization of proteins on nanofibrillated cellulose derivatives and their bioactive film formation. Biomacromolecules 13(3):594–603. https://doi.org/10.1021/bm201676q

Bajer D, Janczak K, Bajer K (2020) Novel starch/chitosan/aloe vera composites as promising biopackaging materials. J Polym Environ 28(3):1021–1039. https://doi.org/10.1007/s10924-020-01661-7

Byun Y, Kim YT, Whiteside S (2010) Characterization of an antioxidant polylactic acid (PLA) film prepared with α-tocopherol, BHT and polyethylene glycol using film cast extruder. J Food Eng 100(2):239–244. https://doi.org/10.1016/j.jfoodeng.2010.04.005

Callegarin F, Gallo JAQ, Debeaufort F, Voilley A (1997) Lipids and bio packaging. J Am Oil Chem Soc 74(10):1183–1192. https://doi.org/10.1007/s11746-997-0044-x

Castro KADF, Moura NMM, Fernandes A, Faustino MAF, Simões MMQ, Cavaleiro JAS, Nakagaki S, Almeida A, Cunha Â, Silvestre AJD, Freire CSR, Pinto RJB, Neves MGPMS (2017) Control of listeria innocua biofilms by biocompatible photodynamic antifouling chitosan-based materials. Dyes Pigments 137:265–276. https://doi.org/10.1016/j.dyepig.2016.10.020

Cazón P, Velazquez G, Ramírez JA, Vázquez M (2017) Polysaccharide-based films and coatings for food packaging: a review. Food Hydrocoll 68:136–148. https://doi.org/10.1016/j.foodhyd.2016.09.009

Cazón P, Vázquez M, Velazquez G (2018) Composite films of regenerate cellulose with chitosan and polyvinyl alcohol: evaluation of water adsorption, mechanical and optical properties. Int J Biol Macromol 117:235–246. https://doi.org/10.1016/j.ijbiomac.2018.05.148

Cazón P, Vázquez M, Velazquez G (2020) Characterization of mechanical and barrier properties of bacterial cellulose, glycerol and polyvinyl alcohol (PVOH) composite films with eco-friendly UV-protective properties. Food Hydrocoll 99:105323. https://doi.org/10.1016/j.foodhyd.2019.105323

Coma V (2013) Polysaccharide-based biomaterials with antimicrobial and antioxidant properties. Polimeros 23(3):287–297. https://doi.org/10.4322/polimeros020ov002

van Crevel R (2016) Bio-based food packaging in sustainable development: challenges and opportunities to utilize biomass residues from agriculture and forestry as a feedstock for bio-based food packaging. FAO report. FAO, Rome

Crini G (2005) Recent developments in polysaccharide-based materials used as adsorbents in wastewater treatment. Prog Polym Sci (Oxf) 30(1):38–70. https://doi.org/10.1016/j.progpolymsci.2004.11.002

de Oliveira AD, Beatrice CAG (2018) Polymer nanocomposites with different types of nanofiller. Nanocomposites-recent evolutions 103–104

Debeaufort F, Voilley A (2009) Lipid based edible films and coatings. In: Edible films and coatings for food applications. Springer, Dordrecht, pp 135–168. https://doi.org/10.1007/978-0-387-92824-1

Du H, Hu Q, Yang W, Pei F, Kimatu BM, Ma N, Fang Y, Cao C, Zhao L (2016) Development, physiochemical characterization and forming mechanism of Flammulinavelutipes polysaccharide-based edible films. Carbohydr Polym 152:214–221. https://doi.org/10.1016/j.carbpol.2016.07.035

Dutta PK, Tripathi S, Mehrotra GK, Dutta J (2009) Perspectives for chitosan based antimicrobial films in food applications. Food Chem 114(4):1173–1182. https://doi.org/10.1016/j.foodchem.2008.11.047

Erdohan ZÖ, Çam B, Turhan KN (2013) Characterization of antimicrobial polylactic acid-based films. J Food Eng 119(2):308–315. https://doi.org/10.1016/j.jfoodeng.2013.05.043
Escamilla-García M, Calderón-Domínguez G, Chanona Pérez JJ, FarreraRebollo RR, AndracaAdame JA, Arzate Vázquez I, Mendez-Mendez JV, Moreno Ruiz LA (2013) Physical and structural characterisation of zein and chitosan edible films using nanotechnology tools. Int J Biol Macromol 61:196–203. https://doi.org/10.1016/j.ijbiomac.2013.06.051
Farris S, Unalan IU, Introzzi L, Fuentes-Alventosa JM, Cozzolino CA (2014) Pullulan-based films and coatings for food packaging: Present applications, emerging opportunities, and future challenges. J Appl Polym Sci 131(13):1–12. https://doi.org/10.1002/app.40539
Feki A, Amara I, Ben Bardaa S, Hajji S, Chabchoub N, Kallel R, Boudawara T, Zghal S, Ben Salah R, Nasri M, Ktari N (2019) Preparation and characterization of polysaccharide-based films and evaluation of their healing effects on dermal laser burns in rats. Eur Polym J 115(2): 147–156. https://doi.org/10.1016/j.eurpolymj.2019.02.043
Ferreira ARV, Alves VD, Coelhoso IM (2016) Polysaccharide-based membranes in food packaging applications. Membranes (Basel) 6(2):1–17. https://doi.org/10.3390/membranes6020022
Francis Suh JK, Matthew HWT (2000) Application of chitosan-based polysaccharide biomaterials in cartilage tissue engineering: a review. Biomaterials 21(24):2589–2598. https://doi.org/10.1016/S0142-9612(00)00126-5
Freitas F, Alves VD, Reis MA, Crespo JG, Coelhoso IM (2014) Microbial polysaccharide-based membranes: current and future applications. J Appl Polym Sci 131(6):1–11. https://doi.org/10.1002/app.40047
Galgano F, Condelli N, Favati F, Di Bianco V, Perretti G, Caruso MC (2015) Biodegradable packaging and edible coating for fresh-cut fruits and vegetables. Ital J Food Sci 27(1):1–20. https://doi.org/10.14674/1120-1770/ijfs.v70
Gómez-Guillén MC, Pérez-Mateos M, Gómez-Estaca J, López-Caballero E, Giménez B, Montero P (2009) Fish gelatin: a renewable material for developing active biodegradable films. Trends Food Sci Technol 20(1):3–16. https://doi.org/10.1016/j.tifs.2008.10.002
Gontard N, Guilbert S (1994) Bio-packaging: technology and properties of edible and/or biodegradable material of agricultural origin. J Food Process Preserv 11:159–181. https://doi.org/10.1007/978-1-4615-2173-0_9
Guerrero P, Retegi A, Gabilondo N, De La Caba K (2010) Mechanical and thermal properties of soy protein films processed by casting and compression. J Food Eng 100(1):145–151. https://doi.org/10.1016/j.jfoodeng.2010.03.039
Guilbert S, Gontard N, Cuq B (1995) Technology and applications of edible protective films. Packag Technol Sci 8(6):339–346. https://doi.org/10.1002/pts.2770080607
Guilbert S, Cuq B, Gontard N (1997) Recent innovations in edible and/or biodegradable packaging materials. Food Addit Contam 14(6–7):741–751. https://doi.org/10.1080/02652039709374585
Gustafsson J, Landberg M, Bátori V, Åkesson D, Taherzadeh MJ, Zamani A (2019) Development of bio-based films and 3D objects from apple pomace. Polymers 11(2):289. https://doi.org/10.3390/polym11020289
Han JH (2013) Edible films and coatings: a review. In: Innovations in food packaging, 2nd edn. Elsevier, London, pp 213–255. https://doi.org/10.1016/B978-0-12-394601-0.00009-6
Han K, Liu Y, Liu Y, Huang X, Sheng L (2020) Characterization and film-forming mechanism of egg white/pullulan blend film. Food Chem 315(1):126201. https://doi.org/10.1016/j.foodchem.2020.126201
Junter GA, Thébault P, Lebrun L (2016) Polysaccharide-based antibiofilm surfaces. Acta Biomater 30(11):13–25. https://doi.org/10.1016/j.actbio.2015.11.010
Kumar N, Neeraj (2019) Polysaccharide-based component and their relevance in edible film/coating: a review. Nutr Food Sci 49(5):793–823. https://doi.org/10.1108/NFS-10-2018-0294
Kuorwel KK, Cran MJ, Sonneveld K, Miltz J, Bigger SW (2011) Antimicrobial activity of biodegradable polysaccharide and protein-based films containing active agents. J Food Sci 76(3):R90–R102. https://doi.org/10.1111/j.1750-3841.2011.02102.x

Lacroix M, Le Tie C (2005) Edible films and coatings from non-starch polysaccharides. In: Han JH (ed) Innovations in food packaging. Elsevier, London, pp 239–262

Leceta I, Etxabide A, Cabezudo S, de la Caba K, De Guerrero P (2014) Bio-based films prepared with by-products and wastes: environmental assessment. J Clean Prod 64:218–227. https://doi.org/10.1016/j.jclepro.2013.07.054

Ljungberg N, Wesslén B (2005) Preparation and properties of plasticized poly (lactic acid) films. Biomacromolecules 6(3):1789–1796. https://doi.org/10.1021/bm050098f

Mikkonen KS, Rita H, Helén H, Talja RA, Hyvönen L, Tenkanen M (2007) Effect of polysaccharide structure on mechanical and thermal properties of galactomannan-based films. Biomacromolecules 8(10):3198–3205. https://doi.org/10.1021/bm700538c

Miller KS, Krochta JM (1997) Oxygen and aroma barrier properties of edible films: a review. Trends Food Sci Technol 8(7):228–237. https://doi.org/10.1016/S0924-2244(97)01051-0

Mlalila N, Hilonga A, Swai H, Devlieghere F, Ragaert P (2018) Antimicrobial packaging based on starch, poly(3-hydroxybutyrate) and poly(lactic-co-glycolide) materials and application challenges. Trends Food Sci Technol 74(1):1–11. https://doi.org/10.1016/j.tifs.2018.01.015

Murata Y, Isobe T, Kofuji K, Nishida N, Kamaguchi R (2010) Preparation of fast dissolving films for oral dosage from natural polysaccharides. Materials 3(8):4291–4299. https://doi.org/10.3390/ma3084291

Nesic A, Castillo C, Castaño P, Cabrera-Barjas G, Serrano J (2020) Bio-based packaging materials. In Biobased products and industries (pp. 279–309). Elsevier

Nesic AR, Seslija, SI (2017) The influence of nanofillers on physical–chemical properties of polysaccharide-based film intended for food packaging. In: Grumezescu AM (ed) Food packaging. Elsevier, London, pp 637–697. https://doi.org/10.1016/b978-0-12-804302-8.00019-4

Nešić A, Cabrera-Barjas G, Dimitrijević-Branković S, Davidović S, Radovanović N, Delattre C (2020) Prospect of polysaccharide-based materials as advanced food packaging. Molecules 25(1):135. https://doi.org/10.3390/molecules25010135

Pandit P, Nadathur GT, Maiti S, Regubalan B (2018) Functionality and properties of bio-based materials. In: Ahmed S (ed) Bio-based materials for food packaging: green and sustainable advanced packaging materials. Springer, Singapore, pp 81–103. https://doi.org/10.1007/978-981-13-1909-9_4

Prokopovich P, Perni S (2009) An investigation of microbial adhesion to natural and synthetic polysaccharide-based films and its relationship with the surface energy components. J Mater Sci Mater Med 20(1):195–202. https://doi.org/10.1007/s10856-008-3555-6

Rhim JW, Mohanty AK, Singh SP, Ng PKW (2006) Effect of the processing methods on the performance of polylactide films: thermocompression versus solvent casting. J Appl Polym Sci 101(6):3736–3742. https://doi.org/10.1002/app.23403

Rhim JW, Park HM, Ha CS (2013) Bio-nanocomposites for food packaging applications. Prog Polym Sci 38(10–11):1629–1652. https://doi.org/10.1016/j.progpolymsci.2013.05.008

Ribeiro DML, Júnior ARC, de Macedo GHRV, Chagas VL, Silva LDS, da Silva Cutrim B, Santos DM, Soares BLL, Zagmignan A, de Cássia Mendonça de Miranda R, de Albuquerque PBS, da Silva LCN (2020) Polysaccharide-based formulations for healing of skin-related wound infections: lessons from animal models and clinical trials. Biomol Ther 10(1):63. https://doi.org/10.3390/biom10010063

Rogovina SZ, Vikhoreva GA (2006) Polysaccharide-based polymer blends: methods of their production. Glycoconj J 23(7–8):611–618. https://doi.org/10.1007/s10719-006-8768-7

Rojas-Graü MA, Tapia MS, Martín-Belloso O (2008) Using polysaccharide-based edible coatings to maintain quality of fresh-cut Fuji apples. LWT Food Sci Technol 41(1):139–147. https://doi.org/10.1016/j.lwt.2007.01.009

Sánchez-González L, Chiralt A, González-Martínez C, Cháfer M (2011) Effect of essential oils on properties of film forming emulsions and films based on hydroxypropyl methylcellulose and chitosan. J Food Eng 105(2):246–253. https://doi.org/10.1016/j.jfoodeng.2011.02.028

Shaw NB, Monahan FJ, O'Riordan ED, Sullivan MO (2002) Effect of soya oil and glycerol on physical properties of composite WPI films. J Food Eng 51(4):299–304. https://doi.org/10.1016/S0260-8774(01)00071-1

Shukla A, Maiti P (2019) Biodegradable polymer-based nanohybrids for controlled drug delivery and implant applications. In: Katiyar V, Gupta R, Ghosh T (eds) Advances in sustainable polymers, Materials horizons: from nature to nanomaterials. Springer, Singapore, pp 3–19. https://doi.org/10.1007/978-981-32-9804-01

Song W, Gaware VS, Rúnarsson ÖV, Másson M, Mano JF (2010) Functionalized superhydrophobic biomimetic chitosan-based films. Carbohydr Polym 81(1):140–144. https://doi.org/10.1016/j.carbpol.2010.01.041

Souza BWS, Cerqueira MA, Teixeira JA, Vicente AA (2010) The use of electric fields for edible coatings and films development and production: a review. Food Eng Rev 2(4):244–255. https://doi.org/10.1007/s12393-010-9029-x

Phan TD, Debeaufort F, Luu D, Voilley A (2005) Functional properties of edible agar-based and starchbased films for food quality preservation. J Agric Food Chem 53(4):973–981

Tariq S (2013) Success factors for the adoption of bio-based packaging in EU food industry. KTH Industrial Engineering and Management, Stockholm

Thakur M, Modi V (2020) Emerging technologies in food science. Springer, Singapore. https://doi.org/10.1007/978-981-15-2556-8

Tongdeesoontorn W, Mauer LJ, Wongruong S, Sriburi P, Rachtanapun P (2011) Effect of carboxymethyl cellulose concentration on physical properties of biodegradable cassava starch-based films. Chem Cent J 5(1):1–8. https://doi.org/10.1186/1752-153X-5-6

Trinetta V, Cutter CN (2016) Pullulan: a suitable biopolymer for antimicrobial food packaging applications. In: Antimicrobial food packaging. Elsevier, Amsterdam, pp 385–397. https://doi.org/10.1016/B978-0-12-800723-5.00030-9

Umaraw P, Verma AK (2017) Comprehensive review on application of edible film on meat and meat products: an eco-friendly approach. Crit Rev Food Sci Nutr 57(6):1270–1279. https://doi.org/10.1080/10408398.2014.986563

Vieira MGA, Da Silva MA, Dos Santos LO, Beppu MM (2011) Natural-based plasticizers and biopolymer films: a review. Eur Polym J47(3):254–263. https://doi.org/10.1016/j.eurpolymj.2010.12.011

Vijayendra SVN, Shamala TR (2014) Film forming microbial biopolymers for commercial applications-a review. Crit Rev Biotechnol 34(4):338–357. https://doi.org/10.3109/07388551.2013.798254

Wang LZ, Liu L, Holmes J, Kerry JF, Kerry JP (2007) Assessment of film-forming potential and properties of protein and polysaccharide-based biopolymer films. Int J food Sci Technol 42(9):1128–1138. https://doi.org/10.1111/j.1365-2621.2006.01440.x

Wang LF, Rhim JW, Hong SI (2016) Preparation of poly(lactide)/poly (butylene adipate-co-terephthalate) blend films using a solvent casting method and their food packaging application. LWT Food Sci Technol 68:454–461. https://doi.org/10.1016/j.lwt.2015.12.062

Whitehead KA, Verran J (2015) Formation, architecture and functionality of microbial biofilms in the food industry. Curr Opin Food Sci 2:84–91. https://doi.org/10.1016/j.cofs.2015.02.003

Wu Y, Weller CL, Hamouz F, Cuppett SL, Schnepf M (2002) Development and application of multicomponent edible coatings and films: a review. Adv Food Nutr Res 44:347–394. https://doi.org/10.1016/S1043-4526(02)44007-7

Yang L, Paulson AT (2000) Mechanical and water vapour barrier properties of edible gellan films. Food Res Int 33(7):563–570. https://doi.org/10.1016/S0963-9969(00)00092-2

Yoshida CMP, Maciel VBV, Mendonça MED, Franco TT (2014) Chitosan biobased and intelligent films: monitoring pH variations. LWT Food Sci Technol 55(1):83–89. https://doi.org/10.1016/j.lwt.2013.09.015

Zafar R, Zia KM, Tabasum S, Jabeen F, Noreen A, Zuber M (2016) Polysaccharide based bionanocomposites, properties and applications: a review. Int J Biol Macromol 92:1012–1024. https://doi.org/10.1016/j.ijbiomac.2016.07.102

Zheng Y, Monty J, Linhardt RJ (2015) Polysaccharide-based nanocomposites and their applications. Carbohydr Res 405:23–32. https://doi.org/10.1016/j.carres.2014.07.016

Chapter 2
Starch and Starch Derivative-Based Films

Saadia Zainab, Xianqing Zhou, Saira Tanweer, Tariq Mehmood, Kashif Ameer, Kappat Valiyapeediyekkal Sunooj, and Abdur Rehman

2.1 Introduction and Sources

Starch is a biological polymer based on glucose that is assumed to be an appropriate option for the biodegradable polymer products. It is thought to be the key storage carbohydrate of green plants. During the production of chlorophyll, starch is formed in chloroplasts that act as energy source and is further used for metabolic process. Starch extracted from various plants is obtained in the form of granules having various size ranges depending upon the resource (Sugih et al. 2009; Bucknow et al. 2009). These granules are not soluble in cold water and swell causing the gel formation on dissolving if outermost membrane broke by any aggressive action

S. Zainab (✉) · X. Zhou
College of Food Science and Engineering, Henan University of Technology, Zhengzhou, China

S. Tanweer
Department of Food Science and Technology, The Islamia University of Bahawalpur, Bahawalpur, Pakistan

T. Mehmood
Institute of Food Science and Technology, Khwaja Fareed University of Engineering and Technology, Rahim Yar Khan, Pakistan

K. Ameer
Department of Integrative Food, Bioscience and Biotechnology, Chonnam National University, Gwangju, South Korea

K. V. Sunooj
Department of Food Science and Technology, Pondicherry University, Kalapet, Puducherry, India

A. Rehman
State Key Laboratory of Food Science and Technology, Jiangnan University, Jiangsu, Wuxi, China

T. Amin et al. (eds.), *Polysaccharide Based Films for Food Packaging: Fundamentals, Properties and Applications*,
https://doi.org/10.1007/978-981-99-4898-7_2

such as grinding process. In contrast if granules are dissolved in warm water, the soluble part of the granule absorbs water by diffusion through wall, while the rest of the granule swells and spurts (Bahram et al. 2016).

There are enormous sources of starch among which the common ones are derived from cereal grains like wheat, corn, and rice. Wheat produces hazy and dense mixture, while corn starch yields more transparent mixture. Another source of starch is legumes, namely soybeans or garbanzo beans. Fruits like banana can also be a source of starch. As powder, starch is also obtained from the branches and trunk of sago palm found in tropical Asia (Vaclavik 2021). Industrial starches are mostly attained through maize, potato, wheat, rice, and cassava. Usage of starch in food products is mainly because of its physicochemical and functional characteristics like water-holding capacity, viscosity, and gel formation. On the other hand, characteristics of native starch are restricted that can be improved by using chemical, physical, or enzymatic approaches to advance their effectiveness (Perez and Acevedo 2017).

Numerous native biopolymers such as starch (Jih-Guan et al. 2010), chitosan (Pereira et al. 2015), pectin (Liu et al. 2008a, 2008b), and konjac glucomannan (Chen et al. 2016) have extensive usage in food packaging (Bin et al. 2017). Starch is extensively used as a packaging material because of its characteristics like inexpensive, effective processability accompanied by its extraordinary film-forming attributes (Aswathy et al. 2019; Zhang et al. 2018). Starch plays vast roles in the food products as a binding agent owing to its adhesiveness; film-forming, gelling, glazing, stabilizing, and moisture-holding capacity; and thickening functions (Shah et al. 2015). Nevertheless, starch is extensively used in emerging food packaging films, but it has quite inadequate mechanical and thermal features. To overcome this problem, starch is mixed with additional polymers such as PVA, PLA, and lignin (Raj and Siddaramaiah 2004) among which PVA has been observed to have extraordinary property as thermal and mechanical property enhancer. For the enhancement of mechanical characteristics, starch polymer can also be cross-linked chemically by sodium trimetaphosphate (STMP) (Gao et al. 2014).

Undoubtedly, starch-based polymers are incredibly appealing as a cheap base for the novel biodegradable polymers because of their mediocre material price and aptitude to be progressed with the traditional plastic processing approaches (Wang et al. 2016; Jimenez et al. 2012). The enhancement of biodegradable starch-based film for the general usage has been provoked by keen attention to the ease of environment load because of the excessive usage of petrochemically originated polymers (Teixeira et al. 2009; Ali et al. 2018).

2.1.1 Cereals

2.1.1.1 Corn Starch

For the industrial intention, most of the starch is an excerpt from corn; however, significant quantity of starch is also collected from other sources. Ordinary corn starch consists of 25% amylose, while remaining is amylopectin. Plants can be transformed to generate starch with amylose and amylopectin out of the normal range that is 25% amylose and 75% amylopectin. Take the example of high-amylose maize in which amylose content is comparatively higher, about 85%, and waxy corn has comparatively lower amylose content, about 0% amylose. Preceding investigations have narrated the waxy, normal, and high-amylose maize starch covering 2%, 23%, and 85% amylose matter (Tester et al. 2000).

Ordinary and waxy maize starches consist of different grains of a vast variety of shapes differing from small spherical to large polyhedral, while the grain size ranges from 3 to 20 μm. There are clearly apparent apertures with variable size and depth on the exterior of several starch grains (Wang et al. 2014a, 2014b). Distinct to the waxy and usual maize starches, two types of granules exist for the high-amylose maize starch that is small oval grains and large elongated grains. It was reported that high-amylose maize starch consisted of 7–32% elongated grains made by the synthesis of some small grains (Jiang et al. 2010). For the waxy and usual maize gelatinization, temperature ranges from 60 °C to 80 °C. Conversely, high-amylose maize starch represents a wide endothermic conversion that concludes at a higher temperature (Wang et al. 2014a, 2014b).

2.1.1.2 Wheat Starch

Wheat is one of the abundantly used and merchandized food granules worldwide. Due to its agronomic flexibility, storage conditions, nutritional values, and the use of its flour in a diverse range of recipes of palatable and satisfying foods, wheat is a main fraction of human diet (Guo et al. 2018). Wheat is frequently divided into two main types: *Triticum aestivum* (also referred to as bread wheat) and *Triticum durum* (also known as pasta wheat). *T. aestivum* is the mostly found variety of wheat, recorded for 90 to 95% of yearly production, while the other one, *T. durum*, accounts for approximately 40 million tons (Gelinas and McKinnon 2016; Sobota et al. 2015; Wang et al. 2015a, 2015b). By wet milling and non-starch constituent removal by mashing, starch is produced. Starches that are isolated from *T. aestivum* and *T. durum* contain 26.3–30% total amylose and 19.3–25.1% apparent amylose, respectively (Soulaka and Morrison 2010; Ao and Jane 2007; Hyunseok and Kerryc 2010).

The bimodal dispersion of grain size can only be seen from Triticeae starches (barley, rye and triticale), not by other cereal, root, tuber, or legume starches (Wilson et al. 2007). By using image analysis and laser diffraction approach, wheat starch granules can be categorized on the basis of size ranges as A-type granules (>15 mm), B-type granules (5–15 mm), and C-type granules (<5 mm) (Wilson et al. 2007).

Efficiency and in vitro assimilation of bread wheat starch grains can be drastically changed by using alkali with elimination of exterior proteins and lipids instead of appreciably changing the interior arrangements of starch granules (Wang et al. 2014a, 2014b). A-type granules display lesser quantity of phosphorus, lipid complex amylose, and obvious amylose in comparison to B-type granules. Furthermore, wheat starch is considered to be characterized to A/B-type granule ratio because of its pasting manner (Shinde et al. 2003). As compared to the most of the botanical resource starches, lenticular or disk-shaped A-granules from *Triticeae* starches are unique apart from slighter recognized starches (Jane et al. 1994).

2.1.1.3 Rye Starch

For the preparation of sour dough breads and crisp breads, rye is frequently consumed. Due to the elevated pentosan matter and inadequate gluten-forming properties of rye flours, it is extremely difficult to isolate rye starch (Autio and Eliasson 2009). Rye starch represents an A-type X-ray diffraction prototype with the projected degree of crystallinity of 15–17% (Schierbaum et al. 1991). Just like wheat, triticale, and barley, rye starch generally has about 22–25% amylose that comprises large A-type and small B-type grains of diameter 23–40 μm and less than 10 μm, respectively.

2.1.1.4 Barely Starch

Barely starch subsidizes 75 to 80% of the whole barley endosperm kernel bulk within dense barely protein matrix. Depending on the genotype, amylose content of barely starches has a wide range. From non-waxy and high-amylose starches, amylose content was recorded as 24.6 and 48.7%, respectively, while only a trace 0.04% of amylose was contained by waxy starch that was determined by using calorimetrical approach (Czuchajowska et al. 1998). Amylose content of hull-less barley varied from 23.1 to 30%, while enlargement power and water solubility index varied from 12.8 to 19.9 g/g and 12.7 to 23.7 g/g, respectively (Kong et al. 2016). Barely grains just like wheat exhibited bimodal dispersal and consisted of large disc modeled grains and small sphere modeled grains (Ao and Jane 2007). Normally, average molecular weight of barely is 1.15×10^6 g/mol, comparatively lower compared to maize amylopectin that is 19.3×10^6 g/mol (Luisarturo et al. 2009). Amylopectin unit chain varied from 9.83 to 16.66% whose DP value is greater than 36 among 14 barley starches (Kong et al. 2016). Among six barely genotypes, average amylopectin chain length varied from 16.2 to 20.4% glucosyl remainder; next to large grains have more amylopectin unit chains of DP value more than 25 and comparatively large chains (Ao and Jane 2007). When barely starches are compared to the maize and rice, they represent less quantity of long chains and more quantity of small chains. The degree of crystallinity in starch grains has been proved to be in the range of 37–44.3% (6 genotypes) (Waduge et al. 2006), 10.7–43.2% (7 genotypes) (Li et al. 2014), and 10.1–13.3% (14 genotypes) (Kong et al. 2016).

2.1.1.5 Oat Starch

Russia, Canada, the United States, Finland, Australia, and China are the countries that are famous for cultivation of oats that is of significant for humans in their diet (Xu et al. 2014). The two mainly grown types are *Avena sativa* L. (hulled oat) and *Avena nuda* L. (naked oat). The latter one is mainly cultivated in North China such as Shanxi, Gansu, Jilin, and Hebei provinces as well as used as conventional Chinese medicine (Hu et al. 2014). Opposite to the first one, hulled oats, which are extensively grown in Western countries, are used as rolled oats and steel cut groats. Subjected to the directional protocols and cooking traditions, oat is known as a low or medium glycemic index (GI) food. Starch, because of the main glycemic carbohydrate, represents more than 60% of dry content of granules together with amylose content varying from 25.2 to 29.4% (Gudmundsson and Eliasson 2009) with A-type glassy polymorphs illustrative of cereal starches (Hoover et al. 2003). In single granules, oat starches incline in the form of clusters. Shape of grains varies from irregular to the polygonal, and their diameter is about 7–8 μm (Hoover et al. 2003) that is comparatively smaller than wheat, barley, corn starch, and rye grains.

2.1.1.6 Rice Starch

Rice is one of the abundantly cultivated crop which is widely used as a staple crop in several countries, while damaged rice granules are mostly milled or ground into flour that is further consumed as the constituent of baby food, noodle puddings, and most of the Asian cuisines. Major constituent of rice grain is starch that is more than 80% dry weight (Hasjim et al. 2013). Rice is significant in modulating the stage of postprandial glucose in blood stream because it is the main dietary source of carbohydrate. Because of the rapid digestion of starch, rice is thought to be of high GI that made its digestibility a topic of nutritional interest. From the industrial viewpoint, rice starch cannot be fully exploited because of expensive manufacturing cost as compared to that of other commercial starches.

Asian rice is categorized into Indica, Japonica, and waxy rice depending upon the granule size and texture. Commonly, Indica rice contain higher amounts of amylose as compared to Japonica rice resulting in poor cooking quality which makes it unpopular among the customers (Tian et al. 2005). To enhance the quality of Indica rice, hybrid rice with less amylose content has been developed in China (Kuang et al. 2016; Ni et al. 2011). Dimension of rice starch grains varies from 3 to 10 μm, while the shape can be round, angular, and polygonal (Ashogbon and Akintayo 2012). Japonica and waxy rice starches have a larger ratio of short A chains and a low ratio of medium B1 chains and long B3 chains regarding to Indica hybrid rice starch (Wang et al. 2017).

2.1.2 *Tubers and Roots*

2.1.2.1 Potato Starch

After maize, wheat, and rice, potato is on the fourth place as a significant crop in the world that has an important role for ensuring universal food security (Alvani et al. 2011; Tong et al. 2018). In the Western world, potatoes are the main sources of carbohydrates, while for the developing countries, its use is continuously increasing day by day. Potatoes are used as vegetables and can also be utilized as fresh ingredient in the production of various items like snacks, breads, starch, and their derivatives (Ek et al. 2014a, 2014b).

Potato tubers have the main content, starch that resident almost 66–80% of dry weight (Chung et al. 2014). The chemical proportion of potato starches gain from 10 cultivators under the identical conditions demonstrate the ranges of protein, amylose, and phosphorus matter varies from 0.30% to 0.34%, 25.2% to 29.1%, and 52.6 to 66.2 mg/100 g (db) correspondingly (Alvani et al. 2011). The range of phosphorus varies from 59 to 71 mg/100 mg (db) segregated starch from various locations of Canada (Chung et al. 2014). Potatoes whose phosphorylation amount is higher are more wanted because of their useful practical characteristics of native potato starch for manufacturing purposes with ecosystem responsive (Singh et al. 2008).

Potato starch is unique due to its characteristics as compared to the industrial starches, like enlarged grain size, transparency, long chains of amylose and amylopectin, existence of phosphate ester groups on amylopectin, comparatively higher swelling and pasting viscosity, clear starch paste, and impartial flavor, and forms gels of thick viscoelastic character when heating and successive cooling are provided (Alvani et al. 2011). For example, potato starch already has been used to the great extent in food as well as nonfood manufacturing companies such as thickening agents, food resources, pharmaceutical plaster, or excipients (Sriroth et al. 1999). Furthermore, fresh potato starch showed higher confrontation against acid and enzymatic hydrolysis in contrast to the other usual starches that is linked to the higher phosphorous content, B-type crystallites, and longer chains of amylopectin (Jane et al. 1997).

2.1.2.2 Cassava Starch

Cassava is a perpetual woody bush consisting of tuberous roots in the spurge family (Euphorbiaceae) that is also called tapioca, manioc, or yuca. Constituents of typical mature roots are 30–40% dry content, 30–35% carbohydrates, 1–2% fat, 1–2% fiber, and 1–2% protein with less quantity of vitamins and minerals. Because of its possible growth in severe innate circumstances, cassava is thought to be a significant crop for food security and commercial uses (Alves 2002). Toxicity of cassava is also a concern which is attributed to the presence of cyanogenic glucosides. Therefore, appropriate directional approach is needed to detoxify cassava efficiently when used in human diet or as animal feed.

Dried cassava roots contribute 70% of starch (Olomo and Ajibola 2010), as a major constitute. Amylose content varies significantly in the range of 0% to 30% of which example is 4050 cassava genotypes whose amylose content varies greatly from 15.2% to 26.5% (Rollandsabate et al. 2012). Grain size varies from 2 to 32 μm, while the shape can be oval, truncated, and round (Rolland-Sabate et al. 2013; Moorthy et al. 2015). The surface of granules is smooth as compared to the potato's surface (Leslaw et al. 2003).

Starches obtained from cassava mainly display A- or C_A-type X-ray diffraction prototypes (Rolland-Sabate et al. 2013), however, a minor amount of B-type crystalline structure is also found that's uncommon (Rollandsabate et al. 2012). Momentous assortment in gelatinization characteristics has been recorded, that is, the ranges of T_o, T_p, T_c, and ΔH for 12 genotypes were 61.1–71.3 °C, 66.8–74.9 °C, 78.4–85 °C, and 15.1–16.4 J/g, respectively (Charoenkul et al. 2011).

2.1.2.3 Yam Starch

Yam (*Dioscorea* spp.) is considered as a significant root and tuber crop that is an eatable starchy tuber, cultivated mainly in tropical and subtropical regions of the world due to its cultural, economic, and nutritional significance. Mostly grown varieties contain water yam (*D. alata*), yellow yam (*D. cayenensis*), lesser yam (*D. esculenta*), Chinese yam (*D. opposita*), white yam (*D. rotundata*), and cush-cush yam (*D. trifida*) (Zhu 2015).

Yam starch constitutes 60–80% of dry content of yam tuber, and that is also considered as the key factor in assessing the physicochemical, rheological, and textural properties of food products manufactured by various yam species (Otegbayo et al. 2011; Amani et al. 2004). Individual yam starch grain size ranges from 1 μm up to 90 μm having monomodal size dissemination. Yam starch exhibits round, oval, polyhedral, and polygonal grains along with minor crevices on the surface. Genotypes from numerous yam varieties exhibit different amylose contents ranging from 1.40% to 50% (Perez et al. 2011; Rolland-Sabate et al. 2003), with less quantity of amylose (<20%) perceived for *D. dumetorum* and *D. esculenta*, than for *D. alata*, *D. rotundata*, *D. cayenensis-rotundata*, and *D. cayenensis* whose amylose content is relatively higher (>20%) (Amani et al. 2004; Otegbayo et al. 2014). Research studies have indicated that yam starches have B- or C-type polymorph (*D. alata*, *D. cayenensis-rotundata*), while some have A-type polymorph (*D. dumetorum*, *D. zingiberensis*) (Amani et al. 2004; Zhou et al. 2013; Jayakody et al. 2007). The degree of crystallinity ranges widely from 24 to 53% across the different species and with the identical species (Zhu 2015). As compared to other tuber and root starches like potato, sweet potato, and cassava starch, yam starches have higher gelatinization temperatures (Zhu 2015).

2.1.2.4 Sweet Potato

Sweet potato (*Ipomoea batatas* L. Lam) is a tuberous rooted perpetual plant in the family of Convolvulaceae. Although sweet potato has a relatively higher amount of starch, its commercial use is still inadequate. Starch is the main constituent of sweet potato responsible for approximately 50–80% of dry content of roots. Degree of polymerization of amylose in sweet potato differs from 3400 to 4400 whereas amylopectin with median chain lengths of DP 21–22. A-type, C-type, or a mixture of both crystalline prototypes are found in sweet potato. Absolute crystallinity for sweet potato starch was recorded at 38% (Moorthy 2015).

2.1.3 Pulses

Pulses are basically comestible kernels of leguminous plants counting lentil (*Lens culinaris* L.), bean (*Phaseolus vulgaris* L.), pea (*Pisum sativum* L.), and chickpea (*Cicer arietinum* L.) (He and Wei 2017). After cereals, pulses are the second main food resource for human diet that is of great significance especially in the developing countries (Kaur et al. 2004; Tharanathan and Mahadevamma 2003). The shape of pulse grains can be varied like they can be round, spherical, oval, or irregular. With no crevice and perforation, grain surfaces are found to be smooth (Hoover and Ratnayake 2002; Chung et al. 2008). Size of pulse starch granules generally ranges from 1 to 100 μm, while most of them have distribution in the range of 5 to 30 μm (Stevenson et al. 2006a, 2006b). Amylose content ranges from 17 to 52% (Wani et al. 2016).

Median branch chain length of amylopectin ranges from DP 20.4 to 20.9 that can be verified by anion exchange chromatography (Stevenson et al. 2006a, 2006b). The structure and physicochemical characteristics of pulse starches obtained from diverse varieties such as faba bean, black bean, and pinto bean are stated to have apparent amylose, phosphorus, and bound lipid contents in the range of 25.8–33.6%, 0.004–0.009%, and 0.13–0.15%, respectively, as has been reported by various researchers (Ambigaipalan et al. 2011). Amylose content of starches from four (04) cultivars of field pea (*Pisum sativum* L.) ranges from 48.8 to 49.6% of that 10.9 to 12.3% usually is complexed with native lipid. The degree of polymerization for amylose varies from 1300 to 1350. Dissemination of chain length of shrewd amylopectin of field pea starches exhibits that fraction of short branch chains of DP 6 to 12 ranges from 16.2 to 18.6% (Ratnayake et al. 2001).

Deflection prototype of C_A- and C_B-type starches is identical to the C_C type with the shoulder peak at about 18° and a powerful peak at 23° for C_A type, while for the C_B type 2 shoulder peaks are obtained at 22° and 24° (He and Wei 2017). A thorough research on the starches obtained from faba beans, black beans, and pinto beans exhibits that B-type polymorphic matter in C-type spreading prototype varied from 20.6 to 26.3%, 15.4 to 17.7%, and 7.9 to 17.3%, respectively. However, particular pea mutants exhibit various ratios of A- and B-type polymorphs within the starch grain that can be further consumed as valuable models to comprehend the enhancement of various polymorphs (Hedley et al. 2015).

2.1.4 Others

2.1.4.1 Fruit Starch

In contrast to the starches obtained from cereals, tubers, roots, and pulses, fruit starches have been a topic of keen interest for the scientists expansively over the decades because of the affiliation of sugar and starch and fruit softening, in sequence to present foundation for the emerging improved storage expertise for post-harvested fruits like apple, tomato, winter squash, and banana (Schaffer et al. 2000). Lately, because of the use of starches from fruit in the food industry, their configuration and efficiency have been studied extensively. Usually, in fruits, starch content varies considerably that may be characterized to the dissimilar harvesting time and vitiation of starch during the ripening process. According to the conducted researches, tomato was stated to comprise up to 20% of dry basis starch after 20 days of anthesis; as for the winter squash, the value is up to 65% (Stevenson et al. 2005). Furthermore, in various cultivars, starch storing tissue possesses a crystalline structure. Take the example of columella and pericarp of tomato fruits that are the first sites of starch loading in immature fruits having various starch properties that are at the identical functional phase, starch properties of columella, and pericarp alter. Tomato starch is considered as C type. As far as degree of crystallinity, columella and pericarp comprise 40.1% and 30.7%, respectively (Luengwilai and Beckles 2008).

Apples, just as other fruits, mount up the starch at initial phase of maturation, and then to enhance the sweetness of fruit, degrade it increasingly during the maturation process. The value of starch augmented up to nearly 20–30 mg per fruit basis after flowering period of 110 and 130 days for the apple specie of “Royal Gala” and “Fuji,” respectively. For various cultivars, starch content ranges from 44 to 53.2% on dry basis (Carrin et al. 2004). Granule dialectology of apple is identical to that of squash fruit starch (Stevenson et al. 2005) and kiwi fruit starch, which exhibit mostly spherical or dome-shaped and split granules with the diameter varying from 2 to 12 μm (Carrin et al. 2004).

Banana is a highly significant fruit subsequent to citrus, comprising approximately 16% of the total world food manufacturing. In green bananas, starch is mainly present in the form of carbohydrate that ranged from 69 g/100 g to 82 g/100 g on dry basis (Kumar et al. 2019). Preceding research has showed that for various banana starches, amylose content ranges from 15 g/100 g to 23 g/100 g (db) and crystallinity from 2 to 12.2% (Kumar et al. 2019). Concerning the variety of banana, mounting environment, and segregation approaches, starch can have A-, B-, or C-type crystalline prototype (Zhang et al. 2005; Zhang and Hamaker 2012), while starch granules happened to seem elongated ovals with ridges (Zhang et al. 2005).

Amylopectin represents lesser ratio of short chains DP 6 to 12 (21.6%) and relatively elevated ratio of long chains of higher value of DP > 36 (21.4%) and comparatively higher ratio of extremely long chains that appreciably vary from corn and potato amylopectin (Zhang and Hamaker 2012). Dissemination of short chains (DP 6–12) is describing the amylopectin and known as the fingerprint of the starches (Zhang and Hamaker 2012).

Natural banana starch is extremely resilient to enzymatic hydrolysis partly ascribed to the exterior dense layer of granules obstructing enzyme dispersion into granules (Cummings and Englyst 1991). Native banana starch has possible benefits for different food applications because of its excellent characteristics analogous to the imperceptibly cross-linked starch (Eggleston et al. 1992).

2.1.4.2 Medicinal Plant Starch

Generally, starches obtained mainly from resources have been studied to a great extent, but the starches acquired from medical plant resources are normally superfluous during the segregation approaches of small molecule bioactive composites. With the improvement of starch commercial usage, novel starches with definite characteristics are of main interest.

For several years, in conventional Chinese medicine, beimu, bulbs of *Fritillaria* species, have been used in a wide range of drugs that are especially effective against cough, phlegm, and high blood pressure (Wang et al. 2006). Bulbs of *Fritillaria* species have starch as the key constituent that is responsible for about 80% of the entire weight (Gao et al. 1999).

2.2 Structure of Starch and Starch-Derived Components

Starch grains obtained from different sources vary in size and shape that range from 2 to 150 microns and being round or polygonal correspondingly. Starch mainly consists of two major constituents that are amylose and amylopectin. Portions of those constituents are associated by glycosidic linkages. One-fourth portion of starch is made up of amylose that is a large linear chain arranged of several glucose elements with linkage of carbon 1 and carbon 4 of glucose component and hence comprises α-1,4 glycosidic linkage. Upon cooling, molecules correlate by forming three-dimensional system that accounts for the gel development of cooked, cooled starch pastes (Vaclavik 2021).

Starches having higher content of amylose have the gel-forming ability or they can resume their shape upon molding, while the starches with low or no amylose content do not form gel. Rough estimation of amylose content for various sources is 26–28%, 17–23%, and 0 with no gel formation for the cereal grains, roots and tubers, and waxy diversities of starches, respectively.

Bound by the vegetal resources, starches exhibit various properties, some of which are gelatinization temperature, paste viscosity and clarity, gelling ability, and retrogradation rate. All these characteristics are due to the relationship of structure and functional traits of starch. These are used for the development of novel and diverse starches with improved functional features that can be of commercial use for different purposes.

2.2.1 *Effects of Structures on Starch Properties*

Because of the exceptional functional characteristics, various varieties of starches have diverse applications in food sector. These uses can be extended after modification by chemical, physical, and enzymatic approaches. In food sector, starch is mainly consumed as thickening agent, gelling agent, stabilizer, emulsifier, fat replacer, encapsulation agent anti-caking agent, and resistant starch (Chung et al. 2010; Bai et al. 2011; Ai and Jane 2016). These ample uses are because of the conversion of starch from partially crystalline grains to amorphous phase by trailing the natural crystalline structure (gelatinization); after engrossing water, a viscous paste is developed (pasting), subsequently developing structure to form a gel (gelling) and to recrystallize an insoluble and cumulative phase (retrogradation). These applications are further explained in detail.

2.2.1.1 Starch Gelatinization

When granules are soaked in surplus water at normal temperature, grains of starch can take in up to 30% (w/w) moisture that mostly circulates in the amorphous part of the starch grains. Crystalline structure of starch grains damages after heat treatment when provided to a specific temperature and becomes amorphous as a result of vanishing of Maltese cross. This phenomenon is known as starch gelatinization, and that specific temperature on which crystalline structure is lost is called gelatinization temperature. Gelatinization temperature is one of the most important characteristics of starches usually increased by increasing the branch chain length of amylopectin due to the formation of double-helical crystallites of starch grains (Jane 2006). Starches having short branch chains like waxy and sweet rice starches can be gelatinized at comparatively lower temperature. Some more altered crops like japonica rice and su2 maize have also reported to exhibit the similar results (Ai et al. 2016; Perera et al. 2001). Opposite to the waxy and high-amylose maize, starches that have much higher amount of amylopectin content with much longer branch chain show much higher gelatinization temperature.

Amylose content and transitional constituents of high-amylose starch have higher thermal steadiness forming double helical crystallites (Jiang et al. 2010a). Consequently, extraordinarily higher final gelatinization temperature could be seen by high-amylose maize starch than that of the waxy starches. In boiling water, high-amylose starches can merely be partly gelatinized. It is of quite interest that potato starch represent lower temperature of gelatinization regardless of higher branched chains owing to the presence of linked phosphate monoester derivatives which undermine the crystallize network by revulsion (Jane et al. 1999).

2.2.1.2 Starch Pasting

During the gelatinization process, starch pallets swell more vigorously in the excess of water during heat treatmentcausing the amylose to leach out resulting in the dissolving of them in an aqueous solution. Viscosity formation during this process is called as pasting characteristics of the starch that can be verified by using amylograph like Brabender Amylograph and Rapid Visco Analyzer (RVA). Some of the starch analyses are shown by using RVA. Amylopectin is a key constituent that broadly considers the swelling power of starch grains and succeeding viscosity manufacture (Tester and Morrison 1990). Consequently, waxy starch normally shows the higher peak viscosity compared to normal and high-amylose starch parts.

Amylose content particularly with the existence of polar lipids has the tendency to oppose the swelling of starch grains in the process of heating. Amylose-lipid complex (ALC) developed in starch increases the predicament of amylose and amylopectin molecules that oppose swelling of granules (Ai et al. 2013). This leads to the drop in peak viscosity and enhances the pasting temperature and consequences of impervious representation of starch paste. In comparison to the cereal starches, wheat starch has the maximum value of pasting temperature and minimum value of peak viscosity due to great meditation of phospholipids, a cluster of polar lipids that can promptly form presence complex with amylose in starch (Debet and Gidley 2006). Potato starch exhibits minimum value of pasting temperature and maximumvalue of peak viscosity among all other starches because of the association of phosphate monoester groups andenlarged molecular size.

The pasting profile of high-amylose maize starches exhibits low or no noticeable viscosity as starch can be partly gelatinized by using amylograph (maximum heating temperature 95 °C) (Fig. 3.5a) (Ai et al. 2013; Debet and Gidley 2006). Transformation of the high-amylose resultant in the amylose exempted waxy starch exhibiting relatively highviscosity with a slightly declined breakdown (Kubo et al. 2010). Only sole trait that can be responsible is the length of branched chains interacting to each other that averts dispersal of starch grains and collapse of viscosity. Fascinatingly, rice starch tends to show a negative trend. Rice starches have long-chain amylopectin branched chains but that can be negatively associated to the breakdown of viscosity and positive to setback viscosity that can be characterized to the long chains avoid the dispersal of swollen starch grain, enhancing predicaments in pasting process (Inouchi et al. 2005; Han and Hamaker 2001).

2.2.1.3 Starch Gelling

Starch granules [pastes of concentration of 5% and 15% (w/w)] upon cooling and followed by subsequent storage, swell to fill up space and distort each other resulting in amylose leaching and upholding the shape of mold to formthe gel. To preserve the volume of swell starch grains, amylose and amylopectin obliged to interrelate to each other to precede a linkage that can help in grasping the integrity of swollen grains (Klucinec and Thompson 1999). Typical starches from cereals and pulses like maize, wheat, rice, and pea are recognized as the starches that hold rigid swollen grains and form powerful gel. In comparison to prior waxy starch, potato and tapioca starches could not form a gel or form feeble gels of identical starch concentration due to the ability of starch to swell freely and vanish after cooking because of the deficiency of amylose and endogenous lipids to form amylose-lipid complex (ALC). Before cooking, tapioca starch is completely mixed with lipids (e.g., soy lecithin, oleic acid, and linoleic acid) that can form gels (Ai et al. 2013). Molecular interrelation can be observed to be enhanced by ALC formation in starch grains preventing swelling and sustaining the incorporation of swollen grains for the formation of gel. Potato starch can also form a powerful gel after heat treatment, moisture treatment, or chemical cross association reactions. Texture analyzer or rheometer can be used for the determination of power of gel (Ai et al. 2013).

Swollen starch grains can simply drop incorporation in the process of heat treatment and vanish in the aqueous solution. Because of these circumstances, with the increase in amylose content, the strength of starch gel also increases due to the presence of long and powerful junction zones formed by linear amylose molecules in the gel network (Klucinec and Thompson 1999). So, starch or flour that has comparatively larger concentration of amylose (like mung bean starch, indica rice flour) is selected for the manufacturing of rich noodles and rice sticks, the products which need strong gelling capability of starch (Kasemsuwan et al. 1998).

2.2.1.4 Retrogradation of Starch and Syneresis

When storage is prolonged, in the aqueous solution, amorphous starch molecules start to recrystallize to develop a double-helical crystallite that tends to lose water-holding ability. This transformation is termed as starch retrogradation that is usually perceived in the stalling of bread and syneresis of soups and sauces along with starch as thickening agent. Starch retrogradation can be increased by enhancing the concentration of starch in the paste, upsurge amylose content, enhancing the amylopectin branch chain length (Jane et al. 1999) and stowing the starch paste at the temperature of 0–5 °C (Wang et al. 2015a, 2015b). The paste or gel formed by native cereal starches has comparatively higher tendency to retrogradation than that of tuber and root starches due to partial swelling of native cereal starch grains and gives adjacent propinquity within the starch chains for increased retrogradation. For the reduction of retrogradation rate of starch, several approaches have been used in food products, among of which are explained as follows (Yongfeng and Jay-lin 2018):

1. Chemical derivatives addition in starch like hydroxypropylation and acetylation.
2. To reduce the starch molecules flexibility, starch foods stowed at a temperature less than glass transition temperature.
3. To form annexation helical complex with amylose and amylopectin, long branch chain complex agents like monoglycerides, lecithins and sodium stearoyl lactylate can be added.
4. Length of glucan chains can be decreased by the hydrolysis process of starch by amylolytic enzymes (Goesaert et al. 2009).

2.3 Production/Preparation of Starch and Starch Derivative-Based Films

Normal starches obtained from various plant resources have special characteristics, though they are not very suitable for several functions and have absence of adaptability for an ample variety of produces like foods, drugs, plastics, and cosmetics. These drawbacks are concerned to the low shear confrontation, thermal decay, and elevated propensity to retrogradation, resulting in reduction in industrial relevance (Sandhu and Singh 2007). To avoid this issue, new altered starches are developed by enzymatic, physical, and chemical approaches as substitutes that have been customized to enhance the starch characteristics, advancing its usage.

To improve the practical functions of starches, both physical and chemical properties of normal ones have been altered. Predominantly, chemical alteration consists of the initiation of functional congregates in the starch molecules that change their physicochemical characteristics like gelatinization, pasting, and retrogradation (Sandhu and Singh 2007).

Responses are generally carried out in aqueous medium, characteristically 30 to 45% solids by mass, frequently at pH of 7–12 with continuous rousing and regulated temperature, but gelatinization temperature should also be avoided. In this suspect, to remove the unresponsive reagents, salt, and extra soluble produces with the help of water wash, polysaccharide can be regained in their dry appearance earlier to the end recovery (Rutenberg and Solarek 1984). Reaction is ended by neutralizing the medium and obtaining clear and dry starch with maximum competence of 70%.

Most of the chemical alterations include responses linked to the hydroxyl groups (-OH) of starch polymer. The alternation significance can be estimated by the degree of substitution (DS), a ratio of -OH clusters median number present in each hydroglucosyl unit (GSU), that is supplementary exchanged by functional groups. Each GSU, for the replacement, has three -OH present so maximum number of DS becomes 3. The response can produce polymeric alternatives if substituent cluster (acetate phosphate) powerfully reacts with reagent. Other representation to exhibit alternation stage is molar substitution (MS) that conveys the moles of alternative groups per mole of GSU (Rutenberg and Solarek 1984).

Generally, industrial product's imitative has DS smaller than 0.2. Chemically, altered starches are nontoxic and safe for human health that they can be used without any worry, inadequacy, or measurable limits. Chemically altered starches can be acquired by derivatization and adaptation. Viscosity of normal starches can be lessened by adaptation approaches that are of importance in some applications like salad dressings. Furthermore, altered starches by adaptation can be practiced in some functions at elevated concentrations (greater than 6%) because of their high solubility than its native counterpart.

Conversion approaches consist of acid hydrolysis, oxidation, dextrinization, and enzymatic treatment, providing starch produces having various unique functionalities (Tharanathan 2005). Gelatinization and cooking properties can be altered while retrogradation propensity and gelling of amylose of starch can be reduced by derivatization process. In terms of importance, at depleted temperature, water-holding capacity is enhanced resulting in the reduction of syneresis because of the increment of starch hydrophilic property. Furthermore, the replacement conveys hydrophobic character, and ionic substitution can be initiated. Thickening, gelling, binding characteristics, and film production can also be persuaded by derivatization (Rutenberg and Solarek 1984). The approaches that can be used for the alternation includes cross-linking, oxidation, esterification or etherification, acid or enzymatic hydrolysis, and dextrinization (Light 1990; Rutenberg and Solarek 1984).

For complex film preparation, oxidized starch has been of great interest that tends to enhance the hydrophilic property of starch resulting in elevation of water vapor penetrability and solubility. Upon enhancing the oxidation echelon, the tensile power of film also increased (Zamudio-Flores et al. 2006, 2009). In the meantime, the hydrophobic property enhanced by acetylation incrementation resulting in derivatives that are perfect for the film and coating substantial (Larotonda et al. 2003). Water vapor penetrability and solubility of film can be reduced by acetylation as hydrophilic property of starch decreases. Dissolving capacity of acetylated starch films reduces at the largest stowage period because of the restructuring of starch main constituents, that is, amylose and amylopectin, present in the film complex (Zamudio-Flores et al. 2009). Starch film mechanical and impediment characteristics were experienced to advance by using double alteration (oxidation and acetylation) approach.

The purpose of this double alteration was to attain films with an amphipathic property, altering their characteristics. A few characteristics of the films derived from double alternation approach are dissimilar from that of the oxidized or acetylated ones (Zamudio-Flores et al. 2009, 2011). Predominantly, the films obtained by double alternation tend to have low water vapor penetrability and comparatively high elastic modulus than their oxidized equivalents. Moreover, to be cooperated to the various produces and contents, starch can be physically altered to gain the appropriate characteristics that result in changing the configuration of normal starches. Heat treatment, high pressure, microwaves, ultraviolet light, radiation, ohmic heating, wet means, and mechanical stress (extrusion) are some of the approaches that are used for the physical alternation. The variations in the normal starch configuration modify its practical characteristics, and the attained derivatives can be of great importance in the various applications like paper manufacturing, clothing, plastics, chemicals, and food industry.

Altered starches obtained from these approaches are arguably innocuous, and natural materials that are favored over the derivatives are attained by the chemical alternation. Predominantly, modification of waxy corn was done to obtain the desired rheological characteristics especially highviscosity and shear power. Addition of xanthan gum results in minor alteration of physicochemical and rheological characteristics of oven dried starch caused by carboxymethylcellulose and guar gum. These small alterations can be related to the changes in amorphous zone integral because of drying consequences (Lim et al. 2007). There are a small number of studies that evidently influence the physical processing approaches on the practical characteristics of starch.

2.4 Properties

2.4.1 Thermal Properties

In the food industry, special focus was given to the thermal attributes of starch for the production of starch-based constituents. Thermal stability as well as degradation of starch during processing is prime factor from both industrial and scientific perspectives. The inimitable microstructure of starch along with a number of phases of transitions during the thermal processing of starch gives a great opportunity to illustrate the pragmatic approach of relationship between structural polymers and processing of starch. This understanding of polymers and processing relationship increased the knowledge of processing conditions of natural polymers, production of pyrolyzed starch (Liu et al. 2013), and preparation of starch-enriched foods via frying, baking, and cooking (Aguilera et al. 2001).

To determine the thermal decomposition and stability of starch polymers, thermogravimetric analysis (TGA) has been used (Soares et al. 2005; Guinesi et al. 2006). In an open system, decomposition and dehydration are two prime processing steps that are associated with the degradation mechanism of starch (Liu et al. 2008a, 2008b, 2009a, 2009b, 2009c). Following the same pattern, Liu et al. (2010a, 2010b, 2010c) used differential scanning calorimetry (DSC) technique to determine the thermal degradation of starch with a stainless steel high-pressure pan in a sealed system. In this sealed system, two temperatures were noted as decomposition temperature as the lower temperature depicted the scission of long chains of starch, and at the higher temperature, the glucose ring decomposition occurred (Xie et al. 2012). The production and development of biodegradable starch-based material have been urged due to the oil deficiencies and the rising attention in facilitating the environmental load of petrochemically derived polymers. The thermal stability of starch is one of the vital issues in the thermal processing of starch-enriched materials in shear stress specifically for the extraction technology under sealed conditions (Liu et al. 2009a, 2009b, 2009c, 2010a, 2010b, 2010c). Moreover, the process of extrusion technology for the modifications in starch and starch-based products has increased in the previous two decades (Xie et al. 2006; Moad 2011).

2.4.2 Barrier Properties

Barrier properties have the ability to enhance the shelf life from deterioration such as humidity, oxidation, and bacteria by protecting the internal side of the product. Any starch-based material having good barrier properties should have excellent water vapor barrier although it will have lower barrier properties for oxygen. The pervasion mechanism of starch depends on the penetrant characteristics as well as on chemical structure of the polymers (Bras 2004).

Humidity and temperature are the prime external factors and have great effect on the structure as well as behavior of both the penetrant and polymer. Hence, tortuosity is mainly responsible for the enhanced migration of molecules and consequently limited permeability. The tortuosity can be obtained by the help of nanofillers and defines why nanocomposition is essential for barrier properties (Chen et al. 2008).

2.4.3 Mechanical Properties

Starch films have wide applications in both food and non-food fields. In starch-based materials, improvements of mechanical properties are still an ongoing challenge owing to the lower level of mechanical properties especially tensile strength (Yu et al. 2006; Le et al. 2010). To enhance the mechanical strength, a number of techniques have been introduced that involved compositional and blending of starch-based materials (Gao et al. 2012; Noshirvani et al. 2017), such as reinforcing of starch with natural fillers or minerals and/or blending of starch with different boardable polyesters (Wilhelm et al. 2003; Aldana et al. 2015). In this context, different cross-linking agents and compatibilizers have been used to increase the interface between the starch and other added components. However, the augmentation of any additive is a subtle case for the development of edible packaging or biomaterial owing to the safety concern and toxic issues (Yu et al. 2006; Chaichi et al. 2017; Tongdeesoontorn et al. 2011). For the optimization of mechanical properties of starch-based biodegraded composites, polysaccharide-based crystals and natural fibers at nanoscale have been under keen observation (Liu et al. 2009a, 2009b, 2009c, 2010a, 2010b, 2010c).

The mechanical properties of starch can be momentously increased by the reinforcement of crystals as the mechanical properties of specific polymer at nanocomposite level can be affected by versatile parameters such as aspect ratio, matrix stiffness, volume fraction, and stiffness of the material (Ali et al. 2018). To secure the barrier properties of starch derivatives, mechanical properties should be improved. Suitable mechanical capability confirms the integrity of starch-based film and its resistance to abrasion and breakage as well as suppressed the possibilities of defects, for example, fissures and pinholes which damage the barrier properties. The adequate flexibility ensures appropriate level of plasticity to adapt probable deformation without breaking during filling process (Guilbert et al. 1995).

The mechanical properties of starch and starch derivatives are mainly dependent on the structural cohesion of material used for the film formation. The plasticizer molecules are inserted into the polymer matrix that further modify the organization of the 3D structure of film and the energy involved in the formation of these films. The augmentation of these plasticizers should be compatible with the composites of film polymers to ensure the distribution of 3D structure in a homogenous way. However, the selection of the plasticizers can affect the mechanical properties of starch derivative films. Some polyols and monosaccharides can be used as plasticizers in starch-based films (Zhang and Han 2006a).

2.4.4 *Dielectric Properties*

Dielectric characteristics are of great importance because of their quality to absorb and convert microwave electromagnetic power to thermal power. They can be affected by the microwave frequency, network constitution, density, temperature, and many more features. Concerned to the starch-based content, these characteristics are unique because of their consistent constitution specifically moisture content and metal ion attentiveness. Furthermore, porosity, temperature, and remaining factors of starch-based network seldom alter during various approaches processes like kneading and puffing, prominent to the alteration in dielectric characteristics of the system. Fine composition of starch regulates its physical and chemical characteristics, deviously disturbing the processability and nutritional characteristics. Real-time intensive care of dielectric characteristics of a starch scheme in the process of microwave treatment permits the effect of microwaves to be implicit exhaustively that endow a hypothetical source for controlling the factors of microwave approach in follow-up levels. Dielectric characteristics of starch-water combination in the process of microwave and conduction heating approaches along with the identical heating rate were recorded (Fan et al. 2017d). In contrast to conduction heating, sample going through microwave heating had powerful dielectric reaction especially at the trough temperature range of 45 to 65 °C (Fan et al. 2017d), representing the elevated contact of microwaves and starch tasters could possibly happen to be in follow-up levels. Consequently, it is proposed to lessen the microwave strength density and dismiss the unwanted harm of starch constitution (Tao et al. 2020).

2.4.5 Tensile Properties

The starch and cellulose nanocrystals have tensile properties as well as Young's modulus (Mittal et al. 2016; Li et al. 2015a, 2015b). These both properties can be enhanced by the addition of crystals as crystals do not dissolve in water, and their inflexible particles have the ability to act as reinforcement mediators. These strengthening mechanisms via the addition of rigid compounds into medium are renowned, and all the crystals act similarly in the same pattern. It was further declared that by increasing the concentration of crystals, the tensile strength as well as Young's modulus increased momentously (Kaushik et al. 2010; Muller et al. 2009; Dai et al. 2015). The starch derivative films that are enriched with cellulose crystals have higher young's modulus because the cellulose crystals have greater thermal stability and modulus as compared to the starch crystals. Besides this, the starch derivative films in which monosaccharide crystals were incorporated depicted higher tensile strength due to the smaller size and spherical shape of monosaccharides as compared to the cellulose crystals (Siqueira and Dufresne 2009; Spirido et al. 2015).

Tensile strength is always represented as Kg force per millimeter square (Kg force/mm^2) measured by autograph. Basically, it is the mechanical property of any starch-derived edible film which exhibits maximum stress that films normally bear before breakage. The concentration of nano-emulsion less than 3% and glycerol less than 30% did not impart any significant effect on the tensile strength of films. Generally, the higher concentration of nano-emulsion depicted lower tensile strength and vice versa. This inverse relation is due to the capabilities of glycerol to increase flexibility of film and lower intermolecular forces in polymer chains of films (Intan and Aizan 2011). By the incorporation of essential oils in starch-derived films, the elongation properties of film increased; however, the tensile strength of films decreased. In this context, Zhang and Han (2006a) concluded that tensile strength of the starch-derived film decreased in which different concentration of lemon grass oil was added. It is concluded that the decrease in tensile strength was due to the cover of lipids on the starch and polysaccharide molecules which have greater tensile strength compared to oil. Moreover, the added glycerol started the breakdown of bonds between polysaccharides, hence forming free units. The molecular interaction decreased, so the tensile strength declined. Glycerol has the ability to penetrate in the polysaccharide molecules and release hydrogen bonding (Madigan et al. 2003).

On the other side, Asiyah et al. (2018) concluded that the amount of glycerol and essential oils such as nutmeg in the form of nano-emulsion had great effect on the characterization of edible films that were prepared from the corn starch. Nearly about 1% nutmeg was added in the corn starch-based film along with 30% glycerol, and it proved the 69.44% elongation, 1873 kg force/mm^2 strength, and 0.0840 mm thickness. Furthermore, the addition of 1% nutmeg inhibits the growth of *Escherichia coli* and *Staphylococcus aureus* in starch-derived films.

2.4.6 Heat Sealing Properties

Sealing property is indicative of the sealing quality. Heat sealability is one of the vital factors for the polymerization of starch-derived films especially edible coatings and flexible packaging films (Kim and Ustunol 2001; Tongnuanchan et al. 2016; Liu et al. 2020). During thermal bonding, to provide sealing ability, two films are pressed for a specific time period between hot metal bars and then immediately cooled down to facilitate the interfacial interaction of melted polymers between the films. Generally, thermal adhesion occurs on their onset temperature. In addition to the strength, the information about the failure of sealing of two films is also an important factor to determine the packaging performances. The integrity of packaging depends upon the sealing, and it is considered as good quality if the interaction between molecules of individual layer is enough to eradicate interface, hence leading to the formation of a homogenous layer (Lopez et al. 2011). By the addition of plasticizer, the melting point of starch-derived films started to decrease by weakened bonding among the molecules; hence, mobility of starch molecules increased. Therefore, sealing can be done at temperature lesser than required for the conformist thermal bonding (Jimenez et al. 2013a, 2013b). From the previous era, it is believed that the addition of water is supposed to be the best-known plasticizer that resulted in the higher elongation of film as well as lowered tensile strength (Matveev et al. 2000; Heljo et al. 2012).

In the last few years, Tongnuanchan et al. (2016) investigated that the sealing properties and their qualities are affected either by the water or not. For this purpose, the heat-sealed starch films were prepared from for different botanical sources by the application of heat gradient tester. Before testing, they preserved every film in a controller relative humidity chamber and maintained the RH for a week by using saturated salt solution. On the conclusion of heat gradient test, they classified the results into three categories as adhesive seal failures, failure at seal edge, and cohesive seal failures.

2.4.7 Flexibility Properties

The thermoplastic starch (TPS) films with lower retrogradation issues and higher flexibility can be obtained by the blending of starch-derived films with other compounds. Besides on the blending of existing polymers and then improving their compatibilities by adding more significant compounds or by chemical modifications, a new tailor-made polyester PBCEA (poly-butylene cyclohexanedicarboxylate co-adipate) copolymer has been synthesized to get all the required properties of starch-derived films (Dominici et al. 2020).

To prepare this molecule to enhance the flexibility of starch-derived films, two subunits, their polymer architecture, and molar ratio were specifically selected. PBCE (poly-butylene cyclohexanedicarboxylate) can be characterized by the availability of cycloaliphatic ring present in the repeating unit. The structure of cycloaliphatic ring is similar to the glucose molecule and, hence, has improved compatibility with starch. This cycloaliphatic ring has good resistance to moisture and heat and higher melting point, making it perfect to be used in starch-derived packaging films (Gigli et al. 2013; Gigli et al. 2014). Furthermore, limonene-derived terephthalic acid and succinic acid, both are renewable resources for the preparation of PBCE (Genovese et al. 2016). Besides this, adipic acid is generally biobased (Skoog et al. 2018) and has been incorporated in the PBCE chain to reduce the melting point of crystalline domains. Hence, the blending temperature can be reduced without starch degradation by the combined effect of shear stress and heat with a subsequent decline in properties (Liu et al. 2010a, 2010b, 2010c).

2.4.8 *Electrical Properties*

Electrical properties can be easily described in the form of electrical conductivity which has vital part in ohmic heating. In ohmic heating, electricity is transformed into thermal energy via the circulation of alternating current through the food. Mostly, ohmic heating is used for the pasteurization of fluids where the main factor is resistance of the liquid part in the mixture during the electrical conductivity. The measurement of electrical conductivity is dependent on the characteristics of material to be used in food such as chilling tolerance, freezing properties, frost sensitiveness, seeds germination, moisture content, pasteurization, mechanical stress, and other many properties of seeds, grains, sugar, meat, fruits and vegetables, milk, soil, wood, maturity of fruits, radio frequency heating, freshness of eggs, and infected foods. The electrical conductivity is also a measurement of electrical current that passes through the food having cross-sectional area (A), with resistance (R) up to a specific unit length (L). It also has inverse relation of electrical resistance and is expressed as s ¼L = ð Þ ð AR 1Þ.

The properties that are linked to measure the interaction of food with the electromagnetic fields are known as dielectric properties (Ahmed et al. 2007). The dielectric loss factor, relative permittivity, alternate current conductivity, and loss tangent are the main concerns in drying, heating, and storing of grains. The concept of dielectric capabilities of foods was first introduced five decades ago.

Besides this, electrical permittivity is another dielectric property which deals with the matter and electromagnetic waves and describes about the electrical filed and the charge density applied on that field (Barbosa-Canovas et al. 2006). In gases, liquids, and solids, the electrical permittivity is linked with the electric constant (ε') which is further associated with the capability of any food material, its storage ability, and the dielectric losses (ε''). The dielectric loss is depended on the energy loss in food when an alternating current filed is applied such as ionic conduction and dielectric relaxation (Jha et al. 2011).

2.4.9 Optical Properties

The polarized microscopic optical images of starch-derived films and films in which starch is bonded with the cellulose crystals depicted higher optical activities as compared to the starch films alone. The gelatinized starch suspension has dark background spots under optical microscopes; however, it is only visible when the film is prepared from pure starch film. Both the cellulose crystals and starch crystals can be identified by using polar optical microscope along with the categorization of starch matrix as homogenous distribution. The optical microscopic images of starch matrix depicted the particle size, geometric shape, and brightness of the crystals which were distributed in the starch suspension and then dried to form a film. This concluded that two types of crystals were stable during film formation and this mechanism can be used to determine the reinforcement of the system (Viguie et al. 2007; Vu-Bac et al. 2015). Mostly starch-based films are transparent; however, this transparency rate decreased with the augmentation of crystals that were formed in double-phase system. The transparency changes with the concentration of cellulose and/or starch crystals as the transparencies decreased with the increase of crystal contents. The transparency of film at wavelength of 206 nm can be used to protect the food from the effect of UV radiation.

2.4.10 Water Solubility Properties

The increased concentration of glycerol in starch-derived films has the ability to increase the initial contact angle of wettability as the surface of the starch-derived film becomes more hydrophilic by the addition of glycerol. However, by the augmentation of sodium montmorillonite (Na-MMT) in the starch matrix, the contact angle of starch matrix as well as hydrophobic condition of biopolymer increased. When a water molecule is adsorbed on the surface of starch-derived film, the attraction between water and the starch molecule occurs, and the strength of that attraction depends on the properties of liquid and solids. The reduced contact angle between water and polymer proved higher attraction strength; hence, they adhere to each other more strongly (Heydari et al. 2013).

2.4.11 Viscous Properties

To determine the viscosity of any starch material viscometer including orifice, capillary flow, falling ball, and/or rotational type, amylograph can be used (Sahin et al. 2006). The viscosity of the starchy material showed non-Newtonian feature as the dose of shear stress does not increase proportionally as increase in shear rate. The shear rate of any starchy material can be expressed as a property of shear stress to be fit in different models as Herschel-Bulky, power law, and Bingham model (Che et al. 2008). Viscosity for starchy material is also a shear thinning (thixotropic) as it displays suppressed viscosity regarding time and shear rate (Biliaderis et al. 2009). Moreover, the viscosity of starch increased with the starch concentration; however, it decreased with higher amylose content and temperature (Ai and Jane 2015).

2.4.12 Pasting Properties

Use of pressure on starch results in the constrain of leakage of amylose and amylopectin, hence altering the structure and elevation in temperature and decreasing viscosity (Liu et al. 2016b). Pasting temperature can be related to the temperature at which gelatinization process of starch is initiated (Schirmer et al. 2015); when gone through pressure, buckwheat, red adzuki bean, sorghum, and proso millet starches are observed to have comparatively higher pasting temperatures than that of 0.1 MPa treated starch presenting the maximum estimation at 600 MPa (Liu et al. 2015, 2016a, 2016b, 2016c, 2018). For both lotus seed and mung bean starches, pasting temperatures recorded to be enhanced under the conditions of 0.1 to 500 MPa and 0.1 to 600 MPa, respectively (Li et al. 2015a, 2015b; Guo et al. 2015). At these elevated stages of pressure, pasting temperature recorded was at the maximum stage. Liu et al. (2018) stated to have no significant alteration in the pasting temperature of pea starch when pressure value alters from 150 to 450 MPa, but when this value elevates up to 600 MPa, the minimum value of pasting temperature is recorded at that for the pea starch. Identical observations were recorded by Li and Zhu (2018) for the maize and quinoa starches, while for the lentil starches, it is shown that pasting temperature reduced from 0.1 to 600 MPa, grasping the minimum value of 600 MPa.

Starch that undergoes pressurization process up to 360 to 480 MPa is described to have comparatively drastically poorer peak time than that of native starches but elevated with more, enhancing the pressure process until the maximum peak value is obtained and was accessed at the value of 600 MPa (Jiang et al. 2015a, 2015b; Li et al. 2011). When lotus seeds and red adzuki bean starches undergo pressure of 600 MPa, the maximum peak is obtained in comparison to the other approaches (Guo et al. 2015; Li et al. 2015a, 2015b). As regards pea starch, Leite et al. (2017) reported that there was no momentous alteration in peak time with the pressure approach ($p > 0.05$).

Maximum peak value was acquired at 600 MPa, but it is not accurately substantially linked to the control. With the increase in pressure of buckwheat and sorghum starches, there is reduction in viscosity value; in contrast to the normal starches, also the minimum value is recorded at 600 MPa (Liu et al. 2016a, 2016b, 2016c). These outcomes are partly in compliance to the lotus seed starches (Guo et al. 2015). When the pressure value changes from 100 to 500 MPa, there is a depression in peak, and then final viscosity value is enhanced proportionate to the normal starch; nonetheless, breakdown and setback viscosity values are reduced.

At the pressure of 600 MPa, viscosity is reduced and had a minimum value. When the pressure value is altered from 0.1 to 600 MPa, breakdown and setback viscosity values of chestnut starches tend to reduce drastically (Ahmed and Al-Attar 2017). Red adzuki bean starches are found to have a minimum viscosity value at the pressure of 600 MPa in contrast to that of 0.1 MPa and other approaches (Li et al. 2015a, 2015b). Overall, at the pressure of 120 to 480 MPa, mung beans' viscosity value elevated comparatively to that of 0.1 MPa and 600 MPa, but the momentous reduction in value recorded at 600 MPa ($p < 0.05$) (Jiang et al. 2015a) rises; the viscosity value of pea starch can be observed at the pressure of 150 to 450 MPa, but a reduction can also be recorded at the value of 600 MPa (Liu et al. 2018). In contrast to the 0.1 MPa, viscosity value at 600 MPa increases, but breakdown value was not noteworthy (Jiang et al. 2015b). In comparison to pressurized mung starch and normal starch, the value of mung starch represented increment, but breakdown value is reduced (Li et al. 2011). Leite et al. (2017) also studied the rise in viscosity value from 0.1 to 600 MPa.

2.4.13 Aroma Retention Properties

Starch and starch-derived biopolymers can be used as stabilizers, textural and gelling agents, and thickeners as well as carrier for aromatic compounds. The composition and nature of starch molecules, granule shape, size, porosity, specific area, composition of starch-polysaccharide matrices, and amorphous or crystalline characteristics have major factor in the binding of volatile compounds (Boutboul et al. 2002; Baranauskiene et al. 2007; Savary et al. 2007; Zafeiropoulou et al. 2010). The binding of aromatic compounds as flavoring agents in starch is well-known as inclusion complex formation via hydrophobic bonding in polar interaction of amylose helix that is further involved in the hydrogen bonding of odorant and hydroxyl groups of starches (Arvisenet et al. 2002; Boutboul et al. 2002). The binding of low molecular weight and short chain compounds with starch is based on the non-specific sorption to granule agglomerates and starch powder (Misharina 2004; Conde-Petit et al. 2006).

When the permeable assembly of the starch extracted from corn is treated with the glucoamylase enzyme, the ability to retain peppermint oil increased. The porous starch molecules have the ability to retain 29% of the peppermint oil when it was kept in open air without any standard procedure for a time period of 2months; however, the ability to retain the oil decreased to 12% on native starch molecules when it was kept in open dish for a time period of 3days. During the succinylation process, the volatile compounds of meat flavors are incorporated in barely and corn-based starches (Jeon et al. 2003). After the storage interval of 4 weeks, the flavoring compounds present in barely and corn starches were greater than the β-cyclodextrin that is most common material used for encapsulation. Further, it was noted that when the starches which were modified hydrophobically were used as encapsulation material, it enhanced the flavoring compound's retention up to 50% (Baranauskiene et al. 2007). Moreover, the emulsified starch used as HiCap was also depicted as an effective matrix for encapsulation of caraway oil and reported to be heat stable during encapsulation process (Partanen et al. 2002).

The research work conducted previously proved that the type of corn starch (Amilogel G) used for the formation of film has effect on the retention of strawberry-based aromatic compounds and been found to be the best carrier for the delivery of majority of analyzed compounds (Vidrith et al. 2009). The retention time of the aromatic compounds present in cross-linked polysaccharide gels and/or waxy corn starches was attributed to the physicochemical interaction between gelling compounds and the low molecular compounds (Savary et al. 2007). Besides this physicochemical and structural attributes of polymer matrix, the retention of aromatic compounds was linked with the chemical and structural properties of starches (Misharina 2004).

2.4.14 Antimicrobial Properties

Primary function of edible films and coatings is to shelter the exterior of the food produces so that the shelf life of the produces can be enhanced along with substantially their nourishing value and sensory quality (Krasniewska and Gniewosz 2012). Significant elements that are the main reason of dilapidation of food are spoilage microorganism and foodborne pathogens. Existence of spoilage microorganism especially involved with the food can cause several unwanted alterations like that can fasten the lipid oxidation that partakes in the alteration of organoleptic characteristics of the produce. In sequence, pathogenic microorganisms can cause foodborne illness in humans (Atares and Chiralt 2016).

Yam starch films are also studied by Mali and Grossmann (2003) as a packaging material to enhance the shelf life of strawberries. The study concluded that starch film decreased deterioration and microbiological adulteration of supplied tasters. Moreover, shelf life of the film covered strawberries extended up to 21 days comparatively higher than that of 14 days without film coverage. The main reason could be linked to the characteristics of the film like oxygen barrier that helps in introversion of development of aerobic pathogens.

Contradictory to the different bacteria and fungi, gum Arabic showed extraordinary antimicrobial activity (Bnuyan et al. 2015). That is the main cause that many efforts are made to use gum Arabic as an eatable covering to sustain the fresh fruits and vegetables. For instance, gum Arabic has been employed to apples for the purpose of sustainability of quality and safety in the process of cold storage. Comparatively, coating appreciably lowers the deterioration (El-Anany et al. 2009). Extreme interest has been shown in the increment and enhancing the antimicrobial characteristics of the eatable packaging content. Subsequently, researches have been conducted to analyze the chances of introduction of active compounds along with hypothetical biology action of films and coating. Coating of apricot gum along with the extraction from Satureja intermediately propels antifungal impression next to *Penicillium citrinum*, *Fusarium oxysporum*, *Aspergillus flavus*, and *Alternaria alternata*. Furthermore, use of these working substances has already been verified for the actual food. Aforementioned coating efficiently constrains the microbial pollution because of the fungi on wild almond kernel during stowage of 60 days (Hashemi and Raeisi 2018).

Antifungal influences of gum Arabic including lemongrass oil and cinnamon oil have been studied extensively by Maqbool et al. (2011) for the regulating postharvest anthracnose on papaya and banana. That exhibits significance reduction in the development of reaction to anthracnose in tropical fruit. Similarly, potato peels combined with oregano vital oil efficiently restrain the development of *Listeria monocytogenes* for 30 days on cold smoked salmon (Tammineni et al. 2013).

2.4.15 Moisture Sorption Properties

In mostly hydrophilic films, water has the ability to work as plasticizers, and the water adsorption capacities of these films depend on the relative humidity level in surrounding environment. In starch-derived films, the water content acting as plasticizer is more hygroscopic as compared to the starch molecules. The difference in water adsorption ability of starch-derived films is mainly dependent on the type and quantity of plasticizer, while the starch content in the film remains constant (Zhang and Han 2006b). On the comparison of type of plasticizer used and the moisture content of the film, the films having glycerol with contestant level of starch showed higher concentration of adsorbed water. This may be due to the better plasticizer potential of glycerol than any other plasticizer. Furthermore, the glycerol molecules have greater effect in lowering the attractive forces between the polymers of starch films, hence enhancing the segmental motions and volume; ultimately water molecule penetrates in the starch film more easily resulting in increased water content (Zhang and Han 2008). Similarly, Zhang and Han (2006b, 2008) suggested that the addition of glycerol as plasticizer in starch-derived films has the capability to hold more moisture content as compared to other polyols used as plasticizer in films, and it was attributed to have higher polarity of glycerol. They further reported that glycerol serves as the water-holding agent hence, higher concentration of water molecules within the network of starch molecules.

Beside this, Talja et al. (2007) depicted that the water adsorption of starch films increased with increase in glycerol content owing to the lower molecular weight of glycerol. The water content value of monolayer glycerol plasticize film was 15.7 to 21.4 g/100 g dry weight. The values obtained by the addition of glycerol as plasticizer were comparable with the monolayers of potato starch (Talja et al. 2007), corn starch films (Bertuzzi et al. 2007), ethyl and methyl cellulose-based films (de la Cruz et al. 2001), and pea starch films (Zhang and Han 2008).

Furthermore, the monomer layer of glycerol plasticizer film is used to express the sequence and number of active sites available in starch water for the water adsorption (Inchuen et al. 2009). Consequently, it was suggested that the addition of glycerol into the starch-derived film leads to increase in the free sorption sites. However, this result was in line according to the molecular weight of glycerol compounds. Owing to the smaller molecular weight, the constant plasticizer effects of starch, and the addition of glycerol, the maximum available hydroxyl groups increased the moisture content of monolayer film as compared to other plasticizers. Conclusively, the availability of glycerol monomers in starch-derived films can increase the hygroscopic properties (Cho and Rhee 2002; Mali et al. 2005a, 2005b; Martelli et al. 2006).

2.4.16 Release Kinetics of Active Packaging

The antioxidant active packaging films can be divided into two categories depending upon the incorporation of antioxidants in the polymer surface or in the matrix of film. During the manufacturing of starch-derived film-based antioxidant packaging material, the active moieties are intimately mixed with the polymer matrix by melting the starch polymer and then incorporating into melted antioxidant level via extrusion process or by dissolving in appropriate solvent through casting procedure. Currently, other techniques comprise immobilization of bioactive moieties in the surface of polymer that is further treated chemically or physically by coating method (Gomez-Estaca et al. 2016; Fang et al. 2017).

Regardless of the process that was adopted for the preparation of starch-derived film, the direct contact of film with meat products was adjusted according to the availability of the active compounds present in packaging film and the time required to release the bioactive ingredient from the film to the food product (Gomez-Estaca et al. 2016). In this context, the release of antioxidant compounds from starch-derived film is a complex process that depends on a number of factors such as the compound used as antioxidant, its quantity, characteristics of matrix, as well as food (Lopez-De-Dicastillo et al. 2012). Another vital factor to be considered is the number of layers during manufacturing of film; the film is either monolayered or multilayered (Busolo and Lagaron 2015).

2.4.17 Cost-Effectiveness

Starch is one of the most auspicious and adaptable biopolymers for the production of active films (Vargas Junior et al. 2015). All around the globe, the starch-derived films are produced on larger scale by offering the higher availability of films, hence decreasing the market price of these films (Carvalho et al. 2001). The main advantage of the starch-derived films is the taste, and it has an excellent barrier properties for CO_2 and O_2. These films are tasteless, transparent, and odorless (Chiumarelli and Hubinger 2012). Although these films did not provide barrier properties for water vapors owing to the higher hydrophilicity (Suput et al. 2015).

The active packaging films having antioxidant as bioactive molecules are a promising future for the packaging of meat and meat-related products. It is obvious that the utilization of these active packaging films has the potential to increase the shelf of minimally processed meat. Hence, the concurrent application of natural antioxidants in addition to biopolymers added in the monolayer or multilayer of the film not only satisfies the demands of consumer for the replacement of artificial polymers and preservatives but also offers highest protective effects on the prevention of lipid peroxidation in meat and meat-derived products. Therefore, the manufacturing and implementation of these packaging films will be dependent on the cost and acceptance for consumer as well as by the industry (Dominguez et al. 2018).

2.4.18 Safety and Regulations

Starch-derived films and edible coatings are the integral part of the edible portion linked with the food products; hence, it should follow all the essential safety requirements and fulfill the regulation standards as pertinent to food ingredients (Guilbert et al. 1995). Previously, Druchta and Jonhston (1997) concluded that the starch-based edible films also have safety issues and health-related problems, and to maintain the eating quality and safety level of food product, all the ingredients required for the manufacturing of starch-derived films along with the functional additives should be generally recognized as safe (GRAS) and should be used within the limits issued by the Food Drug Administration (FDA). If any of the ingredients of starch-based films have not gained the status of GRAS, and however the manufacturer claimed it as safe to be used, then the manufacture should proceed to the market under the mark of self-determination procedure or file the GRAS affirmation petition to the FDA. On the other hand, all the films and edible coatings contain a number of nutritious/functional ingredients such as coloring compounds, flavoring compounds, antioxidants, and antimicrobial agents in the form of food additives. In developed countries, the added food additives should be given an E number to recognize its GRAS status. Although a few ingredients used in starch-derived films can cause allergies in some individuals. The well-known allergens are mostly from wheat (gluten), milk (casein), peanuts, and soybean. Therefore, the film having any of these ingredients should be mentioned clearly on the label even if used in minute quantity (Franssen and Krochta 2003).

Generally, to get the commercial success, all the ingredients to be used in film formation over and above to the use of additives to enhance the shelf life and quality of the food product should be accepted via sensory evaluation trails to explore the response of consumer to check the preferences for coated and non-coated food products (Wan et al. 2007). Besides this, another important parameter that should be counted for the safety and regulations of starch and starch-derived films is the allergenicity of coatings and films along with the toxicity level. In this context, essential oils having antimicrobial compounds also have allergic effects in some persons even though the oils are registered as generally recognized as safe. Higher dose of that allergic compounds can cause serious health-related issues via oral toxicity, so it is necessary to use in balanced quality to have the effectiveness of natural essential oils along with the risk level of these plant extracts (Ribeiro-Santos et al. 2017).

2.4.19 Methods to Improve Properties

Presently, usage of starch as natural content for the production of biodegradable film has been of great interest together with strong clues of conceivable and inadequacy obliged by the usage of native material for the present intention (Garcia et al. 2000; Mali et al. 2005a, 2005b; Veiga-Santos et al. 2008; Muller et al. 2008). Assuming the hydrophilic character of starch, in comparison to the artificial contents, the produced materials have higher value of water vapor penetrability and lower value of mechanical characteristics. By adding up other contents into the polymer structure, these characteristics can be developed (Muller et al. 2011).

Nowadays, customers have keen interest in each and every aspect of food product with elevated quality and shelf life along with the reduction of disposable packaging resulting in enhancing the recyclability. This interest with so much stress on various points results in the advancement of keen attention towards the eatable and biodegradable contents that operate as mass transmission obstacles to the water, oxygen, taste, smell, fat, and carbon dioxide among the constituents of food and surrounding environment that further prolong the shelf life and enhance the whole quality of food network. Those contents have the capability of reducing the quantity of non-reusable traditional artificial polymer packaging materials and practice elements of agriculturally derived produces. Eatable coatings and films are of great interest for scientists because of their ability to extend the shelf life of various produces (Nunes et al. 2013; Wu et al. 2013). Eatable films can be described as a fine coating of material that can be consumed as an element of the entire food produce that also support in hindrance to the mass transfer (water, taste, and gas) among food and its surroundings and itself too (Broumand et al. 2011).

Amid of biopolymers, bionanocomposite content has a wide range of functions as a packaging material in the food industry (Tang et al. 2008) as starch and its derivative films are safe to eat as a food packaging material. However, starch does not have such good mechanical characteristics that can be enhanced by introducing additives like plasticizers and nanofillers (Sorrentino et al. 2007). Among the improved properties, there are still some disadvantages of biopolymers as food packaging material as compared to the non-biodegradable material consisting of inadequate mechanical characteristics (low tensile strength), adequate water absorbing measurements, stiffness, brittleness, and barrier characteristics like water penetrability (Othman 2014). Specifically, because of some brittleness, depleted heat distortion temperature, elevated gas, and vapor penetrability, deprived confrontation during extended organizational process mainly restrains their use (Rhim et al. 2013).

Starch is mixed with additional polysaccharides and polymers to overawe these mentioned disadvantages that helps in appropriate film-forming process. It is stated by the literature that mechanical characteristics like tensile strength and elongation of artificial or biodegradable are observed to be good, whereas mixing of starch with additional polysaccharides, plasticizers, and additives meets the insistent mechanical characteristics of packaging.

2.4.19.1 Plasticizer

Glycerin is of great importance as a plasticizer agent because of the absence of retrogradation process that usually happens to have with glycerin and more plasticizers (Battegazzore et al. 2015). Film-molding moisture can be preserved by using glycerin. Glycerol is of great significance in food packaging films as it is a bio-diesel remainder making it an appealing plasticizer. In the step-by-step procedure of film making, drying is one of the most vital steps that can directly influence the characteristics of biofilm.

Reduction in moisture diffusion can be seen by adding glycerol in the process of drying plasticized eatable films. It is extremely important during thermal processing of normal starch to add plasticizer at precise time; decomposition temperature is lower than that of melting sooner than gelatinization. Addition of water results in weak mechanical characteristics and changes with changing value of humidity, whereas through thermal processing, viscosity decreases. The influence of water relies on different restrictions and existing extra additives. To acquire the desired characteristics for appropriate gelatinization, different plasticizers have been consumed in mixture with water.

2.4.19.2 Polyols and Sugars

Polyols can also be used as an additive to acquire the good mechanical characteristics and impart momentous alteration in elasticity and evade pores and extension of polymeric network. Flexibility of a film can be decreased by adding a plasticizer because of the hydrogen bonding among polymer chains, hence enhancing intermolecular space (Liu et al. 2009a, 2009b, 2009c). Transparency of film is influenced by the categorization of amylose and quantity of hydrogen bonding formed within the film. Most important property is solubility that plays a vital role in encapsulation of food or additive and film preparation. Glycerin is generally recognized as safe (GRAS) by FDA due to its non-sweet and cost-effective character. The best advantage to use it is the absence of retrogradation that can only occur in the presence of glycerol and another plasticizer agent (Battegazzore et al. 2015).

Glycerol in combination with urea proved to form steady and powerful hydrogen bonds with water and starch as compared to glycerol alone (Rahman et al. 2010). In comparison of sucrose, sorbitol, and glycerol, sucrose happened to have more effectiveness as plasticizer than the other two. To achieve the minimal surface microbial infection of eatable films, sorbates considered as GRAS can be added because of their activity antagonistic molds, yeast, and bacteria (Barzegar et al. 2014). Generally, films derived from starches are considered to be strong, and this power completely relied on the moisture quantity of film and relative humidity of the surroundings in that they are deposited. As plasticizers badly influence the barrier and mechanical characteristics of film, therefore the appropriate amount should be used precisely (Shah et al. 2015).

During the process of starch alteration, addition of plasticizer reduces the intermolecular magnetism, enhances polymer motion, elevates prolongation, and declines tensile strength with the increase of glycerol content. Coating of starches with combination of plasticizer is prone to be uniform and shelter the whole exterior of fruit, while without plasticizer proved to be brittle and retain adverse fractures. Starch-based film with combination of glycerol that undergoes oxidization process demonstrates the hydrophilic character (Kuakpetoon and Wang 2006). Extraordinary film formation and depleted viscosity characteristics can be fabricated by oxidized starch. In the process of crystallization, movement of chain and capability of amylose chain ascertain the quantity of crystallinity in the starch films. Crystallinity portion of films can be seen by slow process of drying, and rapid water vaporization can enhance the amorphous stage that can bound in the existence of plasticizer (sorbate) and also helps to avoid the recrystallization process of starch (Fama et al. 2005). Isosorbide plays an important role as a natural plasticizer because of the characteristics of absence of retrogradation that generally happened in the presence of glycerol and extra plasticizer agents (Battegazzore et al. 2015).

2.4.19.3 Lipids and Waxes Matrix

Enhancement in the characteristics of barrier, mechanical, and optical characteristics of film is done by adding lipid-causing changes in interior structure and exterior layer (Jimenez et al. 2014), while wax addition improved moisture barrier of the film. Combination of lipid nanolayer (sunflower) and hydrophilic film (starch) enhanced the tensile strength and reduced water dispersion coefficient; subsequently, network and hydrophilic character of lipids and oils bound the transfer of gas and vapors (Slavutsky and Bertuzzi 2015).

2.4.19.4 Wetting Agents and Emulsifiers

Lipids have comparatively less vapor penetrability than that of native waxes making it a good choice for the film preparation while generally used are palmitic acid, sunflower oil, stearic acid, and soybean acid. Introducing lipid nanolayer in film amended the water barrier characteristics due to managed penetration process by diffusion across hydrophobic nanolayer (Slavutsky and Bertuzzi 2015). Waxes of high melting point have proved to be effectual in bettering the moisture barrier characteristics of film, while shelf life can be lengthened by totaling 20% eatable oil to starch source layer in oily foods as configuration and hydrophobic character hinder gas and vapor transfer (Slavutsky and Bertuzzi 2015). Mixture of lipid nanolayer (sunflower) and hydrophilic film (starch) imparts high tensile strength, trough water diffusion coefficient resulting in reduction in diffusivity, and Young's module of formed films (Slavutsky and Bertuzzi 2015). Hydrophilic films are water responsive that factor can be reduced by adding hydrophobic materials (Shah et al. 2015). Reduction of optical characteristics of film can be achieved by introducing long chain saturated fatty acid along with fatty alcohols. Lipids also have some disadvantages like they can affect coating clarity and vapor penetrability resulting along with after taste issue. These shortcomings are because of the low lipid melting points, emulsion drop dimension, solvent volatilization from configuration system, polarity, extent of saturation, and polymorphism of lipid constituents on water barrier and mechanical characteristics of films and coatings.

2.4.19.5 Antioxidants

Addition of antioxidants in starch helps to extend produce constancy to lengthen the shelf life of the product, while the native antioxidants used are vitamins, tocopherols, carotenoids, terpenes, phenolic acid, organic acids, etc. (Ashwar et al. 2014; Inam u Nisa et al. 2015). Use of citric acid with starch resulting in film of thick networking involving decline in water loss and penetrability by cross-linking responses. Shah et al. (2014) also reported the betterment of characteristics of film by adding *Nymphaea mexicana* as an antioxidant. Buckwheat is considered as one of the appropriate antioxidants for film making as it is rich in phenolic content (Jan et al. 2015). Also, green tea has higher phenolic content that also proved to be a good choice to be used in film formation (Ahmad et al. 2014).

2.5 Limitations and Future Trends

The future trends of researches towards the starch properties and its applications in food processing include the following:

- Exposure and development of new starch types having desirable physical, chemical, and functional properties along with modifications according to the application.
- The continuous research to determine the structural sequence, molecular organization, and order of starch molecules at different magnitude levels.
- He enhanced understanding of the links between starch and other constituents of foods and the resultant effects on the nutritional value as well as sensory properties of food product.

- To develop the fast and new methods to observe the minute structural and functional changing in starch molecules due to temperature during processing, storage, and distribution.
- To develop the novel methods for the enzymatic, physical, and chemical modifications of starch molecules to enhance the properties from all aspects.

2.6 Conclusion

Starch is considered as one of the significant materials as food packing because of the low cost, low toxicity, biocompatibility, and biodegradability, but low mechanical properties can be improved by introducing plasticizers, polyols and sugars, lipids and waxes, wetting agents and emulsifiers and antioxidants, etc. There are various resources of starch that are cereals, tubers and roots, pulses, and many more. Heat-sealing property of starch films can be increased by the addition of plasticizers that further improves the barrier properties of starch-derived films. On the other side, the optical property of starch describes about the gelatinization of starch as the gelatinized starch showed black background under optical microscopes. When a water molecule is adsorbed on the surface of starch-derived film, the attraction between water and the starch molecule occurs, and the strength of that attraction depending on the properties of liquid and solids is used in starch-derived films. Furthermore, the composition and nature of starch molecules, granule shape, size, porosity, specific area, composition of starch-polysaccharide matrices, and amorphous or crystalline characteristics are major factors in the binding of volatile compounds to enhance the aroma retention properties of starch-based films. On the comparison of type of plasticizer used and the moisture content of the film, the films having glycerol with contestant level of starch showed higher concentration of adsorbed water. The release of antioxidant compounds from starch-derived film is a complex process that depends on a number of factors such as the compound used as antioxidant, its quantity, characteristics of matrix, as well as food. From the cost-effectiveness point of view, the starch-derived films are produced on larger scale by offering the higher availability of films, hence decreasing the market price of these films. For the safety and regulation of starch-derived films, all the ingredients required for the manufacturing of starch-derived films along with the functional additives should be generally recognized as safe (GRAS) and should be used within the limits issued by the Food Drug Administration (FDA). The starch applications include microwave cake manufacturing, baking, fat replacement, frozen sauces, gelling candies, and snacks. The appearance of these food in which starch is added depends on the choice of starch that is being used. Viscosity of the starch increases with the increase of pressure in general trend, but for some starches, this trend is quite opposite. Dielectric properties link the microwaves with the food samples and play an important role in the process of microwave heating. Edible films and coating protect the food entity from food spoilage causing and pathogenic microorganism.

References

Aguilera JM, Cadoche L, Lopez C, Gutierrez G (2001) Microstructural changes of potato cells and starch granules heated in oil. Food Res Int 34:939–947

Ahmad M, Baba WN, Shah U, Gani A, Gani A, Masoodi FA (2014) Nutraceutical properties of the green tea polyphenols. J Food Process Technol 5(11):1000390

Ahmed J, Al-Attar H (2017) Structural properties of high-pressure-treated chestnut flour dispersions. Int J Food Prop 20(1):S766–S778

Ahmed J, Ramaswamy HS, Raghavan VGS (2007) Dielectric properties of Indian basmati rice flour slurry. J Food Eng 80:1125–1133

Ai Y, Jane J (2015) Gelatinization and rheological properties of starch. Starch/Stärke 67:213–224

Ai Y, Jane J (2016) Macronutrients in corn and human nutrition. Compr Rev Food Sci Food Saf 15: 581–598

Ai Y, Hasjim J, Jane J (2013) Effects of lipids on enzymatic hydrolysis and physical properties of starch. Carbohydr Polym 92:120–127

Ai Y, Gong LX, Reed M, Huang JQ, Zhang Y, Jane J (2016) Characterization of starch from bamboo seeds. Starch-Starke 68:131–139

Aldana SD, Andrade-Ochoa S, Aguilar CN, Contreras-Esquivel JC, Nevarez-Moorillon GV (2015) Antibacterial activity of pectic-based edible films incorporated with Mexican lime essential oil. Food Control 50:907–912

Ali A, Xie F, Yu L, Liu H, Meng L, Khalid S, Chen L (2018) Preparation and characterization of starch-based composite films reinforced by polysaccharide-based crystals. Compos B Eng 133: 122–128

Alvani K, Xin Q, Richard FT, Colin ES (2011) Physico-chemical properties of potato starches. Food Chem 125:958–965

Alves AAC (2002) Cassava botany and physiology. In: Hillocks RJ, Thresh JM, Bellotti AC (eds) Cassava biology production and utilization. CABI Publisher, New York, pp 67–89

Amani N, Buleon A, Kamenan A, Colonna P (2004) Variability in starch physicochemical and functional properties of yam (Dioscorea sp.) cultivated in Ivory Coast. J Sci Food Agric 84: 2085–2096

Ambigaipalan P, Hoover R, Donner E, Liu Q, Jaiswal S, Chibbar R (2011) Structure of faba bean, black bean and pinto bean starches at different levels of granule organization and their physicochemical properties. Food Res Int 44:2962–2974

Ao Z, Jane J (2007) Characterization and modeling of the A- and B-granule starches of wheat, triticale, and barley. Carbohydr Polym 67:46–55

Arvisenet G, Voilley A, Cayot N (2002) Retention of aroma compounds in starch matrices: compositions between aroma compounds toward amylose and amylopectin. J Agric Food Chem 50:7345–7349

Ashogbon AO, Akintayo ET (2012) Morphological, functional and pasting properties of starches separated from rice cultivars grown in Nigeria. Int Food Res J 19:181–187

Aisyah Y, Irwanda L P, Haryani S, Safriani, N (2018) Characterization of corn starch-based edible film incorporated with nutmeg oil nanoemulsion. In IOP conference series: materials science and engineering (Vol. 352, No. 1, p. 012050). IOP Publishing.

Ashwar BA, Shah A, Gani A, Shah U, Gani A, Wani ID, Masoodi FA (2014) Rice starch active packaging films loaded with antioxidants-development and characterization. Starch-Starke 66: 1–9

Aswathy J, Heera KV, Sumi TS, Joseph M, Mathew S, Praveen G, Nair IC, Radhakrishnan EK (2019) Starch-PVA composite films with zinc-oxide nanoparticles and phytochemicals as intelligent pH sensing wraps for food package application. Int J Biol Macromol 136(1):395–403

Atares L, Chiralt A (2016) Essential oils as additives in biodegradable films and coatings for active food packaging. Trends Food Sci Technol 48:51–62

Autio K, Eliasson AC (2009) Rye starch. In: BeMiller J, Whistler R (eds) Starch, 3rd edn. Academic Press, New York, pp 579–587

Bahram K, Niazi MBK, Samin G, Jahan Z (2016) Thermoplastic starch: a possible biodegradable food packaging material—a review. J Food Process Eng 40(3):e12447

Bai YJ, Shi YC, Herrera A, Prakash O (2011) Study of octenyl succinic anhydride-modified waxy maize starch by nuclear magnetic resonance spectroscopy. Carbohydr Polym 83:407–413

Baranauskiene R, Bylaite E, Zukauskaite J, Venskutonis PR (2007) Flavor retention of peppermint (Mentha piperita L.) essential oil spray-dried in modified starches during encapsulation and storage. J Agric Food Chem 55:3027–3036

Barbosa-Canovas GV, Juliano P, Peleg M (2006) Engineering properties of foods, in food engineering. In: Encyclopaedia of life support systems (EOLSS). Developed under auspices of the UNESCO. EOLSS Publishers, Oxford

Barzegar H, Azizi MH, Barzegar M, Hamidi-Esfahani Z (2014) Effect of potassium sorbate on antimicrobial and physical properties of starch–clay nanocomposite films. Carbohydr Polym 110:26–31

Battegazzore D, Bocchini S, Nicola G, Martini E, Frache A (2015) Isosorbide, a green plasticizer for thermoplastic starch that does not retrogradate. Carbohydr Polym 119:78–84

Bertuzzi MA, Vidaurre EFC, Armada M, Gottifredi JC (2007) Water vapor permeability of edible starch based films. J Food Eng 80(3):972–978

Biliaderis CG et al (2009) In: BeMiller JN, Whistler RL (eds) Starch chemistry and technology. Academic Press, New York, pp 293–372

Bin L, Xu H, Zhao H, Liu W, Zhao L, Li Y (2017) Preparation and characterization of intelligent starch/PVA films for simultaneous colorimetric indication and antimicrobial activity for food packaging applications. Carbohydr Polym 157:842–849

Bnuyan I, Hindi NKK, Jebur MH, Mahdi MA (2015) In vitro antimicrobial activity of gum arabic (Al Manna and Tayebat) prebiotics against infectious pathogens. Int J Pharm Pharm Res 3:77–85

Boutboul A, Giampaoli P, Feigenbaum A, Ducruet V (2002) Influence of the nature and treatment of starch on aroma retention. Carbohydr Polym 47:73–82

Broumand A, Emam-Djomeh Z, Hamedi M, Razavi SH (2011) Antimicrobial, water vapour permeability, mechanical and thermal properties of casein based Zataraia multiflora Boiss. Extract containing film. LWT Food Sci Technol 44:2316–2323

Bucknow R, Jankowiak L, Knorr D, Versteeg C (2009) Pressure–temperature phase diagrams of maize starches with different amylose contents. J Agric Food Chem 57:11510–11516

Busolo MA, Lagaron JM (2015) Antioxidant polyethylene films based on a resveratrol containing clay of interest in food packaging applications. Food Packag Shelf Life 6:30–41

Carrin ME, Ceci LN, Lozano JE (2004) Characterization of starch in apple juice and its degradation by amylases. Food Chem 87:173–178

Carvalho AJF, Curvelo AAS, Agnelli JAM (2001) A first insight on composites of thermoplastic starch and kaolin. Carbohydr Polym 45:189–194

Chaichi M, Hashemi M, Badii F, Mohammadi A (2017) Preparation and characterization packaging applications. Adv Polym Technol 23(1):32–45

Charoenkul N, Uttapap D, Pathipanawat W, Takeda Y (2011) Physicochemical characteristics of starches and flours from cassava varieties having different cooked root textures. LWT Food Sci Technol 44:1774–1781

Che LM, Li D, Wang LJ, Ozkan N (2008) Rheological properties of dilute aqueous solutions of cassava starch. Carbohydr Polym 74:385–389

Chen G, Wei M, Chen J, Huang J, Dufresne A, Chang PR (2008) Simultaneous reinforcing and toughening: new nanocomposites of waterborne polyurethane filled with low loading level of starch nanocrystals. Polymer 49(7):1860–1870

Chen Y, Zhao H, Liu X, Li Z, Liu B, Wu J, Shi M, Norde W, Li Y (2016) TEMPO-oxidized Konjac glucomannan as appliance for the preparation of hard capsules. Carbohydr Polym 143:262–269

Chiumarelli M, Hubinger MD (2012) Stability, solubility, mechanical and barrier properties of cassava starch–carnauba wax edible coatings to preserve fresh-cut apples. Food Hydrocoll 28(1):59–67

Cho SY, Rhee C (2002) Sorption characteristics of soy protein films and their relation to mechanical properties. LWT Food Sci Technol 35(2):151–157

Chung HJ, Liu Q, Donner E, Hoover R, Warkentin TD, Vandenberg B (2008) Composition, molecular structure, properties, and in vitro digestibility of starches from newly released Canadian pulse cultivars. Cereal Chem 85:471–479

Chung HJ, Lee SE, Han JA, Lim ST (2010) Physical properties of dry-heated octenyl succinylated waxy corn starches and its application in fat-reduced muffin. J Cereal Sci 52:496–501

Chung HJ, Li XQ, Kalinga D, Lim ST, Yada R, Liu Q (2014) Physicochemical properties of dry matter and isolated starch from potatoes grown in different locations in Canada. Food Res Int 57:89–94

Conde-Petit B, Escher F, Nuessli J (2006) Structural features of starch-flavor complexation in food model system. Trends Food Sci Technol 17:227–235

Cummings JH, Englyst HN (1991) Measurement of starch fermentation in the human large intestine. Can J Physiol Pharmacol 69:121–129

Czuchajowska Z, Klamczynski A, Paszczynska B, Baik BK (1998) Structure and functionality of barley starches. Cereal Chem 75:747–754

Dai L, Qiu C, Xiong L, Sun Q (2015) Characterization of corn starch-based films reinforced with taro starch nanoparticles. Food Chem 174:82–88

de la Cruz GV, Torres JA, Martin-Polo MO (2001) Temperature effect on the moisture sorption isotherms for methylcellulose and ethylcellulose films. J Food Eng 48(1):91–94

Debet MR, Gidley MJ (2006) Three classes of starch granule swelling: influence of surface proteins and lipids. Carbohydr Polym 64:452–465

Dominguez R, Barba FJ, Gomez B, Putnik P, Kovacevic DB, Pateiroa M, Santos EM, Lorenzo JM (2018) Active packaging films with natural antioxidants to be used in meat industry: a review. Food Res Int 113:93–101

Dominici F, Gigli M, Armentano I, Genovese L, Luzi F, Torre L, Munari A, Lotti N (2020) Improving the flexibility and compostability of starch/poly (butylene cyclohexanedicarboxylate)-based blends. Carbohydr Polym 246:116631

Druchta JM, Jonhston CD (1997) An update on edible films-scientific status summary. Food Technol 51:60–62

Eggleston G, Swennen R, Akoni S (1992) Physicochemical studies on starches isolated from plantain cultivars, plantain hybrids and cooking bananas. Starch-Stärke 44:121–128

Ek KL, Wang S, Copeland L, Brand-Miller JC (2014a) Discovery of a low-glycaemic index potato and relationship with starch digestion in vitro. Br J Nutr 111:699–705

Ek KL, Wang S, Brand-Miller J, Copeland L (2014b) Properties of starch from potatoes differing in glycemic index. Food Funct 5:2509–2515

El-Anany AM, Hassan GFA, Ali FMR (2009) Effects of edible coatings on the shelf-life and quality of Anna apple (Malus domestica Borkh) during cold storage. J Food Technol 7:5–11

Fama L, Rojas M, Ana GS, Gerschenson L (2005) Mechanical properties of tapioca-starch edible films containing sorbates. LWT Food Sci Technol 38:631–639

Fan D, Wang L, Zhang N, Xiong L, Huang L, Zhao J (2017d) Full-time response of starch subjected to microwave heating. Sci Rep 7:3967

Fang Z, Zhao Y, Warner RD, Johnson SK (2017) Active and intelligent packaging in meat industry. Trends Food Sci Technol 61:60–71

Franssen LR, Krochta JM (2003) Edible coatings containing natural antimicrobials for processed foods. In: Roller S (ed) Natural antimicrobials for minimal processing of foods. CRC Press, Cambridge, pp 250–262

Gao W, Fan L, Paek KY (1999) Ultrastructure of amyloplasts and intercellular transport of old and new scales in Fritillaria ussuriensis. J Plant Biol 42:117–123

Gao C, Yu L, Liu H, Chen L (2012) Development of self-reinforced polymer composites. Prog Polym Sci 37:767–780

Gao F, Li D, Bi CH, Mao ZH, Adhikari B (2014) Preparation and characterization of starch crosslinked with sodium trimetaphosphate and hydrolyzed by enzymes. Carbohydr Polym 103:310–318

Garcia MA, Martinho MN, Zaritzky NE (2000) Microstructural characterization of plasticized starch-based films. Starch-Starke 52:118–124

Gelinas P, McKinnon C (2016) Gluten weight in ancient and modern wheat and the reactivity of epitopes towards R5 and G12 monoclonal antibodies. Int J Food Sci Technol 51:1801–1810

Genovese L, Soccio M, Gigli M, Lotti N, Gazzano M, Siracusa V (2016) Gas permeability, mechanical behaviour and compostability of fully-aliphatic bio-based multiblock poly(ester urethane)s. RSC Adv 6(60):55331–55342

Gigli M, Lotti N, Gazzano M, Siracusa V, Finelli L, Munari A (2013) Fully aliphatic copolyesters based on poly (butylene 1,4-cyclohexanedicarboxylate) with promising mechanical and barrier properties for food packaging applications. Ind Eng Chem Res 52(36):12876–12886

Gigli M, Lotti N, Gazzano M, Siracusa V, Finelli L, Munari A (2014) Biodegradable aliphatic copolyesters containing PEG-like sequences for sustainable food packaging applications. Polym Degrad Stab 105:96–106

Goesaert H, Slade L, Levine H, Delcour JA (2009) Amylases and bread firming - an integrated view. J Cereal Sci 50:345–352

Gomez-Estaca J, Gavara R, Catala R, Hernandez-Munoz P (2016) The potential of proteins for producing food packaging materials: a review. Packag Technol Sci 29(4–5):203–224

Gudmundsson M, Eliasson AC (2009) Some physico-chemical properties of oat starches extracted from varieties with different oil content. Acta Agric Scand Sect A 39:101–111

Guilbert S, Gontard N, Cuq B (1995) Technology and applications of edible protective films. Packag Technol Sci 8(6):339–346

Guinesi LS, da Roz AL, Corradini E, Mattoso LHC (2006) Kinetics of thermal degradation applied to starches from different botanical origins by non-isothermal procedures. Thermochim Acta 447:190–196

Guo Z, Zeng S, Lu X, Zhou M, Zheng M, Zheng B (2015) Structural and physicochemical properties of lotus seed starch treated with ultra-high pressure. Food Chem 186:223–230

Guo P, Yu J, Wang S, Wang S, Copeland L (2018) Effects of particle size and water content during cooking on the physicochemical properties and in vitro starch digestibility of milled durum wheat grains. Food Hydrocoll 77:445–453

Han XZ, Hamaker BR (2001) Amylopectin fine structure and rice starch paste breakdown. J Cereal Sci 34:279–284

Hashemi SMB, Raeisi S (2018) Evaluation of antifungal and antioxidant properties of edible coating based on apricot (Prunus armeniaca) gum containing Satureja intermedia extract in fresh wild almond (Amygdalus scoparia) kernels. J Food Meas Charact 12:362–369

Hasjim J, Li E, Dhital S (2013) Milling of rice grains: effects of starch/flour structures on gelatinization and pasting properties. Carbohydr Polym 92:682–690

He W, Wei C (2017) Progress in C-type starches from different plant sources. Food Hydrocoll 73: 162–175

Hedley CL, Bogracheva TY, Wang TL (2015) A genetic approach to studying the morphology, structure and function of starch granules using pea as a model. Starch-Stärke 54:235–242

Heljo VP, Nordberg A, Tenho M, Virtanen T, Jouppila K, Salonen J, Maunu SL, Juppo AM (2012) The effect of water plasticization on the molecular mobility and crystallization tendency of amorphous disaccharides. Pharm Res 29:2684–2697

Heydari A, Alemzadeh I, Vossoughi M (2013) Functional properties of biodegradable corn starch nanocomposites for food packaging applications. Mater Des 50:954–961

Hoover R, Ratnayake WS (2002) Starch characteristics of black bean, chick pea, lentil, navy bean and pinto bean cultivars grown in Canada. Food Chem 78:489–498

Hoover R, Smith C, Zhou Y, Ratnayake RMWS (2003) Physicochemical properties of Canadian oat starches. Carbohydr Polym 52:253–261

Hu XZ, Zheng JM, Li XP, Xu C, Zhao Q (2014) Chemical composition and sensory characteristics of oat flakes: a comparative study of naked oat flakes from China and hulled oat flakes from western countries. J Cereal Sci 60:297–301

Hyunseok K, Kerryc H (2010) Physicochemical properties and amylopectin fine structures of A- and B-type granules of waxy and normal soft wheat starch. J Cereal Sci 51:256–264

Inchuen S, Narkrugsa W, Pornchaloempong P (2009) Moisture sorption of Thai red curry powder. Maejo Int J Sci Technol 3(3):486–497

Inouchi N, Hibiu H, Li T, Horibata T, Fuwa H, Itani T (2005) Structure and properties of endosperm starches from cultivated rice of Asia and other countries. J Appl Glycosci 52:239–246

Intan DH, Aizan A (2011) Tensile and water absorption properties of biodegradable composites derived from cassava skin/polyvinyl alcohol with glycerol as plasticizer. Sains Malays 40:713–718

Jan U, Gani A, Ahmad M, Shah U, Baba WN, Masoodi FA, Wani ID (2015) Characterization of cookies made from wheat flour blended with buckwheat flour and effect on antioxidant properties. J Food Sci Technol 52:6334–6344

Jane J (2006) Current understanding on starch granule structures. J Appl Glycosci 53:205–213

Jane JL, Kasemsuwan T, Leas S, Zobel H, Robyt JF (1994) Anthology of starch granule morphology by scanning electron microscopy. Starch-Stärke 46:121–129

Jane JL, Wong KS, McPherson AE (1997) Branch-structure difference in starches of A- and B-type X-ray patterns revealed by their Naegeli dextrins. Carbohydr Res 300:219–227

Jane J, Chen YY, Lee LF, Mcpherson AE, Wong KS, Radosavljevic M, Kasemsuwan T (1999) Effects of amylopectin branch chain length and amylose content on the gelatinization and pasting properties of starch. Cereal Chem 76:629–637

Jayakody L, Hoover R, Liu Q, Donner E (2007) Studies on tuber starches. II. Molecular structure composition and physicochemical properties of yam (Dioscorea sp.) starches grown in Sri Lanka. Carbohydr Polym 69:148–163

Jeon YJ, Vasanthan T, Temelli F, Song BK (2003) The suitability of barley and corn starches in their native and chemically modified forms for volatile meat flavour encapsulation. Food Res Int 36:349–355

Jha SN, Narsaiah K, Basediya AL, Sharma R, Jaiswal P, Kumar R, Bhardwaj R (2011) Measurement techniques and application of electrical properties for nondestructive quality evaluation of foods—a review. J Food Sci Technol 48(4):387–411

Jiang HX, Campbell M, Blanco M, Jane JL (2010) Characterization of maize amylose-extender (ae) mutant starches: part II. Structures and properties of starch residues remaining after enzymatic hydrolysis at boiling-water temperature. Carbohydr Polym 80:1–12

Jiang HX, Campbell M, Blanco M, Jane JL (2010a) Characterization of maize amylose extender (ae) mutant starches: part II. Structures and properties of starch residues remaining after enzymatic hydrolysis at boiling-water temperature. Carbohydr Polym 80:1–12

Jiang B, Li W, Hu X, Wu J, Shen Q (2015a) Rheology of mung bean starch treated by high hydrostatic pressure. Int J Food Prop 18(1):81–92

Jiang B, Li W, Shen Q, Hu X, Wu J (2015b) Effects of high hydrostatic pressure on rheological properties of rice starch. Int J Food Prop 18(6):1334–1344

Jih-Guan WU, Wang PJ, Chen SC (2010) Antioxidant and antimicrobial effectiveness of catechin-impregnated PVA–starch film on red meat. J Food Qual 33:780–801

Jimenez A, Fabra MJ, Talens P, Chiralt A (2012) Edible and biodegradable starch films: a review. Food Bioprocess Technol 5:2058–2076

Jimenez A, Fabra MJ, Talens P, Chiralt A (2013a) Phase transitions in starch based films containing fatty acids. Effect on water sorption and mechanical behaviour. Food Hydrocoll 30:408–418

Jimenez A, Fabra MJ, Talens P, Chiralt A (2013b) Physical properties and antioxidant capacity of starch–sodium caseinate films containing lipids. J Food Eng 116:695–702

Jimenez A, Sanchez-Gonzalez L, Desobry S, Chiralt A, Tehrany EA (2014) Influence of nanoliposomes incorporation on properties of film forming dispersions and films based on corn starch and sodium caseinate. Food Hydrocoll 35:159–169

Kasemsuwan T, Bailey T, Jane J (1998) Preparation of clear noodles with mixtures of tapioca and high-amylose starches. Carbohydr Polym 36:301–312

Kaur M, Singh N, Sandhu KS, Guraya HS (2004) Physicochemical, morphological, thermal and rheological properties of starches separated from kernels of some Indian mango cultivars (Mangifera indica L.). Food Chem 85:131–140

Kaushik A, Singh M, Verma G (2010) Green nanocomposites based on thermoplastic starch and steam exploded cellulose nanofibrils from wheat straw. Carbohydr Polym 82:337–345

Kim SJ, Ustunol Z (2001) Thermal properties, heat sealability and seal attributes of whey protein isolate/lipid emulsion edible films. J Food Sci 66:985–990

Klucinec JD, Thompson DB (1999) Amylose and amylopectin interact in retrogradation of dispersed high-amylose starches. Cereal Chem 76:282–291

Kong X, Kasapis S, Zhu P, Sui Z, Bao J, Corke H (2016) Physicochemical and structural characteristics of starches from Chinese hull-less barley cultivars. Int J Food Sci Technol 51: 509–518

Krasniewska K, Gniewosz M (2012) Substances with antimicrobial activity in edible films—a review. Pol J Food Nutr Sci 62:199–206

Kuakpetoon D, Wang YJ (2006) Structural characteristics and physicochemical properties of oxidized corn starches varying in amylose content. Carbohydr Res 341:1896–1915

Kuang Q, Xu J, Wang K, Zhou S, Liu X (2016) Structure and digestion of hybrid Indica rice starch and its biosynthesis. Int J Biol Macromol 93:402–407

Kubo A, Akdogan G, Nakaya M, Shojo A, Suzuki S, Satoh H, Kitamura S (2010) Structure, physical, and digestive properties of starch from wx ae double-mutant rice. J Agric Food Chem 58:4463–4469

Kumar PS, Saravanan A, Sheeba N, Uma S (2019) Structural, functional characterization and physicochemical properties of green banana flour from dessert and plantain bananas (Musa spp.). LWT Food Sci Technol 116:108524

Larotonda FDS, Matsui KS, Paes SS, Laurindo JB (2003) Impregnation of Kraft paper with cassava-starch acetate-analysis of the tensile strength, water absorption and water vapor permeability. Starch-Stärke 55(11):504–510

Le CD, Bras J, Dufresne A (2010) Starch nanoparticles: a review. Biomacromolecules 11:1139–1153

Leite TS, Jesus ALT, Schmiele M, Tribst AAL, Cristianini M (2017) High pressure processing (HPP) of pea starch: effect on the gelatinization properties. Food Sci Technol 76:361–369

Leslaw J, Fortuna T, Krok F (2003) Non-contact atomic force microscopy of starch granules surface. Part I: potato and tapioca starches. Starch-Starke 55:1–7

Li M, Peng L, Wei Z, Long Y, Fengwei X, Huayin P, Hongshen L, Chen L (2011) Extrusion processing and characterization of edible starch films with different amylose contents. J Food Eng 106(1):95–101

Li G, Zhu F (2018) Effect of high pressure on rheological and thermal properties of quinoa and maize starches. Food Chem 241:380–386

Li W, Xiao X, Zhang W, Zheng J, Luo Q, Ouyang S (2014) Compositional, morphological, structural and physicochemical properties of starches from seven naked barley cultivars grown in China. Food Res Int 58:7–14

Li W, Tian X, Liu L, Wang P, Wu G, Zheng J (2015a) High pressure induced gelatinization of red adzuki bean starch and its effects on starch physicochemical and structural properties. Food Hydrocoll 45:132–139

Li X, Qiu C, Ji N, Sun C, Xiong L, Sun Q (2015b) Mechanical, barrier and morphological properties of starch nanocrystals-reinforced pea starch films. Carbohydr Polym 121:155–162

Light JM (1990) Modified food starches: why, what, where, and how. Cereal Food World 35(11): 1081–1092

Lim H, Han JA, BeMiller J, Lim ST (2007) Physical modification of waxy maize starch by dry heating with ionic gums. J Appl Glycosci 53(4):281–286

Liu L, Jin T, Liu CK, Hicks K, Mohanty AK, Bhardwaj R, Misra M (2008a) A preliminary study on antimicrobial edible films from pectin and other food hydrocolloids by extrusion method. J Nat Fibers 5:366–382

Liu X, Yu L, Liu H, Chen L, Li L (2008b) In situ thermal decomposition of starch with constant moisture in a sealed system. Polym Degrad Stab 93:260–262

Liu HS, Xie FW, Yu L, Chen L, Li L (2009a) Thermal processing of starch-based polymers. Prog Polym Sci 34:1348–1368

Liu H, Xie FY, Yu L, Chen LL (2009b) Thermal processing of starch-based polymers. Progr Polym Sci 34:1348–1368

Liu X, Yu L, Liu H, Chen L, Li L (2009c) Thermal decomposition of cornstarch with different amylose/amylopectin ratios in open and sealed systems. Cereal Chem 86:383–385

Liu D, Zhong T, Chang PR, Li K, Wu Q (2010a) Starch composites reinforced by bamboo cellulosic crystals. Bioresour Technol 101:2529–2536

Liu HS, Yu L, Tong Z, Chen L (2010b) Retrogradation of waxy cornstarch studied by DSC. Starch-Starke 62:524–529

Liu WC, Halley PJ, Gilbert RG (2010c) Mechanism of degradation of starch, a highly branched polymer, during extrusion. Macromolecules 43(6):2855–2864

Liu X, Wang Y, Yu L, Tong Z, Chen L, Liu H, Li X (2013) Thermal degradation and stability of starch under different processing conditions. Starch-Starke 65:48–60

Liu B, Xu H, Zhao H, Liu W, Zhao L, Li Y (2015) Preparation and characterization of intelligent starch/PVA films for simultaneous colorimetric indication and antimicrobial activity for food packaging applications. Carbohydr Polym 10:842–849

Liu H, Guo X, Li Y, Li H, Fan H, Wang M (2016a) In vitro digestibility and changes in physicochemical and textural properties of tartary buckwheat starch under high hydrostatic pressure. J Food Eng 189:64–71

Liu H, Wang L, Cao R, Fan H, Wang M (2016b) In vitro digestibility and changes in physicochemical and structural properties of common buckwheat starch affected by high hydrostatic pressure. Carbohydr Polym 144:1–8

Liu H, Fan H, Cao R, Blanchard C, Wang M (2016c) Physicochemical properties and in vitro digestibility of sorghum starch altered by high hydrostatic pressure. Int J Biol Macromol 92: 753–760

Liu M, Wu NN, Yu GP, Zhai XT, Chen X, Zhang M (2018) Physicochemical properties, structural properties, and in vitro digestibility of pea starch treated with high hydrostatic pressure. Starch 70(1–2):1700082

Liu C, Huang J, Zheng X, Liu S, Lu K, Tang K, Liu J (2020) Heat sealable soluble soybean polysaccharide/gelatin blend edible films for food packaging applications. Food Packag Shelf Life 24:100485

Lopez OV, Lecot CJ, Zaritzky NE, Garcia MA (2011) Biodegradable packages development from starch based heat sealable films. J Food Eng 105:254–263

Lopez-De-Dicastillo C, Gomez-Estaca J, Catala R, Gavara R, Hernandez-Munoz P (2012) Active antioxidant packaging films: development and effect on lipid stability of brined sardines. Food Chem 131(4):1376–1384

Luengwilai K, Beckles DM (2008) Structural investigations and morphology of tomato fruit starch. J Agric Food Chem 57:282–291

Luisarturo BP, Sandral RA, Edith AA, Mirnam SR (2009) Solubilization effects on molecular weights of amylose and amylopectins of normal maize and barley starches. Cereal Chem 86: 701–705

Madigan E (2003) New uses for starch. Health guidance for better health. http://www.healthguidance.org/pages/Terms-of-Service. Retrieved 20 Jan 2011

Mali S, Grossmann MVE (2003) Effects of yam starch films on storability and quality of fresh strawberries (Fragaria ananassa). J Agric Food Chem 51:7005–7011

Mali S, Grossmann MVE, Garcia MA, Martino MN, Zaritzky NE (2005a) Mechanical and thermal properties of yam starch films. Food Hydrocoll 19(1):157–164

Mali S, Sakanaka LS, Yamashita F, Grossmann MVE (2005b) Water sorption and mechanical properties of cassava starch films and their relation to plasticizing effect. Carbohydr Polym 60: 283–289

Maqbool M, Ali A, Alderson PG, Mohamed MTM, Siddiqui Y, Zahis N (2011) Postharvest application of gum arabic and essential oils for controlling anthracnose and quality of banana and papaya during cold storage. Postharvest Biol Technol 62:71–76

Martelli SM, Moore G, Paes SS, Gandolfo C, Laurindo JB (2006) Influence of plasticizers on the water sorption isotherms and water vapor permeability of chicken feather keratin films. LWT Food Sci Technol 39(3):292–301

Matveev YI, Grinberg VY, Tolstoguzov V (2000) The plasticizing effect of water on proteins, polysaccharides and their mixtures. Glassy state of biopolymers, food and seeds. Food Hydrocoll 14:425–437

Misharina TA (2004) Retention of essential oil components sorbed by native cornstarch during storage. Appl Biochem Microbiol 40:104–107

Mittal V, Saini R, Sinha S (2016) Natural fiber-mediated epoxy composites a review. Compos B Eng 99:425–435

Moad G (2011) Chemical modification of starch by reactive extrusion. Prog Polym Sci 36:218–237

Moorthy SN (2015) Physicochemical and functional properties of tropical tuber starches: a review. Starch-Stärke 54:559–592

Moorthy SN, Wenham JE, Jmv B (2015) Effect of solvent extraction on the gelatinisation properties of flour and starch of five cassava varieties. J Sci Food Agric 72:329–336

Muller CMO, Yamashita F, Laurindo JB (2008) Evaluation of the effect of glycerol and sorbitol concentration and water activity on the water barrier properties of cassava starch films through a solubility approach. Carbohydr Polym 72:82–87

Muller CMO, Laurindo JB, Yamashita F (2009) Effect of cellulose fibers on the crystallinity and mechanical properties of starch-based films at different relative humidity values. Carbohydr Polym 77(2):293–299

Muller CMO, Laurindo JB, Yamashita F (2011) Effect of nanoclay incorporation method on mechanical and water vapor barrier properties of starch-based films. Ind Crops Prod 33:605–610

Ni D, Zhang S, Sheng C, Yong X, Li L, Hao L (2011) Improving cooking and eating quality of Xieyou57, an elite indica hybrid rice, by marker-assisted selection of the Wx locus. Euphytica 179:355–362

Nisa IU, Ashwar BA, Shah A, Gani A, Gani A, Masoodi FA (2015) Development of potato starch based active packaging films loaded with antioxidants and its effect on shelf life of beef. J Food Sci Technol 52:7245–7253

Noshirvani N, Ghanbarzadeh B, Mokarram RR, Hashemi M, Coma V (2017) Preparation and characterization of active emulsified films based on chitosan-carboxymethyl cellulose containing zinc oxide nano particles. Int J Biol Macromol 99:530–538

Nunes C, Maricato E, Cunha A, Nunes A, Silva JAL, Coimbra MA (2013) Chitosan-caffeic acid-genipin films presenting enhanced antioxidant activity and stability in acidic media. Carbohydr Polym 91:236–243

Olomo V, Ajibola O (2010) Processing factors affecting the yield and physicochemical properties of starch from cassava chips and flour. Starch-Starke 55:476–481

Otegbayo B, Bokanga M, Asiedu R (2011) Physicochemical properties of yam starch: effect on textural quality of yam food product (pounded yam). J Food Agric Environ 9:145–150

Otegbayo B, Oguniyan D, Akinwumi O (2014) Physicochemical and functional characterization of yam starch for potential industrial applications. Starch-Starke 66:235–250

Othman SH (2014) Bio-nanocomposite materials for food packaging applications: types of biopolymer and nanosized filler. Agric Agric Sci Procedia 2:296–303

Partanen R, Ahro M, Hakala M, Kallio H, Forssel P (2002) Microencapsulation of caraway extract in β-cyclodextrin and modified starches. Eur Food Res Technol 214:242–247

Pereira VA, Arruda INQD, Stefani R (2015) Active chitosan/PVA films with anthocyanins from Brassica oleracea (red cabbage) as time–temperature indicators for application in intelligent food packaging. Food Hydrocoll 43:180–188
Perera C, Lu Z, Sell J, Jane J (2001) Comparison of physicochemical properties and structures of sugary-2 cornstarch with normal and waxy cultivars. Cereal Chem 78:249–256
Perez L, Acevedo EA (2017) Starch. In: Starch based materials in food packaging: processing, characterization and applications. CRC Press, Ohio, pp 1–18
Perez E, Gibert O, Rolland-Sabate A, Jimenez Y, Sanchez T, Giraldo A (2011) Physicochemical, functional, and macromolecular properties of waxy yam starches discovered from "Mapuey" (Dioscorea trifida) genotypes in the Venezuelan Amazon. J Agric Food Chem 59:263–273
Raj B, Siddaramaiah USK (2004) Low density polyethylene/starch blend films for food packaging applications. Adv Polym Technol 23(1):32–45
Ratnayake WS, Hoover R, Shahidi F, Perera C, Jane J (2001) Composition, molecular structure, and physicochemical properties of starches from four field pea (Pisum sativum L.) cultivars. Food Chem 74:189–202
Rhim JW, Park HM, Ha CS (2013) Bio-nanocomposites for food packaging applications. Progr Polym Sci 38:1629–1652
Ribeiro-Santos R, Andrade M, Rados de Melo N, Sanches-Silva A (2017) Use of essential oils in active food packaging: recent advances and future trends. Trends Food Sci Technol 61:132–140
Rollandsabate A, Sanchez T, Buleon A, Colonna P, Jaillais B, Ceballos H (2012) Structural characterization of novel cassava starches with low and high-amylose contents in comparison with other commercial sources. Food Hydrocoll 27:161–174
Rolland-Sabate A, Georges Amani NG, Dufour D, Guilois S, Colonna P (2003) Macromolecular characteristics of ten yam (Dioscorea spp.) starches. J Sci Food Agric 83:927–936
Rolland-Sabate A, Sanchez T, Buléon A, Colonna P, Ceballos H, Zhao SS (2013) Molecular and supra-molecular structure of waxy starches developed from cassava (Manihot esculenta Crantz). Carbohydr Polym 92:1451–1462
Rutenberg MW, Solarek D (1984) Starch derivatives: production and uses. In: Whistler R, BeMiller J, Paschall E (eds) Starch chemistry and technology. Academic Press, New York, pp 311–366
Sahin S, Sumnu SG (2006) Physical properties of foods. Springer Science & Business Media
Sandhu KS, Singh N (2007) Some properties of corn starches II: physicochemical, gelatinization, retrogradation, pasting and gel textural properties. Food Chem 101(4):1499–1507
Savary G, Lafarge C, Doublier JL, Cayot N (2007) Distribution of aroma in starch–polysaccharide composite gel. Food Res Int 40:706–716
Schaffer AA, Levin I, Oguz I, Petreikov M, Bar M (2000) ADP glucose pyro phosphorylase activity and starch accumulation in immature tomato fruit: the effect of a Lycopersicon hirsutum derived introgression encoding for the large subunit. Plant Sci 152:135–144
Schierbaum DSF, Radosta S, Richter M, Kettlitz B, Dipl Krist CG (1991) Studies on rye starch properties and modification. Part I: composition and properties of rye starch granules. Starch-Stärke 43:331–339
Schirmer M, Jekle M, Becker T (2015) Starch gelatinization and its complexity for analysis. Starch 67:30–41
Shah U, Baba NB, Ahmad M, Shah A, Gani A, Masoodi FA, Ashwar B (2014) In vitro antioxidant and anti-proliferative activities of seed extracts of Nymphaea mexicana in different solvents and GC-MS analysis. Int J Drug Dev Res 6:280–285
Shah U, Gani A, Ashwar BA, Shah A, Ahmad M, Gani A, Wani IA, Masoodi FA (2015) A review of the recent advances in starch as active and nanocomposite packaging films. Cogent Food Agric 1:1115640
Shinde SV, Nelson JE, Huber KC (2003) Soft wheat starch pasting behavior in relation to A- and B-type granule content and composition. Cereal Chem 80:91–98
Singh N, Isono N, Srichuwong S, Noda T, Nishinari K (2008) Structural, thermal and viscoelastic properties of potato starches. Food Hydrocoll 22:979–988

Siqueira GB, Dufresne A (2009) Cellulose whiskers versus microfibrils: influence of the nature of the nanoparticle and its surface functionalization on the thermal and mechanical properties of nanocomposites. Biomacromolecules 10:425–432

Skoog E, Shin JH, Saez-Jimenez V, Mapelli V, Olsson L (2018) Biobased adipic acid—the challenge of developing the production host. Biotechnol Adv 36(8):2248–2263

Slavutsky AM, Bertuzzi MA (2015) Formulation and characterization of nanolaminated starch based film. LWT Food Sci Technol 61:407–413

Soares RMD, Lima AMF, Oliveira RVB, Pires ATN, Soldi V (2005) Thermal degradation of biodegradable edible films based on xanthan and starches from different sources. Polym Degrad Stab 90:449–454

Sobota A, Rzedzicki Z, Zarzycki P, Kuzawinska E (2015) Application of common wheat bran for the industrial production of high-fibre pasta. Int J Food Sci Technol 50:111–119

Sorrentino A, Gorrasi G, Vittoria V (2007) Potential perspectives of bio-nanocomposites for food packaging applications. Trends Food Sci Technol 18:84–95

Soulaka AB, Morrison WR (2010) The amylose and lipid contents, dimensions, and gelatinization characteristics of some wheat starches and their A- and B-granule fractions. J Sci Food Agric 36: 709–718

Spirido I, Anghel N, Bele A (2015) Behavior of biodegradable composites based on starch reinforced with modified cellulosic fibers. Polym Adv Technol 26:1189–1197

Sriroth K, Santisopasri V, Petchalanuwat C, Kurotjanawong K, Piyachomkwan K, Oates CG (1999) Cassava starch granule structure–function properties: influence of time and conditions at harvest on four cultivars of cassava starch. Carbohydr Polym 38(2):161–170

Stevenson DG, Yoo SH, Hurst PL, Jane JL (2005) Structural and physicochemical characteristics of winter squash (Cucurbita maxima D.) fruit starches at harvest. Carbohydr Polym 59:153–163

Stevenson DG, Domoto PA, Jane JL (2006a) Structures and functional properties of apple (Malu domestica Borkh) fruit starch. Carbohydr Polym 63:432–441

Stevenson DG, Doorenbos RK, Jane JL, Inglett GE (2006b) Structures and functional properties of starch from seeds of three soybean (Glycine max (L.) Merr.) varieties. Starch-Stärke 58:509–519

Sugih AK, Picchioni F, Janssen LPBM, Heers HJ (2009) Synthesis of poly-(epsilon)-caprolactone grafted starch co-polymers by ring-opening polymerization using silylated starch precursors. Carbohydr Polym 77:267–275

Suput DZ, Lazic VL, Popovic SZ, Hromis NM (2015) Edible films and coatings: sources, properties and application. Food Feed Res 42(1):11–22

Talja RA, Helen H, Roos YH, Jouppila K (2007) Effect of various polyols and polyol contents on physical and mechanical properties of potato starch-based films. Carbohydr Polym 67(3): 288–295

Tammineni N, Unlu G, Min SC (2013) Development of antimicrobial potato peel waste-based edible films with oregano essential oil to inhibit listeria monocytogenes on cold-smoked salmon. Int J Food Sci Technol 48:211–214

Tang X, Alavi S, Herald TJ (2008) Barrier and mechanical properties of starch-clay nanocomposite films. Cereal Chem 85:428–433

Tao Y, Yan B, Fan D, Zhang N, Ma S, Wang L, Wu Y, Wang M, Zhao J, Zhang H (2020) Structural changes of starch subjected to microwave heating: a review from the perspective of dielectric properties. Trends Food Sci Technol 99:593–607

Teixeira EM, Pasquini D, Curvelo AAS, Corradini E, Belgacem MN, Dufresne A (2009) Cassava bagasse cellulose nanofibrils reinforced thermoplastic cassava starch. Carbohydr Polym 78: 422–431

Tester RF, Morrison WR (1990) Swelling and gelatinization of cereal starches: effects of amylopectin, amylose, and lipids. Cereal Chem 67:551–557

Tester RF, Debon SJJ, Sommerville MD (2000) Annealing of maize starch. Carbohydr Polym 42: 287–299

Tharanathan RN (2005) Starch-value addition by modification. Crit Rev Food Sci Nutr 45(5): 371–384
Tharanathan RN, Mahadevamma S (2003) Grain legumes-a boon to human nutrition. Trends Food Sci Technol 14:507–518
Tian R, Jiang GH, Shen LH, Wang LQ, He YQ (2005) Mapping quantitative trait loci underlying the cooking and eating quality of rice using a DH population. Mol Breed 15:117–124
Tong C, Ahmed S, Pang Y, Xin Z, Bao J (2018) Fine structure and gelatinization and pasting properties relationships among starches from pigmented potatoes. Food Hydrocoll 83:45–52
Tongdeesoontorn W, Mauer LJ, Wongruong S, Sriburi P, Rachtanapun P (2011) Effect of carboxymethyl cellulose concentration on physical properties of biodegradable cassava starch-based films. Chem Central J 5(6):2–8
Tongnuanchan P, Benjakul S, Prodpran T, Pisuchpen S, Osako K (2016) Mechanical, thermal and heat sealing properties of fish skin gelatin film containing palm oil and basil essential oil with different surfactants. Food Hydrocoll 56:93–107
Vaclavik VA (2021) Starches in food. In: Essential of food science, 5th edn. Springer, New York, pp 38–53
Vargas Junior A, Fronza N, Foralosso FB, Dezen D, Huber E, dos Santos JHZ, Quadri MGN (2015) Biodegradable duo-functional active film: antioxidant and antimicrobial actions for the conservation of beef. Food Bioprocess Technol 8(1):75–87
Veiga-Santos P, Suzuki CK, Nery KF, Cereda MP, Scamparin ARP (2008) Evaluation of optical microscopy efficacy in evaluating cassava starch biofilms microstructure. LWT Food Sci Technol 41:1506–1523
Vidrith R, Zlatic E, Hribar J (2009) Release of strawberry aroma compounds by different starch–aroma systems. Czech J Food Sci 27:58–61
Viguie J, Molina-Boisseau S, Dufresne A (2007) Processing and characterization of waxy maize starch films plasticized by sorbitol and reinforced with starch nanocrystals. Macromol Biosci 7: 1206–1216
Vu-Bac N, Rafiee R, Zhuang X, Lahmer T, Rabczuk T (2015) Uncertainty quantification for multiscale modeling of polymer nanocomposites with correlated parameters. Compos B Eng 68:446–464
Waduge RN, Hoover R, Vasanthan T, Gao J, Li J (2006) Effect of annealing on the structure and physicochemical properties of barley starches of varying amylose content. Food Res Int 39:59–77
Wan VCH, Lee CM, Lee SY (2007) Understanding consumers attitudes on edible films and coatings: focus group findings. J Sens Stud 22:353–366
Wang S, Gao W, Chen H, Xiao P (2006) Studies on the morphological, thermal and crystalline properties of starches separated from medicinal plants. J Food Eng 76:420–426
Wang S, Luo H, Zhang J, Zhang Y, He Z, Wang S (2014a) Alkali-induced changes in functional properties and in vitro digestibility of wheat starch: the role of surface proteins and lipids. J Agric Food Chem 62:3636–3643
Wang S, Wang J, Yu J, Wang S (2014b) A comparative study of annealing of waxy, normal and high amylose maize starches: the role of amylose molecules. Food Chem 164:332–338
Wang SJ, Li CL, Copeland L, Niu Q, Wang S (2015a) Starch retrogradation: a comprehensive review. Compr Rev Food Sci Food Saf 14:568–585
Wang S, Li C, Copeland L, Niu Q, Wang S (2015b) Starch retrogradation: a comprehensive review. Rev Food Sci Food Saf 14:568–585
Wang YF, Liu HS, Xie FW, Yu L, Zhang L, Liao L, Chen L (2016) Morphology and properties of thermal/cooling-gel bi-phasic systems based on hydroxypropyl methylcellulose and hydroxypropyl starch. Compo Part B-Eng 101:46–52
Wang S, Li P, Yu J, Guo P, Wang S (2017) Multi-scale structures and functional properties of starches from Indica hybrid, japonica and waxy rice. Int J Biol Macromol 102:136–143
Wani IA, Sogi DS, Hamdani AM, Gani A, Bhat NA, Shah A (2016) Isolation, composition, and physicochemical properties of starch from legumes: a review. Starch-Stärke 68:834–845

Wilhelm HM, Sierakowski MR, Souza GP, Wypych F (2003) Starch films reinforced with mineral clay. Carbohydr Polym 52:101–110
Wilson JD, Bechtel DB, Todd TC, Seib PA (2007) Measurement of wheat starch granule size distribution using image analysis and laser diffraction technology. Cereal Chem 83:259–268
Wu J, Zhong F, Li Y, Shoemaker CF, Xia W (2013) Preparation and characterization of pullulan–chitosan and pullulan–carboxymethyl chitosan blended films. Food Hydrocoll 30:82–91
Xie FW, Yu L, Liu HS, Chen L (2006) Starch modification using reactive extrusion. Starch-Starke 58:131–139
Xie F, Halley PJ, Averous L (2012) Rheology to understand and optimize processibility, structures and properties of starch polymeric materials. Prog Polym Sci 37:595–623
Xu D, Ren G, Liu L, Zhu W, Liu Y (2014) The influences of drying process on crude protein content of naked oat cut herbage (Avena nuda L.). Dry Technol 32:321–327
Yongfeng A, Jay-lin J (2018) Understanding starch structure and functionality. In: Starch in Food: Structure, Function and Applications Woodhead Publishing Series in Food Science, Technology and Nutrition, pp 151–178
Yu L, Dean K, Li L (2006) Polymer blends and composites from renewable resources. Prog Polym Sci 31:576–602
Zafeiropoulou T, Evageliou V, Gardeli C, Yanniotis S, Komaitis M (2010) Retention of trans-anethole by gelatine and starch matrices. Food Chem 123:364–368
Zamudio-Flores PB, Vargas-Torres A, Pérez-González J, Bosquez-Molina E, Bello-Pérez LA (2006) Films prepared with oxidized banana starch: mechanical and barrier properties. Starch Stärke 58(6):274–282
Zamudio-Flores PB, Bautista-Baños S, Salgado-Delgado R, Bello-Pérez LA (2009) Effect of oxidation level on the dual modification of banana starch: the mechanical and barrier properties of its films. J Appl Polym Sci 112(2):822–829
Zamudio-Flores PB, Gutierrez-Meraz F, Bello-Pérez LA (2011) Effect of low and high acetylation degree in the morphological, physicochemical and structural characteristics of barley starch. Starch-Stärke 63(9):550–557
Zhang P, Hamaker BR (2012) Banana starch structure and digestibility. Carbohydr Polym 87:1552–1558
Zhang V, Han JH (2006a) Plasticization of pea starch films with monosaccharides and polyols. J Food Sci 71:253–260
Zhang YC, Han JH (2006b) Mechanical and thermal characteristics of pea starch films plasticized with monosaccharides and polyols. J Food Sci 71(2):109–118
Zhang Y, Han JH (2008) Sorption isotherm and plasticization effect of moisture and plasticizers in pea starch film. J Food Sci 73(7):313–324
Zhang P, Whistler RL, Bemiller JN, Hamaker BR (2005) Banana starch: production, physicochemical properties, and digestibility-a review. Carbohydr Polym 59:443–458
Zhang Y, Zhang Y, Xu F, Li S, Tan L (2018) Structural characterization of starches from Chinese jackfruit seeds (*Artocarpus heterophyllus* lam). Food Hydrocoll 80:141–148
Zhou Q, Shi W, Meng X, Liu Y (2013) Studies on the morphological, crystalline, thermal properties of an under utilized starch from yam Dioscorea zingiberensis CH Wright. Starch-Stärke 65:123–133
Zhu F (2015) Isolation, composition, structure, properties, modifications, and uses of yam starch. Compr Rev Food Sci Food Saf 14:357–386

Chapter 3
Cellulose and Cellulose Derivative-Based Films

Panthavur Nairveetil Anjali, Sowriappan John Don Bosco, Saadia Zainab, and Kappat Valiyapeediyekkal Sunooj

3.1 Introduction

Biodegradable agricultural produce obtained from nature plays an inevitable role in the packaging industry, especially the plant fibers, which are composed mainly of cellulose. There are natural polymers like cellulose, silk, protein, and synthetic polymers such as polyethylene, polypropylene, polystyrene, etc., for packaging applications. However, natural polymers are nontoxic to soil and the environment. Among natural polymers, fibers can be classified into plant, animal, and mineral fibers. Classification of natural fibers is given in Fig. 3.1. Plant fibers have many advantages like good thermal and acoustical insulation characteristics, sustainability, renewability, lightweight (low density), easily recyclable, low cost, easy to separate, energy recovery, CO_2 sequestration, degradability, and availability and have acceptable specific strength properties with widespread applications (Tajeddin 2014; Al-oqla and Rababah 2017). Moreover, plant fibers can be further classified into wood, fruit, seed, leaf, bast, stalk, and root fibers. They are generally composed of hemicellulose, cellulose (holocellulose), and lignin (HPS et al. 2017).

Cellulose is one of the most abundant organic materials, obtained from numerous plant and non-plant resources like wood, bast fiber, seed fiber (cotton), coconut fiber, grasses (bamboo, bagasse), sisal, eucalyptus, algae, bacteria, fungi, marine creatures, etc. It was discovered in 1838 by the French chemist Anselme Payen. It has been used widely in food, pharmaceutical, textile, paint, cosmetics, and paper industries (Saba and Jawaid 2017; Nasir et al. 2017). Being a biodegradable natural polymer,

P. N. Anjali · S. J. D. Bosco · K. V. Sunooj (✉)
Department of Food Science and Technology, Pondicherry University, Kalapet, Puducherry, India

S. Zainab
College of Food Science and Technology, Henan University of Technology, Zhengzhou, China

T. Amin et al. (eds.), *Polysaccharide Based Films for Food Packaging: Fundamentals, Properties and Applications*,
https://doi.org/10.1007/978-981-99-4898-7_3

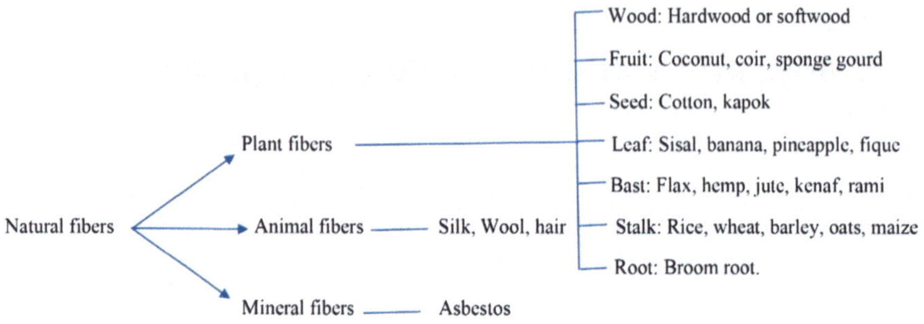

Fig. 3.1 Classification of natural fibers

cellulose attracted the food packaging industry worldwide since our society demands packaging materials with new functionalities and low environmental impact. Plastics was the most commonly used packaging material from a long time back. It became crucial when hundreds of thousands of tonnes of plastic were discarded and accumulated in the environment. Even though the only way to lower the bulk plastics waste is incineration, the high cost and release of harmful chemicals to the environment became the obstacles. Moreover, the hiked prices of crude oil and reserves forced the necessity to find an alternative to oil-derived polymers. The role of packaging is not only product preservation and distribution. It also aims at end-user convenience and effective communication with the consumers. Even though numerous polymers have been exploited to apply in the food packaging industry, cellulose is the most used and well-known renewable and sustainable raw material. It became the ideal candidate for polymer nanocomposites processing owing to the properties like reinforcing capabilities, mechanical properties, low density, and abundance.

3.2 Structure of Cellulose

Cellulose is a fibrous, water-insoluble, natural polysaccharide composed of more than 3000 repeating units of glucose, with a structural formula $(C_6H_{10}O_5)_n$ where n is the degree of polymerization and represents the number of glucose groups. The low water solubility of cellulose provides control over the scaffold design of cellulose-based materials. The average molecular weight of cellulose is between 1,30,000 and 1,90,000, and its degree of polymerization is 800–1200 (Al-oqla and Rababah 2017). It is a well-arranged fibrillar network that contributes mainly to the strength and rigidity of the cell wall of plants, which are not digestible by man. A major portion of the fibrillar cellulose structure is in the crystalline arrangement (ordered), and the remaining is in the amorphous region. The chemical structure consists of a linear chain of D-anhydro-glucose units linked together by β-(1–4)-glycosidic bonds. Figure 3.2 illustrates the chemical structure of cellulose. The linear

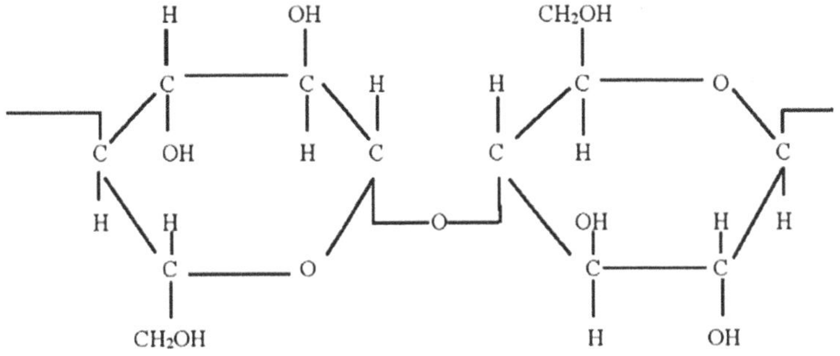

Fig. 3.2 Structure of cellulose

chain of glucose units is the microfibrils with a diameter range of 2–20 nm, and their aggregate forms the fibril. Fibrils are long, thread-like, and compactly packed as bundles by intermolecular hydrogen bonding, and this is responsible for the high degree of crystallinity and low solubility of cellulose (Prusty et al. 2017). It does not exhibit a particular melting point and does not dissolve in common aqueous solvents (Khattab et al. 2017). Three hydroxyl groups at positions C6, C2, and C3 are present in each glucose monomer, which gives a hydrophilic property to cellulose and plays a significant role in governing the physical properties and crystalline packing of cellulose (John and Thomas 2008).

Four types of polymorphs are based on the molecular orientation and hydrogen bonding in the crystalline region of cellulose. They are cellulose I, cellulose II, cellulose III, and cellulose IV. Among all the four types, cellulose I (CI) and cellulose II (CII) are the most common polymorphs of cellulose, and they are named native and regenerated cellulose, respectively. CI is produced by various trees, plants, and tunicates. It can be converted to either CII or CIII. The most stable structure of cellulose is the CII, formed by regeneration or mercerization. CI and CII show different mechanical and hydrophilic characteristics, and the hydrogen bonding in CII is more complicated than CI. Liquid ammonia treatment is used to create CIII, and CIV is developed from it (Mahmud et al. 2019; Fardioui and Bouhfid 2017).

3.3 Cellulose Extraction

Cellulose can be extracted using different mechanical, chemical, or enzymatic methods depending on the requirement. Grinding, high-pressure homogenization, and cryo crushing are the mechanical methods commonly used. Zhao et al. (2013) reported that a combination of two or more methods is adopted for cellulose extraction with higher efficiency. Ultrasonication of plant tissues treated with deionized water is another method to extract cellulose. In chemical methods like alkaline treatment, NaOH solution will be used for digestion to remove hemicellulose and lignin. A mixture of glacial acetic acid and sodium chlorite can be used further for improving the process. Finally, the extracted cellulose will be washed with NaOH solution and deionized water to make the pH neutral. Dilute acid pretreatment is also carried out for efficient removal of hemicellulose and lignin. Even though the structure of lignin is affected by this treatment, the cellulose structure remains the same. Enzymatic treatments involve selective hydrolysis of hemicellulose and lignin (Joseph et al. 2020).

Cellulose extraction was mostly carried out using NaOH treatment, which requires more acid or water to neutralize effluent. It makes the process more hazardous and costly. Corrosion of equipment is also a disadvantage. Recent studies have reported an alternative efficient, mild pretreatment for extracting cellulose microfibers from sugarcane bagasse using Na_2CO_3 (Sankhla et al. 2021). Similarly, acid hydrolysis, especially sulfuric acid hydrolysis, has several drawbacks like very low thermal stability of the final product, the difficult economic recovery of acid, a requirement of high amount of alkali for neutralization, presence of sulfate ion in final nanocellulose, etc. (Chen and Lee 2018). The ionic liquid treatment reported by Sankhla et al. (2021) is a powerful, economically friendly alternate solution for acid hydrolysis pretreatment using sulfuric acid. Ionic liquids can be reused multiple times without affecting the efficiency of the process. Other advantages include high thermal and chemical stability, very low vapor pressure, and easiness in recrystallization of dissolved cellulose.

3.4 Bacterial Cellulose (BC)

Bacteria, algae, and fungi can produce cellulose. Most bacteria produce extracellular polysaccharides like cellulose as a protective covering for the cell. BC is an organic compound secreted by certain types of bacteria such as *Acetobacter*, *Achromobacter*, *Alcaligenes*, *Pseudomonas*, *Rhizobium*, *Sarcina*, *Rhodobacter*, *Agrobacterium*, etc. In contrast to other cellulose, it is free of hemicellulose, lignin, and pectin, increasing purity during extraction. It exhibits properties like strength, moldability, and increased water holding capacity. The water holding capacity of BC is associated with the production of temporary artificial skin to treat skin burns. It also tends to enhance the growth of the human cell. The high tensile strength and uniform fiber distribution help to serve as an acoustic membrane also. Even though the main application of BC is in the biomedical field, the food industry applications are also vast. It acts as a thickener, binder, stabilizer, water-binding additive, and packaging material in the food industry. Packaging films formed from BC are gaining much importance because even disassembled BC can be processed into microfibrils, nanofibrils, and nanocrystals. Polymeric composites are also formed by incorporating BC into various polymeric matrices as a reinforcing agent (Cazón and Vázquez 2020).

3.5 Physically Modified Cellulose

The microfibrillated cellulose (MFC) and microcrystalline cellulose (MCC) are the physically modified cellulose. MFC is created by passing a slurry of cellulose through a small orifice under conditions of high shear and pressure difference. It will disrupt the cellulose structure to form micro fragments. MCC is created by treating cellulose with hydrochloric acid so that the amorphous region will be dissolved and the remaining crystalline region forms fine crystals (Stephen et al. 1995). Both MFC and MCC have broad applications in nanocellulose composite formations. For example, a study reported composite film development by dissolving MCC as a matrix and MFC and MCC as reinforcing phases. The tensile strength of composite decreased due to the presence of MCC, whereas the addition of MFC increased the tensile property of cellulose. At a low content of reinforcement, strong interfacial interaction between the matrix and reinforcing phase was observed. The results exhibited MFC has more reinforcing ability than MCC (Tanpichai 2018).

3.6 Chemically Modified Cellulose Derivatives

Cellulose derivatives are the natural polymers obtained by partial or complete reaction between the three hydroxyl groups present in the anhydro-glucose units with different reagents. The applications of cellulose derivatives are innumerable in the food industry. It is used mainly for food packaging, where the packaging material comes into direct contact with the food. Water-soluble edible films can also be produced from cellulose derivatives. Etherification and esterification are the important derivatizations of cellulose, and their functional, structural, and physical properties vary according to the process involved (Seddiqi et al. 2021). Etherification is the process of converting alcohol or phenol into an ether by the reaction between alkali cellulose and various reagents. Similarly, the chemical reaction involving organic acid (RCOOH) and alcohol (ROH) results in the formation of ester (RCOOR). Table 3.1 summarizes the chemical reactions involved in the formation of cellulose derivatives. Methylcellulose (MC), hydroxypropyl methylcellulose (HPMC), carboxymethylcellulose (CMC), cellulose acetate (CA), ethylcellulose (EC), hydroxyethylcellulose (HEC), etc. are the most commonly synthesized cellulose derivatives.

3.6.1 Methylcellulose (MC)

MC is the simplest and most widely used cellulose ether, which can be synthesized in an alkaline medium using an appropriate methylating agent like methyl chloride.

Table 3.1 The chemical reactions involved in the formation of cellulose derivatives

Cellulose derivatives	*R*	Chemical reaction
Methylcellulose	H or CH_3	Cellulose+NaOH+methyl chloride or dimethyl sulfate
Carboxymethylcellulose	H or CH_2COOH	Cellulose+NaOH+monochloroacetic acid
Ethylcellulose	H or CH_2CH_3	Cellulose+NaOH+ethyl chloride
Hydroxyethylcellulose	H or CH_2CH_2OH	Cellulose+NaOH+ethylene oxide
Hydroxypropylcellulose	H or $CH_2CH(OH)CH_3$	Cellulose+NaOH+1,2-propylene oxide
Cellulose acetate	H or $(C{=}O)CH_3$	Cellulose+aceticanhydride+sulfuric acid (catalyst) +acetic acid (solvent)
Cellulose nitrate	H or NO_2	Cellulose+nitric acid
Cellulose sulfate	H or SO_3H	Cellulose+sulfuric acid or chlorosulfonic acid or sulfur trioxide

In the presence of alkali (mostly NaOH) and water, cellulose structure will get activated by disrupting the crystalline structure. It will promote the access of polymer hydroxyl groups to react with the alkylating agent that further leads to etherification. MC is completely soluble in cold water to form a clear solution, whereas on heating between 48 °C and 64 °C, it becomes gel based on the grade. Various properties of MC can be altered depending on the degree of substitution (DS), either by the time of substitution or methylating agent used (Viera et al. 2007). For example, MC is soluble in some solvents, if the DS is in the range of 2.4–2.8, and if the degree is less than that of 2.0, it will be soluble in water only (Nasatto et al. 2015). Researchers have reported that MC is a flexible, tasteless, nontoxic, transparent film with good oxygen barrier properties but poor water vapor barrier properties (Mura et al. 2011).

3.6.2 Carboxymethylcellulose (CMC)

CMC is synthesized by replacing some of the hydroxyl groups in the chemical structure of cellulose with carboxymethyl groups. Along with the hydrophilic carboxylic groups, it is composed of hydrophobic polysaccharides, which together show an amphiphilic character (Toğrul and Arslan 2004). It is used widely for film-forming formulations owing to unique features like high chemical stability, biocompatibility, hydrophilicity, film-forming capacity, biodegradability, thermal gelatinization, and water solubility. It dissolves readily in cold water but does not dissolve in some organic solvents like ethanol. The solubility rate depends on the DS and the degree of the polymerization reaction. The DS is 0.4–1.4; DS more than 0.6–0.8 provides a good water solubility (Sahoo and Jena 2018). CMC can act as a multifunctional agent due to its ability to perform as an addictive and other property, including solubility in water, immiscibility in oil, etc. Therefore, it plays a significant role in industries as a thickening agent, binding agent, or stabilizer (Tajeddin 2014).

3.6.3 Hydroxypropylmethylcellulose (HPMC)

HPMC is a cellulose ether that is synthesized further from MC. Structurally MC and HPMC are the same, except for the substitution of hydroxypropyl groups in place of some of the remaining free hydroxyl groups, enhancing the open up of the cellulose molecule. Propylene oxide is the alkylene oxide added to MC to form HPMC. The films developed using HPMC are strong, but it is not that much strong compared to MC. It is also soluble in cold water like MC. Studies have reported that a 3% solution of HPMC can synthesize a strong and clear film of thickness 20 mils (0.02 in.). The film can be peeled out from the casting surface easily at a stretch indicating its strength (Embuscado and Huber 2009).

3.6.4 Cellulose Acetate (CA)

A collection of acetate esters of cellulose is generally termed as derived cellulose acetate (Malhotra et al. 2015). It is widely used in the food packaging industry, along with cellulose diacetate and triacetate, to manufacture film wrappings and membranes. Different researchers have carried out several studies related to cellulose acetate, diacetate, and triacetate. CA films with antimicrobial properties can be developed by the addition of potassium salt of sorbic acid, lysozyme, and sodium propionate. It slowly releases the active agents from film to dissolve on the surface of food in contact so that the growth of pathogens can be prevented and enhance the extended shelf life of a product (Uz and Altinkaya 2011; Gemili et al. 2009; Soares et al. 2002). Another work was carried out by Suvorova et al. (2000) related to cellulose diacetate combined with potato or corn starch. Immobilization of CA films and entrapping behavior of active agents in pores of CA films have also been studied (Gemili et al. 2010).

3.7 Cellulose and Cellulose Derivatives-Based Films

Packaging materials can have a significant impact on the food industry because they reduce food loss due to deterioration and spoilage. It acts as a barrier for providing protection and improving the quality of food material and hence extends the shelf life of the product. A good packaging material prevents the food from direct contact with atmospheric gases, physical contaminants, microbial attack, water, oil, etc. At the same time, the disposal of used packaging materials can create a negative impact on the environment. Therefore, packaging materials should possess specific quality characteristics, most important being biodegradability and compostability.

Cellulose and cellulose derivatives have wide acceptance in the food packaging industry as they are biodegradable and can produce films and coatings that can be consumed along with the food (Dara et al. 2021). Even though the terms film and coating are used interchangeably, the film is a stand-alone wrapping material. In contrast, a coating can be applied directly on the product's surface either by the spraying, dipping, panning, or fluidized bed method. The most preferred method is spray coating that offers uniform coating, thickness control, and easiness to perform in a production environment (Andrade et al. 2015). The coating can regulate physiological, biochemical, or oxidation processes that enhance the shelf life of the product. It also improves product appearance by shining surfaces, hiding scars, and delayed decaying. The nutritional and sensory attributes can also be enhanced by various active ingredients like colorings, flavorings, or sweeteners (Panahirad et al. 2021).

The packaging film, either clear or opaque, is a dried interacting polymer network of three-dimensional gel structures, having a thickness of less than 0.3 mm. It has four major components: basic material, plasticizers, additives, and solvents (Avramescu et al. 2020).

- Cellulose derivatives like CMC, HPC, and HPMC are widely used as basic materials for film preparation.
- Plasticizers like glycerol, mannitol, sorbitol, and sucrose are added for polymer rigidity at a proportion of 10–65%. It improves polymer formation, reduces brittleness, and increases the flexibility and hardness of the film.
- Additives include various compounds like antioxidants, antimicrobial agents, colorings, flavorings, sweeteners, fragrances, nutraceuticals, probiotics, emulsifiers, and browning inhibitors.
- Dispersion of all film-forming materials takes place in a solvent medium, most commonly an aqueous solution of ethanol with different concentrations. The main benefit of using ethanol as a solvent is its rapid drying power after the film formation and its ability to reduce microbial growth.

The film-forming process is a conversion of a wet gel-structured medium (basic film-forming material and solvent) into a dry film by eliminating the excess solvent in the gel solution with drying. Generally, the film-forming material will be capable of forming gel solution by gelatinization. Plasticizers and additives are then added to the solution for giving specific characteristics to the film. The next step is to process the liquid by casting, pouring the solution to a solid smooth surface either by lamination or extrusion. The solution has to be dried carefully to remove the excess solvent and form a dried film till it reaches a certain humidity. One of the important factors to be considered in the wet process of film making is the selection of solvent because the film should be edible and biodegradable. Ethanol and water are the most preferred solvents. It should disperse homogenously in the gel solution, and when applied to a smooth flat surface, it should form a thin film (Embuscado and Huber 2009; Tajeddin 2014).

3.8 Properties of Cellulose Derivative-Based Films

3.8.1 Thermal Properties

The thermal properties of cellulose derivative-based films have a high impact on their functional behavior. It determines the viability of biodegradable films in food packaging because the developed films might have to undergo several heat treatments even after application on the food products (Moradi and Kalanpour 2019). The process of film making involves several heat treatments where the structure and behavior of cellulose derivatives change. For example, the clear solution of MC will form a gel by losing water molecules when heated between 48 °C and 64 °C. The heated MC can hold this gel or fall out entirely based on the concentration. Similarly, the hydrated gel of HPMC solution (2%) will fall out completely when heated to 82 ° C or above. Both MC and HPMC redissolves when cooled down to room temperature (Embuscado and Huber 2009). Therefore, depending on the thermal reactions involved in the film-making process, alterations will be seen in the stability and behavior of the film formed.

Thermogravimetric analysis (TGA) and differential scanning calorimetry (DSC) are the techniques used for assessing the thermal stability of films. TGA is a measure of the change in mass relative to temperature, and the output can be analyzed by a TGA curve, which shows the polymer deformations according to temperature range (Yaradoddi et al. 2020). Zhou and Wang (2021) reported that the cellulose film prepared from durian rind (DC) exhibited high thermal stability compared to commercial cellophane film (CC). The TGA curve implicated two distinct mass loss stages around 270 °C for CC and 345 °C for DC, which may be due to the decomposition of cellulose matrix and degradation of salt, respectively. The first weight loss at around 150 °C was observed for CC, which may be attributed to the added additives resulting in lower thermal stability of the film. Another study based on the all cellulose nanocomposite film (ACNC) developed from sugarcane bagasse nanofiber revealed lower thermal stability for ACNC (174.1 °C) than cellulose nanofiber (176.9 °C), which might be due to the poor crystallinity of ACNC (Ghaderi et al. 2014). In a recent study on the water-insoluble and flexible transparent film based on CMC, a weight-loss stage before 150 °C was observed for the soluble CMC film. It might be due to the moisture content present. The insoluble flexible transparent film developed exhibited greater thermal stability than the soluble CMC film (Wang et al., 2021).

DSC measures the difference in heat flow rate between a sample and an inert reference as a function of time and temperature. i.e., During a change in temperature, the heat flow either emitted or absorbed by the sample on the basis of temperature difference in the sample and reference material will be measured. The result will be a DSC curve, with temperature or time dependence on heat flow (per time unit). Endo or exothermic peaks will be formed based on the heat flow, showing either positive or negative deviation from a baseline (baseline is recorded when there is no reaction in the system) (Gill et al. 2010).

Viera et al. (2007) synthesized MC from sugarcane bagasse cellulose and characterized the prepared cellulose derivative in comparison with original cellulose. Thermal properties were analyzed in a temperature range of 25 °C to 300 °C, and the glass transition temperature (T_g) observed was 184.24 °C. The two endotherm peaks at 211 °C and 219 °C indicate the fusion of crystalline regions and MC degradation. Another study was conducted involving the extraction of CMC from agricultural waste for packaging applications. The film formed from derived CMC, along with gelatin, agar, and different concentrations of glycerol (A-1.5%, B-2%, C-2.5%), were taken for the study. The T_g value obtained for the three samples, A, B, and C, is 50.25 °C, 76.08 °C, and 50.25 °C, respectively. The higher $T_g$value indicates greater mobility and a tendency to form a crystal structure. The observed heat fusion of samples was 223.9 J/g, and they attained the crystallization temperature at 63.65 °C (Yaradoddi et al. 2020).

3.8.2 *Barrier Properties*

The packaging films should meet certain quality criteria for particular food products and the conditions in which they will be used. For example, the packaging material for a liquid product should be waterproof, and the film used for a fat-based product should possess gas barrier properties to prevent oxidation reactions. In general, the quality and shelf life of food products on storage and transportation depends mainly on the barrier properties of the packaging material used. It includes water vapor/oil permeability, light barrier, gas permeability (O_2, CO_2, N_2), barrier to aroma and flavor compounds, etc. The transfer of water vapor, gas, or flavor compounds takes place gradually by three steps, adsorption on the film surface, diffusion, and desorption. Temperature, pressure, and relative humidity also have a role in determining barrier properties along with the nature of the film. The transmission rate varies according to the thickness of the film, partial pressure of permeant, and the relative humidity of the environment (Wang et al. 2021).

3.8.2.1 Water Vapor Permeability (WVP)

The prevention of moisture transfer between food and the surrounding is a function of packaging material. Therefore, the amount of water that can be easily penetrated through the film has to be tested to check the barrier property of the film (Di Filippo et al. 2021). The hydrophilic character of cellulose imparts WVP to cellulose-based films (Wang et al. 2021). Several researchers have reported that the addition of lipid can reduce WVP, but mechanical properties will also get reduced (Bravin et al. 2004). It was observed that the addition of gelatin also improved the moisture barrier at a certain pH level, whereas mechanical properties decreased regardless of pH (Chambi and Grosso 2011). The DS in cellulose derivative-based films like CMC will directly affect the hydrophilic character and thus water vapor transmission rate.

Spence et al. (2011) mentioned another method of surface coating and mineral filler addition to decrease the WVP of films. The solubility of water vapor decreased due to mineral filling. A three-layer surface coating with paraffin, beeswax, and cooked starch also enhanced the moisture barrier property of cellulose-based films. Another study based on the characteristics of HPMC-based edible film reported that the WVP of a film depends on the storage conditions like temperature and relative humidity, composition, and thickness of the film (Osorio et al. 2011). The results showed that at a storage temperature of 5 °C and relative humidity of 100%, the WVP of HPMC film observed was in a range of ($1.67 \times 10^{-8} \pm 0.37 \times 10^{-8}$) and ($2.61 \times 10^{-7} \pm 0.36 \times 10^{-7}$) (g mm h^{-1} cm^{-2} Pa^{-1}), whereas WVP of HPMC film without storage condition was observed to be 3.79×10^{-8} (g mm h^{-1} cm^{-2} Pa^{-1}).

3.8.2.2 Oxygen Barrier Properties

The oxygen barrier properties of films are very important in food packaging. It is reported that material is considered to be an excellent oxygen barrier, if the permeability of oxygen is <10 cm^3 μm m^{-2} day^{-1} kPa^{-1} (Edlund et al. 2010). Most polysaccharide-based films have a good oxygen barrier capacity due to the complex structure with a high number of hydrogen bonds limiting free volume for oxygen transmission. Studies have reported that oxygen transmission rates will be increased upon increasing relative humidity. It might be due to the weakening of hydrogen bonds that causes a change in the cellulose compactness (Lyytikäinen et al. 2021). A recent study on the incorporation of nano-fibrillated cellulose and nisin on cold plasma treated BOPP/LDPE laminates, performed by Lu et al. (2018), showed improved oxygen barrier properties in packaging applications. The observed oxygen transmission rate was 24.02 cc/m^2.day that is very low compared to the non-coated film (67.03 cc/m^2.day). Another study revealed the excellent oxygen barrier properties of cellulose nanomaterial films (CNM), challenging even the most commonly used petroleum-based polymers. CNM films come under high oxygen barrier classification with a relative humidity range of up to 65%. It was found to be the best-suited film for dry foods stored at low temperature and relative humidity conditions (Wang et al. 2018).

3.8.3 Mechanical Properties

Mechanical properties of packaging films are attributed to the strength and flexibility properties characterized by deformation at the break, percentage elongation at break, tensile strength, and elastic modulus (Embuscado and Huber 2009). The surrounding conditions like temperature, partial pressure, and relative humidity will also influence the mechanical properties during film production, storage, and its use. Cellulose and cellulose derivative-based films can impart a good mechanical property owing to their intermolecular network interactions providing structural integrity to film.

Additionally, plasticizers like glycerol, polyethylene glycol, sorbitol, etc. influence mechanical behavior.

In a recent study, Zhou and Wang (2021) reported the mechanical properties of zein and MC composite film developed by adding polyethylene glycol (PEG) and oleic acid (OA) as plasticizers. MC exhibited a high tensile strength (TS) of 59.6 $\pm$ 1.5 MPa. It slightly decreased when PEG was added, but elongation increased from 13.5 $\pm$ 1.5 to 45.9 $\pm$ 8.9 cm/100 cm. According to the increased percentage of MC added to the composite film, elongation increased from 12.4 $\pm$ 1.3 (zein-MC-50) to 28.5 $\pm$ 2.5. Another study revealed the excellent mechanical properties of CNF, in which the mechanical strength and stability are strongly associated with the dispersal of fiber in the polymer matrix. The high specific strength and aspect ratio of CNF composite film upholds a conversion from continuous polymer phase to discontinuous fiber phase (Somvanshi and Gope 2021).

3.8.4 Dielectric Properties

Electrical insulators that can be polarized by the applied electric field are called dielectric materials. The stored energy during polarization is represented by the dielectric constant (ε), and dissipated energy converted to heat is the dielectric loss. It is characterized by the loss tangent, tan δ (Yang et al. 2019; Nelson 2015). The dielectric materials with high dielectric constant and low dielectric loss are considered appropriate for various applications. Conventional dielectric polymers are non-renewable thermoplastics like polypropylene, polystyrene, polyethylene terephthalate, polyethylene naphthalate, etc. Researchers have found that renewable, biodegradable cellulose nanomaterials can replace conventional dielectric polymers to some extent (Tao et al. 2021). Cellulose nanomaterials are the nanostructured fibers of cellulose, either CNC, CNF, or BNC. The increase in surface area due to nanostructure enhances the interaction between cellulose and the surrounding particles like water, low-molecular-weight compounds, etc.

Cellulose-based derivatives like CMC with high viscosity, cyanoethyl cellulose (CEC) with high dielectric constant, and heat stability can be introduced to nanocellulose to improve dielectric properties. Also, high dielectric fillers like ceramic fillers and conductive fillers can be incorporated into nanocellulose for enhancing dielectric properties (El Sayed et al. 2015; Jia et al. 2016). A recent study on the dielectric properties of nanofibrillated cellulose thin films developed from bagasse pulp reported decreased dielectric constant with increasing frequency due to orientation polarization. It might be due to the freedom of the dipole molecular chain when there is a rise in temperature (Abdel-karim et al. 2018). Diantoro et al. (2017) reported that the addition of ZnO in CA to form nano CA-ZnO composite film reduced the crystallinity of CA-based film and increased its dielectric constant. It might be due to the greater polarization effect of the composite film under the influence of the external field.

3.8.5 Tensile Properties

Tensile strength is the ability of a film to withstand external force without breaking. It is the maximum stress a material can hold without failure. For example, the tensile strength of MC is 60–70 MPa, and commercial CA is 41–87 MPa (Paunonen 2013). Several techniques can be applied to improve the tensile strength of the film, like the addition of a filler or a cross-linking agent, physical treatments like ultrasound methods, or modifying the manufacturing process of the film. Borges et al. (2004) studied the tensile properties of cross-linked and uncross-linked microcrystalline cellulose fibers incorporated composite films developed from HPC. The results showed that even with a small amount of fiber, the strength of cross-linked polymer increased due to the improved fibril network structure. Another study reported the enhanced mechanical properties of regenerated cellulose by incorporating various plasticizers. It was observed that the tensile strength of control cellulose film was comparatively lower than plasticized films. The addition of sorbitol, glycerol, and CMC increased tensile strength by 43.5%, 46.2%, and 58%, respectively (Pang et al. 2013).

3.8.6 Heat Sealing Properties

Heat sealing is the method for closure and protection of packaging that was primarily used for flexible packaging. Care should be taken during heat sealing because moisture, gas or aroma, and flavor compounds transfer through the tiny pores if not properly sealed. Pressure, energy, and time are the factors to be considered for heat sealing. The heat sealability of a film is the maximum force required per width unit to separate a heat seal. It is tested using a sealing strength tester (Willberg-Keyriläinen et al. 2017). Heat sealing is done by introducing a particular amount of heat to melt the surface of the films. The interfacial interactions result in the melting and sealing of the contact surface (Nilsuwan et al. 2018). It was observed that the sealing temperature and seal strength increased according to the chain length of the polymer, and the contact surface of films diffuses as a function of temperature (Cooper 2014). Studies based on the three-layer CNF film, coated with molar mass controlled cellulose esters with different side-chain lengths, were carried out by Willberg and coworkers. The sealing strength of the three-layer film measured by a 20 s delay time was found to be <100 N/m. The thickness of the ester coating also affects the heat sealability of the film (Willberg-Keyriläinen et al. 2017).

3.8.7 *Flexibility Properties*

Cellulose and its derivatives have several applications owing to their shape flexibility. It is suitable to use as a flexible substrate or develop flexible electronics other than packaging applications (Zhao et al. 2020). The intrinsic stiffness of the film, which defines the flexibility properties, is measured by elastic modulus. It is the ratio of stress to strain when a body is subjected to deformation on applied force. A comparative study on the mechanical properties of cellulose derivative films reported that the flexibility decreased when the number of bonds between molecules increased. The result showed smaller young modulus (YM) values for pectin and HPC films, indicating that these structures are more fragile than other films with more intra- and intermolecular interaction. The CMC films interact with water without losing the intermolecular interactions to retain the elastic properties (Espinoza-Herrera et al. 2011).

Several studies have reported that adding a small molecule or polymer to a CNCs colloidal solution without any modification allows it to retain the chiral nematic arrangement with some flexibility in the final films. For example, the final product obtained by adding PEG to the CNCs solution was humidity responsive. Exposure to humidity increased the pitch values of the film in a reversible process and became flexible as the PEG content increased (Saraiva et al. 2020).

3.8.8 *Electrical Properties*

Cellulose nanofiber films have become a promising material for use as substrates for electronic devices even though CNF has low electrical conductivity due to the non-availability of free electrons. The friction or rubbing of electrical insulators generally provides static electricity (Shimizu et al. 2019). Therefore, to increase the electrical conductivity of devices, inorganic particles can be incorporated into them, or they can be coated with antistatic materials.

Nontoxic, biocompatible conductive fibers have gained much attention in recent decades for their applications in flexible electronic devices and wearable electronic appliances like nano-generators, supercapacitors, batteries, and flexible solar cells. Cellulose is considered to be an inexhaustible source to replace chemically derived compounds. The substantial intra-intermolecular hydrogen bonds make cellulose less soluble, which hinders its wide applications. However, cellulose can dissolve in ionic liquids effectively. The productive and promising green solvents, ionic liquids, have several advantages like superior dissolving capacity, recyclability, environmentally friendly behavior, recoverability, and variable structures. Different polymers have produced heterostructure composites in ionic liquid systems such as films and fibers (Zhang et al. 2018).

A recent study on the development of nanofibrillated thin films from bagasse pulp revealed the electrical conductivity of the prepared film. It was observed that the dc electrical conductivity followed Arrhenius behavior, and the highest conductivity correlated to less activation energy. The ac electrical conductivity followed the power law equation, and it increased with frequency. The study revealed that conductivity observed for thin films depends on several factors like polar functional groups and free charges, higher water sorption properties, and compactness of film (Abdel-karim et al. 2018).

3.8.9 Optical Properties

The optical property of films has greater importance in the packaging industry since they are used in food applications. The surface color of the film is commonly measured using a colorimeter and opacity measured with a spectrophotometer. Cellulose fiber dimensions play a significant role in improving the optical characteristics of the film. A study was conducted based on the light scattering of fibers and micro-scaled pores that result in the opacity of regular papers constituted with micro-sized wood fibers. TEMPO oxidation and homogenization were used to modify the morphological behavior of wood fiber. The results showed limited light scattering and high transparency. The total transmittance and transmittance haze were analyzed using a UV spectrometer in wavelength of 400–1100 nm. It was observed that tempo oxidized cellulose nanofibrils (TOCN) have the highest transmittance (88%) and lowest optical haze (3.8%) at a wavelength of 550 nm. These results are similar to those of flexible glass and PET films. When the tempo oxidized wood fiber (TOWF) was incorporated, the TOCN films exhibited an optical haze of 62.3%. The total transmittance was found to be 83% from 88% at a wavelength of 550 nm. These results signify the potential applications of transparent substrates to fabricate next-generation green optoelectronic devices (Yang et al. 2018).

The study on the effect of PVA and copper oxide nanoparticles on the optical properties of CMC film showed that blending the film with PVA increased optical transmittance from 87% to 89%. The CuO nanoparticles decreased the optical transmittance to 77% after 0.5 wt% doping. The refractive index of CMC film was lowered by the addition of PVA (1.645–1.576) but increased after the blending of CuO nanoparticles (1.852) (El Sayed et al. 2015).

3.8.10 Water Solubility

Water solubility is an important factor of packaging material for food applications. Cellulose is insoluble in water and many organic solvents because of the strong inter and intramolecular hydrogen bonds, but the water solubility of cellulose derivatives varies according to the substitution reactions. The most commonly used cellulose derivatives, MC, CMC, HPMC, HPC, NaCMC, etc., are soluble in water.

A recent study on the use of snail mucus extract in the preparation of cellulose derivatives-based film by Di Filippo et al. (2021) showed that HPMC-based films immediately solubilize in water irrespective of their composition. The CMC-based film is also soluble in water, but the presence of snail extract preserved the structure for more than 2 weeks. The change in solubility of CMC film might be due to the protonation of the COO^- groups into the COOH group in an acidic pH (Qiu et al. 2014). Another study based on the development of water-insoluble, transparent CMC film by the freeze-thaw method and acid adjustment was carried out by Wang et al. (2021), which promises the approach of cellulose derivatives as substitutes for functional films. The ionized carboxyl groups of CMC that contribute to the water solubility of the film are protonated to form hydrogen bonds. The result showed that the freeze-thaw method, along with acid adjustment, resulted in the gelation of CMC solution and insoluble CMC film.

3.8.11 Viscous Properties

The viscous properties of a film are the function of the molecular weight of a polymer. According to the cellulose derivatives involved, the required viscous characteristics of an end product can be achieved by controlling many factors, including the DS, homogeneity of chemical substitution, or molecular weight, by varying the amount of chain oxidation. Factors like temperature, pH, and solute concentration also affect the viscous properties of the solution. The viscosity of most of the cellulose ether solutions decreases as the temperature increases. Generally, minor changes have been noted in a pH range of 5.0 to 9.0. In the case of CMC solution, precipitation of the polymer occurs below pH 4.0 due to the free acid production, and above pH 10.0, the viscosity will decrease, leading to polymer degradation. The solutes like salt or polar nonsolvents minimize the viscosity drop of a CMC solution according to the nature of addition.

Nonionic MC solution exhibits a different viscous property compared to CMC solution. The effect of pH is reduced, and the temperature has a significant role. On heating, an aqueous solution of MC and HPMC forms a gel, and upon cooling, it will return to the original state. The addition of solute like phosphate, sulfate, and carbonate salt causes changes in hydrogen bonding strength involved in the cellulose structure. The viscous properties of EMC are also similar to MC. EMC will precipitate upon heating above 60 °C in an aqueous solution instead of the gelling nature of MC. Similarly, HPC also shows precipitation above 45 °C (Stephen et al. 1995).

Arrieta et al. (2014) reported a study on the development of nanocomposite films based on the CNC blended with poly(lactic acid)-poly(hydroxybutyrate), PLA-PHB. The results showed that the addition of CNC improved the thermal stability of the film, and a reduction in viscosity molecular weight of PHB was observed. The viscosity molecular weight of PLA was unaffected.

3.8.12 Printing Properties

Cellulose, being a biodegradable polymer, has a wide scope in printing technologies. Three-dimensional printing (3D) has received good acceptance because it can produce customer-specific design and rapid on-demand fabrication at a low cost (Guvendiren et al. 2016). The printing ink should have specific features like well-controlled viscoelastic properties to improve the resistance during extrusion. They should possess a stable network and lower shrinkage property during drying to avoid deformations (Shao et al. 2015). Cellulose and its derivatives can be used as inks for excellent printing properties. The chemical structure of cellulose contains abundant functional groups, mainly OH, that can bind other materials. The swelling behavior also enhances high sensitivity, increasing the concentration of analyte molecules (Gu and Huang 2013).

Cellulose is insoluble in water and most of the solvents. Therefore, derivatives like cellulose ethers and esters are also widely used in printing applications for performing multiple roles. Cellulose ether (CMC, HPC) is mainly involved in viscosity enhancement. It provides low viscosity at the high shear rate, and when the shear rate stops, the viscosity will be increased. Different cellulose ethers can be mixed to adjust ink viscosity. Cellulose derivatives like MC are used as plasticizers in printing ink to adhere the adsorbent particles with the binder. The rheological properties of CMC find application in using the derivative as matrix material. Nanocellulose also offers a wide application in printing technology (Rees et al. 2015). A recent study focusing on the printability of CNF-based material analyzed the uniformity, precision, and stability of the printed pattern. The result showed a clogginess appearance when the CNF concentration was greater than 0.8%, and it might be due to the larger fiber particles in CNF (Lille et al. 2018).

3.8.13 Aroma Retention Properties

The aroma is one of the characteristics that determine the freshness and quality of food. In fact, a particular aroma or flavor recognized for a food product is the result of interactions between a large number of compounds so that even a minute variation can diminish the quality of food. The aroma gives the first impression about food to the consumers. The main reason for the loss of aroma from food is their volatility and sensitivity to air, light, heat, and moisture (Saifullah et al. 2019). The presence of oxygen or moisture content in the environment reacts readily with aroma compounds and affects food quality. Therefore, the packaging materials should be a barrier to aroma loss from food. Cellulose-based films are used widely for retaining the aroma and flavor of foods. Among them, HPMC films exhibit excellent aroma barrier properties. The study on the effect of HPMC-beeswax (BW)-based film on the postharvest quality of coated "Angeleno" plums showed that the coating reduced the softening, bleeding, and aroma-flavor loss, which indicates the ability of packaging material to create a modified atmosphere in the fruit (Navarro-Tarazaga et al. 2011).

3.8.14 Antimicrobial Properties

The presence of microorganisms in fresh and processed food products is a severe concern of safety and quality. Microbes invade and harbor mainly on the surface of fruits, vegetables, meat, fish, poultry, and dairy products. Antimicrobial films are developed to reduce the microbial activity on food products and hence extend the shelf life of the product. The antimicrobial films offer innumerable advantages, which includes the following:

- Preservative agents are directly applied to packaging film. Therefore, reduced amount of preservatives will be added to food.
- Steps involved in the process can be simplified by combining packaging and the addition of preservatives simultaneously.
- The interaction between various food components like protein and lipid with the preservative agents can be reduced by the slow migration of preservative agents from packaging material to food.
- Efficient protection is provided even during transportation and storage of food products.

A vast number of antimicrobial agents are present, either natural or synthetic, to incorporate in films for retarding microbial growth. Bacteriocins, enzymes, plant extracts, chitosan, polypeptides, spices, etc. are some of the biological antimicrobial agents. The common synthetic antimicrobial agents are organic acids, including benzoic acid, acetic acid, propionic acid, citric acid, malic acid, etc. (Irkin and Esmer 2015; Valencia-Chamorro et al. 2011).

Cellulose and cellulose derivatives have wide applications in the production of antibacterial packaging films. Different techniques like incorporating inorganic nanoparticles, grafting reactive groups into cellulose nanofibers, tethering antimicrobial polymers by coupling reactions, and surface-initiated polymerization of monomers having antimicrobial properties can be employed for film production (Wu et al. 2016). It has been reported that nisin has good antimicrobial properties in HPMC films against *Listeria monocytogenes*, *Staphylococcus aureus*, *Aspergillus niger*, etc. *L. monocytogenes* and *S. aureus* activity was inhibited to 70% and 40%, respectively, by adding 15% stearic acid to HPMC films (Valencia-Chamorro et al. 2011).

A study on the antimicrobial activity of edible MC-based film incorporated with *Pimpinella affinis* oil (PAO) reported the film with PAO inhibited most of the organisms. PAO is an essential oil that can destruct the cell wall and membranes of bacteria and fungi and inhibit the synthesis of DNA, RNA, polysaccharides, proteins, and enzymes. Therefore, the film incorporated with PAO exhibited a clear inhibition zone, indicating the efficient antibacterial activity (Ariaii et al.

2014). Another study aimed at the production of the optically transparent film from ginger cellulose nanofiber, with good antimicrobial properties by chemical and ultrasonication treatment, revealed the high antibacterial properties of the ginger fiber. The diameter of the inhibition zone obtained was greater than 6 mm, due to the action of bioactive compounds present in the ginger fiber (Abral et al. 2020).

3.8.15 Moisture Sorption Properties

The microstructure of polymers like cellulose can be studied using sorption properties. Water vapor is the main sorbate used for studying the sorption properties of cellulose and cellulose derivatives. Vapors of inert nonpolar liquids can analyze the porous structure and surface area of hydrophilic cellulose derivatives, whereas water vapor can be used for hydrophobic derivatives. The sorption properties are described using isotherms that exhibit the relative amount of sorbate in the sorbent (Ioelovich and Leykin 2011).

In a study on the water vapor sorption properties of CNC and CNF, the sorption behavior of four nanocellulose samples (CNF and CNC with cellulose I and II structures) was analyzed using the DVS technique. The effect of the nanocellulose preparation procedure on equilibrium moisture content, sorption hysteresis, and sorption kinetics was studied in detail. Parallel exponential kinetics (PEK) model was used for explaining the sorption kinetics behavior, which divided the sorption kinetic curve into fast and slow regimes (Guo et al. 2017). Another study based on the water vapor sorption behavior of two types of microfibrillated cellulose films in a wide range of temperature and water activity was carried out. The sorption isotherm exhibited an initial downward curve followed by a linear trend. Two models, namely Park and Gab, were used for the study. At a temperature below 35 °C, an endothermic sorption process was noted. There was no effect on solubility above 35 °C. At a lower temperature, the sorption process was diffusive, but when the temperature and moisture content in the films increased, another process was observed longer. It might be due to the structural relaxation of interfibrillar bonds in the matrix (Meriçer et al. 2017).

3.8.16 Release Kinetics of Active Packaging

Active packaging can be defined as the technology of incorporating certain additives into packaging films or containers to maintain and extend the shelf life of products. The additives mainly used in active packaging include oxygen scavengers, moisture-carbon dioxide-ethylene adsorbers, flavor/odor enhancers, antioxidants, or other preservatives. The shelf life depends on several intrinsic and extrinsic factors like composition of food, water activity (a_w), acidity (pH), nutrients present, respiration rate, storage temperature, relative humidity, and atmospheric pressure. Active packaging is applied to a product by considering these factors (Coles et al. 2003).

The mechanism behind active packaging is the slow release of additives from film to dissolve directly in the surface of the food. Therefore, the release rate of additives is crucial in active packaging in which different techniques can be applied to achieve a controlled release. Porosity is the main factor of packaging film that decides the release kinetics of active packaging. Films with controlled permeability and morphology can be developed using different phase inversion techniques. CA film formed by the dry phase inversion technique is an example. The difference in porosity of films was achieved by changing the initial casting composition, wet casting thickness, and drying temperature. Potassium sorbate was used as an antimicrobial agent in the study. The crystallization of potassium sorbate was found to be affected during drying. The degree of crystallization was low in dense film. Here the release rate of antimicrobial agent is controlled by the dissolution of emerged crystals. The entrapping behavior of active agents to pores in films varies according to the properties of film and the molecules of additives used. In asymmetric films where one side is denser than the other, release direction can be controlled easily. When multiple layers are present in the film, one layer can act as a barrier, and an additional layer will contain the active agents. The release kinetics can also be controlled by changing the chemical linking of the active agent into the matrix (Paunonen 2013).

3.8.17 Cost-Effectiveness

One of the problems regarding non-degradable plastics is resource availability. The primary source of cellulose is the plant cell wall, wood pulp, cotton, hemp, jute, rice straw, wheat straw, corn, sisal, etc. that are abundantly present in nature. Therefore, it is very cheap and readily available. Wood pulp is used chiefly for paper manufacture and cotton for the textile industry. Generally, cotton fibers which are almost pure cellulose, are not preferred for food-grade applications. Cost-effectiveness is one of the reasons behind the usage of different sources for various purposes.

Waste management is another crucial criterion of concern in the packaging industry. Synthetic plastics persist in the environment without degradation, limiting the landfill space. Since the food industry requires a large number of packaging materials, even a minute reduction in the number of plastics will have a positive effect. Cellulose being biodegradable offers a reduction in landfill space during waste management that can add to the cost-effectiveness of cellulose-based films and composites.

3.8.18 Safety and Regulations

The primary function of packaging is product safety and quality. Therefore, the packaging material also should be safe to use and should follow the regulatory norms. The cellulose and cellulose derivative films assure safety and compatibility with the environment because of their biodegradability. The film-forming components, including the solvents, plasticizers, and functional additives, also should be safe and nontoxic. All the film-forming components should be food grade in edible film production. Otherwise, there is a possibility of migration of toxic compounds to food in contact during storage or transport. Any reaction from packaging material or surroundings will directly affect the quality of food. Studies related to the migration of natural base nanoparticles like microcrystalline cellulose in polylactic acid (PLA) have been reported. It was observed that the migration level of CNC in isooctane was higher than in 10% (v/v) ethanol (Bumbudsanpharoke and Ko 2015). CNC might expose humans and the environment to unknown risks during production because of its unknown properties (Mu et al. 2019).

The functional ingredients like additives, antioxidants, preservatives, or color in packaging should be adequately mentioned in the product label. The food additives commonly used are stabilizers, emulsifiers, thickeners, flavor enhancers, gelling agents, etc. Another aspect regarding films is the labeling of ingredients that can cause allergic reactions to consumers. Allergens inherent to milk protein-casein, peanuts, wheat (gluten), soy, corn, nuts, oats, etc. are some of them which should be mentioned clearly. The equipment, process, and premises should strictly follow current good manufacturing practices (GMP).

3.8.19 Methods to Improve Properties

The primary function of packaging is to prevent food degradation and spoilage that enhances the quality and shelf life of products sold to the consumers. Therefore, the properties of the packaging material are very crucial. It decides the efficiency in packaging by evaluating mechanical, thermal, barrier, water-soluble, heat sealing, electrical, optical, viscous, and antimicrobial properties.

Cellulose and cellulose derivative films have excellent mechanical properties owing to their fibrillar structural behavior. The important characteristics that describe mechanical properties are tensile strength (TS), tensile modulus (TM), and elongation at break (EB), which indicate the ability of packaging films to maintain integrity under stress during processing, handling, and storage (Rhim et al. 2006). Studies have reported that nanocomposite materials like CNC are sensitive to relative humidity and exposure to water. It was observed that the TS of the CNC system improved over storage by the formation of hydrogen bonding due to water absorption (Salmieri et al. 2014). The addition of lipid can improve barrier properties like WVP. Other methods to enhance the barrier property are the addition of gelatin at controlled pH, surface coating, and mineral filler techniques. Oxygen barrier properties can be improved by controlling the relative humidity, as mentioned in Sect. 3.8.2.2. The water solubility of cellulose and its derivatives is determined by the hydroxyl groups present. Therefore, the DS of hydroxyl groups can improve their solubility.

The electric and dielectric properties of cellulose and nanocellulose matrix can be enhanced by incorporating high conductive fillers like ceramic fillers and conductive fillers. Barium titanate ($BaTiO_3$) and titanium dioxide (TiO_2) are the nanosized ceramic particles with relatively high dielectric constant. They also have excellent piezoelectric and pyroelectric properties. In the preparation of polymer composites, carbon-based fillers like graphene and carbon nanotubes (CNTs) are introduced to elevate the electrical and dielectric properties due to their high conductivity, high surface area, and large aspect ratio (Tao et al. 2021).

Cellulose and cellulose derivatives have tailored improved optical properties for optoelectric devices. Cellulose nano papers prepared by the solution casting method will produce a smooth surface that results in lower transmission haze (Yang et al. 2017). Cellulose nanofibers can fill the mesoporous wood fiber-based paper to produce a lower-cost transparent substrate of higher optical haze. These features are altered with cellulose fiber diameter and nano papers packing density (Yang et al. 2018).

3.9 Applications

Various parameters like the physicochemical properties, chemical structure, molecular weight, processing method, cellulose fiber dimensions, and presence of other active ingredients in the formulation determine the applications of cellulose and cellulose derivatives in the food packaging industry (Kenji Kamide 2005). Their commercial food applications other than packaging include the following:

- Cellulose: Cereals, bread, pastry, pasta, cookies
- MC: Fried foods, soup, sauce, bread, foams, restructured foods, stabilization in fruit juices
- HPMC: Baked products, bakery fillings, whipped toppings, icings, sauces, fried foods, frozen desserts
- HPC: Extruded foods, whipped toppings, mousses
- MEC: Egg white substitute, whipped toppings
- Microcrystalline cellulose: Baked products, salad dressings, sauces, beverages, whipped toppings, low-calorie foods, desserts, ice cream
- CMC: Stabilization of emulsion in condensed milk and chocolate milk, stabilization of ice cream against crystal formation, water binding, yield increase of cheese preparations, thickening of pudding
- NaCMC: Dressings, sauces, syrup, frozen desserts, baked goods, beverages, animal foods

Cellulose and its derivatives have versatile applications in the nonfood industry, like pharmaceuticals, cosmetics, fiber and textile, wood and paper, etc. It is a major constituent of the paper and textile industry. CNF is widely used in pharmaceuticals for applications like optoelectronic conversion, reinforcement of drug tablets, energy conversion, nanofillers, bioimaging, filtration media, protective coatings, antimicrobial activity, barrier membranes, etc. (Menon et al. 2017). Cellulose in the form of nitrocellulose is used in smokeless gun powder manufacture. Electrical equipment like cables and transformers use cellulose as insulation. It also provides good strength for use as a building material as an alternative to plastics and resins. The applications of cellulose derivatives in the nonfood industry are given in Table 3.2.

3.10 Future Trends and Limitations

The role of biopolymer-based packaging materials from renewable resources has become a vast area of interesting research. The current trend follows impressive advances in nanopolymer composites, composite films, and multifunctional composites towards an environmentally friendly approach. The excellent mechanical properties, reinforcing abilities, low weight, biodegradability, and wide availability are the reasons for the advancement in the cellulose field (Shaghaleh et al. 2018). It is estimated that from 2016 to 2025, the growth of flexible packaging will be 4.6% at an annual rate. The rapid increase of stand-up pouches for the food and beverage industry is a sign of their expected growth (Wang et al. 2018). Future trends demand an increase in the bio-contents of products to meet consumer sustainability consciousness. Therefore, the CNM-based films which can provide the required stiffness, rigidity, and oxygen barrier properties to stand-up pouches are of great significance in the near future. The petroleum-based polymers and biopolymers together form a biobased economic hybrid system that can be marketable and practical. However, research involving more business-friendly methods has to be developed for the sustainable application of cellulose in large-scale processing (Joseph et al. 2020). Future research demands the following trends:

- Research based on morphology, crystallinity, and surface functionality of CNC and CNF films to control the processing conditions for obtaining optimum pore size, nanoparticle packing, free volume, and sufficient interactions between permeants and polymers.
- New techniques have to be adopted for hybridizing or composting cellulose derivative films with other materials to overcome moisture barrier limitations and improve other barrier properties like mechanical and thermal stability.
- Conventional techniques like corona and plasma treatments can be applied for improved adherence between polymers and cellulosic films.

Even though there is a promising future for the applications of cellulose biopolymers, the main limitation lies in the production of these films and composites from native cellulose. Alternative cellulose extraction methods have to be introduced

Table 3.2 Applications of cellulose derivatives

Cellulose derivatives	Applications
Methylcellulose	• Thickener, emulsifier
	• Variable viscosity personal lubricant
	• Used in the preparation of solutions for tears or saliva substitute
	• Capsule production in nutritional supplements
	• Used as a vegetarian alternative to gelatin, which is commonly derived from collagen of the animal body part
	• Used as mild glue that can be washed with water, sizing in the paper manufacture and textile
	• Used to treat irritable bowel syndrome, diverticulosis, hemorrhoids, and constipation
Ethylcellulose	• Water-insoluble polymer used for the coating of dosage forms
	• Used as a carrier material in matrix preparation
	• Preparation of ethylcellulose microparticles by different microencapsulation methods
	• Thickener and binding agent in tablets
	• Ability to sustain the release of drugs
Hydroxypropylmethylcellulose	• Used in coated and uncoated matrix systems
	• Used as polymer blends
	• Used in aqueous and solvent film coating
Hydroxyethylcellulose	• Used as a gelling and thickening agent
	• Used in cosmetics for the manufacture of cleansing solutions and shampoo
	• Used in textile and paper industry
	• Used in film forming
	• Used as a pesticide
Hydroxypropylcellulose	• Used for artificial tears
	• Help in the treatment of conditions like keratoconjunctivitis sicca, where the insufficient tear is produced, corneal erosions occur frequently, corneal sensitivity lowers
	• Used for lubrication in artificial eyes
	• Food additive, thickener, low-level binder, and emulsion stabilizer
	• Used as a matrix for sieving in DNA separation by different electrophoresis methods
Carboxymethylcellulose	• Used as a thickener to modify viscosity
	• Emulsion stabilizer in products like ice cream
	• Constituent of jelly, laxatives, diet tablets, paints, textile sizing, paper products, toothpaste, and detergents
	• Lubricant in nonvolatile eye drops
	• Used as an ingredient of drilling mud in the oil drilling industry
	• Used for protein purification by ion-exchange chromatographic technique

(continued)

Table 3.2 (continued)

Cellulose derivatives	Applications
Sodium carboxymethylcellulose	• Fine filming properties
	• Thickening and binding agent
	• Used in textile and paper industry
	• Reduce water loss by acting as a protective colloid
	• Used as an insecticide
	• Used in operations like oil well drilling
	• Used in the manufacture of products like shampoo, foamed product syrups, toothpaste, bulk laxative ointments, cream, lotion

since it requires pretreatment of lignocellulose to separate cellulose from lignin. The biopolymer synthesis also requires deconstruction, derivatization, or nanoparticle formation. For example, the conversion of wood pulp to film requires a series of chemical reactions to form the viscous solution by breaking down the pulp. It has to undergo further filtration processes to avoid waste and impurities to develop a good quality film with maximum purity. The viscous solution has to extrude and cast along a series of washing and rolling steps to obtain a soft, clear transparent film with desired optical and mechanical properties. The organic solvents used for cellulose extraction also have to be replaced with some cost-effective, green solvents. Solvent-free methods of extraction are also under research. Another limitation lies in the hydrophilicity of cellulose-based films. The uncoated films can be used only for foods that do not require a complete barrier to moisture and gas.

3.11 Conclusion

The potential of cellulose and cellulose derivatives in the field of food packaging is inevitable. The utilization of renewable biological resources to develop edible films, coatings, and composites marked a transitional change in creating a sustainable environment. The combination of cellulose biopolymers and their modified derivatives together competes with the commercially available synthetic polymer-based packaging materials. Since cellulose is the most abundant organic compound and primarily found in plant cell walls, paper, wood, and cotton, it is very cost-effective. Therefore, cellulose films produce a sound effect economically and on an environmental basis. The broad scope of cellulose film composites opens up a promising future that belongs to a complete biodegradable green plastics environment.

References

Abdel-karim AM, Salama AH, Hassan ML (2018) Electrical conductivity and dielectric properties of nanofibrillated cellulose thin films from bagasse. J Phys Org Chem 31:e3851. https://doi.org/10.1002/poc.3851

Abral H, Ariksa J, Mahardika M et al (2020) Transparent and antimicrobial cellulose film from ginger nanofiber. Food Hydrocoll 98:105266. https://doi.org/10.1016/j.foodhyd.2019.105266

Al-oqla FM, Rababah M (2017) Challenges in design of nanocellulose and its composites for different applications. In: Khalil A, Jawaid M, Boufi S (eds) Cellulose-reinforced nanofiber composites: production, properties and applications. Woodhead Publishing, London, pp 113–127

Andrade R, Skurtys O, Osorio F (2015) Drop impact of gelatin coating formulated with cellulose nano fibers on banana and eggplant epicarps. LWT Food Sci Technol 61:422–429. https://doi.org/10.1016/j.lwt.2014.12.035

Ariaii P, Tavakolipour H, Rezai M, Rad AHE (2014) Properties and antimicrobial activity of edible methylcellulose based film incorporated with Pimpinella affinis oil. Eur J Exp Biol 4:670–676

Arrieta MP, Fortunati E, Dominici F et al (2014) PLA-PHB/cellulose based films: mechanical, barrier and disintegration properties. Polym Degrad Stab 107:139–149. https://doi.org/10.1016/j.polymdegradstab.2014.05.010

Avramescu SM, Butean C, Popa CV et al (2020) Edible and functionalized films/coatings—performances and perspectives. Coatings 10:687

Borges JP, Godinho MH, Martins AF et al (2004) Tensile properties of cellulose fiber reinforced hydroxypropylcellulose films. Polym Compos 25:102–110. https://doi.org/10.1002/pc.20008

Bravin B, Peressini D, Sensidoni A (2004) Influence of emulsifier type and content on functional properties of polysaccharide lipid-based edible films. J Agric Food Chem 52:6448–6455. https://doi.org/10.1021/jf040065b

Bumbudsanpharoke N, Ko S (2015) Nano-food packaging: an overview of market, migration research, and safety regulations. J Food Sci 80:R910–R923. https://doi.org/10.1111/1750-3841.12861

Cazón P, Vázquez M (2020) Bacterial cellulose as a biodegradable food packaging material: a review. Food Hydrocoll 113:106530. https://doi.org/10.1016/j.foodhyd.2020.106530

Chambi H, Grosso C (2011) Effect of surfactants on the functional properties of gelatin-polysaccharide-based films. Eur Food Res Technol 232:63–69. https://doi.org/10.1007/s00217-010-1361-0

Chen YW, Lee HV (2018) Revalorization of selected municipal solid wastes as new precursors of "green" nanocellulose via a novel one-pot isolation system: a source perspective. Int J Biol Macromol 107:78–92. https://doi.org/10.1016/j.ijbiomac.2017.08.143

Coles R, McDowell D, Kirwan MJ (2003) Food packaging technology. Blackwell, Oxford

Cooper R (2014) Microanalysis of polymer chain diffusion in heat seals. All Theses. Paper 2039

Dara F, Primadona I, Srikandace Y, Karina M (2021) Development of bacterial cellulose/chitosan films: structural, physicochemical and antimicrobial properties. J Polym Res 28:1–8. https://doi.org/10.1007/s10965-020-02328-6

Di Filippo MF, Dolci LS, Liccardo L et al (2021) Cellulose derivatives-snail slime films: new disposable eco-friendly materials for food packaging. Food Hydrocoll 111:106247. https://doi.org/10.1016/j.foodhyd.2020.106247

Diantoro M, Mustikasari AA, Wijayanti N et al (2017) Microstructure and dielectric properties of cellulose acetate-ZnO/ITO composite films based on water hyacinth. J Phys Conf Ser 853: 012047. https://doi.org/10.1088/1742-6596/853/1/012047

Edlund U, Zhu Ryberg Y, Albertsson A-C (2010) Barrier films from renewable forestry waste. Biomacromolecules 11:2532–2538. https://doi.org/10.1021/bm100767g

El Sayed AM, El-Gamal S, Morsi WM, Mohammed G (2015) Effect of PVA and copper oxide nanoparticles on the structural, optical, and electrical properties of carboxymethyl cellulose films. J Mater Sci 50:4717–4728. https://doi.org/10.1007/s10853-015-9023-z

Embuscado ME, Huber KC (2009) Edible films and coatings for food applications. Springer, New York

Espinoza-Herrera N, Pedroza-Islas R, San Martín-Martinez E et al (2011) Thermal, mechanical and microstructures properties of cellulose derivatives films: a comparative study. Food Biophys 6: 106–114. https://doi.org/10.1007/s11483-010-9181-0

Fardioui M, Bouhfid R (2017) Cellulose nanocrystal–based nanocomposites. In: Khalil A, Jawaid M, Boufi S (eds) Cellulose-reinforced nanofiber composites: production, properties and applications. Woodhead Publishing, London, pp 373–389

Gemili S, Yemenicioğlu A, Altinkaya SA (2009) Development of cellulose acetate based antimicrobial food packaging materials for controlled release of lysozyme. J Food Eng 90:453–462. https://doi.org/10.1016/j.jfoodeng.2008.07.014

Gemili S, Yemenicioğlu A, Altınkaya SA (2010) Development of antioxidant food packaging materials with controlled release properties. J Food Eng 96:325–332

Ghaderi M, Mousavi M, Yousefi H, Labbafi M (2014) All-cellulose nanocomposite film made from bagasse cellulose nanofibers for food packaging application. Carbohydr Polym 104:59–65. https://doi.org/10.1016/j.carbpol.2014.01.013

Gill P, Moghadam TT, Ranjbar B (2010) Differential scanning Calorimetry techniques: applications in biology and nanoscience. J Biomol Technol 21:167

Gu Y, Huang J (2013) Colorimetric detection of gaseous ammonia by polyaniline nanocoating of natural cellulose substances. Colloid Surf A Physicochem Eng Asp 433:166–172. https://doi.org/10.1016/j.colsurfa.2013.05.016

Guo X, Wu Y, Xie X (2017) Water vapor sorption properties of cellulose nanocrystals and nanofibers using dynamic vapor sorption apparatus. Sci Rep 7:1–12. https://doi.org/10.1038/s41598-017-14664-7

Guvendiren M, Molde J, Soares RMD, Kohn J (2016) Designing biomaterials for 3D printing. ACS Biomater Sci Eng 2:1679–1693. https://doi.org/10.1021/acsbiomaterials.6b00121

HPS AK, Saurabh CK, Asniza M et al (2017) Polymer composites. In: Khalil A, Jawaid M, Boufi S (eds) Cellulose-reinforced Nanofiber composites: production, properties and applications. Woodhead Publishing, London, pp 1–24

Ioelovich M, Leykin A (2011) Study of sorption properties of cellulose and its derivatives. Bioresources 6:178–195. https://doi.org/10.15376/biores.6.1.178-195

Irkin R, Esmer OK (2015) Novel food packaging systems with natural antimicrobial agents. J Food Sci Technol 52:6095–6111. https://doi.org/10.1007/s13197-015-1780-9

Jia C, Shao Z, Fan H et al (2016) Barium titanate as a filler for improving the dielectric property of cyanoethyl cellulose/antimony tin oxide nanocomposite films. Compos Part A Appl Sci Manuf 86:1–8. https://doi.org/10.1016/j.compositesa.2016.03.016

John MJ, Thomas S (2008) Biofibers and biocomposites. Carbohydr Polym 71:343–364. https://doi.org/10.1016/j.carbpol.2007.05.040

Joseph B, Sagarika VK, Sabu C et al (2020) Cellulose nanocomposites: fabrication and biomedical applications. J Bioresour Bioprod 5:223–237. https://doi.org/10.1016/j.jobab.2020.10.001

Kamide K (2005) Cellulose and cellulose derivatives. Elsevier, London

Khattab MM, Abdel-Hady NA, Dahman Y (2017) Cellulose nanocomposites: opportunities, challenges, and applications. In: Khalil A, Jawaid M, Boufi S (eds) Cellulose-reinforced nanofiber composites: production, properties and applications. Woodhead Publishing, London, pp 483–516

Lille M, Nurmela A, Nordlund E et al (2018) Applicability of protein and fiber-rich food materials in extrusion-based 3D printing. J Food Eng 220:20–27. https://doi.org/10.1016/j.jfoodeng.2017.04.034

Lu P, Guo M, Xu Z, Wu M (2018) Application of Nanofibrillated cellulose on BOPP/LDPE film as oxygen barrier and antimicrobial coating based on cold plasma treatment. Coatings 8:207. https://doi.org/10.3390/coatings8060207

Lyytikäinen J, Morits M, Österberg M et al (2021) Skin and bubble formation in films made of methyl nanocellulose, hydrophobically modified ethyl(hydroxyethyl)cellulose and microfibrillated cellulose. Cellulose 28:787–797. https://doi.org/10.1007/s10570-020-03557-0
Mahmud M, Perveen A, Jahan RA et al (2019) Preparation of different polymorphs of cellulose from different acid hydrolysis medium. Int J Biol Macromol 130:969–976. https://doi.org/10.1016/j.ijbiomac.2019.03.027
Malhotra B, Keshwani A, Kharkwal H (2015) Antimicrobial food packaging: potential and pitfalls. Front Microbiol 6:611. https://doi.org/10.3389/fmicb.2015.00611
Menon MP, Selvakumar R, Kumar PS, Ramakrishna S (2017) Extraction and modification of cellulose nanofibers derived from biomass for environmental application. RSC Adv 7:42750–42773. https://doi.org/10.1039/C7RA06713E
Meriçer Ç, Minelli M, Giacinti Baschetti M, Lindström T (2017) Water sorption in microfibrillated cellulose (MFC): the effect of temperature and pretreatment. Carbohydr Polym 174:1201–1212. https://doi.org/10.1016/j.carbpol.2017.07.023
Moradi Z, Kalanpour N (2019) Kefiran, a branched polysaccharide: preparation, properties and applications: a review. Carbohydr Polym 223:115100. https://doi.org/10.1016/j.carbpol.2019.115100
Mu R, Hong X, Ni Y et al (2019) Recent trends and applications of cellulose nanocrystals in food industry. Trends Food Sci Technol 93:136–144. https://doi.org/10.1016/j.tifs.2019.09.013
Mura S, Corrias F, Stara G et al (2011) Innovative composite films of chitosan, methylcellulose, and nanoparticles. J Food Sci 76:N54–N60. https://doi.org/10.1111/j.1750-3841.2011.02295.x
Nasatto PL, Pignon F, Silveira JLM et al (2015) Methylcellulose, a cellulose derivative with original physical properties and extended applications. Polymers (Basel) 7:777–803. https://doi.org/10.3390/polym7050777
Nasir M, Hashim R, Sulaiman O, Asim M (2017) Nanocellulose: preparation methods and applications. In: Khalil A, Jawaid M, Boufi S (eds) Cellulose-reinforced nanofiber composites: production, properties and applications. Woodhead Publishing, London, pp 261–276
Navarro-Tarazaga ML, Massa A, Pérez-Gago MB (2011) Effect of beeswax content on hydroxypropyl methylcellulose-based edible film properties and postharvest quality of coated plums (cv. Angeleno). LWT Food Sci Technol 44:2328–2334. https://doi.org/10.1016/j.lwt.2011.03.011
Nelson SO (2015) Dielectric properties of agricultural materials and their applications. Academic Press, London. https://doi.org/10.1016/C2014-0-02694-9
Nilsuwan K, Benjakul S, Prodpran T (2018) Physical/thermal properties and heat seal ability of bilayer films based on fish gelatin and poly(lactic acid). Food Hydrocoll 77:248–256. https://doi.org/10.1016/j.foodhyd.2017.10.001
Osorio FA, Molina P, Matiacevich S et al (2011) Characteristics of hydroxy propyl methyl cellulose (HPMC) based edible film developed for blueberry coatings. Procedia Food Sci 1:287–293. https://doi.org/10.1016/j.profoo.2011.09.045
Panahirad S, Dadpour M, Peighambardoust SH et al (2021) Applications of carboxymethyl cellulose- and pectin-based active edible coatings in preservation of fruits and vegetables: a review. Trends Food Sci Technol 110:663–673. https://doi.org/10.1016/j.tifs.2021.02.025
Pang J, Liu X, Zhang X et al (2013) Fabrication of cellulose film with enhanced mechanical properties in ionic liquid 1-allyl-3-methylimidazolium chloride (AmimCl). Materials (Basel) 6:1270–1284. https://doi.org/10.3390/ma6041270
Paunonen S (2013) Strength and barrier enhancements of cellophane and cellulose derivative films: a review. BioResources 8:3098–3121. https://doi.org/10.15376/biores.8.2.3098-3121
Prusty K, Sahu D, Swain SK (2017) Nanocellulose as a template for the production of advanced nanostructured material. In: Khalil A, Jawaid M, Boufi S (eds) Cellulose-reinforced nanofiber composites: production, properties and applications. Woodhead Publishing, London, pp 427–454

Qiu L, Shao Z, Liu M et al (2014) Synthesis and electrospinning carboxymethyl cellulose lithium (CMC-Li) modified 9,10-anthraquinone (AQ) high-rate lithium-ion battery. Carbohydr Polym 102:986–992. https://doi.org/10.1016/j.carbpol.2013.09.105

Rees A, Powell LC, Chinga-Carrasco G et al (2015) 3D bioprinting of carboxymethylated-periodate oxidized nanocellulose constructs for wound dressing applications. Biomed Res Int 2015: 925757. https://doi.org/10.1155/2015/925757

Rhim J-W, Mohanty AK, Singh SP, Ng PKW (2006) Effect of the processing methods on the performance of polylactide films: thermocompression versus solvent casting. J Appl Polym Sci 101:3736–3742. https://doi.org/10.1002/app.23403

Saba N, Jawaid M (2017) Recent advances in nanocellulose-based polymer nanocomposites. In: Khalil A, Jawaid M, Boufi S (eds) Cellulose-reinforced nanofiber composites: production, properties and applications. Woodhead Publishing, London, pp 89–112

Sahoo PK, Jena DK (2018) Synthesis and study of mechanical and fire retardant properties of (carboxymethyl cellulose-g-polyacrylonitrile)/montmorillonite biodegradable nanocomposite. J Polym Res 25:1–10. https://doi.org/10.1007/s10965-018-1659-3

Saifullah M, Shishir MRI, Ferdowsi R et al (2019) Micro and nano encapsulation, retention and controlled release of flavor and aroma compounds: a critical review. Trends Food Sci Technol 86:230–251. https://doi.org/10.1016/j.tifs.2019.02.030

Salmieri S, Islam F, Khan RA et al (2014) Antimicrobial nanocomposite films made of poly(lactic acid)–cellulose nanocrystals (PLA–CNC) in food applications—part B: effect of oregano essential oil release on the inactivation of Listeria monocytogenes in mixed vegetables. Cellulose 21:4271–4285. https://doi.org/10.1007/s10570-014-0406-0

Sankhla S, Sardar HH, Neogi S (2021) Greener extraction of highly crystalline and thermally stable cellulose micro-fibers from sugarcane bagasse for cellulose nanofibrils preparation. Carbohydr Polym 251:117030. https://doi.org/10.1016/j.carbpol.2020.117030

Saraiva DV, Chagas R, de Abreu BM et al (2020) Flexible and structural coloured composite films from cellulose nanocrystals/hydroxypropyl cellulose lyotropic suspensions. Crystals 10:122. https://doi.org/10.3390/cryst10020122

Seddiqi H, Oliaei E, Honarkar H et al (2021) Cellulose and its derivatives: towards biomedical applications. Cellulose 28:1–39. https://doi.org/10.1007/s10570-020-03674-w

Shaghaleh H, Xu X, Wang S (2018) Current progress in production of biopolymeric materials based on cellulose, cellulose nanofibers, and cellulose derivatives. RSC Adv 8:825–842. https://doi.org/10.1039/c7ra11157f

Shao Y, Chaussy D, Grosseau P, Beneventi D (2015) Use of microfibrillated cellulose/lignosulfonate blends as carbon precursors: impact of hydrogel rheology on 3D printing. Ind Eng Chem Res 54:10575–10582. https://doi.org/10.1021/acs.iecr.5b02763

Shimizu M, Kusumi R, Saito T, Isogai A (2019) Thermal and electrical properties of nanocellulose films with different interfibrillar structures of alkyl ammonium carboxylates. Cellulose 26: 1657–1665. https://doi.org/10.1007/s10570-018-2155-y

Soares NFF, Rutishauser DM, Melo N et al (2002) Inhibition of microbial growth in bread through active packaging. Packag Technol Sci 15:129–132. https://doi.org/10.1002/pts.576

Somvanshi KS, Gope PC (2021) A review on properties of Nano biocomposite film for packaging applications from cellulose nano fiber. Int J Eng Res Appl 11:29–39. https://doi.org/10.9790/9622-1101032939

Spence KL, Venditti RA, Rojas OJ et al (2011) Water vapor barrier properties of coated and filled microfibrillated cellulose composite films. Bioresources 6:4370–4388

Stephen AM, Phillips GO, Williams PA (1995) Food polysaccharides and their applications. Taylor & Francis, London

Suvorova AI, Tyukova IS, Trufanova EI (2000) Biodegradable starch-based polymeric materials. Russ Chem Rev 69:451. https://doi.org/10.1070/rc2000v069n05abeh000505

Tajeddin B (2014) Cellulose-based polymers for packaging applications. In: Thakur VK (ed) Lignocellulosic polymer composites: processing, characterization, and properties. Wiley, London, pp 477–498

Tanpichai S (2018) A comparative study of nanofibrillated cellulose and microcrystalline cellulose as reinforcements in all-cellulose composites. J Met Mater Miner 28:10–15. https://doi.org/10.14456/jmmm.2018.3

Tao J, Jiao L, Deng Y (2021) Cellulose- and nanocellulose-based dielectric materials. In: Thomas S, Pottathara YB (eds) Nanocellulose based composites for electronics. Elsevier, Oxford, pp 73–100

Toğrul H, Arslan N (2004) Extending shelf-life of peach and pear by using CMC from sugar beet pulp cellulose as a hydrophilic polymer in emulsions. Food Hydrocoll 18:215–226. https://doi.org/10.1016/S0268-005X(03)00066-3

Uz M, Altinkaya SA (2011) Development of mono and multilayer antimicrobial food packaging materials for controlled release of potassium sorbate. LWT Food Sci Technol 44:2302–2309. https://doi.org/10.1016/j.lwt.2011.05.003

Valencia-Chamorro SA, Palou L, Delfio MA, Pérez-Gago MB (2011) Antimicrobial edible films and coatings for fresh and minimally processed fruits and vegetables: a review. Crit Rev Food Sci Nutr 51:872–900. https://doi.org/10.1080/10408398.2010.485705

Viera RGP, Filho GR, de Assunção RMN et al (2007) Synthesis and characterization of methylcellulose from sugar cane bagasse cellulose. Carbohydr Polym 67:182–189. https://doi.org/10.1016/j.carbpol.2006.05.007

Wang J, Gardner DJ, Stark NM et al (2018) Moisture and oxygen barrier properties of cellulose nanomaterial-based films. ACS Sustain Chem Eng 6:49–70. https://doi.org/10.1021/acssuschemeng.7b03523

Wang M, Jia X, Liu W, Lin X (2021) Water insoluble and flexible transparent film based on carboxymethyl cellulose. Carbohydr Polym 255:117353. https://doi.org/10.1016/j.carbpol.2020.117353

Willberg-Keyriläinen P, Vartiainen J, Harlin A, Ropponen J (2017) The effect of side-chain length of cellulose fatty acid esters on their thermal, barrier and mechanical properties. Cellulose 24:505–517. https://doi.org/10.1007/s10570-016-1165-x

Wu Y, Luo X, Li W et al (2016) Green and biodegradable composite films with novel antimicrobial performance based on cellulose. Food Chem 197:250–256. https://doi.org/10.1016/j.foodchem.2015.10.127

Yang W, Jiao L, Min D et al (2017) Effects of preparation approaches on optical properties of self-assembled cellulose nanopapers. RSC Adv 7:10463–10468. https://doi.org/10.1039/C6RA27529J

Yang W, Jiao L, Liu W et al (2018) Morphology control for tunable optical properties of cellulose nanofibrils films. Cellulose 25:5909–5918. https://doi.org/10.1007/s10570-018-1974-1

Yang W, Jiao L, Liu W, Dai H (2019) Manufacture of highly transparent and hazy cellulose nanofibril films via coating TEMPO-oxidized wood fibers. Nano 9:107. https://doi.org/10.3390/nano9010107

Yaradoddi JS, Banapurmath NR, Ganachari SV et al (2020) Biodegradable carboxymethyl cellulose based material for sustainable packaging application. Sci Rep 10:1–13. https://doi.org/10.1038/s41598-020-78912-z

Zhang S, Zhang F, Pan Y et al (2018) Multiwall-carbon-nanotube/cellulose composite fibers with enhanced mechanical and electrical properties by cellulose grafting. RSC Adv 8:5678–5684. https://doi.org/10.1039/c7ra11304h

Zhao J, Zhang W, Zhang X et al (2013) Extraction of cellulose nanofibrils from dry softwood pulp using high shear homogenization. Carbohydr Polym 97:695–702. https://doi.org/10.1016/j.carbpol.2013.05.050

Zhao D, Zhu Y, Cheng W et al (2020) Cellulose-based flexible functional materials for emerging intelligent electronics. Adv Mater 2000619:1–18. https://doi.org/10.1002/adma.202000619

Zhou L, Wang Y (2021) Physical and antimicrobial properties of zein and methyl cellulose composite films with plasticizers of oleic acid and polyethylene glycol. LWT 140:110811. https://doi.org/10.1016/j.lwt.2020.110811

Chapter 4
Pullulan-Based Films

Md. Nahidul Islam, Md. Aslam Ali, Md. Ahiduzzaman, Md. Arifur Rahman Khan, and Md. Shofiul Azam

4.1 Introduction and Sources

Pullulan is a nonionic water-soluble microbial exopolysaccharide derived from the fermentation of a medium supplied with carbon, nitrogen, and other critical ingredients from *Aureobasidium pullulans*, a polymorphic fungus-like black yeast (Farris et al. 2014; Prajapati et al. 2013). Bernier (1958) was the first to report it, and Bender et al. (1959) expounded on its structure. It is a glucan made up of maltotriose units connected by α-(1 → 6) glycosidic connections (Leduy and Boa 1983;Shin et al. 1989).The existence of maltotetraosyl units was reported by Catley and Whelan (1971). Polymaltotriose units are connected together mostly via α-(1 → 6) connections.

Aureobasidium pullulan is commonly referred to as "dark yeast" because of its ability to produce dark pigment "melanin." This well-known microbe is a dimorphic species with considerable plasticity that can grow in a variety of environments. In the

M. N. Islam · M. A. Ali · M. Ahiduzzaman
Department of Agro-Processing, Bangabandhu Sheikh Mujibur Rahman Agricultural University, Gazipur, Bangladesh

Institute of Food Safety and Processing, Bangabandhu Sheikh Mujibur Rahman Agricultural University, Gazipur, Bangladesh
e-mail: nahidul.islam@bsmrau.edu.bd

M. Arifur Rahman Khan (✉)
Department of Agronomy, Bangabandhu Sheikh Mujibur Rahman Agricultural University, Gazipur, Bangladesh
e-mail: arif@bsmrau.edu.bd

M. S. Azam (✉)
Department of Food Engineering, Dhaka University of Engineering and Technology, Gazipur, Bangladesh
e-mail: shofiul@duet.ac.bd

T. Amin et al. (eds.), *Polysaccharide Based Films for Food Packaging: Fundamentals, Properties and Applications*,
https://doi.org/10.1007/978-981-99-4898-7_4

food and pharmaceutical industries, it is commonly used. *Aureobasidium pullulans*, *Aureobasidium pullulans* var. *melanogenum*, *Aureobasidium pullulans* var. *aubasidani* Yurlova, *Aureobasidium pullulans/subglaciale*, and *Aureobasidium pullulans* var. *namibiae* are some of the species' variants (Duan et al. 2008). In different strains, the molecular weight ranges from 50,000 to 4×10^6 (Israilides et al. 1999).

Aureobasidium pullulan can be found in a variety of living environments, including soil, water, ice, fuel, and even rockets and nuclear reactors. It is mostly found on leaves and other surfaces, such as solid, limestone, wood, dirt, and barks. Sweet potato, coconut byproducts, deproteinized whey, jiggery (a concentrated sugar stick juice), carob pod, and Jerusalem artichoke are also sources of fermentable sugars, which can also be separated (Singh et al. 2019; Wu et al. 2009; Sharma et al. 2013). It can also be found in agro-industrial wastes such as grape peel and flesh, starch waste, potato starch waste, olive oil waste effluents, beet molasses, and brewery wastes (Sugumaran and Ponnusami 2017a; Barnett et al. 1999; Singh et al. 2019; Sugumaran and Ponnusami 2017b).

Pullulan has gained lots of industrial interest and commercial relevance and is being used in a variety of industries (Krochta and Mulder-Johston 1997; Katagiri et al. 1990). The most common application of pullulan standards is to determine the molecular mass distribution of cellulosic samples. It can be used as a low-calorie viscosity enhancer, binder, and food quality improver as a partial replacement for starch components. It has been used as an oxygen-impermeable food packaging material that prevents oxidation and preserves the aroma, flavor, and freshness of food while also extending shelf life (Israilides et al. 1999). Food applications have been the primary market for pullulan over the past two decades. Pullulan with desirable properties may be achieved as a result of technological advancements.

A number of scientists in this area are making contributions to the pharmacological, medical, and environmental remediation fields. Pullulan's α-connections provide it with physical features, including water solubility and fiber flexibility, allowing it to form fibers and films akin to other petrol-derived polymers (Haniffa et al. 2016). As a non-ionic, non-hygroscopic, non-immunogenic, nontoxic, noncarcinogenic, and non-mutagenic polymer, it can be made into a transparent and edible film (Li et al. 2017; Leathers 2003). Furthermore, the Food and Drug Administration in the United States designated pullulan as "Generally Recognized as Safe" (GRAS), and regulations in many nations certified it safe for food use. As a result, it can be used in the food, pharmaceutical, cosmetic, and biomedical industries. The aim of this chapter is to present updated information regarding the production, properties, application, and future challenges of pullulan-based biofilm.

4.2 Structure of Pullulan and Pullulan-Derived Components

The structure of pullulan is shown in Fig. 4.1. Pullulan is made up of α-(1 → 4) linkage repeating trimer maltotriose linked together by α-(1 → 6) glycosidic linkage (Leathers 2003; Angelo et al. 2017). Pullulan polymer has particular physical features due to this unique linking arrangement, including flexibility and high mechanical strength, allowing it to form films and fibers (Singh et al. 2008; Angelo et al. 2017).

The backbone of Pullulan, which has nine hydroxyl groups per repeating unit, can be easily altered by grafting other chemical structures on it (Wang et al. 2018). Esterification, etherification, sulfation, oxidation, co-polymerization, amidification, and other methods can be used to derivatize the hydroxyl groups (Prajapati et al. 2013). Pullulan esterification modifies their solubility and other attributes by reducing their hydrogen bond network and changing their solubility. Highly water-soluble polymer carriers tend to release pharmaceuticals quickly for drug delivery and sustained-release applications, whereas polymers that simply swell or disperse in water have the ability to give gradual drug release (Pereira et al. 2014). For example, pullulan acetate is a key pullulan derivative with promising applications in drug administration via self-aggregation nanoparticles in water and pullulan-modified materials (Tang et al. 2010; Ganie et al. 2021; Bala and Kale 2019).

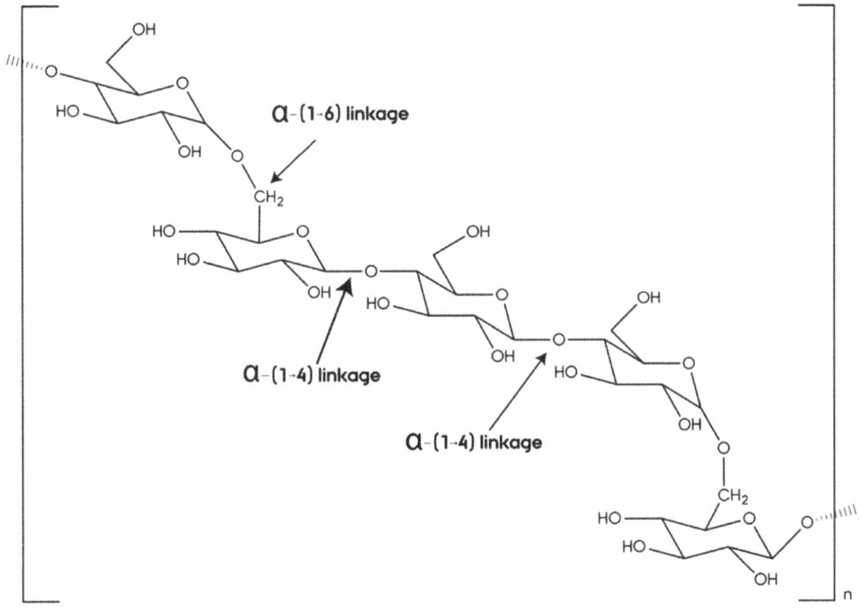

Fig. 4.1 Structure of pullulan (Tabasum et al. 2018)

4.3 Production/Preparation of Pullulan and Pullulan Derivative-Based Films

The flow diagram in Fig. 4.2 summarizes the main procedures for pure pullulan synthesis.

Pullulan production is influenced by a variety of factors, including the fungal strain used, the carbon source, the nitrogen source, the culture temperature, the dissolved

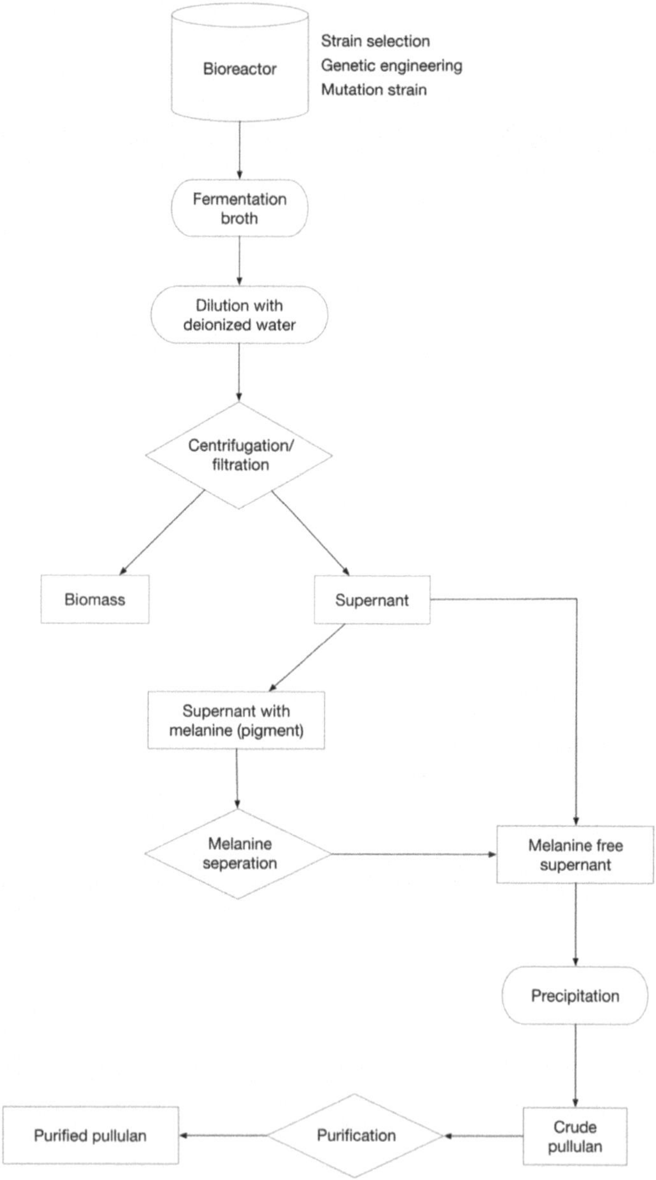

Fig. 4.2 Flow diagram of pullulan production and downstream processing. Adapted from Singh and Saini (2012)

oxygen level, the fermenter and design, the fermentation temperature, the aging duration, and the starting pH. The greatest results for pullulan manufacturing have been reported with a C:N component ratio of 10:1 (Singh and Saini 2012). Several studies have determined that the optimal pH for pullulan production is between 5.5 and 7.5 (Sugumaran et al., 2013; Jiang 2010). Pullulan accumulation is inhibited by a somewhat low pH, although insoluble glucan production is boosted. The optimal temperature for pullulan production varies depending on the strain; however, it is usually between 25 °C and 30 °C. For cells to manufacture pullulan, aeration is a critical factor. Pullulan is not developed or produced in anaerobic circumstances. Due to the collection of pullulan-synthesizing cells, excellent air circulation (up to 2.0 volume of air per volume of medium per minute during fermentation) causes an expansion in pullulan focus (Shingel 2004; Singh et al. 2008).

Pullulan manufacturing using submerged fermentation has been successfully implemented for commercial use. The commercialization of *Aureobasidium pullulans*-produced pullulan is dependent on development characteristics such as bioreactor structure design, which can be batch, fed-batch, or continuous. The combination of increasing cell population and continuous fermentation, on the other hand, is a more practical strategy for long-term production (Sugumaran et al., 2013; Sugumaran and Ponnusami 2017b). Pullulan fermentation numerical models provide information not only on the kinetic and metabolic nature of pullulan but also on how to better control and improve pullulan production during fermentation. Forecasting of pullulan production based on biomass generation would be more economical because biomass can be predicted faster in a continuous system. The logistic model, Luedeking-Piret model, and Gompertz model are three attempts to design a model that depicts the development of pullulan producers and pullulan production (Sugumaran and Ponnusami 2017b).

The exopolysaccharide provided by microorganisms from the culture broth solution requires downstream processing. The technique for pullulan involves the removal of cells and melanin pigments produced during the growth, as well as the precipitation of exopolysaccharides using a suitable solvent system. Melanin was one of the impediments to pullulan formation, resulting in a dull green or dark color in the soup. This problem can be remedied by developing new pullulan-producing strains that are pigment-reduced or pigment-free. Cross-flow membrane filtering can be used to remove the biomass of *Aureobasidium pullulans* from the fermented stock in advance. Pullulan recovery requires the precipitation stage, which can be accomplished with natural organic solvents. For both safety and cost considerations, a 2:1 (v/v) mixture of 95% ethanol and culture supernatant is commonly used. Ultrafiltration and ion-exchange resins can be used to purify pullulan even more. After separation, pullulan is dried and finely milled into powder (Sugumaran and Ponnusami 2017a; Ju et al. 2015; Wang et al. 2016; Goswami et al. 2021).

Natural pullulan has limited usage in the food packaging industry due to its lack of antibacterial action. Adding other antibacterial agents, such as nanoparticles and/essential oils, is currently the most popular option (Youssef et al. 2015; Abdollahi et al. 2012). However, they are frequently obtained through a series of time-consuming preparatory procedures. Film preparation becomes more challenging as a result of these processes.

Pullulan can be chemically changed or combined with other polymers, resulting in new functions and a broader range of applications. Pullulan-sorbitol-sucrose fatty ester films, for example, have been shown to extend the shelf life of coated strawberries while decreasing the shelf life of kiwifruits via changing the microclimate and hence ripening (Diab et al. 2001). The physical addition of a hydrophobic moiety (chemical bonding) to pullulan reduces the water vapor permeability of the films, hence expanding the range of its food preservation coating applications.

Chemical modification of polysaccharides to enhance their intrinsic antibacterial activity has proven to be a worthwhile approach to improving polysaccharide antibacterial capabilities. In fact, 3-aminopropyltrimethoxysilane was employed to functionalize pullulan in order to make antibacterial transparent pullulan films with better antibacterial properties and thermal stability (Abdollahi et al. 2012). However, single-component polysaccharide films have weak mechanical properties and are challenging to meet the requirements of packaging materials.

4.4 Properties

4.4.1 Thermal Properties

With the addition of polyvinyl alcohol, the thermal stability and mechanical properties of pullulan membranes have been improved (Islam et al. 2012). Because of the quick solidification process of stretched chains during electrospinning, the majority of chains are in a non-crystalline condition. Pure pullulan shows a significant melting transition thermogram peak at 95 °C (Karim and Islam 2011).

Differential scanning calorimetry is a useful tool for determining the compatibility of polymer mixtures and determining the likelihood of interactions between matrix components. Before and after electrospinning, pullulan and pullulan/hydroxypropyl-β-cyclodextrin mixes showed two endothermic events (Poudel et al. 2020; Leyva-Porras et al. 2020). Poudel et al. (2020) used peak area to calculate the magnitude of the enthalpy change associated with endothermic transitions. The first transition, measured between 92 °C and 105 °C with a peak temperature of 96.18 °C and an enthalpy change of 2.921 J/g for electrospun pullulan fibers and pullulan powder, is most likely the result of the elimination of coupled water. Karim and Islam (2011), on the other hand, investigated the thermal characteristics of pullulan/montmorillonite and pullulan/polyvinyl alcohol nanofibers and concluded that the endothermic peaks in the thermograms of pure pullulan spun fibers (detected at 95–99 °C) were due to a melting transition temperature.

4.4.2 Barrier Properties

The lowest water vapor permeability and oxygen permeability values have been found in pure pullulan films (7.053106 g.m/Pa.h.m^2 and 0.48 × 106 mL.mm/m^2day.pa, respectively). The absorbance bands of COO^- groups changed dramatically to higher wavenumbers (1640 and 1413 cm^{-1}, respectively) when pullulan was included in alginate. This could be due to the addition of pullulan, which disrupted the intermolecular hydrogen connections that existed between the carboxylic groups at the beginning. The pullulan films diffractogram revealed that pullulan was a totally amorphous polymer (Xiao et al. 2015; Kristo et al. 2007; Wu et al. 2021).

4.4.3 Mechanical Properties

Commercially available pullulan films without any plasticizer are brittle because they are made by evaporating a pullulan solution at a high temperature (such as 100 °C) to save casting time. Casting the solutions at a lower temperature resulted in stiffer and more flexible films; however, this requires a longer evaporation duration (Kawahara et al. 2003; Vuddanda et al. 2017). The effects of cross-linking degree on tensile strength in addition to other physical properties, such as water absorption, swelling behavior, and water vapor permeability, were assessed by Liang et al. (2019). The inclusion of glycerol as a plasticizer improved the extensibility and hydrophilicity of films, resulting in increased water vapor permeability. Glutaraldehyde can also be used as cross-link enzymes with alginate beads (Bhushan et al. 2015). The best tensile strength of pullulan films without glycerol was found when the glutaraldehyde concentration was around 2% (w/w) (Chen et al. 2017). This indicates that films made from high molecular weight pullulan had higher tensile strength (94.43 MPa when 2% (w/w) glutaraldehyde was added) than those made from low molecular weight pullulan.

Pullulan modification and beeswax integration improved the water-proof features such as water solubility, water vapor permeability, and water contact angle of the pullulan biofilms. Octenylsuccination of pullulan had a negative impact on the mechanical properties of the films produced, whereas beeswax insertion into pullulan-octenyl succinic anhydride films solved this issue (Duan et al. 2018; Omar-Aziz et al. 2021).

Tensile strength and elongation at break are two significant and relevant parameters for describing thin film mechanical properties. Furthermore, chain length, molecular weight, and the number of branched chains in the biopolymer, as well as how they are positioned, have a major impact on the mechanical properties of the finished film (Khanzadi et al. 2015; Luís et al. 2021). In general, octenylsuccination of pullulan decreased the film tensile strength, which likely limited pullulan's use in certain fields, while beeswax incorporation into the pullulan octenyl succinic anhydride film matrix increased pullulan's tensile strength (Omar-Aziz et al. 2021).

Moreover, pullulan film elasticity can be influenced by the interaction between xanthan gum and glycerin, and the combination of xanthan gum and locust bean showed a detrimental impact on mechanical properties (Guo et al. 2014).

4.4.4 Dielectric Properties

The addition of cyanoethyl pullulan to the polyvinylidene fluoride terpolymer and high-temperature thermal treatment alters not only the microstructure of the polymer but also its dielectric characteristics (Yi et al. 2021). It was investigated whether cyanoethyl pullulan material could be used in electrowetting on dielectric devices. This material is easily manufactured using the spin-coating method and has the highest dielectric performance when annealed at 100 °C in an environment with a dielectric constant of about 18 (100 kHz) (Jianfeng et al. 2013). The incorporation of chlorotrifluoroethylene units into the crystalline phase is favored by the high-temperature thermal treatment process, which induces a less polar crystal structure in the terpolymer, resulting in a more relaxor ferroelectric–dielectric behavior, a higher dielectric constant, and a lower dielectric loss (Yi et al. 2021).

4.4.5 Tensile Properties

Pullulan's brittle to ductile transition occurs at the same time as its glass to rubber transition. At moisture levels larger than 8%, the component present in higher proportion dominated the tensile characteristics of bilayer films, but films created with biopolymer blends exhibited mechanical behavior similar to plain pullulan films (Kristo et al. 2007). Local stress concentrations in mixed systems are higher than in multilayer films. Furthermore, because the interface of stacked films is aligned along the tensile direction, no significant stress concentration occurs near the interface (Graziano et al. 2019; Lee et al. 2005). Glutaraldehyde and glycerol were added to a pullulan solution to make pullulan films. The tensile strength of pullulan films could be greatly improved when the amount of glutaraldehyde was within 1–5%; the addition of glycerol improved the extensibility, hydrophilicity, and water vapor permeability (Chen et al. 2017).

4.4.6 Heat Sealing Properties of Pullulan Film-Based Packaging

The heat sealability of a packaging film is one of the most important properties when considering its use, and the integrity of the resultant seal is of paramount importance to the ultimate integrity of the package. A number of factors are involved in determining the quality of a heat seal, and they can be conveniently summarized as machine factors (dwell or clamp time; temperature and pressure), resin factors (molecular weight and additives in the resin), film factors (gauge; style or form, e.g., whether gusseted or not), and treatment for printing (Robertson 2011).

Heat sealable films are defined as those that can be sealed together using regular heat and pressure, such as conductance from a heavy heat-resistant metal bar carrying a heating element. Non-heat-sealable films cannot be sealed in this manner, although heat-sealable coatings can typically be used to make them heat sealable. By applying heat and pressure for the requisite dwell time, the two facing coated surfaces get bound to each other. Pullulan has a number of advantages over other polysaccharides, including high gloss and transparency, heat sealability, and the ability to combine with other biopolymers, in addition to the fact that its derivatives are tasteless and colorless (Li et al. 2015).

4.4.7 Flexibility Properties

Pullulan demonstrates a linear, unbranched structure made up of maltotriose units linked by α-(1 $\rightarrow$ 6)-glycosidic linkages. The pullulan molecule's distinguishing qualities, such as structural flexibility and excellent solubility, are determined by its regular pattern and linear structure (Chen et al. 2012; Leathers 1993). The tensile strength of several films was compared (Wu et al. 2013), where the elongation at break is a measure of its flexibility. Compared to the chitosan film, which had a higher film thickness (69.8 m) and elongation (19.2%), pullulan had a film thickness of 47.3 m and elongation of 1.27%.

The addition of sakacin A, on the other hand, showed a rise in elasticity noted and antibacterial activity. Pullulan and glycerin considerably alter the properties of pullulan films, according to several studies. Pullulan-glycerin and xanthan gum were found to have strong interactions (xanthan gum-glycerin). With the addition of sakacin A to the film mixture, there was no discernible effect on the optical properties of the film, but there was an increase in flexibility (Trinetta et al. 2011).

4.4.8 Electrical Properties

With the application of high-voltage DC (5 kV) at 100–120 °C, poling of the composite cyanoethyl pullulan film produced increased pyroelectricity and ferroelectricity. The piezoelectricity, pyroelectricity, and ferroelectricity of the cyanoethyl pullulan composite film can be improved by poling. Several studies also discovered ferroelectric phenomena in cellulose, such as a D–E hysteresis loop (Suzuki et al. 2003; Wang et al. 2019). The cyanoethyl pullulan film showed a pyroelectric constant of about 50 μC/m^2 at 35 °C; the value for the composite cellulose and $BaTiO_3$ film was about five times higher (Suzuki et al. 2003).

4.4.9 Optical Properties

In terms of the optical properties of the biocomposite films, it was discovered that when apple fiber was added, color coordinates (L^*, a^*, and b^*) increased significantly (p-value 0.05), which is compatible with their yellowish appearance. Furthermore, a substantial reduction in transparency (p-value 0.05) was seen in the pullulan–apple fiber films, which could be seen visually (Luís et al. 2021). No significant differences in the optical properties of pullulan biofilm were found with the addition of sakacin A (Trinetta et al. 2011; Luís et al. 2021).

To determine the particular molecular features of pullulan, optical properties of pullulan films were investigated before and after uniaxial stretching. Uniaxial stretching showed no effect on the specific optical rotation of pullulan film, which was the same as that of pullulan in an aqueous solution. On the basis of the modified stress-optical rule, the stress-optical coefficient for pullulan was determined as a material constant when combined with the mechanical measurement. During the stress relaxation of uniaxially stretched pullulan film, the birefringence was positive and increased with time (Horinaka 2018).

4.4.10 Water Solubility Properties

Pullulan dissolves quickly in water, resulting in a viscous, stable solution. In an environment with a relative humidity of less than 70%, pullulan has an equilibrium moisture content of 10–15% but no hygroscopicity or tackiness. Xu et al. (2018) looked into the impact of drying temperature on pullulan's water solubility. With increasing drying temperature, the time for dissolution is reduced. The evaporation rate of film moisture accelerates, and the moisture content drops as the drying temperature rises, resulting in reduced hydrogen bonds between molecules. As the film's tightness deteriorated, water was able to penetrate it easily. Pullulan films disintegrated faster and were more hydrophilic in water than hydroxypropyl methylcellulose films. A considerable quantity of hydroxyl in pullulan molecular chains contributed to the enhanced hydrophilicity. In general, as the concentration of pullulan and hydroxypropyl methylcellulose films grows, the time to dissolve into water increases as well (Tsujisaka and Mitsuhashi 1993).

In water, a composite film with the ratio (1/2) dissolves significantly slower than a composite film with the other ratios (1/1, 1/1.5, 1/2, and 1/3). The compatibility of pullulan and hydroxypropyl methylcellulose has an impact on the water solubility of composite films. Pullulan molecules can enter the gap between hydroxypropyl methylcellulose chains, and in the ratio (1/2), the gap fills completely. The film's compatibility with the ratio (1/2) is the best. Pullulan is insoluble in all organic solvents except dimethylformamide and dimethyl sulfoxide (Chaen 2009; Tomasik 2003). It may become insoluble in water and soluble in acetone or ethanol when converted into ethers and esters, depending on the degree of substitution. Pullulan has a lower viscosity than other hydrocolloids.

4.4.11 Viscous Properties

Over a wide range of pH values (from 2 to >11), pullulan solution viscosity is generally unaffected (Trinetta 2017). Though pullulan is fairly heat resistant, its viscosity can be affected by temperature. When heated at 90 °C for 1 h at a pH of 4–6, the viscosity of pullulan with a molecular weight of 3×10^6 daltons reduced by around 10%, whereas the viscosity of a 105 daltons pullulan decreased a little bit (Jindal and Singh Khattar 2018). Some ions, such as borate, create a combination with the hydroxyl groups of pullulans, increasing viscosity, but such solutions do not gel (Popović et al. 2018; Jindal and Singh Khattar 2018; Tsujisaka and Mitsuhashi 1993).

4.4.12 Printing Properties

The biodegradable, edible film made of pullulan is thermally stable, elastic, and antistatic and can be printed (Cheng et al. 2011). They are readily compressible in heat and humidity. Transparency, solubility in water, and insolubility in organic solvents are some of the other qualities. Furthermore, the addition of a plasticizer is not required due to its filmogenic peculiarities (Lima et al. 2017; Singh et al. 2008; Islam et al. 2012; Wu et al. 2013).

4.4.13 Aroma Retention Properties

Pullulan-based coatings do not have any taste or aroma and are colorless and shiny. Active compounds can be retained at mucosal locations for longer periods of time with mucoadhesive polysaccharides, and formulations can be devised to manage their release and bioavailability. Research has been conducted to look at how different polysaccharides with different physicochemical features (e.g., functional groups and molecular weight) affect taste component release and perception from films. High or low viscosity carboxymethyl cellulose, pullulan, or hydroxypropyl methylcellulose was used to make polysaccharide films. Some of the polysaccharides evaluated have a mucoadhesive character, which affected flavor delivery over time since those attach to the oral cavity would continue to release taste, whereas those that are not mucoadhesive would be swallowed (Cook et al. 2018; Dinu et al. 2019).

4.4.14 Antimicrobial Properties

Pullulan inhibits the growth of microorganisms in fresh baby carrots (Gniewosz et al. 2013). Pathogenic microorganisms such as *Listeria monocytogenes* and *Staphylococcus aureus*, which contribute to meat deterioration, were inactivated by Ag (nanoparticles), ZnO (nanoparticles), rosemary, and oregano containing pullulan films (Khalaf et al. 2013). The antibacterial activity of pullulan films with integrated meadowsweet flower extracts as well as the efficiency of these coatings in extending apple stability in varied storage settings was investigated (Leimann et al. 2018). The meadowsweet flower extracts coated pullulan evened and unified the surface of apple skins. Pullulan coating with meadowsweet flower extracts—water–ethanol prevented natural mesophilic microflora growth more efficiently, with the amount of microflora decreasing by two logarithmic cycles at the end of storage. Therefore, pullulan coatings containing meadowsweet flower extracts may help to preserve apple quality during storage and turnover at varying temperatures and times. Pullulan films containing meadowsweet flower extract showed antibacterial activity, suppressing the development of bacteria more effectively than fungi. Pullulan coatings using meadowsweet flower extracts may help to preserve apple quality during storage and rotation at varying temperatures and times. These coatings are simple to apply to the fruits, and they give them a shinier, greener appearance (Gniewosz et al. 2014; Luís et al. 2021).

4.4.15 Moisture Sorption Properties

The gravimetric method has been used to determine the moisture sorption isotherms of pure pullulan, alginate, and mix films at 25 °C, 35 °C, and 45 °C. At high water activity, the hydrophilic group $^{-}COONa^{+}$ is more easily exposed than the hydrophilic group ^{-}OH. Within the tested temperature range, the microstructures of all films were stable and did not alter during moisture sorption. The moisture sorption process was non-spontaneous in pure pullulan films, whereas it was spontaneous in alginate and mix films. Furthermore, the sorption process for all investigated films was enthalpy-controlled, according to the enthalpy-entropy compensation theory (Xiao and Tong 2013). The ability of hydroxypropyl methylcellulose or pullulan capsules to absorb moisture is less than that of gelatin capsules. Hydroxypropyl methylcellulose capsules have a lower ability to retain moisture than pullulan or gelatin capsules (Yang et al. 2020).

4.4.16 Release Kinetics of Active Packaging

When it comes to modified release and instant release dosage forms, the drug release from the formulations plays an important role. Drug release from dosage forms is influenced by a variety of factors, including physicochemical properties of pharmaceuticals, excipients, dosage form design, manufacturing process variables, and design. The release pattern of medicine from a dosage form is crucial since it determines the dosage form's efficacy. Several models aid in the study of release patterns, allowing for the development of successful formulations (Paarakh et al. 2018).

The effect of increasing potassium sorbate concentrations (0%, 0.5%, 1%, and 2% w/w) on the release and antibacterial characteristics of edible films based on pullulan, gelatin, and their mixes (75/25, 50/50, and 25/75) were examined. The dissolution and swelling of the polymer matrix are the primary causes of potassium sorbate release from the pullulan and gelatin-rich carriers, respectively. The pullulan 50/gelatin 50 films' large microspheres produced a release profile comparable to the mixing system with a higher gelatin content. It was discovered that the potassium sorbate diffusion mechanism is Quasi-Fickian. The Weibull models were effective in stimulating the release of potassium sorbate from the films (Kowalczyk et al. 2020).

4.4.17 Cost-Effectiveness

Pullulan costs between \$25 and \$30 per kilogram, which is significantly more than other biopolymers derived from polysaccharides and proteins. The high cost of pullulan is primarily due to its manufacturing. Though pullulan has several beneficial characteristics, its use in culinary applications is often hampered by its high cost. As a result, pullulan is routinely combined with other biopolymers to lower costs and improve material characteristics (Tong et al. 2013;Luís et al. 2021).

When compared to petroleum-based products, the cost of producing pullulan is quite high. As a result, various studies have identified low-cost carbon and nitrogen sources that can support *Aureobasidium pullulan's* growth, as well as determining the rate of polysaccharide synthesis. Chen et al. (2019) used hydrolyzed sweet potato as a fermentation medium and found that continuous hydrolysis with pullulanase and α-amylase resulted in high pullulan yields (up to 30 g/L). Roukas and Biliaderis (1995) observed similar results when using carob pods as a substrate for pullulan synthesis. Although the use of these substrates appears to be cost-effective, Trinetta and Cutter (2016) recommended additional research to enable the transition from laboratory to industrial production.

4.4.18 Safety and Regulations

Pullulan is a water-soluble, nontoxic, and non-mutagenic edible biopolymer with excellent film-forming abilities and adhesive properties (Kraśniewska et al. 2019). It was recognized as safe by the FDA in the United States in 2002. It has recently been approved as a food additive (E 1204) in the European Union for use in capsules, tablets, and films under Directive 2006/52/EC. It is also legal to use in the food sector in some Asian countries, Russia, and certain South American countries (Popović et al. 2018; Wüstenberg 2015; Venugopal 2016).

In a study, the toxicity of pullulan and proposed use as a texturizer and flavor substrate for meat and meat substitutes was assessed (Kimoto et al. 1997). Hematological and clinical chemistry levels of pullulan-treated animals showed comparable to control values. In terminal organs and body weights, there was no evidence of pullulan-related toxicity. Treatment did not cause any harmful lesions, according to macroscopic and microscopic investigations. In another study, pullulan's mutagenicity was tested in *Salmonella typhimurium* with and without metabolic activation (TA100, TA1535, TA98, and TA1537). Pullulan did not increase the number of revertants per plate in any strain at any dose, even 10,000 micrograms/plate with or without metabolic activation suggests that it is not carcinogenic or mutagenic (Wolf et al. 2003). These findings lead to the conclusion that pullulan has no major toxicological activity. In addition, the European Food Safety Authority (EFSA) has declared the active substance *Aureobasidium pullulans* (strains DSM 14940 and DSM 14941) as a safe and efficient fungicide and bactericide on fruits and vegetables (Authority 2013; Kimoto et al. 1997).

4.4.19 Methods to Improve Properties

Pullulan–whey protein isolate–montmorillonite nanocomposite films were generated and characterized as a function of integrating varying amounts of montmorillonite nanoparticles (0%, 1%, 3%, and 5% weight). When the nano-montmorillonite content was raised, the water vapor permeability, moisture content, moisture absorption, and water solubility decreased. With increased montmorillonite content, tensile strength improved while elongation at break decreased. Polymer intercalation could be influenced by the polymer's molecular weight as well as polar interactions between the montmorillonite and the polymer (Almasi et al. 2010). The intercalation of biopolymer chains into the montmorillonite layers' galleries was most likely caused by strong polar contacts between the hydroxyl groups present in the whey protein isolate and pullulan chains and the silicate layers.

Montmorillonite improved the mechanical characteristics, thermal properties, and water resistance of whey protein isolate-pullulan film specimens without causing any visual or color issues. The tortuous structure of montmorillonite in the polymer matrix improved the water-barrier qualities of the nanocomposite films. Whey protein isolate-pullulan-montmorillonite nanocomposites demonstrated a homogeneous dispersion of montmorillonite in the polymer matrix in scanning electron micrographs. The homogeneous dispersion and exfoliated structure of montmorillonite are responsible for the majority of the nanocomposites' properties (Hassannia-Kolaee et al. 2016).

4.5 Applications

Pullulan is an exopolysaccharide created by a fungal fermentation process that is used in industry. It has unique qualities that make it suitable for use in edible coatings. It is edible, odorless, tasteless, nontoxic, and extremely soluble in water and acts as an oxygen barrier in food packaging.

This biopolymer has outstanding thermal stability, mechanical strength, and moisture barrier qualities, making it an ideal film former. It can be utilized as an active packaging for food and pharmaceuticals, as well as a carrier for drug or gene delivery. As a result, researchers and manufacturers are particularly interested in pullulan and its blends and composites with other natural chemicals. Listerine was the first commercial product to employ pullulan, which is a breath freshener made out of pullulan strips that disintegrate swiftly. Pullulan has been combined with natural polymers, including cellulose, chitosan, and starch, as well as synthetic polymers like polyvinyl alcohol and polyethyleneimine, to diversify its potential applications (Goswami et al. 2021; Kimoto et al. 1997; Yan et al. 2012). Various applications in different fields of pullulan are shown in Fig. 4.3.

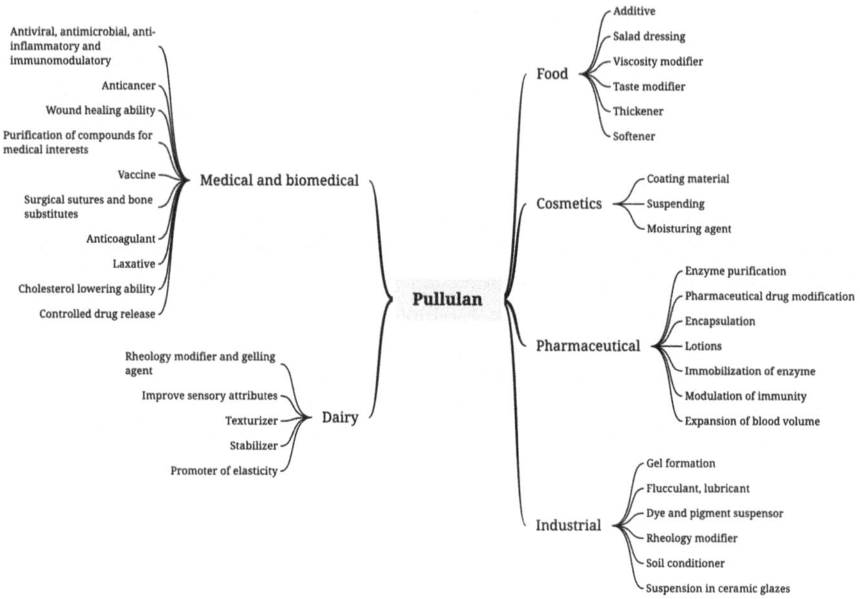

Fig. 4.3 Various applications of pullulan. Adapted from Coltelli et al. (2020)

4.5.1 Pullulan-Based Coatings on Foods

Initial attempts to use pullulan in packaged foods are delayed behind its established use as a food additive (e.g., thickening agent, binder, and stabilizer), with the very first works going back to the early 1990s. Certain unusual features of pullulan, including its high-water solubility and barrier property against oxygen and carbon dioxide, were thought to provide significant benefits at that time. Pullulan edible films that were water-soluble were first offered as edible pouches for premeasured quantities that could be progressively dissolved in water or hot food. But even so, the first widespread use of pullulan in food packaging was as an edible coating, which is a relatively thin layer of material applied and formed directly on the surface of a food product and may be eaten together with it (Trinetta and Cutter 2016).

4.5.2 Edible Coatings

It has long been recognized that edible coatings can be used to protect meals. Edible coatings can assist in preventing food deterioration by preventing dehydration, inhibiting respiration, enhancing textural quality, allowing volatile flavor compounds to be retained, and decreasing microbial development. Because of this potential, interest in developing and using edible coatings to extend the shelf life and improve the quality of fresh, frozen, and prepared foods has exploded in the recent two decades. Pullulan edible coatings have been recommended for fresh and barely processed fruits and vegetables in particular. The necessity for high quality and minimal processing and storage technologies led to the introduction of pullulan-based edible coatings for these types of food products (Biliaderis et al. 1999).

4.5.3 "Active" Films and Coatings

One of the most remarkable breakthroughs in the field of food packaging in the last two decades is "active packaging." Active packaging refers to packaging that interacts with the food and/or the environment in order to improve food quality, safety, and convenience. Absorbing and releasing systems are examples of active packaging technology. Oxygen and carbon dioxide scavengers, humidity absorbers, exuded absorbers, and other items fall into the first priority. However, the majority of active packaging research for food and beverage applications focuses on the development of antimicrobial and antioxidant systems; hence, pullulan plays a significant role (Diab et al. 2001).

4.5.4 Pharmaceutical Industry Applications

Pullulan's widespread use in the pharmaceutical business is due to the fact that it complies with regulatory requirements. It is utilized in the pharmaceutical industry as an alternative to marine algae or plant-derived natural gums. It is distinguished from other polysaccharides by its ability to create a stable solution in aqueous conditions without forming gels, even at high concentrations. Furthermore, when compared to other similar biopolymers, this biopolymer solution's film-forming capacity is exceptional. Soft and hard capsule shells, as well as mucosal film formulations, benefit from these superior film-forming characteristics in the pharmaceutical industry (Kumar et al. 2012).

4.5.5 Biomedical Applications

Several investigations at the molecular level have recently uncovered the underlying causes of cell damage and tissue degeneration. These investigations revealed the possibility for selected nanomaterials to be employed in tissue engineering in situ to treat diseases using the body's natural healing power. Pullulan has been applied as scaffolds for the attachment and development of transplanted cells (Cutiongco et al. 2014). Pullulan plays an important role in targeted drug and gene delivery. Due to selective accumulation, drug delivery vehicles exert direct biological effects on the target areas. Hence, two separate approaches have been used to build them with appropriate aims. The first step is to create a carrier with a suitable design and then tweak it with other appropriate ligands or moieties to lower the drug delivery vehicle's or carrier's hematological toxicity (Bae and Park 2011). Gene therapy has enormous potential for treating a variety of inherited and acquired human genetic problems. It works by delivering antisense RNA or short interference RNA genes to target genomic regions in order to turn on the synthesis of the desired protein for appropriate functioning or turn off the production of damaging genes (Jo et al. 2007).

4.5.6 Environmental Application: Waste Remediation

Due to increased groundwater contamination from industrial effluents comprising heavy metals, dyes, and hydrocarbons, environmental pollution has become a severe concern. Even at very low concentrations, all of these are nonbiodegradable and have significant toxicity, carcinogenicity, or mutagenicity. Pullulan has shown to be a viable candidate for heavy metal removal from wastewater due to its strong biosorption ability, which is mediated by contact between heavy metal ions and acidic functional groups of exopolysaccharides (Shuhong et al. 2014;Ye et al. 2014).

4.6 Limitations and Future Trends

Pullulan has not gotten nearly as much attention as it deserves, despite its many intriguing features. Indeed, there are currently very few commercial goods on the market, with the majority of recommended uses being at the laboratory or semi-industrial scale. The fundamental cause of such commercial underdevelopment is the high cost of doing business. Pullulan is substantially more expensive than most biopolymers of both polysaccharide and protein origin, costing between 25 and 30 US$/kg. The high cost of pullulan is primarily due to its manufacturing.

Carbohydrates are used as carbon sources in the commercial bioprocess for the manufacture of pullulan because they allow for high productivities and yields (Singh and Saini 2008; Bajaj et al. 2007). However, the selling prices for sugar and starch, which are the most often used substrates for the manufacture of most microbial polysaccharides, are currently around 350–518 US$/ton and 279–310 US$/ton, respectively. Because the cost of the substrate contributes up to 40% of the entire cost of manufacturing microbial polymers, it seems evident that using low-cost raw materials, such as wastes and byproducts, is a good place to start. Despite the fact that many efforts are being made in this direction, some obstacles appear to make the path very difficult, such as the difficulty in ensuring their supply in terms of quantity and quality, the different nutrient composition of raw materials, the presence of contaminants in the substrates, and the accumulation of non-reacted components in the broth (Freitas et al. 2014).

Another factor that adds to the high price of pullulan is the removal of the melanin pigment during the fermentation broth's downstream processing. Activated charcoal is typically used to remove melanin, following which pullulan purification can take place. Selection of melanin-free, high-yield strains could thus be another useful method for expanding pullulan applications, along with innovative alternative fermentation schemes (De Baets et al. 2004; Gniewosz and Duszkiewicz-Reinhard 2008; Ravella et al. 2010; Choudhury et al. 2012).

Another factor that adds to the high price of pullulan is the removal of the melanin pigment during the fermentation broth's downstream processing. Activated charcoal is typically used to remove melanin, following which pullulan purification can take place. Selection of melanin-free, high-yield strains could thus be another useful method for expanding pullulan applications, along with innovative alternative fermentation schemes.

Some technological concerns should be addressed in the coming years to open up new market prospects, particularly in the food packaging industry. Several uses are hampered by drawbacks related to pullulan's strong affinity for water molecules. Although pullulan's hydrophilic characteristic can be advantageous in a variety of applications, it is also the cause of its extreme sensitivity to damp surroundings. If protective layers (e.g., polyolefin films) are placed on top of it, inherent or increased qualities (e.g., the oxygen barrier) are not impaired; this would certainly preclude it from being used in typical containers. Pullulan's neutrality is another factor that may rule it out for some food packaging uses. Because it has no charge on its backbone, it cannot form ionic connections with countercharged polymers and/or small molecules, which is necessary for the formation of certain release systems, polyion–complex structures, and complex coacervates.

4.7 Conclusion

Optimization of the food packaging system and the reduced use of fossil-fuel-derived plastics are the trends in food packaging research. Biopolymers are clearly a viable solution for achieving this goal. Because of its unrivaled features, pullulan has significant benefits over other polysaccharides and proteins. However, due to high cost, it can only be used in experimental or pilot-scale applications. Improvement of the production processes and improvement of the functionality are two factors that primarily drive the market expansion of already-existing pullulan-based technologies.

Unique industrial applications can be achieved if academic and industrial researchers can find inspiration to develop new pathways and tactics that enable the development of new high-performance pullulan materials with a greater level of sophistication. It is always critical to focus on solutions that are recyclable, eco-friendly, energy-efficient, dependable, and cost-effective in this attempt. Despite the high yield of pullulan, there is an urgent need to explore and adapt the pullulan synthesis route in order to increase production. Genetic engineering approaches and parameter optimization can be used to improve the current situation. Pullulan has a lot of application possibilities as an exopolysaccharide. More research in this area could help lower the cost of pullulan, making it more competitive with related products like dextran, xanthan, and other microbially generated polysaccharides. Although the particular cell type responsible for pullulan synthesis is still unknown, a bioreactor design involving cell immobilization and biomass buildup has attracted a lot of interest as a way to boost pullulan output. In addition, novel pullulan separation and purification procedures should be researched because they may reduce labor and capital input.

Acknowledgments The authors acknowledge the financial support from Ministry of Science and Technology, Peoples Republic of Bangladesh (Project # SRG-221157), Research Management Wing, Bangabandhu Sheikh Mujibur Rahman Agricultural University, Gazipur, Bangladesh (37.01.3303.133.34.002.22.1765-003), and University Grants Commission of Bangladesh (CropScience-09/2021-2022).

Conflict of Interest The authors declare no conflict of interests.

References

Abdollahi M, Rezaei M, Farzi G (2012) Improvement of active chitosan film properties with rosemary essential oil for food packaging. Int J Food Sci Technol 47:847–853

Almasi H, Ghanbarzadeh B, Entezami AA (2010) Physicochemical properties of starch–CMC–nanoclay biodegradable films. Int J Biol Macromol 46:1–5. https://doi.org/10.1016/j.ijbiomac.2009.10.001

Angelo M, Cerqueira PR, CoutoTeixeira JA, AugustoVicente A (2017) Edible packaging today. In: Edible food packaging. CRC Press, Boca Raton, pp 1–8. https://doi.org/10.1201/b19468-1

Authority EFS (2013) Conclusion on the peer review of the pesticide risk assessment of the active substance Aureobasidium pullulans (strains DSM 14940 and DSM 14941). EFSA J 11:3183

Bae YH, Park K (2011) Targeted drug delivery to tumors: myths, reality and possibility. J Control Release 153:198–205. https://doi.org/10.1016/j.jconrel.2011.06.001

Bajaj IB, Survase SA, Saudagar PS, Singhal RS (2007) Gellan gum: fermentative production, downstream processing and applications. Food Technol Biotechnol 45:341–354

Bala P, Kale SN (2019) Self-assembled pullulan acetate nanoparticles for pH-dependent controlled drug delivery application. Adv Sci Eng Med 11:758–765

Barnett C, Smith A, Scanlon B, Israilides CJ (1999) Pullulan production by Aureobasidium pullulans growing on hydrolysed potato starch waste. Carbohydr Polym 38:203–209

Bender H, Lehmann J, Wallenfels K (1959) Pullulan, an extracellular glucan from Pullularia pullulans. Biochim Biophys Acta 36:309–316. https://doi.org/10.1016/0006-3002(59)90172-6

Bernier B (1958) The production of polysaccharides by fungi active in the decomposition of wood and forest litter. Can J Microbiol 4:195–204. https://doi.org/10.1139/m58-020

Bhushan B, Pal A, Jain V (2015) Improved enzyme catalytic characteristics upon glutaraldehyde cross-linking of alginate entrapped xylanase isolated from Aspergillus flavus MTCC 9390. Enzyme research, 2015.

Biliaderis CG, Lazaridou A, Arvanitoyannis I (1999) Glass transition and physical properties of polyol-plasticised pullulan–starch blends at low moisture. Carbohydr Polym 40:29–47. https://doi.org/10.1016/S0144-8617(99)00026-0

Catley BJ, Whelan WJ (1971) Observations on the structure of pullulan. Arch Biochem Biophys 143:138–142. https://doi.org/10.1016/0003-9861(71)90193-7

Chaen H (2009) Pullulan. In: Imeson A (ed) Food stabilisers, thickeners and gelling agents. Wiley, London, pp 266–274. https://doi.org/10.1002/9781444314724.ch14

Chen J, Wu S, Pan S (2012) Optimization of medium for pullulan production using a novel strain of Aureobasidium pullulans isolated from sea mud through response surface methodology. Carbohydr Polym 87:771–774

Chen CT, Chen KI, Chiang HH, Chen YK, Cheng KC (2017) Improvement on physical properties of pullulan films by novel cross-linking strategy. J Food Sci 82:108–117. https://doi.org/10.1111/1750-3841.13577

Chen G, Zhu Y, Zhang G, Liu H, Wei Y, Wang P, Wang F, Xian M, Xiang H, Zhang H (2019) Optimization and characterization of pullulan production by a newly isolated high-yielding strain Aureobasidium melanogenum. Prep Biochem Biotechnol 49:557–566. https://doi.org/10.1080/10826068.2019.1591988

Cheng K-C, Demirci A, Catchmark JM (2011) Pullulan: biosynthesis, production, and applications. Appl Microbiol Biotechnol 92:29–44. https://doi.org/10.1007/s00253-011-3477-y

Choudhury AR, Sharma N, Prasad G (2012) Deoiled jatropha seed cake is a useful nutrient for pullulan production. Microb Cell Factories 11:1–9

Coltelli M-B, Danti S, De Clerck K, Lazzeri A, Morganti P (2020) Pullulan for advanced sustainable body-and skin-contact applications. J Funct Biomater 11:20. https://doi.org/10.3390/jfb11010020

Cook SL, Methven L, Parker JK, Khutoryanskiy VV (2018) Polysaccharide food matrices for controlling the release, retention and perception of flavours. Food Hydrocoll 79:253–261

Cutiongco MFA, Tan MH, Ng MYK, Le Visage C, Yim EKF (2014) Composite pullulan–dextran polysaccharide scaffold with interfacial polyelectrolyte complexation fibers: a platform with enhanced cell interaction and spatial distribution. Acta Biomater 10:4410–4418. https://doi.org/10.1016/j.actbio.2014.06.029

De Baets S, Vandamme E, Steinbuchel A (2004) Polysaccharides II: polysaccharides of eukaryotes. Chemical Industry Press, Shanghai

Diab T, Biliaderis CG, Gerasopoulos D, Sfakiotakis E (2001) Physicochemical properties and application of pullulan edible films and coatings in fruit preservation. J Sci Food Agric 81:988–1000

Dinu V, Gadon A, Hurst K, Lim M, Ayed C, Gillis RB, Adams GG, Harding SE, Fisk ID (2019) An enzymatically controlled mucoadhesive system for enhancing flavour during food oral processing. NPJ Sci Food 3:11. https://doi.org/10.1038/s41538-019-0043-y

Duan X, Chi Z, Wang L, Wang X (2008) Influence of different sugars on pullulan production and activities of α-phosphoglucose mutase, UDPG-pyrophosphorylase and glucosyltransferase involved in pullulan synthesis in Aureobasidium pullulans Y68. Carbohydr Polym 73:587–593. https://doi.org/10.1016/j.carbpol.2007.12.028

Duan X, Ouyang Q, Tao N (2018) Effect of applying cinnamaldehyde incorporated in wax on green mould decay in citrus fruits. J Sci Food Agric 98:527–533. https://doi.org/10.1002/jsfa.8490

Farris S, Unalan IU, Introzzi L, Fuentes-Alventosa JM, Cozzolino CA (2014) Pullulan-based films and coatings for food packaging: present applications, emerging opportunities, and future challenges. J Appl Polym Sci 131:40539

Freitas F, Alves VD, Reis MA, Crespo JG, Coelhoso IM (2014) Microbial polysaccharide-based membranes: current and future applications. J Appl Polym Sci 131:40047

Ganie SA, Rather LJ, Li Q (2021) A review on anticancer applications of pullulan and pullulan derivative nanoparticles. Carbohydr Polym Technol Appl 2:100115. https://doi.org/10.1016/j.carpta.2021.100115

Gniewosz M, Duszkiewicz-Reinhard W (2008) Comparative studies on pullulan synthesis, melanin synthesis and morphology of white mutant Aureobasidium pullulans B-1 and parent strain Ap-3. Carbohydr Polym 72:431–438

Gniewosz M, Kraśniewska K, Woreta M, Kosakowska O (2013) Antimicrobial activity of a Pullulan–caraway essential oil coating on reduction of food microorganisms and quality in fresh baby carrot. J Food Sci 78:M1242–M1248. https://doi.org/10.1111/1750-3841.12217

Gniewosz M, Synowiec A, Kraśniewska K, Przybył JL, Bączek K, Węglarz Z (2014) The antimicrobial activity of pullulan film incorporated with meadowsweet flower extracts (Filipendulae ulmariae flos) on postharvest quality of apples. Food Control 37:351–361. https://doi.org/10.1016/j.foodcont.2013.09.049

Goswami S, Sharma A, Anjana, Bhattacharya M (2021) Pullulan films and natural compounds: applications and perspectives. In: Rai M, dos Santos CA (eds) Biopolymer-based nano films. Elsevier, London, pp 19–41. https://doi.org/10.1016/B978-0-12-823381-8.00006-5

Graziano A, Jaffer S, Sain M (2019) Review on modification strategies of polyethylene/polypropylene immiscible thermoplastic polymer blends for enhancing their mechanical behavior. J Elastom Plast 51:291–336. https://doi.org/10.1177/0095244318783806

Guo J, Ge L, Li X, Mu C, Li D (2014) Periodate oxidation of xanthan gum and its crosslinking effects on gelatin-based edible films. Food Hydrocoll 39:243–250. https://doi.org/10.1016/j.foodhyd.2014.01.026

Haniffa M, Ching Y, Abdullah L, Poh S, Chuah C (2016) Review of bionanocomposite coating films and their applications. Polymers 8:1–33

Hassannia-Kolaee M, Khodaiyan F, Shahabi-Ghahfarrokhi I (2016) Modification of functional properties of pullulan-whey protein bionanocomposite films with nanoclay. J Food Sci Technol 53:1294–1302. https://doi.org/10.1007/s13197-015-1778-3

Horinaka J-I (2018) Optical and mechanical properties of pullulan films studied by uniaxial stretching. Int J Biol Macromol 118:584–587. https://doi.org/10.1016/j.ijbiomac.2018.06.127

Islam MS, Yeum JH, Das AK (2012) Effect of pullulan/poly (vinyl alcohol) blend system on the montmorillonite structure with property characterization of electrospun pullulan/poly (vinyl alcohol)/montmorillonite nanofibers. J Colloid Interface Sci 368:273–281

Israilides C, Smith A, Scanlon B, Barnett C (1999) Pullulan from agro-industrial wastes. Biotechnol Genet Eng Rev 16:309–324. https://doi.org/10.1080/02648725.1999.10647981

Jianfeng C, Yuhua Y, Xiangyu Z, Jian L, Jia Z (2013) Evaluation of cyanoethyl pullulan material as the dielectric layer for EWOD devices. IEEE 10th international conference on ASIC, 28–-31 October 2013. https://doi.org/10.1109/ASICON.2013.6812063

Jiang L (2010) Optimization of fermentation conditions for pullulan production by Aureobasidium pullulan using response surface methodology. Carbohydr Polym 79:414–417

Jindal N, SinghKhattar J (2018) Microbial polysaccharides in food industry. In: Grumezescu AM, Holban AM (eds) Biopolymers for food design. Academic Press, London, pp 95–123. https://doi.org/10.1016/B978-0-12-811449-0.00004-9

Jo J-I, Ikai T, Okazaki A, Yamamoto M, Hirano Y, Tabata Y (2007) Expression profile of plasmid DNA by spermine derivatives of pullulan with different extents of spermine introduced. J Control Release 118:389–398. https://doi.org/10.1016/j.jconrel.2007.01.005

Ju X-M, Wang D-H, Zhang G-C, Cao D, Wei G-Y (2015) Efficient pullulan production by bioconversion using Aureobasidium pullulans as the whole-cell catalyst. Appl Microbiol Biotechnol 99:211–220. https://doi.org/10.1007/s00253-014-6100-1

Karim MR, Islam MS (2011) Thermal behavior with mechanical property of fluorinated silane functionalized superhydrophobic pullulan/poly(vinyl alcohol) blends by electrospinning method. J Nanomater 2011:979458. https://doi.org/10.1155/2011/979458

Katagiri Y, Naora K, Ichikawa N, Hayashibara M, Iwamoto K (1990) High-performance liquid chromatographic determination of ciprofloxacin in rat brain and cerebrospinal fluid. Chem Pharm Bull 38:2884–2886. https://doi.org/10.1248/cpb.38.2884

Kawahara M, Mizutani K, Suzuki S, Kitamura S, Fukada H, Yui T, Ogawa K (2003) Dependence of the mechanical properties of a pullulan film on the preparation temperature. Biosci Biotechnol Biochem 67:893–895. https://doi.org/10.1271/bbb.67.893

Khalaf H, Sharoba A, El-Tanahi H, Morsy M (2013) Stability of antimicrobial activity of pullulan edible films incorporated with nanoparticles and essential oils and their impact on Turkey deli meat quality. J Food Dairy Sci 4:557–573

Khanzadi M, Jafari SM, Mirzaei H, Chegini FK, Maghsoudlou Y, Dehnad D (2015) Physical and mechanical properties in biodegradable films of whey protein concentrate–pullulan by application of beeswax. Carbohydr Polym 118:24–29. https://doi.org/10.1016/j.carbpol.2014.11.015
Kimoto T, Shibuya T, Shiobara S (1997) Safety studies of a novel starch, pullulan: chronic toxicity in rats and bacterial mutagenicity. Food Chem Toxicol 35:323–329
Kowalczyk D, Kordowska-Wiater M, Karaś M, Zięba E, Mężyńska M, Wiącek AE (2020) Release kinetics and antimicrobial properties of the potassium sorbate-loaded edible films made from pullulan, gelatin and their blends. Food Hydrocoll 101:105539. https://doi.org/10.1016/j.foodhyd.2019.105539
Kraśniewska K, Pobiega K, Gniewosz M (2019) Pullulan–biopolymer with potential for use as food packaging. Int JFood Eng 15:20190030
Kristo E, Biliaderis CG, Zampraka A (2007) Water vapour barrier and tensile properties of composite caseinate-pullulan films: biopolymer composition effects and impact of beeswax lamination. Food Chem 101:753–764. https://doi.org/10.1016/j.foodchem.2006.02.030
Krochta J, Mulder-Johston C (1997) Scientific status summary: edible and biodegradable polymer films: challenges and opportunities. Food Technol 51(2):61–74
Kumar D, Saini N, Pandit V, Ali S (2012) An insight to pullulan: a biopolymer in pharmaceutical approaches. Int J Basic Appl Sci 1:202–219
Leathers TD (1993) Substrate regulation and specificity of amylases from Aureobasidium strain NRRL Y-12,974. FEMS Microbiol Lett 110:217–221
Leathers TD (2003) Biotechnological production and applications of pullulan. Appl Microbiol Biotechnol 62:468–473. https://doi.org/10.1007/s00253-003-1386-4
Leduy A, Boa JM (1983) Pullulan production from peat hydrolyzate. Can J Microbiol 29:143–146
Lee J, Macosko CW, Bates FS (2005) Development of discrete nanopores. II Comparison between layered films and blends of polyolefins. J Appl Polym Sci 95:708–718
Leimann FV, Gonçalves OH, Sakanaka LS, Azevedo AS, Lima MV, Barreiro F, Shirai MA (2018) Active food packaging from botanical, animal, bacterial, and synthetic sources. In: Grumezescu AM, Holban AM (eds) Food packaging and preservation. Elsevier, London, pp 87–135
Leyva-Porras C, Cruz-Alcantar P, Espinosa-Solís V, Martínez-Guerra E, Piñón-Balderrama CI, Compean Martínez I, Saavedra-Leos MZ (2020) Application of differential scanning Calorimetry (DSC) and modulated differential scanning calorimetry (MDSC) in food and drug industries. Polymers 12:5
Li Y, Yokoyama W, Wu J, Ma J, Zhong F (2015) Properties of edible films based on pullulan–chitosan blended film-forming solutions at different pH. RSC Adv 5:105844–105850. https://doi.org/10.1039/C5RA21876D
Li R, Tomasula P, de Sousa AMM, Liu S-C, Tunick M, Liu K, Liu L (2017) Electrospinning pullulan fibers from salt solutions. Polymers 9:32. https://doi.org/10.3390/polym9010032
Liang J, Wang R, Chen R (2019) The impact of cross-linking mode on the physical and antimicrobial properties of a chitosan/bacterial cellulose composite. Polymers 11:491. https://doi.org/10.3390/polym11030491
Lima IAD, Pomin SP, Cavalcanti OA (2017) Development and characterization of pullulan-polymethacrylate free films as potential material for enteric drug release. Braz J Pharm Sci 53:e00002
Luís Â, Ramos A, Domingues F (2021) Pullulan–apple fiber biocomposite films: optical, mechanical, barrier, antioxidant and antibacterial properties. Polymers 13:870
Omar-Aziz M, Khodaiyan F, Yarmand MS, Mousavi M, Gharaghani M, Kennedy JF, Hosseini SS (2021) Combined effects of octenylsuccination and beeswax on pullulan films: water-resistant and mechanical properties. Carbohydr Polym 255:117471. https://doi.org/10.1016/j.carbpol.2020.117471
Paarakh M, Jose P, Setty C, Christoper G (2018) Release kinetics concepts and applications. Int J Pharm Res Technol 8(1):12–20
Pereira JM, Mahoney M, Edgar KJ (2014) Synthesis of amphiphilic 6-carboxypullulan ethers. Carbohydr Polym 100:65–73. https://doi.org/10.1016/j.carbpol.2012.12.029

Popović SZ, Lazić VL, Hromiš NM, Šuput DZ, Bulut SN (2018) Biopolymer packaging materials for food shelf-life prolongation. In: Grumezescu AM, Holban AM (eds) Biopolymers for food design. Academic Press, London, pp 223–277. https://doi.org/10.1016/B978-0-12-811449-0.00008-6

Poudel D, Swilley-Sanchez S, O'Keefe S, Matson J, Long T, Fernández-Fraguas C (2020) Novel electrospun pullulan fibers incorporating hydroxypropyl-β-cyclodextrin: morphology and relation with rheological properties. Polymers 12:2558

Prajapati VD, Jani GK, Khanda SM (2013) Pullulan: an exopolysaccharide and its various applications. Carbohydr Polym 95:540–549

Ravella SR, Quiñones TS, Retter A, Heiermann M, Amon T, Hobbs PJ (2010) Extracellular polysaccharide (EPS) production by a novel strain of yeast-like fungus Aureobasidium pullulans. Carbohydr Polym 82:728–732

Robertson GL (2011) Packaging materials for biscuits and their influence on shelf life. In: Manley D (ed) Manley's technology of biscuits, crackers and cookies, 4th edn. Woodhead Publishing, London, pp 247–267. https://doi.org/10.1533/9780857093646.2.247

Roukas T, Biliaderis CG (1995) Evaluation of carob pod as a substrate for pullulan production by Aureobasidium pullulans. Appl Biochem Biotechnol 55:27–44

Sharma N, Prasad GS, Choudhury AR (2013) Utilization of corn steep liquor for biosynthesis of pullulan, an important exopolysaccharide. Carbohydr Polym 93:95–101. https://doi.org/10.1016/j.carbpol.2012.06.059

Shin YC, Kim YH, Lee HS, Cho SJ, Byun SM (1989) Production of exopolysaccharide pullulan from inulin by a mixed culture of *Aureobasidium pullulans* and *Kluyveromyces fragilis*. Biotechnol Bioeng 33:129–133. https://doi.org/10.1002/bit.260330117

Shingel KI (2004) Current knowledge on biosynthesis, biological activity, and chemical modification of the exopolysaccharide, pullulan. Carbohydr Res 339:447–460. https://doi.org/10.1016/j.carres.2003.10.034

Shuhong Y, Meiping Z, Hong Y, Han W, Shan X, Yan L, Jihui W (2014) Biosorption of Cu2+, Pb2+ and Cr6+ by a novel exopolysaccharide from Arthrobacter ps-5. Carbohydr Polym 101: 50–56. https://doi.org/10.1016/j.carbpol.2013.09.021

Singh RS, Saini GK (2008) Production, purification and characterization of pullulan from a novel strain of Aureobasidium pullulans FB-1. J Biotechnol 136:S506–S507

Singh R, Saini G (2012) Biosynthesis of pullulan and its applications in food and pharmaceutical industry. In: Satyanarayana T, Johri B (eds) Microorganisms in sustainable agriculture and biotechnology. Springer, Dordrecht, pp 509–553. https://doi.org/10.1007/978-94-007-2214-9_24

Singh RS, Saini GK, Kennedy JF (2008) Pullulan: microbial sources, production and applications. Carbohydr Polym 73:515–531

Singh RS, Kaur N, Kennedy JF (2019) Pullulan production from agro-industrial waste and its applications in food industry: a review. Carbohydr Polym 217:46–57. https://doi.org/10.1016/j.carbpol.2019.04.050

Sugumaran KR, Ponnusami V (2017a) Review on production, downstream processing and characterization of microbial pullulan. Carbohydr Polym 173:573–591. https://doi.org/10.1016/j.carbpol.2017.06.022

Sugumaran KR, Ponnusami V (2017b) Conventional optimization of aqueous extraction of pullulan in solid-state fermentation of cassava bagasse and Asian palm kernel. Biocatal Agric Biotechnol 10:204–208. https://doi.org/10.1016/j.bcab.2017.03.010

Sugumaran K, Gowthami E, Swathi B, Elakkiya S, Srivastava S, Ravikumar R, Gowdhaman D, Ponnusami V (2013) Production of pullulan by Aureobasidium pullulans from Asian palm kernel: a novel substrate. Carbohydr Polym 92:697–703. https://doi.org/10.1016/j.carbpol.2012.09.062

Suzuki Y, Kataoka Y, Matsui H (2003) Electrical properties of composite cyanoethylated pullulan film. J Wood Sci 49:0100–0103. https://doi.org/10.1007/s100860300016

Tabasum S, Noreen A, Maqsood MF, Umar H, Akram N, Nazli Z-I-H, Chatha SAS, Zia KM (2018) A review on versatile applications of blends and composites of pullulan with natural and synthetic polymers. Int J Biol Macromol 120:603–632. https://doi.org/10.1016/j.ijbiomac.2018.07.154

Tang H-B, Li L, Chen H, Zhou Z-M, Chen H-L, Li X-M, Liu L-R, Wang Y-S, Zhang Q-Q (2010) Stability and in vivo evaluation of pullulan acetate as a drug nanocarrier. Drug Deliv 17:552–558. https://doi.org/10.3109/10717544.2010.490250

Tomasik P (2003) Chemical and functional properties of food saccharides. CRC Press, Boca Raton

Tong Q, Xiao Q, Lim LT (2013) Effects of glycerol, sorbitol, xylitol and fructose plasticisers on mechanical and moisture barrier properties of pullulan–alginate–carboxymethylcellulose blend films. Int J Food Sci Technol 48:870–878

Trinetta V (2017) The application of edible and active pullulan coatings on foods. In: Reference module in food science. Elsevier, London, pp 1–6. https://doi.org/10.1016/B978-0-08-100596-5.21129-3

Trinetta V, Cutter CN (2016) Pullulan: a suitable biopolymer for antimicrobial food packaging applications. In: Barros-Velázquez J (ed) Antimicrobial food packaging. Academic Press, San Diego, pp 385–397. https://doi.org/10.1016/B978-0-12-800723-5.00030-9

Trinetta V, Cutter CN, Floros JD (2011) Effects of ingredient composition on optical and mechanical properties of pullulan film for food-packaging applications. LWT Food Sci Technol 44: 2296–2301. https://doi.org/10.1016/j.lwt.2011.07.015

Tsujisaka Y, Mitsuhashi M (1993) Pullulan. In: Whistler RL, Bemiller JN (eds) Industrial gums, 3rd edn. Academic Press, London, pp 447–460. https://doi.org/10.1016/B978-0-08-092654-4.50020-6

Venugopal V (2016) Marine polysaccharides: food applications. CRC Press, Boca Raton

Vuddanda PR, Montenegro-Nicolini M, Morales JO, Velaga S (2017) Effect of plasticizers on the physico-mechanical properties of pullulan based pharmaceutical oral films. Eur J Pharm Sci 96: 290–298. https://doi.org/10.1016/j.ejps.2016.09.011

Wang D, Ju X, Zhang G, Wang D, Wei G (2016) Copper sulfate improves pullulan production by bioconversion using whole cells of Aureobasidium pullulans as the catalyst. Carbohydr Polym 150:209–215. https://doi.org/10.1016/j.carbpol.2016.05.035

Wang Z, Xie J, Shen M, Nie S, Xie M (2018) Sulfated modification of polysaccharides: synthesis, characterization and bioactivities. Trends Food Sci Technol 74:147–157

Wang Y, Guo Z, Qian Y, Zhang Z, Lyu L, Wang Y, Ye F (2019) Study on the electrospinning of gelatin/pullulan composite nanofibers. Polymers 11:1424. https://doi.org/10.3390/polym11091424

Wolf BW, Garleb KA, Choe YS, Humphrey PM, Maki KC (2003) Pullulan is a slowly digested carbohydrate in humans. J Nutr 133:1051–1055

Wu S, Jin Z, Tong Q, Chen H (2009) Sweet potato: a novel substrate for pullulan production by Aureobasidium pullulans. Carbohydr Polym 76:645–649. https://doi.org/10.1016/j.carbpol.2008.11.034

Wu J, Zhong F, Li Y, Shoemaker C, Xia W (2013) Preparation and characterization of pullulan–chitosan and pullulan–carboxymethyl chitosan blended films. Food Hydrocoll 30:82–91

Wu F, Misra M, Mohanty AK (2021) Challenges and new opportunities on barrier performance of biodegradable polymers for sustainable packaging. Prog Polym Sci 117:101395. https://doi.org/10.1016/j.progpolymsci.2021.101395

Wüstenberg T (2015) General overview of food hydrocolloids. In: Wüstenberg T (ed) Cellulose and cellulose derivatives in the food industry fundamentals and applications. Wiley, Weinheim, pp 1–68

Xiao Q, Tong Q (2013) Thermodynamic properties of moisture sorption in pullulan–sodium alginate based edible films. Food Res Int 54:1605–1612. https://doi.org/10.1016/j.foodres.2013.09.019

Xiao Q, Lu K, Tong Q, Liu C (2015) Barrier properties and microstructure of pullulan–alginate-based films. J Food Process Eng 38:155–161. https://doi.org/10.1111/jfpe.12151

Xu Q, Yao Y, Zhao T, Shi Q, Li Z, Tian W (2018) Dissolution characteristics of freeze-dried pullulan particles affected by solution concentration and freezing medium. Int JFood Eng 14. https://doi.org/10.1515/ijfe-2018-0073

Yan JJ, Li Z, Zhang JF, Qiao CS (2012) Preparation and properties of pullulan composite films. Adv Mater Res 476-478:2100–2104. https://doi.org/10.4028/www.scientific.net/AMR.476-478.2100

Yang N, Chen H, Jin Z, Hou J, Zhang Y, Han H, Shen Y, Guo S (2020) Moisture sorption and desorption properties of gelatin, HPMC and pullulan hard capsules. Int J Biol Macromol 159: 659–666. https://doi.org/10.1016/j.ijbiomac.2020.05.110

Ye S, Ma Z, Liu Z, Liu Y, Zhang M, Wang J (2014) Effects of carbohydrate sources on biosorption properties of the novel exopolysaccharides produced by Arthrobacter ps-5. Carbohydr Polym 112:615–621. https://doi.org/10.1016/j.carbpol.2014.05.076

Yi K, Liu J, Chen P, Chu B (2021) Miscibility, microstructure, and dielectric properties of the polymer blends of cyanoethyl pullulan and fluoropolymer. J Mater Sci Mater Electron 32:3577–3590. https://doi.org/10.1007/s10854-020-05104-w

Youssef AM, Abou-Yousef H, El-Sayed SM, Kamel S (2015) Mechanical and antibacterial properties of novel high performance chitosan/nanocomposite films. Int J Biol Macromol 76: 25–32. https://doi.org/10.1016/j.ijbiomac.2015.02.016

Chapter 5
Chitosan-Based Films

Aiman Zehra, Tawheed Amin, Sajad Mohd Wani, Haamiyah Sidiq, Iqra Bashir, Sehrish Mustafa, Zahida Naseem, and Seerat Showkat

5.1 Introduction

Currently, food safety and environmental issues have caused several apprehensions of the public and green chemistry demands to many scientists. Chitosan, derived from deacetylation of chitin, is an amino polysaccharide containing D-glucosamine units and N-acetyl-D-glucosamine (Singh et al. 2014). This is a predominant biopolymer in nature comprising shells, insects, and plants. Chitosan is the next highest produce of polysaccharide followed by cellulose. Also, biological activity and structural documentation from investigations have revealed that chitosan holds excellent properties that are unlike from other polysaccharides. Chitosan corresponds to cyclodextrin but matches well with biofilms due to the presence of amino group at position 2. Chitosan has been demonstrated to be beneficial in numerous fields, such as bioengineering, agriculture, medicine, food, environment, and textile, owing to its admirable antimicrobial properties, biocompatibility, biodegradability, biofunctional, chelating ability, etc. (Tian and Liu2020).

Chitosan produces effective membranes and films. The use of chitosan film for packaging, particularly as food packaging, is possible. This is because of its excellent oxygen and carbon dioxide barriers and attractive antibacterial characteristics. Chitosan is a treasured polymer for microspheres, artificial membranes, beads, nanoparticles, and wound dressings since it may be combined with crosslinking substances like sodium tripolyphosphate and sodium citrate. It is a semi-crystalline polysaccharide which is not soluble in water at a neutral pH, unlike many biodegradable polymers. The qualities of chitosan films include durability, toughness, and flexibility. This presents a chance for emerging a diversity of films by different

A. Zehra · T. Amin · S. M. Wani (✉) · H. Sidiq · I. Bashir · S. Mustafa · Z. Naseem · S. Showkat
Sher-e-Kashmir University of Agricultural Sciences and Technology of Kashmir, Shalimar, Srinagar, Jammu and Kashmir, India

T. Amin et al. (eds.), *Polysaccharide Based Films for Food Packaging: Fundamentals, Properties and Applications*,
https://doi.org/10.1007/978-981-99-4898-7_5

chitosans from crawfish and organic acid solvents (Bhuvaneshwari et al. 2011). Chitosan-based films have also been claimed to be excellent transporters of various useful compounds, such as antioxidants and antibacterial agents, in order to improve the quality of food storage(Tripathi et al. 2008). Chitosan films are formed by cross-linking with different substances like ternary chitosan-glucomannan-nisin, aglycone geniposidic acid, ferulic acid combined with starch chitosan (Mathew and Abraham 2008), potassium sorbate, garlic oil and nisin amalgamation, starch/chitosan blend film under the action of irradiation, chitosan-potato starch film using microwave treatment (Tripathi et al. 2008), and amylose and water-soluble chitosan film (Suzuki et al. 2005), but chitosan-based films are stiff and need plasticizers to reduce film upsurge film elasticity and manageability and brittleness. The plasticizers increase the movement of polymer chains and diminish the intermolecular forces, thereby enhancing extensibility and flexibility of the film. However, plasticizers often strengthen solute permeability of the film. Owing to the use of plasticizers in film's formulation, mechanical characteristics and maintenance of the film during prolonged storage periods are thus achieved (Singh et al. 2014).

5.2 Source

Chitin is the polysaccharide that is utmost prevalent after cellulose on Earth. Chitosan consists of three different reactive functional groups, i.e., an amino group at position C-2, primary and secondary hydroxyl groups at positions C-3 and C-6, and both of these groups at position C-3. It is a natural polymer mainly present in insects as a principal element of the exoskeleton, in oceanic invertebrates and fungi as a cell walls constituent. Every year, people consume marine invertebrates including shrimp, oysters, lobster, and crab as rich protein sources from sea food because they contain (1,4)-linked 2-amino-deoxy-b-D-glucan, a deacetylated derivative of chitin. The non-edible portions and external shells of these crustaceans make nearly half of the body mass and are typically thrown away as waste. These leftover materials discarded as wastes are noticeable chitin source (Kumar et al. 2020).

Chitosan derived from crustacean chitin derived from shrimp and crab shell wastes is the most commonly used form of chitosan. This is nearly 70% of the organic complexes in such shells. Pulverized shells are demineralized and deproteinated by progressive treatments of acid and alkali, and then the extracted chitin is deacetylated to chitosan at higher temperature by alkaline hydrolysis for formulating chitosan. Except in certain fungi, chitosan is sporadic in nature. In recent years, the chitosan produced from fungi, by means of fermentation, has attained much importance (Prabaharan 2012).

5.3 Structure of Chitosan and Chitosan-Derived Components

Chitin, a linear cationic polysaccharide, is made of β-(1,4) linked N-acetyl-D-glucosamine. Chitosan is produced by the deacetylation of chitin with concentrated NaOH solution and comprises of a heteropolysaccharide of β-1,4-linked D-glucosamine and N-acetyl-D-glucosamine (Dutta et al. 2009). The level of acetamidation is a defining characteristic of chitin and chitosan. It is greater than 50% in chitin and less than 50% in chitosan (Brasselet et al. 2019).

Chitosan is particularly simple to modify since it contains reactive amino (-NH2) and hydroxyl (-OH) groups. These changes are intended to improve the materials' biological and chemical characteristics and alter their solubility in accordance with the intended uses. The main modifications of chitosan are quaternization, N-acyl modifications, N-alkyl modifications, and C-6 oxidation. Chemical changes to produce new biofunctional materials are of particular interest because such procedure would not alter the essential skeleton of chitin and chitosan. Because these adjustments would not alter the skeleton of chitin and chitosan, chemical modifications to produce new biofunctional materials are of particular interest. The solubility of chitin and chitosan may be altered by the introduction of groups such as phosphonic acid or phosphonate by reactivity of phosphorylating chemical onto the amino groups. There are several methods for getting phosphate derivatives (Jayakumar et al. 2006).

Chemical modification of the amino and hydroxyl groups with sulphate at specific sites can result in pharmaceutical products (Jayakumar et al. 2007).

5.4 Production/Preparation of Chitosan and Chitosan Derivative-Based Films

Chitosan has made the most important progress and has revolutionized the food packaging sector. The most recent improvements among the approaches to formulate chitosan-based films include layer-by-layer assembly, dipping, direct casting, coating, and extrusion. In the food industry, chitosan films have been examined for its non-toxicity, biodegradability, and excellent performance. Food packaging materials made of chitosan can be used to prolong shelf life of food. Films made from chitosan preserve a wide variety of foods like pomegranate, banana, tomato, mango, fish, papaya, wolfberry, and carrot (Kerch 2015).

There is evidence that chitosan is harmless, biocompatible, and biodegradable. Shrimp shell chitosan is a nontoxic, commercially viable waste product of the seafood industry. Chitosan can be converted into edible films by mixing it in diluted acetic acid solution, hydrochloric acid, and formic acid. Chitosan films typically have no pores and are translucent and clear.

In order to create plasticized chitosan films, Suyatma et al. (2005) combined chitosan powder (4.2 g) with 1% acetic acid solution (420 mL). The films were then cast on a polypropylene plate and dried for 12 h at 60 °C. To increase the chain mobility, extendibility, flexibility, and strength, small amounts of the plasticizers sorbitol and glycerol are added. Emulsifiers and surfactant, like Tween-80, Tween-20, Span 20, and Brij 56, are incorporated to increase the properties of chitosan film (Alix et al. 2013). Bhuvaneshwari et al. (2014) developed chitosan films by dissolving chitosan (1&2 g) in 100 mL of known acetic acid solutions (1%, 2%, 3%). The chitosan powder was vigorously mixed with the acetic acid. The solution was allowed to degas for about 2 h. After degassing, the solution was poured into the petri plate. The mixture was dried on a plate in an oven at 72 °C for 20 h. Plates containing dry films were immersed in 2 N sodium hydroxide at room temperature for over 30 min. The wet films were removed from the glass plate, dried at 37 °C for a week, and then washed with distilled water.

5.5 Properties

5.5.1 Thermal Properties

The thermal properties of the chitosan are dependent on its molecular structure, length of the chains of the macromolecules, and crystalline structure. Some important thermal analyzers such as dynamic mechanical analyzer (DMA), thermogravimetric analyzer (TGA), and DSC are used to estimate the thermal properties of the chitosan film. Suyatma et al. (2005) studied the thermal characteristics of the chitosan-based films using differential scanning calorimetry (DSC). The sample was first subjected to the heating cycle of 190 °C for about 1 min, and then the sample was made to cool at 0 °C for 5 min keeping the isothermal conditions. The second heating was done at 300 °C again for 5 min and again cooled. The two cycles for each heating and cooling eliminated the effect of moisture. The various values of Tg were being observed by different researchers. *Tg* of chitosan film was observed around 103 °C by Cheung et al.(2002). Glass transition temp at 195 °C was found for chitosan by Shanta and Harding (2002). Dong et al. (2004) determined the Tg of the chitosan films at temperatures between 140 and 150 °C by different techniques, viz. thermally simulated current spectroscopy (TSC), differential scanning calorimetry (DSC), dynamic mechanical thermal analysis (DMTA), and dilatometry (DIL) measurements. Tg of unplasticized film is usually high at almost about 196 °C as compared to that of plasticized films; usually it is in the range of 184 °C, 178 °C, and 169 °C (Suyatma et al., 2005).

5.5.2 Barrier Properties

The difficulty in commercializing the biodegradable polymers is its barrier properties. Barrier properties are one of the fundamental parameters for food packaging, which include (O_2, CO_2, and N_2) aroma compounds and light. The barrier properties against the water, nitrogen, oxygen, CO_2, UV light, etc. have been investigated widely. The researchers have published studies on the chitosan films that they owe poor barrier properties. Chitosan films, being semipermeable membrane, allow the oxygen in the packaging material to some extent; in the foods like fruits and vegetables, ripening is controlled by the oxygen. To retain aromatic compounds in the packaging material during storage, the barrier properties of the film must be improved in order to delay the rancidity of polyunsaturated fats (Kerch and Korkhov 2011).

The film barrier property against water vapor is one of the significant properties for food packaging. The organic materials (polyphenols, nanoparticles) build up strong connections with hydrophilic groups of chitosan and water molecules resulting in the decrease of the permeability of water vapors. Nanoparticles incorporated in the film fill the voids and hence prevent the moment of water molecules. Moreover, glycerol when added in high concentration yields the film with poor barrier properties (Sun et al. 2017). The shelf life and worth of the food is reduced once it comes in contact with oxygen. Oxygen promotes the growth of the microbes such as yeasts, molds, and bacteria. However, chitosan films form strongly packed hydrogen bonds which inhibits the gas entry (Benbettaieb et al. 2019).

Water vapor permeability and mechanical properties of chitosan films are inversely related to storage temperature and linearly related to molecular weight and storage time. Preparing the composites with nano and microfillers and integrating them into the film matrix has enhanced the barrier properties. Cellulose, cellulose nanocrystals (CNC), nanoclay, nano carbon, lignin, and CS nanoparticles are some of the biopolymers used to study their impact on barrier properties. The barrier property and the mechanical property of the film get enhanced with the increase in the concentration of plasticizer (Kerch and Korkhov 2011). Azerdo et al. (2010) investigated that the incorporation of chitosan-cellulose nano-whisker (CSNW) and glycerol at 15% and 18%, respectively, has enhanced the barrier properties. Chitosan owes resistance with regard to gas permeability to some extent but lacks the resistance to water vapors due to its strong water loving nature (Zotkin et al. 2004).The physical characteristics (barrier properties against oxygen and water) of the chitosan film have been preserved in a more and superior way by infrared drying compared to those films which are dried in oven or room temperature (Srinivasa et al. 2004). The research, which has been investigated by numerous researchers Tual et al. (2000) and Zheng et al. (2000), related with the chitosan films, which have been incorporated with active compounds such as antimicrobial agents, showed an expeditious release of biocides onto food material surface. The biocide property of the chitosan film would get deactivated, when the protonated glucosamine groups are released during the disintegration of the film into any solvent such as water or chemical and would get deactivated. Tang et al. investigated how cross-linking agents incorporated in the film form covalent bonds and prevent the interface between the water and polysaccharide to strengthen the structure of the film (Tang et al. 2003).

5.5.3 Mechanical Properties

Another most important characteristic besides the barrier property of chitosan films is its mechanical property which contributes to a significant role in designing the packaging. The mechanical behavior of chitosan films is determined by interactions between molecules dispersed in the medium, such as plasticizers, water, and additives. However, the mechanical properties of the chitosan film are improved by incorporating it with other biopolymer such as starch, cellulose, and alginate. The mechanical properties of the film are frequently checked using yield strength, Young's modulus, fracture strength, and elongation at yield and break. The mechanical property of the film improves as the concentration of plasticizer increases (Kerch and Korkhov 2011). A research conducted by Zhang and Wang (2009) exhibited that the nanocomposite chitosan films filled with Na+ –montmorillonite (MMT) and multiwalled carbon nanotubes (MWCNT) produce synergistic effects on its mechanical properties. Hassan et al. (2012) observed that nanofiller-reinforced chitosan nanocomposites are used from the rice straw which significantly increased the tensile strength of chitosan nanocomposites. A nanocomposite of chitosan MgO consisting clove as an essential oil has increased the tensile strength (Sanuja et al., 2014). The influence of carboxymethyl cellulose and nanoparticles of zinc oxide, which was investigated by Youssef et al. (2016), revealed that it has a great consequence on the mechanical properties.

5.5.4 Dielectric Properties

Due to its fine mechanical properties, capability to perform as O_2 obstruction, and antimicrobial behavior, chitosan is formed into biocomposite films designed for food packaging. Many remarkable properties exist in polymers. With prompt technical development, aluminum and iron have been substituted in polymers for several purposes that need pressure and elevated temperature (Alturki 2018). Dielectric properties of polymeric substances rely on chemical composition, morphological features, crystallinity, and existence of additives or fillers. The effective dipole moment increases the dielectric constant of a substance (Mansour et al. 1994). For development of an effective packaging, dielectric measurements are used to understand the molecular behavior of polymeric substances.

Sunilkumar et al. (2014) studied dielectric properties of LDPE/chitosan combined films and reported that pure LDPE had a low dielectric constant (ε_r). Alturki (2018) found that different preparation methods affected the particle size, dielectric constant, and antibacterial properties of ZnO nanoparticles and ZnO/chitosan thin films. It was observed that the pure chitosan film had a higher ε_r value than all nanocomposite films of chitosan/ZnO samples. It is possible that Zn ions in samples will "nano impact" the rather swift segmental motion of chitosan to slow it down. Once nanoparticles are added to a polymer composite, the dielectric constant decreases.

5.5.5 Tensile Properties

Mechanical properties are just as critical as barrier properties in edible and biodegradable films. A film's credibility is ensured by its technical efficiency and the absence of minor flaws. Hydrocolloid interactions with small molecules present in the matrix space, such as water, plasticizers, lipids, and other additives, have an impact on the mechanical behavior of films. For the packaging design process, quantitative data on the mechanical parameters of edible films is needed. Tensile properties are the most considered, tested, and used mechanical properties of plastic materials in the industry. Yield power, modulus of elasticity, fracture strength, and elongation at yield and break are all valuable data obtained from tensile research. The maximum fracture strength of a film corresponds to the maximum tensile stress it can withstand. The maximum deformation in length of the test sample before it breaks is called strain. The modulus of elastic is the most ultimate characteristic of film stiffness. The test conditions would be crucial because interactions with the external environment influence the mechanical properties of hydrophilic films. Temperature and relative humidity have a significant impact on the stress-strain behavior of these polymers. Because of the plasticizing impact of the water molecules, films conditioned in a 50% RH setting are poorer than films equilibrated in a lower RH environment. Maximum tensile strength, tensile yield strength, and Young's modulus all tend to decline as temperature rises. The elongation at yield and split, on the other hand, tends to increase. Mechanical properties are also affected by additives and plasticizers. Tensile strength decreases as plasticizer content rises, while elongation rises (Leceta et al. 2013).

5.5.6 Heat Sealing Properties

The essential characteristic of chitosan film is its sealing ability which should be strong enough to manufacture bags, pouches, or sachets to pack liquids or dry food. Like conventional thermoplastic polymer films, chitosan films cannot be heat sealed (Cao et al. 2009). There are high demand and desire to increase the heat-sealing properties. Sealability and seal strength of polymer structure can be increased by adding hydroxyl groups and a copolymer (Abdorreza et al. 2011; Das and

Chowdhury 2016). Gelatin has been used to alter the functional groups and promotes the molecular interdiffusion and enhances the melting properties and heat sealability (Abdelhedi et al. 2017). Prateepchanachai et al. (2019) used CO_2, glycerol, and gelatin to make chitosan-based films for packaging to enhance seal strength and mechanical properties of chitosan film compared with those films which have used only glycerol (Farhan and Hani 2017; Tongnuanchan et al. 2016). The films without any additive, i.e., control films, were hard to heat seal due to their stiff crystalline nature and presence of tough inter- or intramolecular hydrogen bonding results in lower molecular interdiffusion of films during their thermal degradation (Mittal et al. 2017; Caro et al. 2015; Leceta et al. 2013). The functional groups like carboxyl (COOH), aldehyde (CHO), and hydroxyl (OH) are liable for the adhesion power of the chitosan film (Elsabee and Abdou 2013; Hamed et al. 2016).

5.5.7 Flexibility Properties

The chitosan film is fragile despite its outstanding film-forming ability. The fragility of chitosan films is mostly determined by the strong suit of polymer interaction, which can be controlled by adding a plasticizer and changing the polymer chemistry. Small molecules, for instance polyols like, polyethylene glycol (PEG), sorbitol, and glycerol, which are used as plasticizers interpolate and intermingle between the polymer chains, thereby dispersing the chains separately to surge the flexibility, disturbing hydrogen bonding and the gas permeability of the films and water vapor transmission rate. Sorbitol (SOR) and glycerol (GLY) are functional plasticizers for edible films (Suyatma et al. 2005). However, the films plasticized with SOR can also crystallize if kept at low and intermediate relative humidity levels (Ray et al. 2009).

5.5.8 Electrical Properties

Chitosan has drawn huge attentiveness as biomaterial for several applications such as biomedical, pharmaceutical, issue engineering, food package, and wound dressing material (Goenka et al. 2014). Moreover, active bonding sites are produced by the high density of amine groups present in chitosan and help it to preserve its natural properties even after processing into film. Therefore, chitosan has been nominated to be fabricated into film by means of electrochemical deposition and used as a sensing material. Also, the films fabricated from pure chitosan are brittle and rigid. Thus, to enhance the objectionable chitosan properties, various approaches have been proposed to add other polymers as additives in the preparation of chitosan films. CMC, a natural nontoxic, biodegradable, and biocompatible anionic polymer generated by chemical modification of natural cellulose, is one of the forms of plasticizers that can be employed. Since CMC and chitosan share a strong ionic interaction, this structural similarity enables CMC to enhance the properties of chitosan film (Nasution et al. 2017).

The electrical and mechanical properties of the chitosan matrix may also be enhanced by the introduction of graphene, carbon nanotubes (CNTs), and other conductive fillers. CNTs are often unfavorable because of the toxicity of synthesis-related impurities. Also, the poor dispersion in biological media and hydrophobic character leading to agglomeration makes it unfavorable. Thus, to prepare conductive biopolymer-based composites, graphene (Gr) may be used as a conductive filler. It is the thin material in the world and has malleable properties. Additionally, it is more durable than diamond and has the best ability of any material to conduct electricity at ambient temperature (Yenier et al. 2016). Single-layer graphene has greater thermal and electrical conductivities as a result of the low defect density in the crystal lattice. Single-layer, defect-free graphene has thermal conductivities between 4500 and 5200 W/mK and electrical conductivities of 104 S/cm, respectively (Goenka et al. 2014). Moreover, single-layer defect-free graphene has a breaking strength that is 200 times greater than steel. The fracture strength and Young's modulus for defect-free graphene are 130 GPa and 1 TPa, respectively (Li et al. 2006). Due to these extraordinary mechanical, thermal, and electrical properties, graphene has gained huge attention in material science and is preferred as a supporting material in the biopolymer (Lee et al. 2013).

5.5.9 Optical Properties

The terms that define the optical properties of polymeric substances are opacity, transparency, haziness, gloss, and refractive index. Packaging that is clear and shiny is preferred over packaging that is cloudy and opaque. Foods with shorter shelf lives are especially favored by transparent packaging since it allows consumers to visually assess the food's quality. The transparency of a polymer can be influenced by its shape. Transparent polymers do not absorb all visible light wavelengths, and the crystallinity of a polymer has an inverse relationship to transparency. Visible light scattering increases with crystallinity because visible light's wavelength corresponds to the size of the crystals (Bower et al. 2003). Transparency is reduced when scattering increases. There are numerous papers describing how the morphology of polymers can be changed in order to regulate the optical properties of biodegradable polymers. This can be done by changing the crystallinity, processing conditions, or incorporating more fillers. Oksman et al. (2016) investigated the effects of incorporating chitin and cellulose nanocrystals into plasticized polylactic acid using compression molding and extrusion while employing two cooling speeds. A faster rate of cooling resulted in the formation of more amorphous and advanced material. The creation of polymers with UV-barrier properties is a good strategy to increase the shelf life of food or to maintain its sensory qualities (Cazón et al. 2019). In their study of the production, structural, dielectric, and optical characteristics of chitosan-MgO nanocomposite, Khalil et al. (2020) noted that the UV-shielding capacity of

MgO nanoparticles significantly reduced the UV transmission through the chitosan-MgO film, indicating that the MgO nanoparticles were used to create the chitosan's UV protection.

5.5.10 Water Solubility Properties

Due to its robust and stiff crystalline structure, chitosan is a partially crystalline polymer, a weak base that is insoluble in water, alkali, liquid over pH 7, and frequently organic solvents. Chitosan is polydispersed and can dissolve in a wide range of inorganic and organic acids, including HCl, phosphoric acid, lactic acid, propionic acid, succinic acid, acetic acid, tartaric acid, and formic acid. When the pH 6.4 is just right and the mixture is vigorously stirred for a period of time (Chen et al. 2008). The allocation of the CH_3CO groups that are still present next to the chain is what gives chitosan its distinctive properties, but generally, free amino ($-NH_2$) groups are what determine structural characteristics through intra- and/or intermolecular H-bonding. This lowers the associative interactions between chains and makes chitosan soluble in acidic liquids at pH levels less than 6.5. Chitosan is one of the few cationic polyelectrolytes found in nature and is formed by amino groups (pKa ≈ 6.5), compared with other carbohydrates, which can either have a neutral or a negative charge. Chitosan has peculiar features because of its basicity: it is protonated before dissolving in liquid acidic medium at pH less than 6.5, but when it dissolves, it has a strong positive charge on $-NH^{3+}$ groups, and the soluble polysaccharide that forms as a result is positively charged. It sticks to negatively charged surfaces as a result. It chelates heavy metal ions and collects polyanionic components. Chitosan has exceptional gel-forming characteristics due to its solubility in acidic solutions and its accumulation with polyanions (Chenite et al. 2001; Krajewska 2004; Qin et al. 2006). Even though chitosan is recognized to have important functional properties, its poor solubility is the fundamental barrier to its utilization. By breaking down the chitosan polymer chain, soluble chitosan oligomers are produced. Additionally, water-soluble chitosan is produced by chemically altering chitosan in a controlled manner to the degree of replacement (Qin et al. 2006). Due to their shorter chain lengths and free amino groups in D-glucosamine units, incompletely hydrolyzing chitosan has attracted new interest since doing so makes it readily soluble in water (Ilyina et al. 1999; Qin et al. 2006). It was discovered that whereas water-soluble chitosan had little to no impact on bacteria and yeast, the water-insoluble variety showed signs of having an antibacterial effect (Qin et al. 2006). Improved physiological benefits of water-soluble chitosan include anticancer activity and in vivo immune-stimulating effects. Chitosan has a strong ability to form films, and a major component of a film that influences its water resistance quality is how soluble it is in water (Karlsen and Skaugrud 1991). In their study of the preparation and characterization of chitosan film containing thinned young apple polyphenols (YAP) as an active packaging material, Sun et al. (2017) found that the YAP-chitosan films had significantly higher water solubility (p0.05) than the control films, which had the lowest solubility. The higher value of YAP-chitosan films' water solubility can be attributed to the polyphenols' affinity for water, which allows them to easily mix with water molecules (Mathew et al. 2006). The addition of YAP increases the matrix's capacity to mix with water, increasing its hydrophobicity. Furthermore, compared to films made using the casting process, electrospun chitosan films had a significantly better water solubility (Cai et al. 2019).

5.5.11 Viscous Properties

To clarify the processability of the film, steady shear behavior of chitosan film-forming solution has been carried out. The flow properties of the film-forming fluid influence the blend film's thickness, spreadability, and homogeneity. With an increase in shear rate, the blend solution's apparent viscosity drops (Sun et al. 2020). Blend solution, a film-forming solution devoid of or containing various amounts of glycerol, PEG-400, or sorbitol, demonstrated a shear thinning tendency and was regarded as a non-Newtonian fluid. It had been shown that the viscosity of film-forming solutions was due to the entanglement of molecules. Accordingly, the primary cause of shear-thinning behavior was the breakdown of the entanglement structure between molecules during high-speed shear. The structure could not be put back together quickly. The chitosan/zein film without plasticizer displayed higher apparent viscosity when compared to other film solutions. A significant result observed by Sun et al. (2020) was that with increase in the plasticizer concentration, apparent viscosity decreased.

5.5.12 Incorporation of Nanoparticles Using Inkjet Printing

Inkjet printing was used to print chitosan-tripolyphosphate nanoparticles and chitosan-tripolyphosphate nanoparticles loaded with thymol on new active packaging films made from chitosan and a combination of chitosan and quinoa protein (Caro et al. 2015). 20% glycerol was added to the chitosan-tripolyphosphate-thymol nanoparticle suspension prior to printing it into films in order to change the suspension's surface tension and kinematic viscosity. Compared to the control, both printed films demonstrated improved barrier characteristics. The placements of the chitosan-tripolyphosphate-thymol nanoparticles on the films were thought to be the cause of the variations in the thymol-release profiles of the films. The antibacterial activity of the prepared films was stronger against *L. innocua*, *S. aureus*, *E. typhimurium aerogenes*, *E. coli*, and *P. aeruginosa* than the control films, *E. coli*.

5.5.13 Aroma Retention Properties

The undesirable chemical changes due to oxidation and spoilage caused by microbial agents can be inhibited by incorporating the antimicrobial agent obtained from synthetic or natural origin. Carvacrol, eugenol, and limonene are three fragrance molecules found in various EOs and known to possess antibacterial activities packaging. Carvacrol, eugenol, and limonene are three fragrance molecules found in various EOs and known to possess antibacterial activities (Rivera et al. 2015). There should be a proper control on the release of aroma compounds from the packaging material. The aroma compounds should be properly introduced in the polymer matrix for understanding its mechanism of release in the head space of food, which is a challenging requirement during the designing and processing an efficient active packaging (Ali et al. 2010). Low-molecular-weight chitosan, on the other hand, can penetrate the cytosol and bond with cell DNA, impairing the synthesis of mRNA and proteins as well as leading to cell death.

5.5.14 Antimicrobial Properties

The development of safe, wholesome food free of synthetic chemicals is one of the major challenges confronting food producers and the food industry (preservatives). As a result of customers' growing health and environmental concerns, researchers' attention has been drawn more intensely in the last decade to the use of bio-based packaging materials as a replacement for synthetic plastic polymer(s) in food packaging (Kumar et al. 2020). These biodegradable counterparts for plastic polymers include polysaccharides like starch, cellulose, chitosan, proteins, and agar such as casein. When the proper temperature, moisture levels, and oxygen are present, bioplastics breakdown without leaving any harmful residues behind. Depending on the source and kind of the polymer, certain biopolymers disintegrate in just a few weeks as opposed to synthetic polymers, which might take years to breakdown (Barboza et al. 2018).

As a result of the surge in consumer demand for minimally processed, safe food free of artificial preservatives, the food industry has paid close attention to antimicrobial packaging. It includes packaging methods like modified atmosphere packaging used to limit the activities of bacteria and control growth, as well as packaging materials consisting of antimicrobial agents. The primary goal of antimicrobial packaging is to ensure food safety. Several attempts have recently been made to develop biopolymer-based antimicrobial packaging solutions in order to increase food shelf life and quality. Due to its nontoxicity, biodegradability, and biocompatibility, as well as its widespread availability in nature and cheaper cost, chitosan (CS) has emerged as a potential biopolymer material for antimicrobial applications in the food industry (Perinelli et al. 2018; Silberbauerand Schmid 2017).

Chitosan is a naturally occurring biopolymer that exhibits excellent antimicrobial activity against a wide range of pathogens, such as yeast, filamentous fungus, and Gram-positive and Gram-negative bacteria, such *as Pseudomonas aeruginosa, Escherichia coli, Salmonella enteritidis*, and *Listeria monocytogenes, Lactobacillus fructivorans*, and *Staphylococcus aureus* (Zivanovic et al. 2005; Hosseinnejad and Jafari 2016; Pan et al. 2015). There are many potential mechanisms for how antimicrobial effects work, but the most promising ones are as follows: (i) the interaction of chitosan's positively charged amine groups with negatively charged microbial cell membranes changes the properties of the membrane barrier, which leads to leakage of intracellular contents and ultimately cell death (Hosseinnejad and Jafari 2016). (ii) The second hypothesized mechanism is based on chitosan's chelating abilities. Chitosan preferentially binds to metals, preventing the active centers of a variety of metabolic enzymes in microbial cells and inhibiting microbial development. (iii) Chitosan's molecular weight has an impact on its antibacterial properties (Qin et al. 2006).Low-molecular-weight chitosan, on the other hand, can enter the cytosol and bind with cell DNA, causing disruptions in mRNA and protein synthesis as well as cell death. High-molecular-weight chitosan can form an impermeable polymeric layer on the surface of a microbial cell, altering permeability and ultimately blocking nutrient access (Hosseinnejad and Jafari 2016). The most widely accepted theory for how chitosan obtains its polycationic feature in acidic environments ($pH > 6.3$) is the interaction between positively charged amino groups of the chitosan polymer and negatively charged components on microbial cell membranes, which results in cell death. In addition to other physicochemical characteristics, chitosan's antimicrobial activity is greatly influenced by its degree of deacetylation, molecular weight, concentration in solution, acid concentration and type, other compounds in the system, type of microorganisms, pH of the medium, and environmental factors (Kumar et al. 2020).

5.5.15 Moisture Sorption Properties

The mathematical expression of the moisture sorption isotherm describes the water activity and moisture content of the food. The graph of an isotherm can be obtained in two ways: adsorption isotherm and desorption isotherm. To explain the sorption isotherm of edible food packaging films, various models have been developed. To analyze the sorption isotherms, a variety of equation-based mathematical models have been created. In their study, Chirife and Iglesias (1978) discussed 23 typical models and went through a few isotherm equations. Several models are created using theoretical background. Other ideas include a background to explain the sorption mechanism theoretical models that are simplified or only empirical. The sorption isotherm is critical for determining packaging shelf life, food quality, and stability while being transported and stored. Several popular types of edible food film packaging below are described: Brunauer–Emmett–Teller (BET) model, Guggenheim–Anderson–de Boer (GAB) model, Halsey model, Henderson model,

Oswin model, Peleg model, Langmuir model, and Freundlich model. Halsey, Oswin, Henderson, Peleg, BET, GAB are among the well-known models employed to assess the moisture absorption of edible films. GAB is one of these models that is frequently employed since the parameters have a physical meaning. The stability and shelf life of the film used for food packaging are greatly influenced by its capacity to absorb moisture. Modified models are not frequently utilized to calculate moisture sorption studies. In moisture sorption isotherm research, the concentration of polymer should be taken into account in addition to temperature (Loredo et al. 2017). Chitosan films are made of natural materials; they are extremely sensitive to environmental factors including temperature and relative humidity. It is essential to research how films absorb water in order to acquire and create chitosan films with high mechanical and barrier qualities. The mechanical characteristics of a film are influenced by its moisture content since moisture may move through it readily, affecting its physicochemical characteristics and ability to bind moisture. The moisture sorption isotherm describes the relationship between the equilibrium moisture content and the relative humidity of a food product at a given temperature. Knowledge regarding the sorption isotherm is crucial for designing and optimizing drying equipment, designing packaging materials, and predicting stability, quality, and shelf life. According to Muzaffar and Kumar (2016), adsorption isotherms may be described in terms of hysteresis graphs that represent either adsorption or desorption processes. Loredo et al. (2017) explored a number of equations and models to determine how temperature affects adsorption isotherms. He concluded that the experimental results of the moisture adsorption isotherms of biodegradable chitosan films were better suited by the GAB and Oswin models (Souza et al. 2015).

According to research done by Loredo et al. (2017), temperature, chemical composition, and relative humidity levels in the storage space all affect how much moisture the biodegradable films can absorb. The modified BET, GAB, Halsey, Henderson, and Oswin models were investigated for characterizing the behavior of moisture adsorption isotherms of chitosan films at 15 °C, 20 °C, 25 °C, and 30 °C. The experimental sorption data of chitosan films were best fit by GAB, Oswin, and Halsey.

5.5.16 Release Kinetics of Active Packaging

The antimicrobial active agent is being incorporated and immobilized into the polymeric surface. The surface of the packaging material would have need a larger molecular structure to maintain antimicrobial activity. The most propitious system in the polymeric material is its capability to release the active agent when it comes in close proximity with the surface of food. The antimicrobial components incorporated into the films have several limitations. One of the limitations is that most of the time the active compound is not able to specifically target the spoilage reactions on the surface of food. Focus is mostly placed on controlled release systems that might demonstrate the preservation effect towards certain foods in order to achieve the necessary shelf life (Siepmann et al. 2002). Antimicrobial agent release is dependent on its concentration and its diffusion. It is predictable that there should be slow release of antimicrobial agent which will be sufficient enough to inhibit microbial growth. Chandrasekar et al. (2016) prepared the films from the complex of alginate and chitosan and investigated their aqueous release. These biopolymer complexes showed very low degrees of swelling in water (below 10%), indicative of sustained antimicrobial (nisin) release characteristics. Riyandari et al. (2018) studied the eugenol's release from the chitosan and alginate complex and a model of its kinetics. Various factors that effected its release were also been investigated that includes the concentration of eugenol and the composition of chitosan-alginate. Shurshina et al. (2015) studied the kinetics of sustained release of medicinal substances from chitosan-based films. Behera and Mishra (2019) characterized the chitosan films and examined the controlled release of doxorubicin—an anticancer drug.

5.5.17 *Cost-Effectiveness*

The commercial availability of biofilms has been constrained by the fact that they cost around 50 times more than plastic films made of petroleum. Therefore, research is needed to lower the cost of ingredients and the cost of production of edible films. In order to determine sensory qualities, bio-compatibility, as well as possible toxicity of the films, further research is essential (Badii et al. 2014).

5.5.18 *Safety and Regulations*

There is a huge spike in the requirement for safe and healthy food with minimum use of synthetic ingredients. As a result of the negative environmental effects of plastic-based polymers, significant efforts are being undertaken to evaluate alternatives to synthetic plastic packaging. Chitosan has developed into a desirable biopolymer for preservation of food and food packaging applications applied by dipping, coating, spraying, or wrapping by films because of its film-forming characteristic and antibacterial activity (Kumar et al. 2020).

Public and academic experts have expressed a variety of worries about current issues relating to food and environmental safety. Chitosan has various uses in industries including agriculture, medicine, textile, food, and the environment because to its exceptional qualities like antibacterial activity, biocompatibility, non-toxicity, biodegradability, and chelating capacity, among others. Capacity of chitosan to form films is a result of its solubility in acetic acid and hydrochloric acid. Researchers have created chitosan-based films using a variety of techniques, including coating, casting, layer-by-layer assembly, etc. They have also altered the films: barrier, antibacterial, mechanical, antioxidant, optical, thermal stability, and sensing/indicating capabilities. They have also combined additional functional materials with chitosan to create composite films, expanding the combined benefits of the resulting films (Wang et al. 2018). Since its use in the food industry depends on its safety, chitosan particles and solutions have also been the focus of extensive research. Chitosan has been approved for use as a food additive in Japan and Korea since 1983 and 1995, respectively. Its biological safety has also been demonstrated by feeding testing on domesticated animals. It is also crucial to note that, according to Allwin et al. (2015), chitosan was declared and generally recognized as safe (GRAS) by the US FDA in 2005. Therefore, it is believed that chitosan is a raw material that may be utilized on fruits and vegetables and is safe, nontoxic, and biocompatible (Uriarte-Montoya et al. 2010; Xing et al. 2010; Youn et al. 2009).

5.5.19 Methods to Improve Properties

Due to their exceptional performance, chitosan films have been investigated for years, particularly in the food business. Films made of chitosan are used as packaging to keep food fresh longer. By incorporating a small amount of additives to the chitosan solution, the characteristics of the chitosan films can be improved. Plasticizers such as glycerol, emulsifiers, or surfactants like Tween-80 and Span 20 might also be added. However, when compared to synthetic films, biopolymer-produced films have poor mechanical and barrier qualities. Plasticizers (glycerol, polyethylene glycol, etc.) are being added to increase flexibility/plasticity. Growing interest has been shown in incorporating nanofiller into biopolymer films for use in food packaging (Sejidov et al. 2005). Because of their antimicrobial properties and unique properties, such as the capacity to enhance the mechanical, barrier, and thermal characteristics of the film, nano fillers like zinc oxide and carbon clay are being added to food packaging films (Janjarasskul and Krochta 2010). Since microbial deterioration has become a significant issue influencing food quality, the incorporation of antimicrobial chemicals into packing materials has gained more traction. This access can reduce the food additives that are usually incorporated into food (Han and Gennadios 2005). Packaging films are evaluated on the basis of their mechanical, thermal properties, oxygen, and water vapor transmission rate. In addition to these characteristics, knowledge of the film's moisture sorption isotherm is crucial. For package design and optimization, quality, stability, and shelf-life forecasts, as well as calculating moisture changes that occur during storage, it is essential to understand and be familiar with the sorption isotherm (Ariahu et al. 2006).

5.6 Applications

Chitosan is a natural source of biopolymer that has been certified and widely accepted for its antimicrobial activity and wide range of applications in the food packaging industry. Chitosan films have improved the quality of food and prolonged its shelf stability (Chen et al. 2022). Sathivel et al. (2007) observed that chitosan-based film when used as a glazing for fillets of skinless salmon delayed the lipid oxidation as it performed as a tremendous barrier to oxygen, while possessing poor water vapor barrier properties. Chitosan-coated fillets showed eminent thaw yield compared to the control one. Green tea extract (GTE) combined with chitosan film shown good active packaging capabilities, according to research by Siripatrawan and Harte (2010). The polyphenolic content, antioxidant characteristics, mechanical, and water vapor barrier of the films were improved by GTE. In order to prevent spoiling, Estaca et al. (2010) added numerous essential oils to the chitosan films. These oils inhibited several bacterial strains that cause spoilage, and it was discovered that clove essential oil had the strongest inhibiting impact. Essential fatty acids, vitamins, vital amino acids, and minerals are abundant in meat products. The two primary obstacles to the meat products' preservation are microbial contamination and lipid oxidation. Biofilms that limit the development of harmful microbes and minimize lipid oxidation can be used to keep meat products safe. Bioactive compounds have also been added to the films in addition to chitosan to slow the pace of microbial growth (Lorenzo and Pateiro 2013; Lorenzo et al. 2014). Soncu et al. (2018) investigated that chitosan films' capacity to prevent mold growth on dry fermented sausages has been increased by the incorporation of 0.2% (w/w) thyme or rosemary oil without degrading the quality characteristics. Lipid oxidation of frozen pompano can be inhibited by incorporating the citric acid into the chitosan coating (Qiu et al. 2016). The traditional method for preserving the meat products is through temperature storage, radiation, modified atmosphere packaging, and high static pressure which are simple and economical viable, although they have certain limitations too. Irradiation has some safety concern for consumer. Expensive processing equipment and higher production costs are required for modified atmosphere packaging and high static pressure packaging (Mahmoud et al. 2016).

The microbiological growth is not completely inhibited by the low temperature conditions as stated by Ge et al. (2016). Arfat et al.(2014) prepared chitosan-based film from the nanoparticles of zinc oxide with clove essential oil as an antimicrobial agent. This film demonstrated the greatest antibacterial activity against *Listeria monocytogenes* and *Salmonella typhimurium* inoculated in shrimp during chilled storage (Ojagh et al. 2010). Chitosan-based packaging films are used to store fruits and vegetables as the films have excellent antibacterial activities, water vapor and gas barrier properties, adhesion, chemical and physical stability, high stability under high relative humidity conditions, and mechanical properties. As an eco-friendly packaging material, chitosan-based films offer significant benefits that help increase the shelf life of fruits and vegetables. Fresh vegetables and fruits are abundant source of nutrients, vitamins, minerals, and amino acids. During storage, the fruits and vegetables undergo metabolic reactions due to which they have short shelf life (Xing et al. 2011). The loss of nutrients, transpiration of water, and infections by spoilage microorganisms impart negative effect on their quality. Krech (2005) investigated that the chitosan films have delayed enzymatic browning by inhibiting the polyphenol oxidase and inhibited the microorganisms. On several fruits and vegetables, the natural wax layer has been replaced by edible coatings made from chitosan (Corbo et al. 2015).

The major concern for the shelf life of breads is its staling properties and microbial spoilage. The degradation of texture and flavor quality is generally due to staling (Gray and Bemiller 2003). A number of researchers have documented the application of the chitosan for prolonging the shelf life of bread. Films stored in chitosan-based film have reduced weight and retarded hardness and retrogradation of bread (Park et al. 2002).

5.7 Limitations and Future Trends

The biofilms are produced by solution casting method as it is easily been implemented on laboratory scale for academic studies. In order to commercialize the biofilms, more research and systematic examination still needs to be done to optimize their vital specifications that includes concentration of plasticizer, moisture content, drying temperature, etc. Petroleum-based plastic films cannot be competed with the biopolymer-based films as more research is needed to be done to enhance their mechanical strength, appearance, flexibility, and moisture/gas barrier properties. It will be necessary to use appropriate mathematical models to represent and identify the release profiles of bioactive components from biofilms. The major limitation that limits the widespread utilization of the biofilms is their cost, which is higher compared to the petroleum-based plastics. To lower ingredient and production costs for making edible films, research is necessary (Chen et al. 2022). As a result of the inherent properties, chitosan being mechanically as well as physically weak is sensitive to water vapor and oxygen. Chitosan is a biopolymer that can hold a variety of additives, including antifungal agents, antioxidants, coloring agents, antimicrobials, and nutrients. To modify the physical and mechanical properties of chitosan films, a wide range of additives are being used (Marroquin et al. 2013). For future improvement, antimicrobial packaging has proven as one of the best alternative for packaging, preserving, and ameliorating the quality of food during storage. Chitosan, being a versatile and promising polymer, has biodegradable nature as well as antimicrobial properties and is preferable for packaging of food. Many researchers have found that chitosan can be obtained from various sources. To obtain deficient expensive and decisive analytical methods and the standard quality of chitosan films, the researchers can work on improving its functional properties by blending it with other, more water-resistant film-forming materials. Preparing hybrid films made from nanoparticle composites is another chance access. By incorporating the nanoclay, it was seen that water barrier and mechanical properties of chitosan hyped significantly, without altering their biodegradability and antimicrobial properties (No et al. 2007). The functional application of ZnO is its barrier, physical, mechanical, and antimicrobial properties, and possessing high chemical stability and non-toxicity has been used as an additive in biofilm. By incorporating $CaCl_2$, the properties of the film could be improved as it acts as a plasticizer for polysaccharide-based films. PEG acts a plasticizer reduces the intermolecular interactions and alleviates the molecular motion (Orue et al. 2018).

5.8 Conclusion

Chitosan is a polymer derived from the deacetylation of chitin that has antibacterial activity against a wide range of food-borne fungi, bacteria, and yeast. Since 1995 and 1983, chitosan has been approved as a food additive in Korea and Japan. Additionally, it was authorized in the USA and granted GRAS status by the US FDA. The chitosan biofilms as well as chitosan coatings have proven very effective in maintaining sensory and nutritional quality, the microbiological quality. Chitosan has a special property that allows it to develop the mixing property, i.e., combination of components, viz., polysaccharides, plasticizers or lipids, and proteins. Chitosan films, being biodegradable, biocompatible, and bioactive, have been colossally used and researched since decades. Despite its performance, its mechanical, water barrier, and thermal properties need to be ameliorated.

References

Abdelhedi O, Nasri R, Mora L, Toldrá F, Nasri M, Jridi M (2017) Collagenous proteins from black-barred halfbeak skin as a source of gelatin and bioactive peptides. Food Hydrocoll 70:123–133

Abdorreza MN, Cheng LH, Karim AA (2011) Effects of plasticizers on thermal properties and heat sealability of sago starch films. Food Hydrocoll 25(1):56–60

Ali A, Maqbool M, Ramachandran S, Alderson PG (2010) Gum Arabic as a novel edible coating for enhancing shelf-life and improving postharvest quality of tomato (*Solanum lycopersicum* L.) fruit. Postharvest Biol Technol 58:42–47. https://doi.org/10.1016/j.postharvbio.2010.05.005

Alix S, Mahieu A, Terrie C, Soulestin J, Gerault E, Feuilloley MGJ, Gattin R, Edon V, Ait-Younes T, Leblanc N (2013) Active pseudo-multilayered films from polycaprolactone and starch based matrix for food-packaging applications. Eur Polym J 49:1234–1242

Allwin SIJ, Jeyasanta KI, Patterson J (2015) Extraction of chitosan from white shrimp (*Litopenaeus vannamei*) processing waste and examination of its bioactive potentials. Adv Biol Res 9(6): 389–396

Alturki AM (2018) Effect of preparation method on the particles size, dielectric constant and antibacterial properties of ZnO nanoparticles and thin film of ZnO/chitosan. Orient J Chem 34(1):548

Arfat YA, Benjakul S, Prodpran T, Sumpavapol P, Songtipya P (2014) Properties and antimicrobial activity of fish protein isolate/fish skin gelatin film containing basil leaf essential oil and zinc oxide nanoparticles. Food Hydrocoll 41:265–273

Ariahu C, Kaze SA, Achem CD (2006) Moisture sorption characteristics of tropical fresh water crayfish. J Food Eng 75:355–363

Azerdo H, Attoso LM, Bustillos AR, Munford M, Wood D, McHugh TH (2010) Nanocellulose reinforced chitosan composite films as affected by nanofiller loading and plasticizer content. J Food Sci Technol 75:1–7

Badii F, MacNaughtan W, Mitchell JR, Farhat IA (2014) The effect of drying temperature on physical properties of thin gelatin films. Dry Technol 32(1):30–38. https://doi.org/10.1080/07373937.2013.808206

Barboza LGA, Vethaak AD, Lavorante BR, Lundebye AK, Guilhermino L (2018) Marine microplastic debris: an emerging issue for food security, food safety and human health. Mar Pollut Bull 133:336–348

Behera KD, Mishra K (2019) Formulation and evaluation of chitosan-polypyrrole nanocomposites for controlled release of anticancer drug doxorubicin. Int J Appl Pharm 11(2):247–253

Benbettaieb N, Debeaufort F, Karbowiak T (2019) Bioactive edible films for food applications: mechanisms of antimicrobial and antioxidant activity. Crit Rev Food Sci Nutr 59(21): 3431–3455. https://doi.org/10.1080/10408398.2018.1494132

Bhuvaneshwari S, Sruthi D, Sivasubramanian V, Niranjana K, Sugunabai J (2011) Development and characterization of chitosan films. Int J Adv Eng Technol 1(2):292–299

Bhuvaneshwari S, Sruthi D, Sivasubramanian V, Kalyani N, Sugunabai J (2014) Development and characterization of chitosan film. Int J Adv Eng Technol 1(2):292–299

Bower GC, Wright MC, Falcke H, Backer DC (2003) Interferometric detection of linear polarization from Sagittarius A* at 230 GHz. Astrophys. J Suppl Ser 588(1):331

Brasselet C, Pierre G, Dubessay P, Dols-Lafargue M, Coulon J, Maupeu J, Vallet-Courbin A, de Baynast H, Doco T, Michaud P, Delattre C (2019) Modification of chitosan for the generation of functional derivatives. Appl Sci 9(7):1–34. https://doi.org/10.3390/app9071321

Cai L, Shi H, Cao A, Jia J (2019) Characterization of gelatin/chitosan polymer films integrated with docosahexaenoic acids fabricated by different methods. Sci Adv 9(1):1–11

Cao N, Yang X, Fu Y (2009) Effects of various plasticizers on mechanical and water vapor barrier properties of gelatin films. Food Hydrocoll 23(3):729–735

Caro N, Medina E, Díaz-Dosque M, López L, Abugoch L, Tapia C (2015) Novel active packaging based on films of chitosan and chitosan/quinoa protein printed with chitosan tripolyphosphate-thymol nanoparticles via thermal ink-jet printing. Food Hydrocoll 52:520–532

Cazón P, Vázquez M, Velazquez G (2019) Environmentally friendly films combining bacterial cellulose, chitosan, and polyvinyl alcohol: effect of water activity on barrier, mechanical, and optical properties. Biomacromolecules 21(2):753–760

ChandrasekarV, John NCoupland, and sAnantheswaran RC (2016) Release kinetics of nisin from chitosan–alginate complex films. J Food Sci81(10): E2503-E2510.doi: https://doi.org/10.1111/1750-3841.13443

Chen LM, Liu J, Chen JC, Shi S, Tan CP, Zheng KC, Ji LN (2008) Experimental and theoretical studies on the DNA-binding and spectral properties of water-soluble complex . J Mol Struct 881 (1–3):156–166

Chen W, Ma S, Wang Q, McClements DJ, Liu X, Ngai T, Liu F (2022) Fortification of edible films with bioactive agents: a review of their formation, properties, and application in food preservation. Crit Rev Food Sci Nutr 62(18):5029–5055. https://doi.org/10.1080/10408398.2021.1881435

Chenite A, Buschmann M, Wang D, Chaput C, Kandani N (2001) Rheological characterisation of thermogelling chitosan/glycerol-phosphate solutions. Carbohydr Polym 46(1):39–47

Cheung MK, Wan KPY, Yu PH (2002) Miscibility and morphology of chiral semicrystalline poly-(R)-(3-hydroxybutyrate)/chitosan and poly-(R)-(3-hydroxybutyrate-co-hydroxyvalerate)/chitosan blends studied with DSC, 1H T1 and T1F CRAMPS. J Appl Polym Sci 86:1253–1258

Chirife J, Iglesias AH (1978) Equations for fitting water sorption isotherms of foods: part 1: a review. Food Sci Technol 13:159–174

Corbo MRD, Campaniello D, Speranza B, Bevilacqua A, Sinigaglia M (2015) Non-conventional tools to preserve and prolong the quality of minimally-processed fruits and vegetables. Coatings 5(4):931–961. https://doi.org/10.3390/coatings5040931

Das M, Chowdhury T (2016) Heat sealing property of starch based self-supporting edible films. Food Packag Shelf Life 9:64–68

Dong Y, Ruan Y, Wang H, Zhao Y, Bi D (2004) Studies on glass transition temperature of chitosan with four techniques. J Appl Polym Sci 93:1553–1558
Dutta PK, Tripathi S, Mehrotra GK, Dutta J (2009) Perspectives for chitosan based antimicrobial films in food applications. Food Chem 114(4):1173–1182
Elsabee MZ, Abdou ES (2013) Chitosan based edible films and coatings: a review. Mater Sci Eng C 33(4):1819–1841
Estaca GJ, Lopez de Lacey A, Lopez-Caballero E, Gomez-Guillen C, Montero P (2010) Biodegradable gelatin-chitosan films incorporated with essential oils as antimicrobial agents for fish preservation. Food Microbiol 27:889–896
Farhan A, Hani NM (2017) Characterization of edible packaging films based on semi-refined kappa- carrageenan plasticized with glycerol and sorbitol. Food Hydrocoll 64:48–58
Ge L, Xu YW, Xia Q, Jiang X (2016) Differential role of endogenous cathepsin and microorganism in texture softening of ice-stored grass carp (*Ctenopharyngodon idella*) fillets. J Sci Food Agric 96(9):3233–3239. https://doi.org/10.1002/jsfa.7506
Goenka S, Sant V, Sant S (2014) Graphene-based nanomaterials for drug delivery and tissue engineering. J Control Release 173:75–88
Gray JA, Bemiller JN (2003) Bread staling: molecular basis and control. Compr Rev Food Sci Food Saf 2:1–21
Hamed I, Özogul F, Regenstein JM (2016) Industrial applications of crustacean by-products (chitin, chitosan, and chitooligosaccharides): a review. Trends Food Sci Technol 48:40–50
Han JH, Gennadios A (2005) Edible films and coatings: a review. Innov Food Packag 15:239–262
Hassan ML, Fadel SM, El-Wakil NA, Oksman K (2012) Chitosan/rice straw nanofibers nanocomposites: preparation, mechanical, and dynamic thermomechanical properties. J Appl Polym Sci 125:216–222
Hosseinnejad M, Jafari SM (2016) Evaluation of different factors affecting antimicrobial properties of chitosan. Int J Biol Macromol 85:467–475
Ilyina AV, Tatarinova NY, Varlamov VP (1999) The preparation of low-molecular-weight chitosan using chitinolytic complex from *Streptomyces kurssanovii*. Process Biochem 34(9):875–878
Janjarasskul T, Krochta JM (2010) Edible packaging materials. Annu Rev Food Sci Technol 1:415–448
Jayakumar R, Reis RL, Mano JF (2006) Chemistry and applications of phosphorylated chitin and chitosan. e-Polymers 6(1):1–16. https://doi.org/10.1515/epoly.2006.6.1.447
Jayakumar R, Nwe N, Tokura S, Tamura H (2007) Sulfated chitin and chitosan as novel biomaterials. Int J Biol Macromol 40(3):175–181
Karlsen J, Skaugrud O (1991) Excipient properties of chitosan. Manuf Chem 62(6):18–19
Kerch G (2015) Chitosan films and coatings prevent losses of fresh fruit nutritional quality: a review. Trends Food Sci Technol 46:159–166
Kerch G, Korkhov V (2011) Effect of storage time and temperature on structure, mechanical, and barrier properties of chitosan-based films. Eur Food Res Technol 232:17–22
Khalil KD, Bashal AH, Khalafalla M, Zaki AA (2020) Synthesis, structural, dielectric and optical properties of chitosan-MgO nanocomposite. J Taibah Univ Sci 14(1):975–983
Krajewska B (2004) Application of chitin-and chitosan-based materials for enzyme immobilizations: a review. Enzym Microb Technol 35(2–3):126–139
Krech S (2005) Reflections on conservation, sustainability and environmentalism in indigenous north america 107(1):78–86. https://doi.org/10.1525/aa.2005.107.1.078
Kumar S, Mukherjee A, Dutta J (2020) Chitosan based nanocomposite films and coatings: emerging antimicrobial food packaging alternatives. Trends Food Sci Technol 97:196–209
Leceta I, Guerrero P, De La Caba K (2013) Functional properties of chitosan-based films. Carbohydr Polym 93(1):339–346
Lee JH, Marroquin J, Rhee KY, Park SJ, Hui D (2013) Cryomilling application of graphene to improve material properties of graphene/chitosan nanocomposites. Compos Part B 45(1): 682–687

Li JL, Kudin KN, McAllister MJ, Prud'homme RK, Aksay IA, Car R (2006) Oxygen-driven unzipping of graphitic materials. Phys Rev Lett 96(17):176101

Loredo RYA, Hernandez AIR, Velazquez G (2017) Modelling the effect of temperature of the water sorption isotherms of chitosan films. Food Sci Technol 37(1):112–118

Lorenzo JM, Pateiro M (2013) Influence of type of muscles on nutritional value of foal meat. Meat Sci 93(3):630–638. https://doi.org/10.1016/j.meatsci.2012.11.007

Lorenzo JM, Sarries MV, Tateo A, Polidori P, Franco D, Lanza M (2014) Carcass characteristics, meat quality and nutritional value of horsemeat: a review. Meat Sci 96(4):1478–1488. https://doi.org/10.1016/j.meatsci.2013.12.006

Mahmoud BSM, Nannapaneni R, Chang S, Wu Y, Coker R (2016) Improving the safety and quality of raw tuna fillets by X-ray irradiation. Food Control 60:569–574. https://doi.org/10.1016/j.foodcont.2015.08.039

Mansour AA, El-Sabagh S, Yehia AA (1994) Dielectric investigation of SBR-NBR and CR-NBR blends. J Elastomers Plast 26(4):367–378

Marroquin JB, Rhee KY, Park SJ (2013) Chitosan nanocomposite films: enhanced electrical conductivity, thermal stability, and mechanical properties. Carbohydr Polym 92:1783–1791

Mathew S, Abraham TE (2008) Characterisation of ferulic acid incorporated starch–chitosan blend films. Food Hydrocoll 22(5):826–835

Mathew S, Brahmakumar M, Abraham TE (2006) Microstructural imaging and characterization of the mechanical, chemical, thermal, and swelling properties of starch–chitosan blend films. Biopolymers 82(2):176–187

Mittal G, Rhee KY, Park SJ, Hui D (2017) Generation of the pores on graphene surface and their reinforcement effects on the thermal and mechanical properties of chitosan-based composites. Compos Part B 114:348–355

Muzaffar K, Kumar P (2016) Moisture sorption isotherms and storage study of spray dried tamarind pul powder. Powder Technol 291:322–327. https://doi.org/10.1016/j.powtec.2015.12.046

Nasution TI, Asrosa R, Machrina Y, Nainggolan I, Balyan M, Rumansyah R (2017) Improved electrical properties of chitosan based acetone sensor by adding carboxymethylcellulose (CMC). IOP Conf Ser Mater Sci Eng 180(1):012018

No HK, Meyers SP, Prinyawiwatkul W, Xu Z (2007) Applications of chitosan for improvement of quality and shelf life of foods: a review. J Food Sci 72(5):R87–R100

Ojagh M, Rezaei M, Razavi H, Hosseini H (2010) Effect of chitosan coatings enriched with cinnamon oil on the quality of refrigerated rainbow trout. Food Chem 120:193–198

Oksman K, Aitomäki Y, Mathew AP, Siqueira G, Zhou Q, Butylina S, Hooshmand S (2016) Review of the recent developments in cellulose nanocomposite processing. Compos A: Appl Sci Manuf 83:2–18

Orue A, Eceiza A, Arbelaiz A (2018) The effect of sisal fiber surface treatments, plasticizer addition and annealing process on the crystallization and the thermomechanical properties of poly(lactic acid) composites. Ind Crop Prod 118:321–333. https://doi.org/10.1016/j.indcrop.2018.03.068

Pan I, Mate JI, Gardrat C, Coma V (2015) Effect of chitosan molecular weight on the antimicrobial activity and release rate of carvacrol-enriched films. Food hydrocoll 51:60–68

Park IK, Lee YK, Kim MJ, Kim SD (2002) Effect of surface treatment with chitooligosaccharide on shelf-life of baguette. J Chitin Chitosan 7(4):214–218

Perinelli DR, Fagioli L, Campana R, Lam JK, Baffone W, Palmieri GF, Bonacucina G (2018) Chitosan-based nanosystems and their exploited antimicrobial activity. Eur J Pharm Sci 117:8–20

Prabaharan M (2012) Sources of Chitosan. In: Prabaharan M (ed) Chitosan and its derivatives as promising drug delivery carriers. ASME Press, New York

Prateepchanachai S, Thakhiew W, Devahastin S, Soponronnarit S (2019) Improvement of mechanical and heat-sealing properties of edible chitosan films via addition of gelatin and CO2 treatment of film-forming solutions. Int J Biol Macromol 131:589–600

Qin C, Li H, Xia Q, Liu Y, Zhu J, Du Y (2006) Water-solubility of chitosan and its antimicrobial activity. Carbohydr Polym 63(3):367–374

Qiu X, Chen G, Liu H, Lin (2016) Inhibition of lipid oxidation in frozen farmed ovate pompano (*Trachinotus ovatus* L.) fillets stored at −18 _C by chitosan coating incorporated with citric acid or licorice extract. J Sci Food Agric 96(10):3374–3379. https://doi.org/10.1002/jsfa.7517

Ray D, Roy P, Sengupta S, Sengupta S, Mohanty AK, Misra M (2009) A study of physicomechanical and morphological properties of starch/poly (vinyl alcohol) based films. J Polym Environ 17(1):56–63

Rivera J, Crandall PG, Bryan CAO, Ricke SC (2015) Essential oils as antimicrobials in food systems—a review. Food Control 54:111–119. https://doi.org/10.1016/j.foodcont.2014.12.040

Riyandari RA, Suherman S, Siswanta D (2018) The physico-mechanical properties and release kinetics of eugenol in chitosan-alginate polyelectrolyte complex films as active food packaging. Indones J Chem 18(1):82–91

Sanuja S, Agalya A, Umapathy MJ (2014) Studies on magnesium oxide-reinforced chitosan bionanocomposite incorporated with clove oil for active food packaging application. Int J Polym Mater 63:733–740

Sathivel S, Liu Q, Huang J, Prinyawiwatkul W (2007) The influence of chitosan glazing on the quality of skinless pink salmon (*Oncorhynchus gorbuscha*) fillets during frozen storage. J Food Eng 83:366–373

Sejidov FT, Mansoori Y, Goodarzi N (2005) Esterification reaction using solid heterogeneous acid catalysts under solvent-less condition. J Mol Catal A Chem 240(12):186–190

Shanta KL, Harding DRK (2002) Synthesis and characterization of chemically modified chitosan microspheres. Carbohydr Polym 48:247–253

Shurshina A, Kulish E, Lazdin R (2015) Study of kinetics of medicinal substances release from chitosan films. Chem Chem 9(3):319–323

Siepmann J, Streubel A, Peppas NA (2002) Understanding and predicting drug delivery from hydrophilic matrix tablets using the sequential layer model. Pharm Res 19:306–314

Silberbauer A, Schmid M (2017) Packaging concepts for ready-to-eat food: recent progress. J Packag Technol Res 1(3):113–126

Singh TP, Chatli MK, Sahoo J (2014) Development of chitosan based edible films: process optimization using response surface methodology. J Food Sci Technol 52(5):2530–2543

Siripatrawan U, Harte R (2010) Physical properties and antioxidant activity of an active film from chitosan incorporated with green tea extract. Food Hydrocoll 24:770–775

Soncu ED, Arslan B, Ertürk D, Küçükkaya S, Ozdemir N, Soyer A (2018) Microbiological, physicochemical and sensory characteristics of Turkish fermented sausages (sucuk) coated with chitosan-essential oils. Lwt 97:198–204

Souza SJF, Alves AI, Vieira ENR, Vieira JAG, Ramos AM, Romero J (2015) Study of thermodynamic water properties and moisture sorption hysteresis of mango skin. Food Sci Technol 35(1): 157–166

Srinivasa PC, Ramesh MN, Kumar KR, Tharanathan RN (2004) Properties of chitosan films prepared under different drying conditions. J Food Eng 63:79–85

Sun L, Sun J, Chen L, Niu P, Yang X, Guo Y (2017) Preparation and characterization of chitosan film incorporated with thinned young apple polyphenols as an active packaging material. Carbohydr Polym 903(163):81–91

Sun Y, Liu Z, Zhang L, Wang X, Li L (2020) Effects of plasticizer type and concentration on rheological, physico-mechanical and structural properties of chitosan/zein film. Int J Biol Macromol 143:334–340

Sunilkumar M, Gafoor AA, Anas A, Haseena AP, Sujith A (2014) Dielectric properties: a gateway to antibacterial assay—a case study of low-density polyethylene/chitosan composite films. Polym J 46(7):422–429

Suyatma NE, Tighzert L, Copinet A, Coma V (2005) Effects of hydrophilic plasticizers on mechanical, thermal, and surface properties of chitosan films. J Agric Food Chem 53(10): 3950–3957

Suzuki S, Shimahashi K, Takahara J, Sunako M, Takaha T, Ogawa K, Kitamura S (2005) Effect of addition of water-soluble chitin on amylose film. Biomacromolecules 6(6):3238–3242

Tang R, Du Y, Fan L (2003) Dialdehyde starch-crosslinked chitosan films and their antimicrobial effects. J Polym Sci 41:993–997

Tian B, Liu Y (2020) Chitosan-based biomaterials: from discovery to food application. Polym Adv Technol 31(11):2408–2421

Tongnuanchan P, Benjakul S, Prodpran T, Pisuchpen S, Osako K (2016) Mechanical, thermal and heat sealing properties of fish skin gelatin film containing palm oil and basil essential oil with different surfactants. Food Hydrocoll 56:93–107

Tripathi S, Mehrotra GK, Dutta PK (2008) Chitosan based antimicrobial films for food packaging applications. e-Polymers 8(1):1–7

Tual C, Espuche E, Escoubes M, Domard A (2000) Transport properties of chitosan membranes: influence of crosslinking. J Polym Sci 38:1521–1529

Uriarte-Montoya MH, Arias-Moscoso JL, Plascencia-Jatomea M, Santacruz-Ortega H, Rouzaud-Sández O, Cardenas-Lopez JL, Ezquerra-Brauer JM (2010) Jumbo squid (*Dosidicus gigas*) mantle collagen: extraction, characterization, and potential application in the preparation of chitosan–collagen biofilms. Bioresour Technol 101(11):4212–4219

Wang H, Qian J, Ding F (2018) Emerging chitosan-based films for food packaging applications. J Agric Food Chem 66(2):395–413

Xing Y, Li X, Xu Q, Yun J, Lu Y (2010) Antifungal activities of cinnamon oil against *Rhizopus nigricans*, *Aspergillus flavus* and *Penicillium expansum* in vitro and in vivo fruit test. Int J Food Sci Technol 45(9):1837–1842

Xing Y, Xu Q, Li X, Chen C, Ma L, Li S, Che Z, Lin H (2011) Chitosan-based coating with antimicrobial agents: preparation, property, mechanism, and application effectiveness on fruits and vegetables. Int J Polym Sci 2016:4851730

Yenier Z, Seki Y, Şen İ, Sever K, Mermer Ö, Sarikanat M (2016) Manufacturing and mechanical, thermal and electrical characterization of graphene loaded chitosan composites. Compos Part B 98:281–287

Youn DK, No HK, Prinyawiwatkul W (2009) Physicochemical and functional properties of chitosans prepared from shells of crabs harvested in three different years. Carbohydr Polym 78(1):41–45

Youssef AM, Sayed SM, Sayed HS, Salama HH, Dufresne A (2016) Enhancement of Egyptian soft white cheese shelf life using a novel chitosan/carboxymethyl cellulose/zinc oxide bionanocomposite film. Carbohydr Polym 151:9–19

Zhang JP, Wang AQ (2009) Synergistic effects of Na+−montmorillonite and multi-walled carbon nanotubes on mechanical properties of chitosan film. Express Polym Lett 3:302–308

Zheng H, Du YM, Yu JH, Xiao L (2000) The properties and preparation of crosslinked chitosan films. Chem J Chin Univ 21:809–812

Zivanovic S, Chi S, Draughon AF (2005) Antimicrobial activity of chitosan films enriched with essential oils. J Food Sci 70(1):M45–M51

Zotkin MA, Vikhoreva GA, Smotrina TV, Derbenev MA (2004) Thermal modification and study of the structure of chitosan films. Fibre Chem 36:16–20

Chapter 6
Alginate-Based Films

Prastuty, Gurkirat Kaur, and Swati Kapoor

6.1 Introduction and Sources

Food packaging mainly aims at protecting the food products from all the external physical, chemical, and microbial factors to prevent it from any damage, maintaining the safety and product integrity from purchase to consumption of the food. Plastic-based packaging materials have become a topic of debate and need extensive research to be replaced with some natural biobased packaging films as plastic films are nonbiodegradable and are the major causes of pollution causing many adverse health effects. Their excessive use in today's packaging industry is due to their less cost, easy availability and ease to process and reform into different shapes and sizes. Food industry is considered to be a major user of plastic-based packaging materials which helps in retaining the freshness and overall quality attributes of a food product. The adverse effects and challenges of plastic packs are expected to increase tenfold in the next upcoming 20 years affecting the soil, drinking water, human, and animal bodies. Considering these alarming outcomes in the future has increased the demand for alternates to the plastic-based packaging materials in all sectors including the food industry (Ananda et al. 2017).

Bio-based polysaccharide materials were traditionally used as a packaging ingredient for food products. They are known to be inert and therefore do not interact with the biological materials and interfere with the biochemical processes of the

Prastuty · S. Kapoor
Department of Food Science and Technology, Punjab Agricultural University, Ludhiana, Punjab, India

G. Kaur (✉)
Electron Microscopy and Nanoscience Lab, Punjab Agricultural University, Ludhiana, Punjab, India
e-mail: gurkirat@pau.edu

T. Amin et al. (eds.), *Polysaccharide Based Films for Food Packaging: Fundamentals, Properties and Applications*,
https://doi.org/10.1007/978-981-99-4898-7_6

biological tissues. With the start of twentieth century, these natural biopolymers started getting replaced with synthetic polymers to impart better film properties but found to intervene with the biological tissues. The biopolymers were again focused to mimic certain roles of natural materials incorporated in packaging material to increase their functionality and barrier properties (Huebsch and Mooney 2009). The European Bioplastic organization declared the biopolymers to be renewable, compostable, and sustainable (Galgano 2015). A large range of polysaccharides and their derivatives have been used in the area of food packaging such. They possess good barrier properties against oxygen and CO_2 as well as good mechanical properties. Their hydrophilic nature is the major drawback as it makes the polysaccharide-based packaging films to be less resistant to water vapors. Cellulose, starch, chitin, carrageen, pectin, and alginate are majorly used polysaccharides for the formulation of bio-based packaging films for food products. Alginate has been widely used in food industry as a coating material, thickener, stabilizer, and gelling agent to protect the food products against microbial and external deterioration such as in coffee, powered milk, and instant tea powders (Arroyo et al. 2020; Gundewadi et al. 2018).

Alginates are known to be a good source for film making due to their nontoxic, easily degradable, and easily compostable nature. The algal purification of brown seaweeds led to the production of sodium alginate ($C_6H_9NaO_7$), which is known to be the first by-product of the process and possess many interesting properties like the production of environment-friendly packaging films. Out of majority of hydrocolloids been in use for food packaging purposes, alginate is one of the most popular and acceptable polysaccharides as it is approved by the FDA (US Food and Drug Administration) and classified as generally recognized as safe (GRAS) substance (Tabassum and Khan 2020).

6.2 Sources

Alginate occurs in different forms of salts of calcium, magnesium, and sodium in the cell wall of mainly brown algae. Most species of brown algae are the major producers of alginates in one or the other forms. *Laminariales* and *Fucales* are the major alginate producing species in acknowledged quantities of about 30–45% (d.w) (Fig. 6.1). They are mainly found in the areas of many coastal regions of the United States, Japan, Mexico, France, Canada, the United Kingdom, China, Norway, Africa, and coastal areas of Gujarat in India. Brown algae flourished and grown in water at temperatures not more than 20 °C are known to be the best suitable sources for alginate extraction. Although brown algae can also be grown in warm waters, they are not considered to be a good source for alginate production as it is not used as food. The main commercial sources are species *of Ascophyllum*, *Durvillaea*, *Ecklonia*, *Laminaria*, *Lessonia*, *Macrocystis*, *Sargassum*, *and Turbinaria*. Of these the most important are *Laminaria*, *Macrocystis*, *and Ascophyllum*.

Laminaria is commonly found in the regions of Scotland, China, Japan, France, Ireland, Norway, and Korea. It is a popular food in many of the Asian countries,

Fig. 6.1 Different sources of brown seaweeds for extraction of alginates

whereas in Japan and Korea, it is considered to be an exclusive raw material for the production of alginate due to its higher cost. Cultivation of *Laminaria japonica* is very much successful in China (approximately up to one million tons or wet seaweeds per year), out of which about two-thirds are used as their diet source, and the rest available is used for alginate production.

Macrocystis can be found on the west coast of North America, and it has been observed that large quantity of wet seaweed can be cultivated annuallly in the United States. This specie is also grown on the east coast of Tasmania and Australia, although in an insufficient quantity to facilitate the alginate producing industries.

Ascophyllum nodosum is cultivated in the intertidal zones of Scotland and Ireland and are generally harvested by hands. It is also grown in some southern parts of Nova Scotia.

6.3 Structure of Alginates

Alginate is naturally and abundantly available indigestible polysaccharide which is mainly extracted from brown seaweed (*Ascophyllum nodosum*, *Cystoseira barbata*, *Laminaria digitata*, *Laminaria hyperborea*, *Laminaria japonica*, *and Macrocystis pyrifera*, *Padina* sp., *Phaeophyceae* sp., and several *Sargassum* sp.) cell walls and from some bacteria strains (*Azotobacter*, *Pseudomonas*). It is a linear polysaccharide

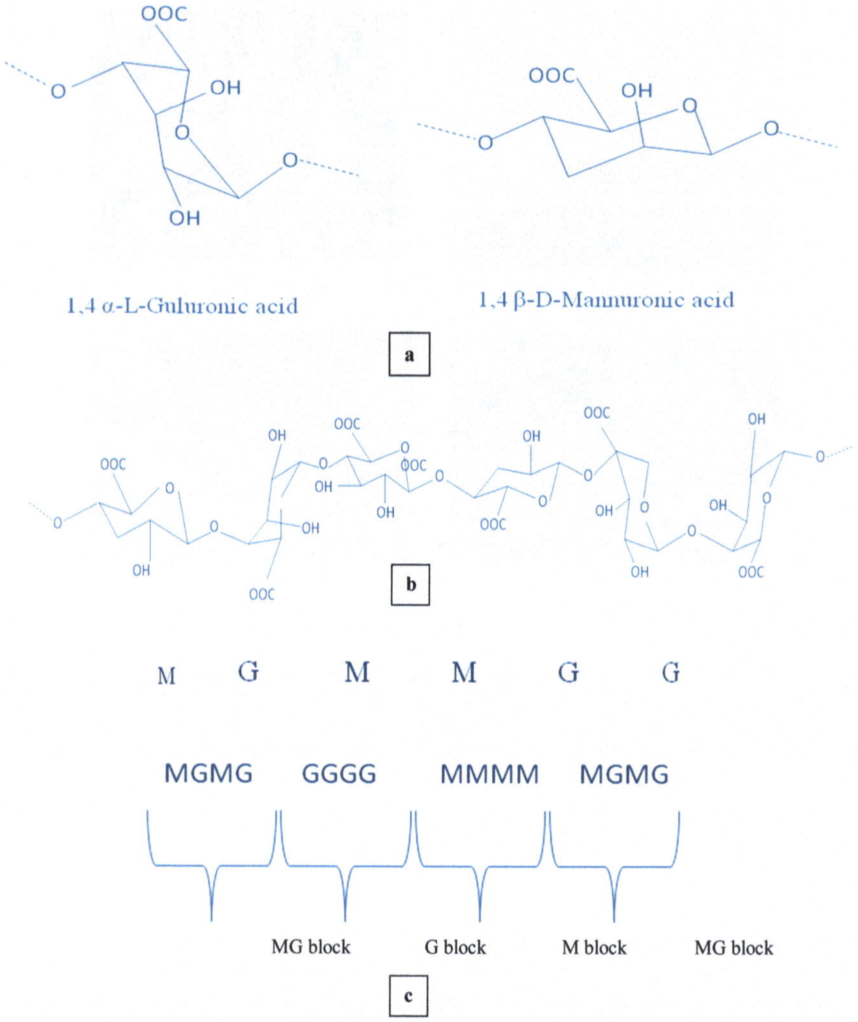

Fig. 6.2 Structure of (**a**) ALG monomers, (**b**) chain confirmation, (**c**) block distribution

which is soluble in water. The molecular structure of alginate constitutes monomeric units of 1–4-linked-D-mannuronate (M blocks) and -L-guluronate (G blocks) along with alternating segments of mannuronic and glucuronic acids known as MG blocks (Fig. 6.2). The extraction source and relative ratio of these three blocks decides the physical properties of the alginates (Pawar and Edgar 2012). Alginate compositions and ratio of mannuronic and glucuronic acid vary in different alginate producing species of seaweeds as well as different parts of the same seaweed (Pereira and Cotas 2019). The extraction source of alginates is the main contributing factor to decide the structural organization; as in the example of *Laminaria hyperborean*, the mannuronic acid is found in higher amounts in leaves, whereas guluronic acid is abundant in the outer cortex. In some other plants, fruiting bodies contain high amounts of mannuronic acid in comparison to higher guluronic acid in old tissues (Galus and Lenart 2019). The linear structure of alginates contributes to the strong nature of films derived from alginates (Tavassoli-Kafrani et al. 2016; Parreidt et al. 2018). Comparatively, bacterial alginates have higher molecular weights compared to the algal polymers.

6.4 Production/Preparation of Alginate and Alginate Derivative-Based Films

Homogeneous and uniform films can be easily formulated by dissolving sodium alginate in water followed by the process of casting or solvent evaporation. Development of sodium alginate films is a difficult and complex task as it relies on various processes and their conditions like cross-linking, drying, and adding of plasticizers to improve the film properties (Bierhalz et al. 2014; Wong et al. 2014).

Many polyvalent cations are used to enhance the film properties like water barrier, rigidity, and mechanical integrity, although the methods and types of cross-linking agents decide the final characteristics of the developed sodium alginate-based films. Alginate is known to form instant gel when it combines with cross-linking agents, which is not known to be good for film making as it hinders the process of casting (Crossingham et al. 2014). However, the use of monovalent ions like Na^+ and K^+ as well as glutaraldehyde has proved to impart better film properties such as enhanced tensile strength when used with cross-linking agents and alginate. Addition of several plasticizers is known to improve the brittleness of the films, improve the flexibility of the films, and reduce their shrinkage while handling and storage. Certain polyols such as xylitol, sorbitol, glycerol, fructose, and sucrose help in imparting the flexibility and enhance the sensory appeal along with the mechanical properties of the developed packaging films. Studies have shown that addition of plasticizers also helps in improving the release properties and bioactive penetration. It was documented by Avella et al. (2007) that a specific ratio between sodium alginate and plasticizers is required to achieve the appropriate effects of plasticizer. Blending of alginates with other polymers has also reported to be effective in improving their film properties. Mechanical properties of alginate-based films can be improved with the addition of cellulose cuoxam. Bacterial cellulose, which is known for its high moisture content and chemical purity, can be blended with alginate-based films to attain its applications in various fields. Films developed with blending of alginate and chitosan have been reported to have better tensile strength as well as the thermostability in comparison to the films prepared from alginic acid and blend of cellulose-alginate. Other known polymer sources which are used to blend with alginates are pectin, starch, cashew tree gum, carrageenan, etc.

Several fillers like cellulose whiskers from cotton or coconut husk fiber, silk fibrion, clays like Montmorillonite and bentonite, grapheme, and grapheme oxides are used to formulate the compost films, which improve their physical properties and versatility. Wet formulations of alginate-based films are required to be dried to get a stable structure for easy handling and storage. Alginate films are dried at temperatures 25, 35, 40, 50, and 60 °C for variable time duration depending on the degree of temperature. Drying methods and conditions are the major deciding factors for the film characteristics like their integrity, thickness, barrier properties, crystalline behavior, and mechanical and tensile properties. It is reported by some researchers that due to the polymer–polymer interaction as induced by the higher drying temperatures, the plasticity of the film tends to decrease at temperature ranging between 40 and –60 °C.

6.5 Properties

6.5.1 Thermal Properties

Moisture trapped inside and on the surface of polymers gets lost and decomposed on heating. During the heating process, components of polymers interact with each other, and this interaction is dependent on the miscibility degree of the components and composition nature. Loss of water is the first stage of heating a polymer and shows small weight loss, while the second step of decomposition shows considerable weight loss, and observable setback in this process is indicative of material stability (McNeill 1997). Giridhar Reddy and Thakur (2018) studied the thermal degradation process of pure sodium alginate and sodium alginate/lignosulfonic acid (LS). Pure alginate degrades in two steps, whereas pure LS degrades in multiple steps. Thus, blending of sodium alginate films with LS improves their thermal stability. Calcium-coated films are the most utilized form of alginate-based films. Sodium alginate films cross-linked using Ca^{+2} lead to formation of calcium alginate films, which possess intrinsic flame retardation property. Exposing calcium alginate-based films to high temperature leads to breakage of glycosidic bond, resulting in thermal instability. In a study by Zhao et al. (2018), application of reduced graphene oxide was done to form thermally stable composites based on sodium alginate. Calcium alginate/reduced graphene oxide (CaAlg/rGO)-based film showed better thermal stability than calcium alginate film (CaAlg). rGO intervention helped in decreasing the hygroscopicity and polarity of CaAlg. Wong et al. (1996) prepared alginate films by using cooling hot calcium alginate gels and pH-controlled calcium release into the alginate solution. pH-controlled alginate gel requires more quantity of energy to dissociate tightly linked structure of calcium and alginate, thus containing increased quantity of high temperature constituent. Cooling hot alginate gels caused relatively less strong cross-linking between calcium with alginate.

6.5.2 Barrier Properties

Water vapor permeability (WVP), moisture content, water vapor transmission rate (WVTR), and water vapor resistance (WVR) comprise the moisture barrier characteristics. Contemplation of moisture barrier properties is essential in formulating the process of coating (Díaz-Mula et al. 2012). Edible coating and films are usually used as a supplementary barrier layer on untreated and raw fruits and vegetables for reducing the rate of transpiration and weight loss. Alginates are alginic acid salt having hydrophilic nature, due to which their moisture barrier property is considered inadequate. To resolve this issue, calcium incorporation with alginate is being practiced for decreasing the water vapor permeability (WVP) and turning the films insoluble. Calcium works by binding with guluronic acid (G block) and forming cross-linkages; thus, WVP properties of alginate films change with varying proportion of G-blocks available in the alginates (Olivas and Barbosa-Cánovas 2008). Vapor-phase diffusion driven by the variation in water vapor pressure between the product and the air is the key mechanism behind the loss of moisture from food products (Díaz-Mula et al. 2012). Various parameters that influence the mass transfer rate are temperature, moisture permeability, relative humidity, and film thickness (Yaman and Bayondrl 2002). Integration of additives with the formed film influences its moisture transfer properties. Different studies reported reduction in moisture transfer with incorporation of lipids along with the coating (Guilbert et al. 1995; Tapia et al. 2008; Oms-Oliu et al. 2008). Pranoto et al. (2005) found that garlic oil (0.4%) incorporation into the alginate films significantly reduced its WVP value. Contrastingly, plasticizer addition in the alginate film formulation enhances the WVP by decreasing the intermolecular bonds within the polymer chains (Tapia et al. 2008). Various biopolymers possessing gel-forming properties also help in improving the water barrier properties of alginate films (Senturk Parreidt et al. 2018).

Capillary diffusion (occurs in porous, imperfect materials) and activated diffusion (gas solubilization in the film) are the primary mechanisms that take place during gas transport. Relative humidity surrounding the product heavily influences the O_2 and CO_2 transport rates. Enhanced interaction between film and water molecules has been reported with increased RH, resulting in plasticization that supported mass transfer (Bonilla et al. 2012). High temperature for storage increases the rate of respiration and is considered another factor influencing the film properties significantly (Baldwin et al. 1995). Jost et al. (2014) revealed that increase in the glycerol concentration increases the oxygen permeability within alginate films, whereas sorbitol did not illustrate any alteration in barrier properties of the alginate films. Hydrocolloidal nature of alginate-based films usually helps in reducing the respiration rate of products during storage, facilitating its quality attribute retention (Senturk Parreidt et al. 2018).

6.5.3 Mechanical Properties

Sodium alginate possesses film-forming capability on solvent evaporation and solution casting, due to which they are utilized as a film-forming material for various applications. Sodium alginate films are hydrophilic in nature, having soluble matrices leading to easy dissolution in water. Polyvalent ion cross-linking process is usually employed to enhance the rigidity, water resistance, cohesiveness, mechanical resistance, and barrier properties of the alginate films (Crossingham et al. 2014). Alginate is a well-recognized polyuronide, naturally possessing ion exchange properties. Charged nature of alginate is responsible for its film formation. Alginate with no bivalent ions can only be utilized for viscosity increment. However, due to the mechanism of ion exchange, incorporation of specific bivalent cations with alginate solution results in formation of gel. Alginate affinity with earth metals (alkaline) in ascending order is as follows: $Ca^{2+} < Sr^{2+} < Ba^{2+}$ (Lee and Rogers 2012). Conversely, various other divalent cations such as Cd^{2+}, Pb^{2+}, Zn^{2+}, Ni^{2+}, Mn^{2+}, and Cu^{2+} can cause gelation, but their utilization is restricted due to their toxic nature (Gombotz and Wee 1998). Alginate hydrogel properties are greatly influenced by length and proportion of the guluronic acid block also known as G-block, gelling ion type, divalent ion binding capacity, and gelling conditions (Tapia et al. 2008). Various studies revealed that high proportion of G-block leads to formation of dense and firm gels while higher mannuronic acid block (M-block) results in porous and flexible gels (Morris et al. 1973; Kierstan et al. 1982). Extension, thickness, and tensile strength are some of the few parameters used to test the mechanical properties of films. Krochta and Mulder-Johnston (1997) reported that standard edible films generally have standard elongation between the range of 10 and 50%. Marismandani and Husni (2020) disclosed that increase in the alginate concentration leads to increase in the extension percentage of the film. The resultant elongation increase is attributable to the rise in the dissolved solids in the film-forming solution which ultimately leads to increase in the thickness of film (Costa et al. (2018) studied the effect of Ca^{2+} (in the form of $CaCl_2$) cross-linking on alginate films. Results demonstrated that cross-linking is the main cause behind the increased tensile strength and young modulus, leading to enhanced film stiffness and resistance. $CaCl_2$ reacts with alginate and forms an egg-like structure, and consequently not enough alginate remains free to bind with water molecules. This process of cross-linking in alginates causes increase in the tensile strength and decrease in the elongation at break. Increase in the bonds between alginate and polyvalent ions induces strong, cohesive, and rigid films (Rhim 2004). Low molecular weight plasticizers such as glycerol easily penetrate inside the H_2-alginate bond and disturb the denseness of alginate matrix by decreasing the intermolecular interaction, leading to enhancement in alginate mobility and tensile strength (Marismandani and Husni 2020).

6.5.4 Dielectric Properties

Microenvironment and structural transition properties of complex polymer system like dipoles, charged groups mobility, and enzyme activity quantity of water can all be determined by determining the dielectric measurements. Quantity of dipole alignment (permanent and induced) is depicted by dielectric constant. Even though dielectric data for alginate beads and solution are accessible, literature related to alginate films linked with polyvalent ions is very limited. In a study conducted by Bekin et al. (2014) on sodium alginate films linked with Ca^{+2} ions, it was seen that rotational motion of polar molecules of sodium alginate and calcium ion-linked film was not rapid enough to change its direction along with electric field (obtained with increase in frequency). Due to this, polarization reduces and ultimately reaches a constant value. Hence, dielectric constant is inversely related to frequency (Zhan et al. 2011). Cross-linking found between alginate films linked with calcium ions causes reduction in the film polarization and leads to higher dielectric constant for sodium alginate films due to a phenomenon called dielectric relaxation, demonstrated by various polymer dielectrics. Polar groups present on the side chain of sodium alginate are the reason behind the beta relaxation of the sodium alginate-based polymer films, leading to dielectric constant (Jean-Mistral et al. 2010). On the contrary, when carboxylate (-COOH $^{-}$) ions of sodium alginate (40%) are linked with Ca^{+2} ions, b-relaxation becomes difficult. Ion mobility decreases in cross-linked films at high frequencies due to their relatively dense structure, suppressing the ions mobility.

6.5.5 Tensile Strength

Tensile strength can be used to determine film breakage and its strength by checking highest attinable film stretching (Krochta et al. 1994). In a study conducted by Fazilah et al. (2011) on tensile strength determination of starch and alginate films, it was found that linear structure of alginate polymer helped in increasing the tensile strength of starch and alginate-based films by 40% and led to formation of flexible and tough films. Alginate application in film formation is restricted by its inferior mechanical strength as it only possesses a negatively charged hydroxyl group, hence forming weak bonds with other molecules and resulting in low tensile strength. Composition of the film also affects the tensile strength of the film. Addition of lipids in alginate films (hydrocolloidal nature) decreases the tensile strength due to their nonpolar nature, making it complicated to be dissolved inside polar water, consequently resulting in weak bonds and low tensile. Plasticizers like glycerol enhance the mobility of polymers and decrease the intermolecular interactions, disordering the compact structure of alginate matrix and increasing the tensile strength of edible film (Marismandani and Husni 2020). Various studies indicated that addition of $CaCl_2$ into alginate films usually causes increase in the tensile strength. $CaCl_2$ works as a cross-linking agent and leads to improvement in the barrier properties, mechanical properties, and cohesion force of the alginate films with the help of Ca^{2+} ions. Ca^{2+} ions form an ionic bridge with neighboring polymer chains, forming cross-links (Cagri et al. 2004).

6.5.6 Flexibility Properties

Cross-linking of films leads to increase in the young modulus and tensile strength which usually leaves the films more stiff and resistant. Cross-linking of alginate films with Ca^{2+} leads to egg-box structure formation, and consequently only few water molecules are able to make bonds, while other molecules of alginate remain unavailable. Hence, formation of more bonds between Ca^{2+} ions and polymer chains further reduces the free volume in the film and results in strong, cohesive film but of less flexibility (Costa et al. 2018). Plasticizers are miscible and nonvolatile substances which are essentially used to improve the mechanical properties of biodegradable materials by providing a plasticization effect within the polymer network. The fundamental purposes of plasticizers are decreasing brittleness, coating material flow regulation, reduction in intermolecular forces, enhancing the polymer mobility at molecular level and free volume, maintaining flexibility, and increasing the resistance of tear impact (Barreto et al. 2003). Water is considered as the most widely used and effective plasticizer, but environmental factors such as temperature and RH influence the plasticizing nature of water with biopolymers (Guilbert et al. 1995). Sorbitol, glycerol, sucrose, and acetylated monoglyceride are also some of the commonly used plasticizers being utilized for food coatings. Plasticizer having hydrophilic property usually enhances the water vapor permeability rate of the film and affects the flexibility property of food coatings and films. Sodium lactate and glycerine result in formation of relatively more elastic alginate films in comparison to sorbitol (Parris et al. 1995).

6.5.7 Electrical Properties

Electrical conductivity of alginate-based films increased by increasing alginate percent (Basavaraja et al. 2010). Chemical and thermal reduction of sodium alginate/graphene oxide fibers leads to improvement in their conductivity (Umar et al. 2018). In a study conducted by Iwaki et al. (2012), polymer electrolytes based on alginate films illustrated a surge in ionic conductivity with increase in the acetic acid content due to an increase in the quantity of mobile charge carriers, whereas an extremely high concentration of salt and acids leads to reduction in ionic activity due to hydrogen bond and ionic aggregate formation.

6.5.8 Optical Properties

Quality and appearance of food products are significantly influenced by the optical properties of film, affecting the product's acceptability (Atarés and Chiralt 2016). Films made with pure alginate are usually found to be transparent in appearance, having high lightness value. Various studies indicated that addition of liquid substances into the films results in decrease in L^* and increase in film's darkness (Kaczmarek 2020). Utilization of emulsions for the formulation of films tends to decrease the lightness of the alginate films. It was also observed that addition of polyvalent ions such as Ca^{++} also leads to decrease in L^* value of the films and turns the films more opaque. Essential oil-containing films illustrate low transmittance at shorter wavelength owing to specific absorption of light by oil compounds. Different refractive index of dispersed phase present in the matrix promotes dispersion of light (Cofeliceet al. 2019).

6.5.9 Water Solubility Property

Calcium alginate and alginic acid are insoluble in water, whereas ammonium alginate, sodium alginate, and potassium alginate-based polymers are soluble in water (Lu et al. 2006). Solubility of the polymers is influenced by pH, demonstrating low solubility at low pH. Roger et al. (2006) revealed in a study that degree of substitutions has a great impact on solubility of alginate films. At low degree of substitution, alginate films are highly soluble and get completely released in the water. In contrast, at high level of calcium concentration, alginate films resist water solubilization in distilled water. Cross-linking calcium ions with alginate films results in development of a very dense structure which limits the movement of alginate out of the film. Plasticizer concentration is another factor affecting the water solubility of the alginate-based films (Rangel-Marrón et al. 2013). As the quantity of plasticizer is increased, it interacts with the available functional group present on the polysaccharide. This leads to reduction in the intermolecular force present within the film, consequently increasing the solubility.

6.5.10 Viscous Property

Various applications of alginates rely on their thickening property. Even with utilization of alginate at low concentration, fair increase in the viscosity can be achieved. Alginate solutions are usually pseudoplastic in behavior, flowing more easily when stirred. Harper et al. (2014) documented that addition of sodium chloride salt (0.47 M) to the film-forming solution of alginate significantly increased the viscosity as at high ionic strength, an intermolecular junction zone is formed between the monovalent ions and the chains, increasing the viscosity. Quantity of calcium ion and sodium alginate viscosity show a linear relationship. Viscosity of polymer possesses a prominent effect on degree of cross-linking. Alginate having high viscosity produces large and heavy droplets on dispersion in an immiscible liquid, providing more chance cross-linking and collision with calcium chloride solution (Lee et al. 2006).

6.5.11 Antimicrobial Properties

Incorporation of antimicrobial agents into films rather than their direct addition into food products is more valuable as films provide a way for releasing bioactive compounds or antimicrobial agents at a steady rate and also helps in retaining a critical concentration for an extended period of time (Zactiti and Kieckbusch 2009). On the contrary, direct application of agents in food products results in instant inhibition of microbial growth, whereas injured cell recovery and growth of survived cells found later might initiate food-borne disease and quality losses (Quintavalla and Vicini 2002). Various studies have been done on incorporation of antimicrobial agents into alginate-based films (Table 6.1). The précis of mentioned results revealed that alginate films serve as an efficient foundation for bioactive or antimicrobial agents for preserving the quality of food product and reducing its microbial load. Essential oils have been gaining the interest of the public and researchers because of their natural food preservation properties. Essential oils are being integrated into alginate films to utilize their antimicrobial properties (present in the form of bioactive agents), but their strong aroma and flavor, initiated from their phenolic compounds, limits their employment into edible films (Cagri et al. 2005).

6.5.12 Release Kinetics of Active Packaging

Edible coatings are known to be an effective means of preservation of active compounds at high level of concentration for a long period of time having a slow rate of migration. Molecular weight of large molecules influences their diffusivity from the matrix, whereas pore size of the gel matrix influences the diffusivity of small molecules (Martinsen et al. 1989). Quantity of migration, well-established rate of release, and control on the process of release are very critical factors, particularly in biotechnology. Consequently, numerous studies have been done for the same. Diffusion rate of lactalbumin, globulin, and glucose and release rates of NAD and hemoglobin were determined from different alginate coating and films (Kierstanet al. 1982). Content of -L-guluronic acid, protein isoelectric point, alginate concentration, and protein charge are some of the factors that affect the rates of diffusion from alginate films and coatings (Gombotz and Wee 1998). Alginate-based microgels have also been found to reduce the release rate of small hydrophilic molecules

Table 6.1 Antimicrobial activity of alginate-based films and coatings

Food	Coating/ cross-linking	Antimicrobial	Result	Source
Strawberry	Alginate with 1% $CaCl_2$ (EC)	*Cryptococcus laurentii*	Reduced microbial growth, weight loss, and loss of firmness along with enhanced storage quality without affecting the color of palatability of the product	Fan et al. (2009)
Pineapple	Alginate (EC)	*Trans*-cinnamaldehyde encapsulate powder	Impacted the sensory parameters of the food but storage life was extended to 15 days at refrigerated conditions	Mantilla et al. (2013)
Nanfeng mandarin	Alginate with 1% sucrose ester (EC)	*Ficus hirta*	Weight loss and fruit decay were reduced considerably in comparison with sodium alginate film (1 > 5%) alone	Chen et al. (2016)
Peach	Sodium alginate (EC)	*Rhubarb*	Inhibited the growth of *Penicillium expansum and maintained flavor and texture of the peaches*	Li et al. (2019)
Mushroom *Pholiota nameko*	Alginate with 5% $CaCl_2$ (EC)	Thyme essential oil, *L*-cysteine, and nisin	Controlled mesophilic bacteria count, quality deterioration and process of senescence	Zhu et al. (2019)
Arbutus unedo	Alginate with $CaCl_2$ (EC)	Eugenol and citral	Alginate (1%) with eugenol (0.2%) and alginate (1%) with eugenol (0.1%), citral (0.15%) gave the best results and with best nutritional and sensory attributes	Guerreiro et al. (2015)
–	Alginate (EF)	Cinnamon essential oil nanoemulsion	Strong antimicrobial effects against *E. coli*, *S. typhimurium*, *S. aureus*, and *B. cereus* were seen with 20% cinnamon nanoemulsion containing film	Frank et al. (2018)
Grapes	Sodium alginate (EC)	Grapefruit seed extract	NaAlg with 1% grapefruit seed extract showed maximum level of antifungal effect	Aloui et al. (2014)
Fish	Alginate and gelatin (EF)	Oregano essential oil	Delayed microbial growth for 15 days of storage	Kazemi and Rezaei (2015)
Carrots	Alginate with 5% $CaCl_2$ (EF)	Silver-montmorillonite nanoparticles	Microbial cell load for active film-coated sample was under chosen threshold level, and shelf life was extended up to 70 days	Mohammed Fayaz et al. (2009)

EC edible coating, *EF* edible film

considerably (Wang and Zhang Newby 2018). Zactiti and Kieckbusch (2006) studied the influence of alginate cross-linking on the release model and permeability of potassium sorbate. It was examined that a reduction in the permeability constant and active compound mobility was seen with increase in the degree of cross-linking, resulting from an increase in concentration of calcium ions in alginate films. Microencapsulation of essential oils has gained popularity in recent times in order to increase antimicrobial properties of food products, and evaluation of its release parameters from alginate matrix has been studied by various researchers (Bustos et al. 2016; Hambleton et al. 2009; Hambleton et al. 2011). Film matrices and aroma compound interaction is the basis for permeability of aroma compounds.

6.5.13 Safety and Regulations

Sodium alginate is categorized as a generally recognized as safe (GRAS) substance by FDA (Food and Drug Administration) in Title 21 of CFR (Code for Federal Regulations) and described its application as a thickener, emulsifier, gelling agent and stabilizer (FDA Title 21). Food Chemical Codex also provides alginic acid specifications. Ammonium, potassium, calcium, and sodium salts of alginic acid and its propylene glycol ester are classified as GRAS substances. ADI or Acceptable Daily Intake of 50 and 25 mg per body weight per day is specified for salts of alginic acid and propylene glycol alginate, respectively, by JECFA. The European Commission (EC) has registered alginic acid and alginic acid-based salts as a certified food additive (E400-E404) (Younes et al. 2017). Numerous industries like beverage, textile, food, pharmaceutical, and printing are utilizing alginate as a source of stabilizing agent, thickening agent, chelating agent, suspending agent, swelling, and encapsulation or for the purpose of membrane, gel, and film production. Sodium alginate is the most widely used salt of alginate (Kim et al. 2000).

6.5.14 Methods to Improve Properties

Sodium alginate is a polysaccharide having a variety of applications in the food industry. Shelf life of various products can be extended with the help of biodegradable coating of films. Very effective results have been attained on meat, fruits, and vegetable-based products when integration of alginate solution was done with various additives. Nevertheless, improved outcomes can be acquired with addition of new antioxidative, anti-browning, and antimicrobial agents to increase quality and safety of food products (Senturk Parreidt et al. 2018). The major limitation associated with this polymer is its high susceptibility against degradation and its reduced stability in wet conditions. Different modification methods have been studied to improve the properties of sodium alginate, which requires addition of metal ions like Cu^{2+}, Pb^{2+}, and Cd^{2+} (Kenawy et al. 2018). Phenolic acid utilization is another approach being used to improve the sodium alginate-based film properties by cross-linking (Kaczmarek 2020). Synergistic effect of various biopolymers like chitosan and cellulose for improving the properties of alginate-based film can be achieved by blending. Incorporation of carbon nanofibers even in a small quantity leads to considerable reduction in water sorption and swelling of alginate owing to their instantaneous cross-linking, resulting in increased modulus and tensile strength (Llorens-Gámez et al. 2020). The influence of various calcium salts (calcium gluconate, calcium chloride, calcium lactate) on the quality of alginate films needs to be discovered with clear working mechanisms.

6.6 Applications

6.6.1 Meat Products

Pseudomonas fluorescens, one of the potential contaminants in poultry, is responsible for food-borne diseases. An attempt was made by Alves et al. (2019) to incorporate ϕIBB-PF7A bacteriophage (phage) into sodium alginate films to prevent microbial meat spoilage. Around $1.52 \pm 2.12 \times 105$ PFU/cm^2 phages were incorporated in alginate-based film that was stable up to 12 week under refrigerated conditions. Application of phage-incorporated alginate film on chicken breast fillet resulted in 2 log reduction of *P. fluorescens* microflora during the first 2 days of application followed by 1 log reduction after 5 days under refrigerated conditions with no significant difference between control and treated films after 7 days suggesting their potential as antimicrobial film in meat products.

Use of alginate-based coating incorporating vitamin C and tea polyphenols for shelf life extension of bream led to promising results as quoted by Song et al. (2011). Alginate-based coating containing 5% vitamin C and 0.3% tea polyphenols resulted in better shelf life extension of fish compared to alginate-calcium coating probably due to antifungal effect (Chidanandaiah et al. 2009) of vitamin C and antimicrobial effect of tea catechin. Additionally, alginate-coated products show lower moisture loss in meat products due to evaporation of moisture in alginate gel prior to product dehydration (Lu et al. 2009). Application of alginate-based coating in fish proved to be superior in delaying the development of thiobarbituric acid value during storage compared to uncoated fish samples. Overall, antioxidant-incorporated alginate films preserved acceptable fish quality till 21 days for vitamin C-alginate, 18 days for tea polyphenols-alginate, and 15 days for alginate-calcium compared to uncoated fish that retained acceptable quality only till 12 days at 4 °C.

Alginate-based coatings offer the ease to incorporate active agents such as antimicrobial or antioxidants. One such application involves incorporation of natamycin, one of the antimicrobial agents in alginate-based films and pectin-alginate composites that showed promising result due to low diffusion coefficient with better physical properties (Bierhalz et al. 2012). In another study, nisin (bacteriocin)-incorporated alginate film at 1000 IU/mL reduced *S. aureus* population in sliced beef gradually till 7 days when stored at 4 °C (Millette et al. 2007). Similarly, embedding nisin in alginate beads resulted in better contact with *S. aureus* and preserved its quality for a longer time. Linking of nisin to alginate occurs due to reaction between aldehyde group and amino group of alginate and nisin, respectively. This type of covalent bonding preserved nisin from proteases and other denaturing factors present in meat (Le-Tien et al. 2004).

Use of essential oils as antimicrobial compounds has interested the consumers due to its natural origin and potential nutritive components. Incorporation of oregano, cinnamon, and savory in alginate-based films at 1% level was done to control the growth of *E. coli* and *Salmonella typhimurium* in beef muscle slices (Oussalah et al. 2006). In order to stabilize essential oils in alginate-based films, polycaprolactone was added as an emulsifying and hydrophobic agent. Among the essential oils used, oregano and cinnamon were more efficient in reducing the growth of microorganism as compared to savory during storage period of 5 days at refrigerated temperature. The major active ingredients for reducing microbial growth in oregano and savory-based alginate films were phenolic compounds and aldehydes in cinnamon-based films. Sample coated with simple alginate films and control samples comparatively had higher growth as compared to essential oil-based films. With respect to calcium chloride concentration, essential oils with 2% $CaCl_2$ alginate films provided better protection against bacteria compared to 20% $CaCl_2$ alginate films as calcium as per Kanipes et al. (2000) develops resistance to environmental stresses in bacteria by modifying lipopolysaccharide through induction of phosphoethanolamine transferase. However, with respect to reducing growth of *Salmonella typhimurium*, the concentration of 2% and 20% $CaCl_2$ used in essential oil-loaded alginate films was at par.

Similar studies were conducted by Matiacevich et al. (2015) for developing alginate-thyme oil-propionic acid edible coating for preservation of chicken breast fillets. Addition of propionic acid in alginate-thyme oil-based film resulted in improved wettability of coating on meat surface compared to only thyme oil-based coating due to hydrophobic nature of oil. The sensory properties revealed nonsignificant differences in sensory properties of coated and uncoated meat samples. Dehydration or weight loss of meat slices was reduced in alginate composite due to synergistic effect of oil and propionic acid as compared to only alginate-propionic acid coating. Inhibition of microbial growth was failed in sole alginate coatings, whereas alginate-propionic acid, alginate-thyme, and alginate-thyme-propionic acid increased meat shelf life by 2, 1.75, and 0.55 days, respectively. The active ingredient in thyme, i.e., phenols and monoterpenes, acts on bacterial membrane and its structure (Ramos et al. 2010), thereby preserving the food integrity.

Alginate-based films have also been used as delivery vehicles for releasing probiotic bacteria to ham slices. Pavli et al. (2017) attempted to incorporate probiotic bacteria in ham slices pretreated with high pressure processing (HPP) at 500 MPa for 2 min through sodium alginate films. The ham slices containing probiotic bacteria had characteristic acidic taste and aroma with no difference in appearance with levels $>10^6$ CFU/g at storage temperature of 4, 8, and 12 °C. HPP treatment helped reduce initial microbial population in ham slices prior to film application, and even after the storage period, the lactic acid bacteria population remained constant $>10^6$ CFU/g, whereas in uncoated and probiotic-free alginate film-coated ham slices, the microbial counts increased at the end of storage period.

6.6.2 *Fruits and Vegetables*

Application of alginate-based coatings in whole and fresh-cut fruits and vegetables is used as potential postharvest treatment to maintain acceptable quality during storage. Some of the applications are discussed below highlighting the significance of alginate-based coatings either alone or in combination with functional additives.

6.6.2.1 Cantaloupe

Whole cantaloupes were treated with 1% sodium alginate coatings containing 2% cinnamon bark oil (CBO) and coating containing combination of 2% cinnamon bark oil and 0.5% soybean oil (SBO) along with 0.5% Tween 80 (Zhang et al. 2015). Control (uncoated) and only alginate-coated cantaloupes displayed increased yellowness and visible mold growth after 15 days of storage compared to CBO coated cantaloupes. Weight loss was found in all the fruits irrespective of any treatment applied; however, flesh firmness was maintained in cantaloupes treated with CBO and SBO till 10 days of storage, whereas firmness reduced after 3 days of storage in uncoated and only alginate-coated fruits. This might be attributed to delaying ripening of fruits by CBO especially in combination with SBO. The volatility of CBO could be lowered by linking it with long chain fatty acid such as SBO and stabilizing its presence in alginate films (Zhang et al. 2015). Alginate coatings with CBO alone proved efficient in reducing *E. coli* population in cantaloupes, whereas for *S. enterica* and *L. monocytogenes* microorganism, alginate coatings containing both SBO and CBO were more effective due to higher stabilization of CBO in SBO.

6.6.2.2 Apple

Demand for ready-to-eat products is increasing rapidly, and in view of this, fresh-cut fruits and vegetables offer an array of products loaded with numerous health benefits. However, higher perishability associated with fresh-cut produce is the biggest limiting factor for its success in the market. Therefore, use of natural edible coatings incorporating active ingredients with antimicrobial potential for quality maintenance has been studied. Application of thymol-alginate (1 mg/mL) films in fresh-cut apple slices as a replacement of cling film was reported to have better tensile strength, lower water vapor permeability, reduced water solubility, and lower swelling ratio than alginate films without thymol supposedly due to hydrogen-bonding between thymol and alginate (Chen et al. 2021). However, higher thymol content (10 mg/mL) is reported to have weakening effect on hydrogen bonding between the two due to higher hydrophobic content of oil. The antioxidant activity (DPPH assay) of thymol-alginate film was higher than sole alginate film, thus preventing browning reactions in sliced apple. Microorganisms such as *S. aureus* and *E. coli* were not inhibited by application of standard sodium alginate films, whereas thymol-incorporated alginate films resulted in significant reduction of these microorganisms particularly *S. aureus*. Quality characteristics such as weight loss were found to be the lowest in polyethylene-wrapped slices followed by thymol-alginate film after 3 days of storage. Hydrophobicity imparted in thymol-alginate film resulted in localization of moisture around apple slices, thereby reducing water loss as compared to untreated slices. Additionally, thymol-alginate films also possessed UV-Vis-blocking capacity to inhibit oxidative reactions in food and prevent its deterioration. Thus, incorporation of thymol helped improved film properties and maintained apple slices quality by extending its shelf life for 2 days at 4 °C.

6.6.2.3 Guava

Utilization of nutraceuticals bearing antimicrobial and antioxidative properties from fruit and vegetable waste is currently gaining momentum in the food industry owing to status of natural plant extract. One such application involves utilization of pomegranate peel extract (PPE) in alginate-based films to improve quality characters of guava fruit (Nair et al. 2018). PPE-incorporated alginate film reduced respiration rate by altering CO_2 and O_2 exchange due to hydrophobicity property of films. Other nutritional properties such as ascorbic acid, total phenols, and antioxidant properties were also better retained in PPE alginate-coated guava compared to standard alginate films.

6.6.2.4 Mangoes

Enzymatic browning in minimally processed fruits displays a limiting factor in acceptability of these products. Use of anti-browning additives or antioxidants in fresh-cut fruits could help alleviate browning reactions and improve aesthetic appeal of produce. Tissue browning in fresh-cut mangoes limits its acceptability in the market; therefore, application of ascorbic acid and citric acid @ 1% as anti-browning agents embedded in alginate films was used to maintain the quality of mango cubes (Robles-Sánchez et al. 2013). It could be inferred from the study that L^* value (brightness) was reduced by 2% in alginate-ascorbic acid film compared to standard alginate (4%) and control samples (7%) at the end of storage period. This could be due to reducing nature of ascorbic acid to inhibit polyphenol oxidase enzyme action. Addition of ascorbic acid and citric acid in alginate films resulted in reduced oxygen diffusion in the cubes favoring higher retention of vitamin C compared to uncoated samples as vitamin C is prone to oxidative degradation. Beta-carotene, flavonoids, and phenols in mango cubes were retained better due to delay in respiration of cubes coated with alginate-ascorbic acid. Edible coated mangoes could be successfully stored at 4 °C for a period of 12 days without major degradation of sensory and nutritive quality.

6.6.2.5 Plum

Plums are one of the preferred delicacies among fruits due to its taste and nutritional content. However, postharvest decay associated with it includes chilling injury under cold store conditions with prominent symptoms of translucency, mealiness, and flesh decay. Therefore, utilization of edible coating prior to cold store could help solve some of these problems. In a study conducted by Valero et al. (2013), four plum cultivars, viz., Songold, Larry Ann, Golden Globe, and Blackamber, were coated with 1% and 3% alginate coatings and stored at 2 and 20 °C for shelf life evaluation. Ethylene production rate in plums remained stable at 2 °C, but when fruits were transferred to 20 °C, there was abrupt rise in ethylene generation rate among which plums coated with 3% alginate showed suppressed behavior of ethylene gas compared to uncoated fruits. This property also helped to lower fruit softening in coated plums as reduced ethylene gas did not activate cell wall hydrolyzing enzymes in fruits and maintained firmness. Alginate coatings act as an additional layer apart from fruit's natural epidermis and cuticle, covering the stomata of cuticle and thus helping in reducing weight loss of coated plums at storage temperature. Application of alginate coating increased optimal shelf life of "Blackamber" and "Larry Ann" by 2 weeks and Songold and Larry Ann by 3 weeks.

6.6.2.6 Melon

Food-borne illness due to *E. coli* O157:H7 and *Salmonella* in watermelon, honeydew, and cantaloupes has been witnessed particularly in salad form, probably due to the contaminated rind part. Therefore, an approach of applying alginate coatings incorporated with malic acid and essential oils to maintain sensory attributes and enhance shelf life has been studied by Raybaudi-Massilia et al. (2008) in fresh-cut melons. Malic acid inclusion in alginate-based coatings resulted in reduced microflora of mesophilic and psychrophilic organisms. However, combined effect of malic acid and essential oils of palmarosa, cinnamon, and lemongrass in alginate coatings resulted in maximum reduction of *E. coli* O157:H7 and *Salmonella*. Among essential oils used, cinnamon and lemongrass had higher efficacy in lowering mesophilic growth than palmarosa essential oil, whereas in suppressing psychrophilic growth, cinnamon was found to be most effective than palmarosa and lemongrass essential oils. Fresh-cut melons coated with only malic acid-alginate coatings resulted in shelf life of 9–10 days as compared to control (3–4 days), and maximum shelf life of 21 days was observed in essential oil-malic acid-alginate films. Overall, considering sensory acceptance by the panelists and quality maintenance of fruits, 0.3% palmarosa essential oil in alginate films was highly acceptable as other essential oils had to be used at higher concentration to achieve microbial efficiency with altered sensory properties owing to strong aroma.

6.6.2.7 Papaya

Edible coating application in fruits is known to create a modified atmosphere inside the fruit, thereby delaying ripeness and increasing shelf life. One such study signifies the semi-permeable properties of alginate coating in fresh-cut papaya for naturally creating modified atmosphere around fruits (Tabassum and Khan 2020). The authors developed edible coatings containing 2% alginate and 0.5–2 mL thyme and oregano essential oil. Papaya cubes were dipped in these coatings and packed in low-density polypropylene trays and kept at 4 °C. The lipophilic nature of essential oils reduced weight loss in papaya cubes to a higher extent as compared to uncoated cubes. The pH value of uncoated papaya cubes was noted as 6.1 at 0 day which reduced to 4.03 at the end of storage period indicating fungal bloom and microbial growth, whereas pH of essential oil-coated fruits ranged from 5.43 to 5.7 at 0 day that slowly reduced to 4.77 to 4.93. Gaseous exchange was limited profusely in essential oil-alginate-coated cubes wherein the samples showed maximum 4% increase in carbon dioxide content at the end of storage and negligible change in oxygen, whereas in control samples, oxygen reduced from 19% at initial day to 7.83% and carbon dioxide increased from 1.10% on zero day to 5.97% on the last day of storage. Microbiological assay as noted for uncoated cubes was highest in uncoated samples 9.67 log CFU/g followed by alginate films without essential oil (6.57 log CFU/g) and least in essential oil-coated cubes (3 log CFU/g). Conclusively, fresh-cut papaya cubes could successfully be stored for a period of 12 days at 4 °C.

6.6.3 Dairy Products

Among various dairy products consumed, cheese is one of the widely consumed products and is quite diverse in nature with respect to aroma, shape, texture, and flavors. However, cheese has a relatively short shelf life with major problem of bacterial, mold, and fungal growth during storage leading to off-flavor development rendering it unfit for human consumption. Sometimes dehydration issues have also been observed in cheese due to lack of proper packaging. Mostly polythene-based packages or paraffin-based coatings are used to preserve cheese for a longer time. However, environmental issues pertaining to use of such packages shift the interest towards application of natural edible coating. Numerous studies have been conducted on application of edible biofilms or coatings to extend shelf life of cheese which has been compiled in Table 6.2.

Table 6.2 Application of alginate-based formulations in cheese

Cheese type	Coating material	Coating technique	Additives	References
Mozzarella cheese	1% Sodium alginate + 0.25% glycerol	Dipping	–	Zhong et al. (2014)
		Enrobing		
		Spraying		
		Electrostatic spraying		
Homemade cheese	3% Sodium alginate + 2.5% glycerol + essential oil 0.1%	Dipping	Essential oil-*O. basilicum* L., *R. officinalis* L., A. *herba alba* *Asso*., M. *pulegium* L.	Mahcene et al. (2021)
Kashar cheese	2% Alginate + 500 μL glycerol	Film application	Natamycin	Kucuk et al. (2020)
Coalho cheese	0.2% Alginate	Nano-laminate	Lysozyme	Medeiros et al. (2014)
		Alg-Lyso-Algi-Lyso-Algi		
Cheese	2% Alginate + 2% calcium lactate	Film application	–	Kampf and Nussinovitch (2000)
Ultra-filtered soft cheese	2% Alginate + glycerol + 2% citric acid	Film application	Antimicrobial microcrystalline cellulose	El-Sayed et al. (2021)

6.6.4 Environmental Applications

Alginate-based coatings are popularly used in food, energy, and environmental sectors owing to their biocompatible, cost-effective, and nontoxic properties. Inorganic or organic contaminants persistent in environment could be successfully absorbed using alginate-based composites based on electrostatic interactions and ion exchange mechanism (Wang et al. 2019). Processed dye waste used in the textile industry is one of the organic contaminants harmful for aquatic ecosystem. Alginate-based composites incorporating calcium, graphene, halloysite, bentonite, iron compounds, polyaspartate, epichlorohydrin have been used to adsorb dyes such as basic black dye, methylene blue, malachite green, methyl orange, and basic Blue9 (Aravindhan et al. 2007; Zhuang et al. 2016; Liu et al. 2012; Benhouria et al. 2015; Mohammadi et al. 2014; Rocher et al. 2010; Mahmoodi 2013).

Heavy metals are one of the potent contaminants found in wastewater due to their toxicity at lower values and harmful effects on body. Alginate majorly comprises hydroxyl and carboxyl functional groups that can either form certain complexes or react with heavy metals through ion exchange mechanism (Babel and Kurniawan 2003; Zhou et al. 2014). Alginate composites could be used for determination of heavy metals, such as biochar for cadmium and lead (Roh et al. 2015; Do and Lee 2013); graphite nanotubes for cobalt and nickel (Jung et al. 2015); alginate chitosan for copper, cadmium, and cobalt (Gotoh et al.2004); iron oxide for arsenic (Zouboulis and Katsoyiannis 2002); orange peel cellulose for zinc (Lai et al. 2010); banana peel cellulose for copper, lead, and zinc (Lai et al. 2010); and montmorillonite for chromium (Olad and FarshiAzhar 2014). The presence of mannuronic acid in alginate from *Laminaria digitata* species is particularly responsible for ion exchange mechanism with maximum adsorption for lead followed by copper and cadmium (Papageorgiou et al. 2006).

Antibiotics are also one of the potent contaminants found in aquatic environment known to possess toxicity at low levels. Antibiotics such as tetracycline and ciprofloxacin were majorly adsorbed using alginate-Fe_3O_4 hydrogel fibers and alginate/graphene double network hydrogel. Anionic interactions between alginate beads and antibiotics lead to hydrogen bonding and subsequent adsorption of antibiotics (Konwar et al. 2015; Zhuang et al. 2017). Insertion of graphene oxide in calcium alginate composites leads to electron donor-acceptor interactions between ciprofloxacin and graphene oxide (Fei et al. 2016). Majorly hydrogen bonding and π-π interaction resulted in major adsorption mechanism for antibiotics on alginate-based composites.

6.7 Limitations and Future Trends

Alginate-based films have been widely explored for food packaging applications either in film form or directly as coating material. Alginate-based films/coatings also offer the ease of incorporation of antimicrobials, antioxidants, anti-browning agents, and various other additives for shelf life extension of fresh and processed food. However, most of the studies have focused on laboratory applications without any success at commercial scale. Elaborate reports on release kinetics of additives incorporated in alginate films are limited, thereby creating a gap to understand actual potential of additives used in films/coatings on food products. Further research could emphasize on the use of newer more powerful antimicrobials and other additives without altering film properties. Also, studies on various type of plasticizers used in film formulations be exemplified for better understanding of film behavior corresponding to specific food product. Poor adhesion of alginate films on hydrophilic food surface necessitates the use of oil in film formulations and also limits its use in some food products. Minimally processed foods usually require postharvest interventions such as coatings/film application for quality maintenance during storage, among which alginate-based films have proved to be quite effective but somewhere has compromised with sensory properties of the food. Therefore, modification of alginate-based formulations at nano-scale must be extensively carried out in a way to maintain natural sensory properties and other quality parameters of fresh-cut fruits or minimally processed food.

6.8 Conclusion

Edible coatings are much in demand owing to the claim of being natural and mostly preferred by consumers. Thickening and stabilizing properties of alginate make it a quite versatile ingredient for food uses. Apart from this, its film formation capability makes it a polymer of interest having adequate water vapor barrier properties, limited gas exchange, and microbial growth prevention and as a carrier for various types of agents. Overall, it could be concluded that alginate-based films hold potential for commercialization at industrial level for handling mass storage of foods, and focus should now be made on pilot-scale production of alginate film/coatings for wide availability worldwide offering sustainable food packaging alternatives.

References

Aloui H, Khwaldia K, Sánchez-González L, Muneret L, Jeandel C, Hamdi M, Desobry S (2014) Alginate coatings containing grapefruit essential oil or grapefruit seed extract for grapes preservation. Int J Food Sci Technol 49(4):952–959

Alves D, Marques A, Milho C, Costa MJ, Pastrana LM, Cerqueira MA, Sillankorva SM (2019) Bacteriophage ϕIBB-PF7A loaded on sodium alginate-based films to prevent microbial meat spoilage. Int J Food Mircobiol 291:121–127

Ananda AP, Manukumar HM, Umesha S, Soumya G, Priyanka D, Kumar ASM, Krishnamurthy NB, Savitha KR (2017) A relook at food packaging for cost effective by incorporation of novel technologies. J Packag Technol Res 1:67–85

Aravindhan R, Fathima NN, Rao JR, Nair BU (2007) Equilibrium and thermodynamic studies on the removal of basic black dye using calcium alginate beads. Colloid Surf A Physicochem Eng 299(1–3):232–238

Arroyo B, Bezerra A, Oliveira L, Arroyo S, De Melo E, Santos A (2020) Antimicrobial active edible coating of alginate and chitosan add ZnO nanoparticles applied in guavas (*Psidium guajava L.*). Food Chem 309:125566

Atarés L, Chiralt A (2016) Essential oils as additives in biodegradable films and coatings for active food packaging. Trends Food Sci Technol 48:51–62

Avella M, Di Pace E, Immirzi B, Impallomeni G, Malinconico M, Santagata G (2007) Addition of glycerol plasticizer to seaweeds derived alginates: influence of microstructure on chemical–physical properties. Carbohydr Polym 69:503–511

Babel S, Kurniawan TA (2003) Low-cost adsorbents for heavy metals uptake from contaminated water: a review. J Hazard Mater 97(1–3):219–243

Baldwin E, Nisperos-Carriedo M, Baker R (1995) Edible coatings for lightly processed fruits and vegetables. Hort Sci 30:35–38

Barreto PLM, Pires ATN, Soldi V (2003) Thermal degradation of edible films based on milk proteins and gelatin in inert atmosphere. Polym Degrad Stab 79:147–152

Basavaraja C, Kim NR, Jo EA, Huh DS (2010) Characterization and conductivity studies of nanocomposite films with polymer networks containing polyaniline-alginate/titanium dioxide. Polym Compos 31:1754–1761

Bekin S, Sarmad S, Gürkan K, Yenici G, Keçeli G, Gürdağ G (2014) Dielectric thermal and swelling properties of calcium ion crosslinked sodium alginate film. Polym Eng Sci 54(6): 1372–1382

Benhouria A, Islam MA, Zaghouane-Boudiaf H, Boutahala M, Hameed B (2015) Calcium alginate–bentonite–activated carbon composite beads as highly effective adsorbent for methylene blue. Chem Eng J 270:621–630
Bierhalz ACK, da Silva MA, Kieckbusch TG (2012) Natamycin release from alginate/pectin films for food packaging applications. J Food Eng 110:18–25
Bierhalz AC, da Silva MA, Braga ME, Sousa HJ, Kieckbusch TG (2014) Effect of calcium and/or barium cross-linking on the physical and antimicrobial properties of natamycin-loaded alginate films. LWT Food Sci Technol Int 57:494–501
Bonilla J, Atarés L, Vargas M, Chiralt A (2012) Edible films and coatings to prevent the detrimental effect of oxygen on food quality: possibilities and limitations. J Food Eng 110:208–213
Bustos R, Alberti F, Matiacevich S (2016) Edible antimicrobial films based on microencapsulated lemongrass oil. J Food Sci Technol 53:832–839
Cagri A, Ustunol Z, Osburn W, Ryser ET (2004) Antimicrobial edible films and coatings. J Food Prot 67(4):833–848
Cagri A, Ustunol Z, Ryser ET (2005) Antimicrobial edible films and coatings. J Food Prot 67:833–848
Chen C, Peng X, Zeng R, Chen M, Wan C, Chen J (2016) *Ficus hirta* fruits extract incorporated into an alginate-based edible coating for Nanfeng mandarin preservation. Sci Hortic 202:41–48
Chen J, Wu A, Yang M, Ge Y, Pristijono P, Li J, Xu B, Mi H (2021) Characterization of sodium alginate-based films incorporated with thymol for fresh-cut apple packaging. Food Control 126: 108063
Chidanandaiah, Keshri RC, Sanyal MK (2009) Effect of sodium alginate with preservatives on the quality of meat patties during refrigerated (4°C) storage. J Muscle Foods 20:275–292
Cofelice M, Cuomo F, Chiralt A (2019) Alginate films encapsulating lemongrass essential oil as affected by spray calcium application. Coll Interfaces 3(3):58
Costa MJ, Marques AM, Pastrana LM, Teixeira JA, Sillankorva SM, Cerqueira MA (2018) Physicochemical properties of alginate-based films: effect of ionic crosslinking and mannuronic and guluronic acid ratio. Food Hydrocoll 81:442–448
Crossingham YJ, Kerr PG, Kennedy RA (2014) Comparison of selected physicochemical properties of calcium alginate films prepared by two different methods. Int J Pharm 473:259–269
Díaz-Mula HM, Serrano M, Valero D (2012) Alginate coatings preserve fruit quality and bioactive compounds during storage of sweet cherry fruit. Food Bioprocess Technol 5:2990–2997
Do XH, Lee BK (2013) Removal of Pb2þ using a biochar-alginate capsule in aqueous solution and capsule regeneration. J Environ Manag 131:375–382
El-Sayed HS, El-Sayed SM, Mabrouk AMM, Nawwar GA, Youssef AM (2021) Development of eco-friendly probiotic edible coatings based on chitosan, alginate and carboxymethyl cellulose for improving the shelf life of UF soft cheese. J Polym Environ 29(4):1–13. https://doi.org/10.1007/s10924-020-02003-3
Fan Y, Xu Y, Wang D, Zhang L, Sun J, Sun L, Zhang B (2009) Effect of alginate coating combined with yeast antagonist on strawberry (*Fragaria ananassa*) preservation quality. Postharvest BiolTechnol 53:84–90
Fazilah A, Maizura M, Abd Karim A, Bhupinder K, Bhat R, Uthumporn U, Chew SH (2011) Physical and mechanical properties of sago starch – alginate films incorporated with calcium chloride. Int Food Res J 18:1027–1033
Fei Y, Li Y, Han S, Ma J (2016) Adsorptive removal of ciprofloxacin by sodium alginate/graphene oxide composite beads from aqueous solution. J Colloid Interface Sci 484:196–204
Frank K, Garcia CV, Shin GH, Kim JT (2018) Alginate biocomposite films incorporated with cinnamon essential oil nanoemulsions: physical mechanical and antibacterial properties. Int J Polym Sci 2018:1519407
Galgano F (2015) Biodegradable packaging and edible coating for fresh-cut fruits and vegetables. Ital J Food Sci 27:1–20
Galus S, Lenart A (2019) Development and characterization of composite edible films based on sodium alginate and pectin. J Food Eng 115:459–465

Giridhar Reddy S, Thakur A (2018) Thermal stability and kinetics of sodium alginate and lignosulphonic acid blends. Iran J Mater Sci Eng 15(3):53–59
Gombotz WR, Wee S (1998) Protein release from alginate matrices. Adv Drug Deliv Rev 31:267–285
Gotoh T, Matsushima K, Kikuchi KI (2004) Preparation of alginate-chitosan hybrid gel beads and adsorption of divalent metal ions. Chemosphere 55(1):135–140
Guerreiro AC, Gago CM, Faleiro ML, Miguel MG, Antunes MD (2015) The effect of alginate-based edible coatings enriched with essential oils constituents on *Arbutus unedo* L fresh fruit storage. Postharvest Biol Technol 100:226–233
Guilbert S, Gontard N, Cuq B (1995) Technology and applications of edible protective films. Packag Technol Sci 8:339–346
Gundewadi G, Rudra S, Sarkar D, Singh D (2018) Nanoemulsion based alginate organic coating for shelf life extension of okra. Food Packag Shelf Life 18:1–12
Hambleton A, Debeaufort F, Bonnotte A, Voilley A (2009) Influence of alginate emulsion-based films structure on its barrier properties and on the protection of microencapsulated aroma compound. Food Hydrocoll 23:2116–2124
Hambleton A, Voilley A, Debeaufort F (2011) Transport parameters for aroma compounds through i-carrageenan and sodium alginate-based edible films. Food Hydrocoll 25:1128–1133
Harper BA, Barbut S, Lim LT, Marcone MF (2014) Effect of various gelling cations on the physical properties of "wet" alginate films. J Food Sci 79(4):E562–E567
Huebsch N, Mooney DJ (2009) Inspiration and application in the evolution of biomaterials. Nature 462:426–432
Iwaki YO, Escalona MH, Briones JR, Pawlicka A (2012) Sodium alginate-based ionic conducting membranes. Mol Cryst Liq Cryst 554(1):221–231
Jean-Mistral C, Sylvestre A, Basrour S, Chaillout JJ (2010) Dielectric properties of polyacrylate thick films used in sensors and actuators. Smart Mater Struct 19(7):075019
Jost V, Kobsik K, Schmid M, Noller K (2014) Influence of plasticiser on the barrier, mechanical and grease resistance properties of alginate cast films. Carbohydr Polym 110:309–319
Jung W, Jeon BH, Cho DW, Roh HS, Cho Y, Kim SJ, Lee DS (2015) Sorptive removal of heavy metals with nano-sized carbon immobilized alginate beads. J Ind Eng Chem 26:364–369
Kaczmarek B (2020) Improving sodium alginate films properties by phenolic acid addition. Materials 13:2895
Kampf N, Nussinovitch A (2000) Hydrocolloid coating of cheeses. Food Hydrocoll 14:531–537
Kanipes MIS, Lin R, Cotter J, Raetz CR (2000) Ca2+ –induced phosphoethanolamine transfer to the outer 3-deoxy-D-manno-octulosonic acid moiety of *Escherichia coli* lipopolysaccharide. A novel membrane enzyme dependent upon phosphatidylethanolamine. Biol Chem 12:1156–1163
Kazemi SM, Rezaei M (2015) Antimicrobial effectiveness of gelatin–alginate film containing oregano essential oil for fish preservation. J Food Saf 35(4):482–490
Kenawy IM, Hafez MAH, Ismail MA, Hashem MA (2018) Adsorption of Cu (II):Cd (II):Hg (II):Pb (II) and Zn (II) from aqueous single metal solutions by guanyl-modified cellulose. Int J Biol Macromol 107:1538–1549
Kierstan M, Darcy G, Reilly J (1982) Studies on the characteristics of alginate gels in relation to their use in separation and immobilized applications. BiotechnolBioeng 24:1507–1517
Kim YJ, Yoon KJ, Ko SW (2000) Preparation and properties of alginate superabsorbent filament fibers crosslinked with glutaraldehyde. J ApplPolymSci 78:1797–1804
Konwar A, Gogoi A, Chowdhury D (2015) Magnetic alginate-Fe3O4 hydrogel fiber capable of ciprofloxacin hydrochloride adsorption/separation in aqueous solution. RSC Adv 5(99): 81573–81582
Krochta JM, De Mulder-Johnston C (1997) Edible and biodegradable polymer films: challenges and opportunities. J Food Technol 51:61–74
Krochta J, Aldwin M, Dan EA, Carriedo MON (1994) Edible coatings and film to improve food quality. Economic Publishing, Boca Raton

Kucuk GS, Celik OF, Mazi BG, Ture H (2020) Evaluation of alginate and zein films as a carrier of natamycin to increase the shelf life of kashar cheese. Packag Technol Sci 33:39–48

Lai YL, Thirumavalavan M, Lee JF (2010) Effective adsorption of heavy metal ions (Cu2+, Pb2+, Zn2+) from aqueous solution by immobilization of adsorbents on Ca-alginate beads. Toxicol Environ Chem 92:697–705

Lee P, Rogers MA (2012) Effect of calcium source and exposure-time on basic caviar spherification using sodium alginate. Int J Gastron Food Sci 1:96–100

Lee HY, Chan LW, Dolzhenko AV, Heng PWS (2006) Influence of viscosity and uronic acid composition of alginates on the properties of alginate films and microspheres produced by emulsification. J Microencapsul 23(8):912–927

Le-Tien C, Millette M, Lacroix M, Mateescu MA (2004) Modified alginate matrices for the immobilization of bioactive agents. Biotechnol Appl Biochem 39:189–198

Li XY, Du XL, Liu Y, Tong LJ, Wang Q, Li JL (2019) Rhubarb extract incorporated into an alginate-based edible coating for peach preservation. Sci Hortic 257:108685

Liu L, Wan Y, Xie Y, Zhai R, Zhang B, Liu J (2012) The removal of dye from aqueous solution using alginate-halloysite nanotube beads. ChemEng J 187:210–216

Llorens-Gámez M, Salesa B, Serrano-Aroca A (2020) Physical and biological properties of alginate/carbon nanofibers hydrogel films. Int J Biol Macromol 151:499–507

Lu JW, Zhu YL, Guo ZX, Hu P, Yu J (2006) Electrospinning of sodium alginate with poly(ethylene oxide). Polymer 47:8026–8031

Lu F, Liu DH, Ye XQ (2009) Alginate-calcium coating incorporating nisin and EDTA maintains the quality of fresh northern snakehead (*Channa argus*) fillets stored at 4°C. J Sci Food Agric 89:848–854

Mahcene Z, Khelil A, Hasni S, Bozkurt F, Goudjil MB, Tornuk F (2021) Home-made cheese preservation using sodium alginate based on edible film incorporating essential oils. J Food Sci Technol 58(6):2406–2419. https://doi.org/10.1007/s13197-020-04753-3

Mahmoodi NM (2013) Magnetic ferrite nanoparticle–alginate composite: synthesis, characterization and binary system dye removal. J Taiwan Inst Chem Eng 44(2):322–330

Mantilla N, Castell-Perez ME, Gomes C, Moreira RG (2013) Multilayered antimicrobial edible coating and its effect on quality and shelf-life of fresh-cut pineapple (*Ananas comosus*). LWT Food Sci Technol 51(1):37–43

Marismandani ADP, Husni A (2020) Development and characterization of biobased alginate/glycerol/virgin coconut oil as biodegradable packaging. Paper presented in E3S web of conferences, EDP Sciences, 2020

Martinsen A, Skjåk-Bræk G, Smidsrød O (1989) Alginate as immobilization material: I correlation between chemical and physical properties of alginate gel beads. BiotechnolBioeng 33:79–89

Matiacevich S, Acevedo N, Lopez D (2015) Characterization of edible active coating based on alginate–thyme oil–propionic acid for the preservation of fresh chicken breast fillets. J Food Process Preserv 39:2792–2801

McNeill IC (1997) Thermal degradation of polystyrene in different environments. Macro Mol Mater Eng 247:179–195

Medeiros BG, Souza MP, Pinheiro AC, Bourbon AI, Cerqueira MA, Vicente AA, Carneiro-da-Cunha MG (2014) Physical characterisation of an alginate/lysozyme Nano-laminate coating and its evaluation on 'Coalho' cheese shelf life. Food Bioprocess Technol 7:1088–1098

Millette M, Tien CL, Smoragiewicz W, Lacroix M (2007) Inhibition of *Staphylococcus aureus* on beef by nisin-containing modified alginate films and beads. Food Control 18:878–884

Mohammadi A, Daemi H, Barikani M (2014) Fast removal of malachite green dye using novel superparamagnetic sodium alginate-coated Fe_3O_4 nanoparticles. Int J Biol Macromol 69:447–455

Mohammed Fayaz A, Balaji K, Girilal M, Kalaichelvan PT, Venkatesan R (2009) Mycobased synthesis of silver nanoparticles and their incorporation into sodium alginate films for vegetable and fruit preservation. J Agric Food Chem 57(14):6246–6252

Morris ER, Rees DA, Thom D (1973) Characterization of polysaccharide structure and interactions by circular dichroism: order-disorder transition in the calcium alginate system. J Chem Soc Chem Commun 7:245–246

Nair MS, Saxena A, Kaur C (2018) Effect of chitosan and alginate based coatings enriched with pomegranate peel extract to extend the postharvest quality of guava (*Psidium guajava* L.). Food Chem 240:245–252

Olad A, FarshiAzhar F (2014) A study on the adsorption of chromium (VI) from aqueous solutions on the alginate-montmorillonite/polyaniline nanocomposite. Desalin Water Treat 52(13–15): 2548–2559

Olivas GI, Barbosa-Cánovas GV (2008) Alginate–calcium films: water vapor permeability and mechanical properties as affected by plasticizer and relative humidity. LWT Food SciTechnol 41(2):359–366

Oms-Oliu G, Soliva-Fortuny R, Martín-Belloso O (2008) Edible coatings with antibrowning agents to maintain sensory quality and antioxidant properties of fresh-cut pears. Postharvest Biol Technol 50:87–94

Oussalah M, Caillet S, Salmieri S, Saucier L, Lacroix M (2006) Antimicrobial effects of alginate-based film containing essential oils for the preservation of whole beef muscle. J Food Prot 69 (10):2364–2369

Papageorgiou SK, Katsaros FK, Kouvelos EP, Nolan JW, Le Deit H, Kanellopoulos NK (2006) Heavy metal sorption by calcium alginate beads from *Laminaria digitata*. J Hazard Mater 137(3):1765–1772

Pavli F, Kovaiou I, Apostolakopoulou G, Kapetanakou A, Skandamis P, Nychas GE, Tassou C, Chorianopoulos N (2017) Alginate-based edible films delivering probiotic bacteria to sliced ham pretreated with high pressure processing. Int J Mol Sci 18:1867

Parreidt ST, Michael S, Markus S, Kajetan M (2018) Effect of presence and concentration of plasticizers, vegetable oils, and surfactants on the properties of sodium-alginate-based edible coatings. Int J Mol Sci 19:2–21

Parris N, Coffin DR, Joubran RF, Pessen H (1995) Composition factors affecting the water vapor permeability and tensile properties of hydrophilic films. J Agric Food Chem 43:1432–1435

Pawar SN, Edgar KJ (2012) Alginate derivatization: a review of chemistry, properties and applications. Biomaterials 33:3279–3305

Pereira L, Cotas J (2019) Alginates: a general overview. In: Alginates: recent uses of this natural polymer. IntechOpen, London, pp 1–17

Pranoto Y, Salokhe VM, Rakshit SK (2005) Physical and antibacterial properties of alginate-based edible film incorporated with garlic oil. Food Res Int 38:267–272

Quintavalla S, Vicini L (2002) Antimicrobial food packaging in meat industry. Meat Sci 62:373–380

Ramos A, Azqueta A, Pereira-wilson C, Collins AR (2010) Polyphenolic compounds from *Salvia* species protect cellular DNA from oxidation and stimulate DNA repair in cultured human cells. J Agric Food Chem 58:7465–7471

Rangel-Marrón M, Montalvo-Paquini C, Palou E, López-Malo A (2013) Optimization of the moisture content, thickness, water solubility and water vapor permeability of sodium alginate edible films using experimental design. In: Proceedings of the 4th WSEAS international conference in recent advances on chemical engineering, biochemistry and computational chemistry, pp 72–78

Raybaudi-Massilia RM, Mosqueda-Melgar J, Martín-Belloso O (2008) Edible alginate-based coating as carrier of antimicrobials to improve shelf-life and safety of fresh-cut melon. Int J Food Microbiol 121:313–327

Rhim JW (2004) Physical and mechanical properties of water resistant sodium alginate films. LWT Food Sci Technol 37(3):323–330

Robles-Sánchez RM, Rojas-Graü MA, Odriozola-Serrano I, González-Aguilar G, Martin-Belloso O (2013) Influence of alginate-based edible coating as carrier of antibrowning agents on bioactive compounds and antioxidant activity in fresh-cut Kent mangoes. LWT Food Sci Technol 50: 240–246

Rocher V, Bee A, Siaugue JM, Cabuil V (2010) Dye removal from aqueous solution by magnetic alginate beads crosslinked with epichlorohydrin. J Hazard Mater 178(1–3):434–439
Roger S, Talbot D, Bee A (2006) Preparation and effect of Ca2+ on water solubility particle release and swelling properties of magnetic alginate films. J Magn Magn Mater 305(1):221–227
Roh H, Yu MR, Yakkala K, Koduru JR, Yang JK, Chang YY (2015) Removal studies of Cd (II) and explosive compounds using buffalo weed biochar-alginate beads. J Ind Eng Chem 26:226–233
SenturkParreidt T, Müller K, Schmid M (2018) Alginate-based edible films and coatings for food packaging applications. Foods 7(10):170
Song Y, Liu L, Shen H, You J, Luo Y (2011) Effect of sodium alginate-based edible coating containing different anti-oxidants on quality and shelf life of refrigerated bream (*Megalobrama amblycephala*). Food Control 22:608–615
Tabassum N, Khan MA (2020) Modified atmosphere packaging of fresh-cut papaya using alginate based edible coating: quality evaluation and shelf life study. Sci Hortic 259:108853–108861
Tapia MS, Rojas-Graü MA, Carmona A, Rodríguez FJ, Soliva-Fortuny R, Martin-Belloso O (2008) Use of alginate- and gellan-based coatings for improving barrier, texture and nutritional properties of fresh-cut papaya. Food Hydrocoll 22:1493–1503
Tavassoli-Kafrani E, Shekarchizadeh H, Masoudpour-Behabadi M (2016) Development of edible films and coatings from alginates and carrageenans. Carbohydr Polym 137:360–374
Umar M, Li Y, Feng Y, Liu X (2018) Development characterization of alginate/Graphene oxide fibers with improved electrical conductivity. J Fiber Bioeng Inform 11(2):99–111
Valero D, Díaz-Mula HM, Zapata PJ, Guillén F, Martínez-Romero D, Castillo S, Serrano M (2013) Effects of alginate edible coating on preserving fruit quality in four plum cultivars during postharvest storage. Postharvest BiolTechnol 77:1–6
Wang Q, Zhang Newby BM (2018) Layer-by-layer polyelectrolyte coating of alginate microgels for sustained release of sodium benzoate and zosteric acid. J Drug Deliv Sci Technol 46:46–54
Wang B, Wan Y, Zheng Y, Lee X, Liu T, Yu Z, Huang J, Ok YS, Chen C, Gao B (2019) Alginate-based composites for environmental applications: a critical review. Crit Rev Environ Sci Technol 49:318–356
Wong DW, Gregorski KS, Hudson JS, Pavlath AE (1996) Calcium alginate films: thermal properties and permeability to sorbate and ascorbate. J Food Sci 61(2):337–341
Wong TW, Ashikin WHNS, Law CL (2014) Evaporation and diffusion transport properties and mechanical properties of alginate dried film. Dry Technol 32:117–125
Yaman Ö, Bayondrl L (2002) Effects of an edible coating and cold storage on shelf-life and quality of cherries. LWT Food Sci Technol 35:146–150
Younes M, Aggett P, Aguilar F, Crebelli R, Filipi c M, Frutos MJ, Galtier P, Gott D, Gundert-Remy U, Kuhnle GG, Lambre C, Leblanc J-C, Lillegaard IT, Moldeus P, Mortensen A, Oskarsson A, Stankovic I, Waalkens-Berendsen I, Woutersen RA, Wright M, Brimer L, Lindtner O, Mosesso P, Christodoulidou A, Horvath Zs, Lodi F, Dusemund B (2017). Scientific opinion on the re-evaluation of alginic acid and its sodium, potassium, ammonium and calcium salts (E 400–E 404) as food additives. EFSA Journal 15(11):5049. pp. 57. https://doi.org/10.2903/j.efsa.2017.5049
Zactiti EM, Kieckbusch TG (2006) Potassium sorbate permeability in biodegradable alginate films: effect of the antimicrobial agent concentration and crosslinking degree. J Food Eng 77:462–467
Zactiti EM, Kieckbusch TG (2009) Release of potassium sorbate from active films of sodium alginate crosslinked with calcium chloride. Packag Technol Sci 22:349–358
Zhan M, Wool RP, Xiao JQ (2011) Electrical properties of chicken feather fiber reinforced epoxy composites. Compos A Appl Sci Manuf 42(3):229–233
Zhang Y, Ma Q, Critzer F, Davidson PM, Zhong Q (2015) Effect of alginate coatings with cinnamon bark oil and soybean oil on quality and microbiological safety of cantaloupe. Int J Food Microbiol 215:25–30
Zhao W, Qi Y, Wang Y, Xue Y, Xu P, Li Z, Li Q (2018) Morphology and thermal properties of calcium alginate/reduced graphene oxide composites. Polymers 10:990

Zhong Y, Cavender G, Zhao Y (2014) Investigation of different coating application methods on the performance of edible coatings on mozzarella cheese. LWT Food Sci Technol 56(1):1–8

Zhou YM, Gao B, Zimmerman AR, Cao XD (2014) Biochar-supported zerovalent iron reclaims silver from aqueous solution to form antimicrobial nanocomposite. Chemosphere 117:801–805

Zhu D, Guo R, Li W, Song J, Cheng F (2019) Improved postharvest preservation effects of *Pholiota nameko* mushroom by sodium alginate–based edible composite coating. Food Bioproc Technol 12(4):587–598

Zhuang Y, Yu F, Chen J, Ma J (2016) Batch and column adsorption of methylene blue by graphene/alginate nanocomposite: comparison of single-network and double network hydrogels. J Environ Chem Eng 4(1):147–156

Zhuang Y, Yu F, Ma J, Chen J (2017) Enhanced adsorption removal of antibiotics from aqueous solutions by modified alginate/graphene double network porous hydrogel. J Colloid Interface Sci 507:250–259

Zouboulis AI, Katsoyiannis IA (2002) Arsenic removal using iron oxide loaded alginate beads. Ind Eng Chem Res 41:6149–6155

Chapter 7
Carrageenan Based Films

Rim Ben Abdallah, Tesnime Ghazouani, and Sami Fattouch

7.1 Introduction and Sources

In recent decades, due to environmental concerns, renewable resources have gained considerable attention. Particularly, bio-based biodegradable natural polymers constitute alternatives to petroleum-based polymer materials in various applications, including food packaging. In fact, in order to meet the industry incessant demand for food packaging, large quantities of synthetic polymers are produced, such as polyvinyl chloride (PVC). Despite their various advantages, most of these artificial organic polymers are resistant to biodegradation and were found to contribute to the increase of environmental problems. These pollutant plastic materials were harmful for both living organisms and ecosystems (Woodall et al. 2014; Rummel et al. 2016; Prata 2018; The Lancet Planetary Health 2017; Sedayu et al. 2019). This hazardous environmental issue attracted the attention of governments, industries, and researchers to come up with biodegradable and sustainable alternatives to plastic materials used in packaging. Thus, many researchers focused their investigations to use different natural sources such as plants, microbes, and animals to produce bioplastics (Sedayu et al. 2019). Because of their low cost and abundance, the marine macroalgae, known as seaweeds, are among the best natural sources that are used to produce bioplastics. These plant-like organisms, which usually live attached to rocks or other hard supports in coastal areas, constitute promising material since they reduce impacts on the food chain and are generally safe from exposure to chemicals or fertilizers (Rajendran et al. 2012; Sedayu et al. 2019). Carrageenan was found as the most promising polysaccharide that can be extracted

R. B. Abdallah · T. Ghazouani · S. Fattouch (✉)
Echochimie Laboratory (LR21ES02), Department of Biological and Chemical Engineering, National Institute of Applied Sciences and Technology (INSAT), University of Carthage, North Urban Center, Tunis, Tunisia
e-mail: sami.fattouch@insat.ucar.tn

T. Amin et al. (eds.), *Polysaccharide Based Films for Food Packaging: Fundamentals, Properties and Applications*,
https://doi.org/10.1007/978-981-99-4898-7_7

from seaweeds since it has an excellent ability for film forming (Blanco-Pascual et al. 2014; Paula et al. 2015; Siah et al. 2015; Sedayu et al. 2019). The first carrageenan was extracted from Irish moss (*Chondrus crispus*); the name "carrageenan" was given to this polysaccharide from Irish population which means "little rock" (Rinaudo 2007; Sedayu et al. 2019). Besides Irish moss, carrageenan was also extracted from various red seaweeds and algae of the *Rhodophyceae* family (Mohamed et al. 2020; Barrett 2018). Nowadays, around 90% of world's carrageenan is derived from the *Kappaphycus* and *Eucheuma* seaweed spices cultivated in Indonesian farming (Valderrama et al. 2013; Hurtado et al. 2015; Sedayu et al. 2019; Rinaudo 2007). This hydrocolloid makes it possible to form hot gels (up to 60 °C) and is therefore of interest compared to traditional animal gelatins. Besides the wide use of carrageenan for its textural functionality as a thickening and stabilizing agent in the food industry, emerging food packaging technologies have exploited it in multiple innovations, especially in the development of composite materials. It carries the code E407 of the classification of food additives.

The present chapter considers the structure of carrageenan and carrageenan-derived components. Methods of producing films based on these molecules in addition to their properties will also be presented. And finally, the application of carrageenan-based films in addition to their limitation and future trends will be discussed.

7.2 Structure of Carrageenan and Carrageenan-Derived Components

Carrageenan is a linear anionic galactan composed of alternate units of α-(1–3)-D-galactose and β-(1–4)-3,6-anhydro-D- or L-galactose as it is illustrated in Fig. 7.1 (Mohamed et al. 2020; Rinaudo2007; dos Santos and Grenha 2015). This polysaccharide is sulfated and hydrophilic and has a high molecular weight (well above 100 kDa) (dos Santos and Grenha 2015; Webber et al. 2012; Sedayu et al. 2019).

Campo et al. (2009) mentioned that carrageenan family can be split into six basic forms, namely: Iota (ι)-, Kappa (κ)-, Lambda (λ)-, Mu (μ)-, Nu (ν)-, and Theta (θ)-

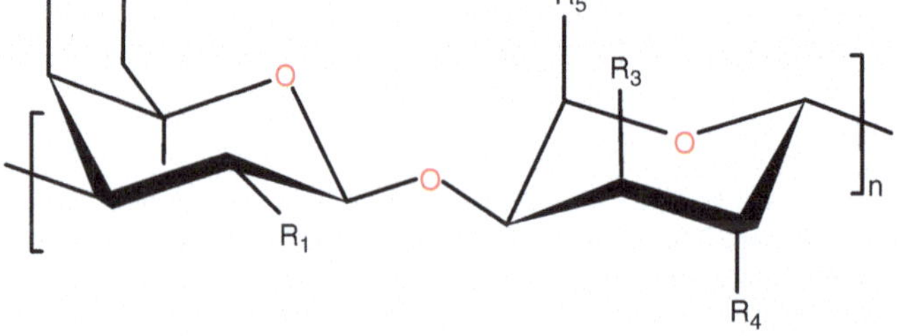

Fig. 7.1 Basic repeat structure of carrageenan

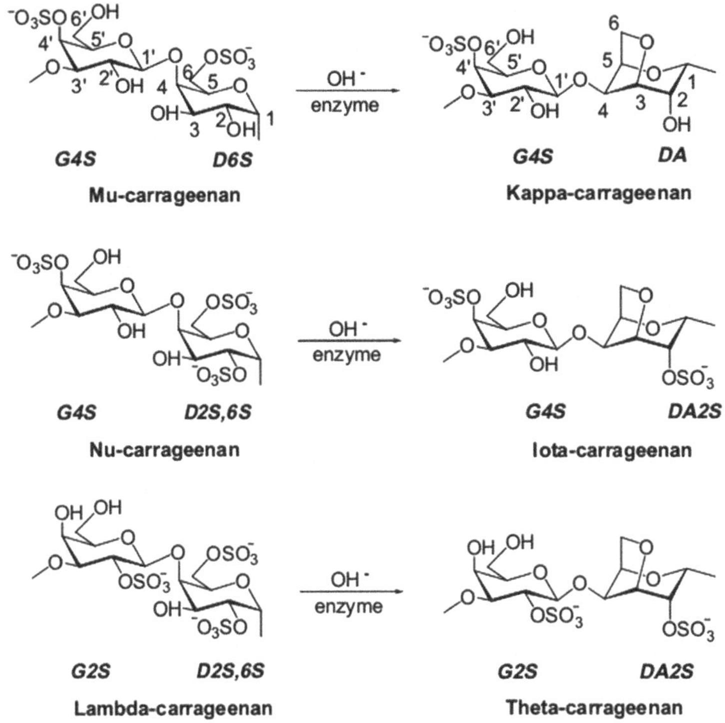

Fig. 7.2 Schematic representation of the different structures of the repeating dimeric units of commercial carrageenans and related structures (Campo et al. 2009)

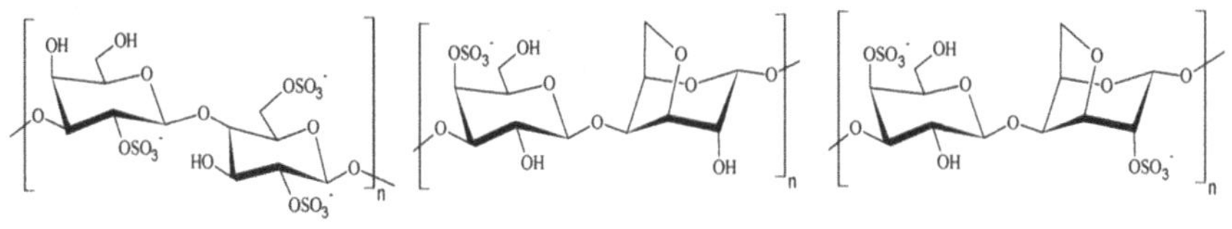

Fig. 7.3 Chemical structures of carrageenans (Cunha and Grenha 2016)

carrageenan. The extraction of carrageenan at high temperatures leads to the formation of the ι- and κ-carrageenans from μ- and ν-carrageenan, respectively (Fig. 7.2).

However, Rinaudo (2007) divided carrageenan into two families depending on whether it is esterified or not. The κ family, composed of ι-, κ-carrageenans, is esterified at the 4-position and the λ family which is non-esterified. Other researchers classified carrageenan on three main types (ι-, κ-, and λ-carrageenan) depending on the sulfate groups on the disaccharide repeating unit as well as the method and the weed from which the molecule is extracted (dos Santos and Grenha 2015; Alvarez-Lorenzo et al. 2013; Necas and Bartosikova 2013; Sankalia et al. 2006; Sedayu et al.

2019; Al-Alawi et al. 2011). Figure 7.3 depicts the position of the sulfate group on the disaccharide unit of the λ-, κ-, and ι-carrageenan.

All the carrageenan has between 22% and 35% of sulfate groups; however, it is known that the higher the ester sulfate levels in the carrageenan molecules, the lower their solubility temperature and gel strength (Sedayu et al. 2019; dos Santos and Grenha 2015; Necas and Bartosikova 2013). κ-Carrageenan has the lowest ester sulfate (25–30%) and between 28% and 35% of anhydrogalactose units. ι–Carrageenan has similar ester sulfate content (28–30%) and about 25–30% anhydrogalactose units as κ-carrageenan. Only the λ-carrageenan has no anhydrogalactose content with higher sulfate amount (32–39%) (Sedayu et al. 2019; dos Santos and Grenha 2015; Barbeyron et al. 2000). Those characteristics make the gel formed from κ-carrageenan strong and rigid and the one formed by ι-carrageenan soft and elastic. Only λ-carrageenan has no ability to gel forming (Rinaudo 2007; Sedayu et al. 2019; dos Santos and Grenha 2015). Yu et al. (2002) characterized the structure of three κ-carrageenan-derived oligosaccharides prepared through mild HCl hydrolysis. Anion exchange high-performance liquid chromatography (HPLC) allowed purifying pentasaccharide, heptasaccharide, and an undecasaccharide, all having a 4-*O*-sulfo-D-galactopyranose residue at both the reducing and non-reducing ends. Depolymerization of κ-carrageenan using carrageenase permitted to obtain low-molecular-weight carrageenans exhibiting anti-tumor activity as validated with the Sarcoma 180 tumor model in mouse when administered orally with a dose of 100 mg kg (Haijin et al. 2003).Special effects on immunological regulation, especially the phagocytosis ratio and phagocytosis index of macrophage, seemed to be at the basis of the observed anti-tumor activity. The anti-tumor capacity of non-sulfonated or light-sulfonated preparations was much greater than that of the high-sulfonated carrageenan. Haijin et al. (2003) reported that the carrageenan-derived products were found to enhance considerably both superoxide dismutase and catalase activities, thus suggesting that the carrageenan-derived oligosaccharides exhibit antioxidant potentials to help eliminate harmful free radicals (Haijin et al. 2003).

7.3 Production/Preparation of Carrageenan and Carrageenan Derivative-Based Films

The production of carrageenan-based films is still in its infancy, compared to other bio-based films. The few reported film fabrication methods until now are solvent casting (Sedayu et al. 2020; Balqis et al. 2017). The first step of the film production is to dissolve the appropriate concentration of carrageenan in deionized water by stirring for 15 min at 80 °C. In a second step, the suitable plasticizer (glycerol, sorbitol, PEG-300, etc.) is added with continuous stirring for about 25 min. Then, the film solution is cooled in a circular polystyrene petri dish plates with the dimension of 140 mm length and 8 mm depth for drying at 25–27 °C for 2days. At the end, the

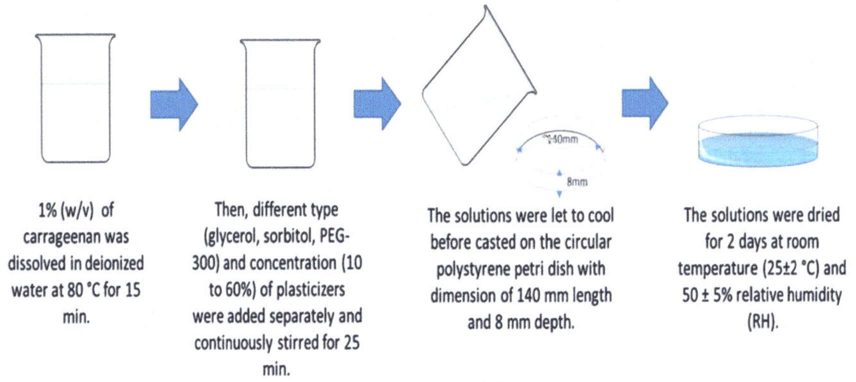

Fig. 7.4 The steps of carrageenan film preparation process (Balqis et al. 2017)

dried film is manually peeled off to be conditioned at a temperature of 25 ± 2 °C and 50 ± 5% relative humidity before characterization. Figure 7.4 illustrates the film preparation process.

7.4 Properties

7.4.1 Thermal Properties

The melting temperature (T_m) and glass transition temperature (T_g) are most important properties to determine thermal properties of the formed film. These two parameters are needed to find out if the polymers are suitable or not in food packaging (Bastarrachea et al. 2011). The impact of the carrageenan onto the thermal properties is highly dependent on the substances used in the blend (Carina et al. 2021). The study of Goonoo et al. (2017) showed that the ratio blend of carrageenan/poly(hydroxybutyrate) modulated the increase in the crystallization temperature and the decrease in the enthalpy of crystallization. Perumal et al. (2018) mentioned that the blend of carrageenan/fucoidan/collagen improved the thermal properties of the final produced biofilm.

7.4.2 Barrier Properties

In order to prevent the migration of chemical compounds, the barrier properties of the film have to be high. Sedayu et al. (2019) reported that carrageenan-based films have a very good oxygen barrier properties compared to other biofilms. Water vapor permeability (WVP) in biofilms has to be as low as possible which give a poor moisture barrier property and a poor water resistance (Cunha and Grenha2016;

Distantina et al. 2013; Roh and Shin 2006; Fabra et al. 2008), and this characterizes almost all bio-based films like the carrageenan-based films.

7.4.3 Mechanical Properties

The mechanical properties of the biofilms, namely the tensile strength (TS), elastic modulus (EM), and elongation at break (EB), indicate the ability of the film to resist to various stresses during the processing, handling, and storing of packaged food without losing its integrity (Carina et al. 2021). In most studies, the carrageenan-based films have a high TS which increases by adding other molecules such as agar (Rhim 2011), alginate (Huq et al. 2012), and nanoclay composite (Martins et al. 2013; Rhim2012). Shahbazi et al. (2016) mentioned that chitosan incorporation to carrageenan-based films enhanced the TS but decreased the EB. Kanmani and Rhim (2014) showed that grapefruit extract mixed with the carrageenan decreased the TS and increased the EB of the developed films. Sedayu et al. (2020) showed that the incorporation of nanoclay into the matrix of the semi-refined carrageenan matrix had no significant effect on the TS but increased the EM of the film by 18.6% and decreased the EB value by 17.2%. In the same study, the poly(caprolactone) incorporation with semi-refined carrageenan increased the TS by 10.7% and the EM by 18.3% but no significant effect on the EB of the carrageenan-based films.

7.4.4 Chemical Properties

The chemical properties, such as hydrophobicity or hydrophilicity, and the composition of the film with functional groups in the packaging play an important role in evaluating the safety and efficiency of food packaging. Due to the different components and their interactions with each other, the way of merging may vary greatly. Goonoo et al. (2017) mentioned that the κ-carrageenan blended with physical properties of polyhydroxy butyrate (PHB) has a high hydrophilicity with an increased surface free energy because the hydrophilic hydroxyl and sulphate groups of the κ-carrageenan take part in polar interactions.

7.4.5 Physical Properties

The type, the nature, and the ratio of the molecules used to prepare the biofilm can change its morphology and its rheological properties. Goonoo et al. (2017) studied the effect of different ratios of the poly(hydroxybutyrate) with the κ-carrageenan. The addition of 10 wt.% κ-carrageenan led to the formation of a porous structure, while a 50/50 blend of PHB and κ-carrageenan showed a spherical protrusion

(Goonoo et al. 2017). The incorporation of the hydrophilic nanoclay in the semi-refined carrageenan film formulation, in the study of Sedayu et al. (2020), gave greater homogenous dispersion in comparison with a less hydrophilic type of nanoclay. The choice of the right ration to incorporate the nanoclay to the semi-refined carrageenan film formulation (6% (w/w) of nanoclay, in this study) gives an optimum mechanical strength.

7.4.6 Optical Properties

Carrageenan-based films are known for their forming ability of excellent optical properties. Regarding the brittleness of its native films that restricts its functionality for food packaging applications, the addition of molecules to enhance the flexibility and workability of the carrageenan-based films is mandatory (Sedayu et al. 2019). Therefore, adjunction of plasticizers is essential to improve film flexibility and workability. These plasticizers, with low molecular weights, are supposed to intercalate and disperse within the carrageenan polymer network, thus disrupting hydrogen bonding and spreading the polymer chains apart, which then increases the flexibility of the resulting film (McHugh and Krochta 1994). Simona et al. (2021) formulated a carrageenan-based edible packaging film to which they incorporated orange essential oil and trehalose. The developed edible packaging was characterized for its optical properties, including UV-Vis spectrum, transparency value, and transmittance. The authors reported that the combination of orange essential oil with trehalose decreases the transmittance value in the UV and Vis regions ($0.14 \pm 0.02\%$ at 356 nm) what was considered an advantage of the film to act as a UV protector (Simona et al. 2021). Relative ratio of combined trehalose, orange essential oils, and carrageenan was found to affect the edible carrageenan-based film properties.

Aji et al. (2008) reported that incorporating SiO_2 nanoparticles to semi-refined ι-carrageenan films slightly improved their optical transparency. Other studies showed that although agar adjunction to carrageenan-based films improved their moisture barrier properties and water resistance, it lowered their optical and tensile properties (Rhim 2012; Rhim and Wang 2013). Additionally, it has been shown that the optical properties of carrageenan-lipid polymer were increased with lipids' addition because of the contribution of the lipid droplets in the polymer matrix to raise the extent of light scattering. The authors highlighted the importance of correlation between the extent of scattering and the size of the droplets as well as the components in the dispersed phase of the polymer matrix (Sánchez-González et al. 2009; Shojaee-Aliabadi et al. 2014).

7.4.7 Cost-Effectiveness

Comparing to the other sources of biofilm materials, especially those derived from terrestrial crops, seaweed may be much more cheaper since it does not need the use of arable lands, fresh water, and chemicals or fertilizers (Konda et al. 2015; Sedayu et al. 2019). However, the whole production chain of carrageenan biofilms, especially the extraction of the carrageenans, which requires a big amount of chemicals and energy, must be significantly and innovatively renewed to reduce the costs of the developed biofilms. Also, the use of the less expensive semi-refined carrageenan may help to reduce the film cots (Sedayu et al. 2019).

It is important to mention that carrageenan-based materials offer environmental benefits compared to the plastic materials and can compensate its higher production costs (DeGruson 2016; Sedayu et al. 2019).

7.4.8 Safety and Regulations

Despite the beneficial properties of seaweed, it can accumulate harmful and/or toxic metals (Circuncisao et al. 2018; Carina et al. 2021). This leads to making of legislation by national or international standardization agencies in regard to the use of the seaweed in foods. Several scientific reports proposed for European Union to set a maximum amount of 3 mg/kg dry weight for cadmium and a maximum level of 10 mg/kg for arsenic with no legislation in place for other toxic metals (Circuncisao et al. 2018; Carina et al. 2021). In South America, the United States, as well as Asian countries, the same lack of legislation can be observed (Circuncisao et al. 2018; Carina et al. 2021). Only France, as a European country, has already set clear regulations on the maximum amount of toxic metals and on 24 seaweeds (Holdt and Kraan 2011; Carina et al. 2021).

7.4.9 Methods to Improve Properties

The improvement of the carrageenan-based film has been investigated to ameliorate its barrier properties and mechanical, optical, and antimicrobial properties. For that, many technical approaches have been adopted: incorporation of antimicrobial agents, nanoparticle reinforcement, blending plasticizing incorporation, and multilayer carrageenan films.

7.4.9.1 Incorporation of Antimicrobial Agents

The aim of the incorporation of antimicrobial agent to the films is to extend the shelf-life and to maintain the quality and the safety of the packaged food. Several types of antimicrobial agents have been incorporated to the carrageenan-based film, namely, essential oils, plant extracts, metallic oxides, and organic acids (Shojaee-Aliabadi et al. 2013, 2014; Martiny et al. 2020; Oun and Rhim 2017; Wibowo et al. 2016).

7.4.9.2 Nanoparticle Reinforcement

Nanotechnology is significantly showing up in the technology science, and it has been developed in the biofilm formation to use small amounts of bioactive molecules. It has been shown that nanoparticles used in an appropriate ratio enhanced the physicomechanical, thermal, and barrier properties of biofilms (Sedayu et al. 2020; Shojaee-Aliabadi et al. 2014).

7.4.9.3 Blending

The carrageenans blending with other polymeric materials is the most simple and effective method to obtain desirable properties of the prepared biofilm (Sedayu et al. 2019). The properties of these films will vary depending on the type, the ratio, and the degree of miscibility of the blended material with carrageenan. Those materials can be lipids, proteins, polysaccharides, and other materials. Even that carrageenan-protein blending results in poor water barrier and water resistance properties, besides the improvement of the mechanical properties of the biofilm; this type of blending is less common (Fabra et al. 2008; Sedayu et al. 2019). Sanchez-Garcia et al. (2010) mentioned that the blending of carrageenan with zein (biopolymer from corn endosperm) highly improved the mechanical properties and decreased the water vapor permeability (17% lower) of the resulted film. Unlike protein blending, carrageenan/polysaccharide blending is the most common. Starch is commonly used as a filler in polymer blending because of its abundance in nature, and it has cheaper price than other molecules (Sedayu et al. 2019). The use of starch in an appropriate ratio enhances the mechanical properties and the moisture sensitivity of the carrageenan-based films (Abdou and Sorour 2014; Ali AkbariGhavimi et al. 2015).

7.4.9.4 Plasticizing Incorporation

The incorporation of the plasticizing is needed for the carrageenan-based films to enhance film flexibility and workability, since the native film shows brittleness that restricts its functionality for food packaging applications. The chosen plasticizer has to be compatible with the polymer system and the processing method used. Moreover, the ratio of both carrageenan and plasticizer must be well studied in the formulation and the enchantment required for the final film products (physical, thermal, and electrical properties); hence, the type of plasticizer has to count as much as its cost (Wittaya 2012; Cao et al. 2009; Farhan and Hani 2017; Sedayu et al. 2019). The increase of the moisture sensitivity and WVP of the carrageenan-based film has been reported as the effect side of the addition of plasticizers (Sedayu et al. 2019).

7.4.9.5 Multilayer Carrageenan Films

The multilayer technique is made to overcome the limitation of the blending method that decreases the barrier properties and water sensitivity as caused by the hydrophilicity of carrageenan and the other molecules (Shojaee-Aliabadi et al. 2013).

7.5 Applications

7.5.1 Edible Films and Coatings

With a rise in consumer commands for more natural foods, and for the environmental protection, a big deal of consideration has been committed to the development of new eco-friendly packaging materials. Edible films and coatings received meaningful interest over the last years due to their possibility to be utilized as edible packaging materials over synthetic ones and to reduce the environmental pollution.

Moreover, development of edible films and coatings can decrease the post-harvest losses and can provide less expensive packaging materials for industry and can lower the costs of the food products. Edible coating technology is a promising process to preserve the quality of food products. Research and development resolutions are leading to the development of novel functional characteristics of coatings, which depends on the properties of the product to be preserved or enhanced. The adoption of edible coating and their application charges are governed by legislation in the country where the coating is being practiced or in the country where the food products are to be exported. In recent years, there has been much concentration on carrageenan as biopolymer in edible film formation (Cian et al. 2014). According to FDA, carrageenans are GRAS (Generally Recognized as Safe) materials; thereby they have been passed the mentioned standards and are considered as edible films and coatings.

Using carrageenan for edible films and coatings involves a wide diversity of food industry applications such as fresh and frozen fish, meat, and poultry (Baldwin et al. 2011; Cha et al. 2002; Seol et al. 2009; Stuchell and Krochta 1995; Wu et al. 2000), several dried food products (Kampf and Nussinovitch 2000), fruit coatings (Bico et al. 2009; Hamzah et al. 2013; Ribeiro et al. 2007), and microencapsulation (Desai and Jin Park 2005).

The main objective of these applications is the limitation of dehydration, unsought chemical reactions (Osorio et al. 2011; Valero et al. 2013), besides the prevention of spoilage and deterioration by microbial contamination (Arvanitoyannis 2010; Soni et al. 2016). Furthermore, carrageenans can similarly be novelized to enhance the transfer and distribution of food additives or to manage the release of active agents, specifically in active packaging applications (Campos et al. 2011; Soni et al. 2016).

Regarding fresh-cut fruit, carrageenan coatings constitute an efficient barrier among the cut surface of the fruit and gas from the environs, thereby dipping respiration and minimalizing discoloration. This polysaccharide-based membrane could modify the internal atmosphere of the fruit and thus extend its shelf life. When used as an edible coating, carrageenan films have the capacity to preserve fruit texture and offer antibacterial protection during storage (Bico et al. 2009; Plotto et al. 2010).

Even if the use of hydrophobic substances such as resins, waxes, or water insoluble proteins accords better moisture barrier properties to edible films or coatings, the application of water-soluble hydrocolloids generally results in better mechanical properties than those offered by the hydrophobic materials (Arvanitoyannis 2010). In addition, water-soluble hydrocolloids similarly exhibit better gas barrier properties (Nussinovitch 2009) owing to their compact and firmly crowded conformation that result from organized hydrogen-bonded network structures (Atarés et al. 2011; Bonilla et al. 2012).

7.5.2 Pharmaceutical and Biomedical Applications

Polymeric films are progressively being considered as an auspicious pharmaceutical form for drug delivery and wound dressings due to advantages, such as their portability, application form, retention time, and easy storage and stability. These films can be used for direct or controlled release depending on the polymer structure of the formulation. The use of carrageenan in pharmaceutical and biomedical applications has been widely investigated (Garcia and Ghaly 2001; Hoffman 2002; Park et al. 2001; Picker 1999; Sankalia et al. 2006; Yamada et al. 2005), and it has been regulated in Europe, the United States, and the United Kingdom (García-González et al. 2011; Li et al. 2014).

In general, the applications of carrageenan-based films in these fields are mainly associated with drug delivery systems and wound dressings. The use of multilayered polymeric film in drug delivery systems is generally appropriate due to its capacity to operate the release of active compounds. Combining pharmaceutical agents into a polymer film pursued by organizing the layers of the film in a specific way may avoid the loss of the drug substances (Wang et al. 2007). Several multilayered films have been investigated to obtain the controlled release of active compounds which can be generated by certain conditions such as temperature, pH, enzymes, or ionic strength (Pinheiro et al. 2012; Quinn and Caruso 2004; Serizawa et al. 2005).

According to Nagai (1985), multilayered drug delivery films characteristically consisted of two principal components: carrier film(s) and barrier film(s). The carrier film includes the pharmaceutical or active compounds that are generally dissolved or dispersed homogenously in the polymer matrix. It was also reported that the better control of the release of the active compounds is correlated with configuring their concentration to aggregate from the outer wall to the inner part of the film. The barrier film is allied to one surface of the carrier film to basically "lock in" the active compounds and preserve stability till the release is activated by certain conditions. This mode of delivery comprises film strips containing active ingredients such as mouth freshener, nutrients or herbs, caffeine, and pharmaceutical ingredients as samples (Siah et al. 2015).

Polyethylene oxide and carrageenan-based solvent cast films were formulated as dressings for drug delivery such as streptomycin and diclofenac. The drug-loaded films showed a high capacity to captivate simulated wound fluid and significant mucoadhesion force and allowed the accurate release of both streptomycin and diclofenac. These films formed higher zones of inhibition against *S. aureus*, *Pseudomonas aeruginosa*, and *E. coli* in comparison to the zones of inhibition of the individual drugs.

Curative strips made from edible films containing drug substances which can be administered without water could represent a good alternative especially for persons who have frustration in swallowing tablets or capsules such as children and the elderly (Siah et al. 2015; Cunha and Grenha 2016). The benefit is that different doses can be obtained by using different lengths of edible film to be convenient for administering prescribed drugs (Nagai 1985).

Furthermore, carrageenan films are used as wound healing aids (Boateng et al. 2013; Fouda et al. 2015; Pawar et al. 2013) with hydrocolloid films due to their distinctive capacity to adhere to both humid and dry surfaces (Heenan 1998). The special preparation of these films based on their cross-linked polymer substances improving their water entrapping ability allows the films the competency to effectively absorb and retain the liquid upon contact with exuding wounds (Boateng et al. 2008). These films have showed required properties in preserving a moist environment, facilitating the drainage and evaporation of wound excretes, in common with permitting adequate gas exchange (Pawar et al. 2013). Another application of carrageenan films is intravaginal film platform, developed by Gu et al. (2015) for focused delivery of small interfering RNA (siRNA)-loaded nanoparticles to dendritic cells as a promising gene therapy for the prevention of sexually transmitted human immunodeficiency virus (HIV) infection. The polymer film was developed with polyvinyl alcohol and carrageenan supplied with the nanoparticles. In general, this application proves that the use of carrageenan improves the condition of the film and allows its application and administration by the vaginal way. This study found a basis for the development of films formed by carrageenan only or in combination for loading drugs for vaginal administration (Gu et al. 2015).

7.5.3 Electrolytes/Electrically Conductive Materials

On behalf of their environmental benefits, some biopolymers, including carrageenan, are considered attractive in the development of electrically conductive materials, taking account their important degree of compatibility with salts (Barbucci et al. 2000; Rozali et al. 2012). The peculiarity of their molecular structure, offering large number of hydroxyl groups, allows carrageenan to establish cross-linking networks with other constituents and substituents with further functional groups in polymer electrolytes (Yang et al. 2011). Thanks to their extensive hydrogen bonding, these polymers exhibit generally an advanced conductivity compared to those with a lesser extent of such bonding (Finkenstadt 2005). Even if liquid electrolytes display better conductivity and the capacity to transmit larger currents, flexible carrageenan-based films can be appropriate for the use in electrically conductive devices (Arof et al. 2010).

According to Prasad and Kadokawa (2010), the application of carrageenan-polymer film as the host polymer of two different ionic components, 1-(3-acryloyloxypropyl)-3-vinylimidazolium bromide and 1-vinyl-3-vinylbenzylimidazolium chloride, engenders a material with good electrical conductivity and mechanical properties.

This compound has, moreover, the competency to be applied for use in semiconductors or organic memory devices. Additionally, the host polymer of the electrolyte composite which is prepared by blending carboxymethyl k-carrageenan and carboxymethyl cellulose could also improve the conductive properties of the material and exposed promising characteristics for fabrication of dye-sensitized solar cells (Rudhziah et al. 2015).

One more innovative technique was reported by Esmaeili et al. (2014). It was founded on electrical conductive device involving carrageenan-based composites in the microbial fuel cell. This process provides economic benefits compared with the commonly used polymer material, poly(pyrrole).

7.6 Limitations and Future Trends

Several studies have reported the transformation of carrageenan into packaging films for a range of diverse requests. Nevertheless, the principal confines of most bio-based packaging materials are their cost, performance, and processing (Sorrentino et al. 2007). The approach of carrageenan film development for commercial applications is still in its infancy considering the quality and economic feasibility of the end-products focuses, face to conventional plastics or further available bio-based plastics in the market. Generally, cost efficiency is the major hurdle to be conquered for the evolution of bio-based plastic materials over the commercialization phase (Mekonnen et al. 2013) since production costs are usually more expensive than those of petroleum-based plastics (Changwichan et al. 2018).

Nevertheless, some unique properties of bio-based plastic materials may allow for future cost reductions, where bio-based materials can be more gainful regarding waste disposal management (van den Oever et al. 2017). Moreover, the bio-based plastic price is relatively more constant than fossil-based plastics because biomass prices typically alter much less than the oil price (van den Oever et al. 2017). In comparison with other sources of bio-based plastic materials such as those derived from corn, cassava, sugarcane, and other terrestrial crops, seaweed may be a less expensive material because it does not necessitate the use of arable lands, fresh water, and fertilizer or chemicals (Konda et al. 2015). However, the extraction process of carrageenan does need a large expense of chemicals and energy (Abdul Khalil et al. 2017) which is the main downside that must be stunned to reach economic feasibility for industrial implementation. However, it has been previously detected that the price of the renewable bio-based plastics becomes more competitive with its synthetic equivalents whenever the economics of large-scale production, conversion into products, and logistics become more promising (van den Oever et al. 2017). Another important reflection that compensates in part for the higher production costs is that carrageenan-based plastics can offer consumers significant inclusive environmental profits compared with some other materials and such perspectives have formerly been related with growing market demands (DeGruson 2016).

A further limitation of the application of pure carrageenan films is their low heat distortion temperature. In addition, their poor resistance to prolonged processing operations and their barrier properties confine their employment. These polymer films still show limitations regarding their water vapor permeability and water resistance since they are naturally hydrophilic (Alves et al. 2011), besides to being inherently brittle (Zarina and Ahmad 2015) which consequently limits their potential use for food packaging applications. Many approaches have been investigated to overcome the shortcomings of carrageenan-based films such as blending with natural or synthetic polymers. Plasticizers are therefore habitually added to the formulations to boost their flexibility (Vieira et al. 2011), and glycerol, a nonvolatile polyol, is one of the more common plasticizers combined with carrageenan films to improve their flexibility and processability (Farhan and Hani 2017; Paula et al. 2015). However, the amount of plasticizer added to carrageenan films must be carefully optimized in order to obtain the best overall film properties. Another approach was applied to overthrow the main properties; constraint of this biopolymer film is the adjustment of the film surface through a UV photo-crosslinking. This process has been investigated in further hydrophilic polymers such as starch-based and poly (lactic acid) polymers and has revealed a desirable enhancement, but these have not yet been applied in the case of carrageenan-based polymers (Sedayu et al. 2019).

The literature shows that carrageenan derivatives are captivating attraction by means of their bioactivities and, thus, could be at the basis of innovative promoting applications. Yamada et al. (2000) succeeded to depolymerize carrageenan using a method based on ultrasonication in the presence of ferrous ions, ascorbic acid, and ethylenediaminetetraacetic acid. The authors reported that the low butanoylated or low hexanoylated derivatives had higher anti-HIV activities as determined in a system in which MT-4 cells were infected with HTLV-III B (Yamada et al. 2000). The anticoagulant activity of the prepared carrageenan derivatives was determined from the measurement of activated partial thromboplastin time and obtained data showed that the butanoylated derivatives of λ-carrageenan with a degree of substitution of 1.1–1.5 had anticoagulant activities of 6.7–8.5 unit/mg, therefore indicating the decreasing effect of the substituted acyl group on anticoagulant activity (Yamada et al. 2000). The application of such findings will certainly push the boundaries of science and technology to focus efforts on this high-value biopolymer, especially with regard to the advances in the nanotechnology field (Shojaee-Aliabadi et al. 2014).

7.7 Conclusion

The incessant growing demand for processed food around the world, new market opportunities are opening for food packaging using bio-based and/or biofunctional materials. Hence, new generations of preservative-free foods and environmentally friendly materials will keep emerging at a significant rate. Particularly, the demand for carrageenan-based films seems to continue, thus boosting basic and applied science and technology research focusing further improvements and the discovery of new functionalities, taking full advantage of carrageenan physicochemical and rheological properties. Nowadays, behind starches, gelatin, and pectin, carrageenan holds the fourth largest hydrocolloid share in the global food textures market, in terms of value, and can rise higher. More focused research is still needed to add more values to carrageenan-based biofilms. The incorporation of bioactive compounds and antimicrobial agents in carrageenan-based films is one of the promoting applications as supposed to be a promising alternative to the use of chemical additives in the food industry.

References

Abdou ES, Sorour MA (2014) Preparation and characterization of starch/carrageenan edible films. Int Food Res J 21(1):189–193

Abdul Khalil HPS, Tye YY, Saurabh CK, Leh CP, Lai TK, Chong EWN, Nurul Fazita MR, Mohd Hafiidz J, Banerjee A, Syakir M (2017) Biodegradable polymer films from seaweed polysaccharides: a review on cellulose as a reinforcement material. Express Polym Lett 11 (4):244–265. https://doi.org/10.3144/expresspolymlett.2017.26

Aji AI, Praseptiangga D, Rochima E, Joni IM, Panatarani C (2008) Optical transparency and mechanical properties of semi-refined iota carrageenan film reinforced with SiO2 as food packaging material. AIP Conf Proc 1927(1):030039. https://doi.org/10.1063/1.5021232

Al-Alawi AA, Al-Marhubi IM, Al-Belushi MSM, Soussi B (2011) Characterization of carrageenan extracted from *Hypnea bryoides* in Oman. Mar Biotechnol 13(5):893–899

Ali AkbariGhavimi S, Ebrahimzadeh MH, Solati-Hashjin M, Abu Osman NA (2015) Polycaprolactone/starch composite: fabrication, structure, properties, and applications. J Biomed Mater Res A 103(7):2482–2498

Alvarez-Lorenzo C, Blanco-Fernandez B, Puga AM, Concheiro A (2013) Crosslinked ionic polysaccharides for stimuli-sensitive drug delivery. Adv Drug Deliv Rev 65(9):1148–1171. https://doi.org/10.1016/j.addr.2013.04.016

Alves VD, Castelló R, Ferreira AR, Costa N, Fonseca IM, Coelhoso IM (2011) Barrier properties of carrageenan/pectin biodegradable composite films. Proc Food Sci 1:240–245

Arof AK, Shuhaimi NEA, Alias NA, Kufian MZ, Majid SR (2010) Application of chitosan/iota-carrageenan polymer electrolytes in electrical double layer capacitor (EDLC). J Solid State Electrochem 14(12):2145–2152

Arvanitoyannis IS (2010) Irradiation of food commodities: techniques, applications, detection, legislation, safety and consumer opinion. Academic Press, London

Atarés L, Pérez-Masiá R, Chiralt A (2011) The role of some antioxidants in the HPMC film properties and lipid protection in coated toasted almonds. J Food Eng 104(4):649–656

Baldwin EA, Hagenmaier R, Bai J (eds) (2011) Edible coatings and films to improve food quality. CRC Press, Boca Rotan, FL

Balqis AI, Khaizura MN, Russly AR, Hanani ZN (2017) Effects of plasticizers on the physicochemical properties of kappa-carrageenan films extracted from *Eucheuma cottonii*. Int J BiolMacromol 103:721–732

Barbeyron T, Michel G, Potin P, Henrissat B, Kloareg B (2000) ι-Carrageenases constitute a novel family of glycoside hydrolases, unrelated to that of κ-carrageenases. J BiolChem 275(45):35499–35505

Barbucci R, Magnani A, Consumi M (2000) Swelling behavior of carboxymethylcellulose hydrogels in relation to cross-linking, pH, and charge density. Macromolecules 33(20):7475–7480

Barrett B (2018) Viral upper respiratory infection. Integr Med 2018:170

Bastarrachea L, Dhawan S, Sablani SS (2011) Engineering properties of polymeric-based antimicrobial films for food packaging: a review. Food Eng Rev 3(2):79–93

Bico SLS, Raposo MFJ, Morais RMSC, Morais AMMB (2009) Combined effects of chemical dip and/or carrageenan coating and/or controlled atmosphere on quality of fresh-cut banana. Food Control 20(5):508–514

Blanco-Pascual N, Montero MP, Gómez-Guillén MC (2014) Antioxidant film development from unrefined extracts of brown seaweeds *Laminaria digitata* and *Ascophyllum nodosum*. Food Hydrocoll 37:100–110

Boateng JS, Matthews KH, Stevens HNE, Eccleston GM (2008) Wound healing dressings and drug delivery systems: a review. J Pharm Sci 97(8):2892–2923

Boateng JS, Pawar HV, Tetteh J (2013) Polyox and carrageenan based composite film dressing containing anti-microbial and anti-inflammatory drugs for effective wound healing. Int J Pharm 441(1-2):181–191. ISSN 0378-5173. https://doi.org/10.1016/j.ijpharm.2012.11.045

Bonilla J, Atarés L, Vargas M, Chiralt A (2012) Edible films and coatings to prevent the detrimental effect of oxygen on food quality: possibilities and limitations. J Food Eng 110(2):208–213

Campo VL, Kawano DF, da Silva JDB, Carvalho I (2009) Carrageenans: biological properties, chemical modifications and structural analysis – a review. CarbohydrPolym 77(2):167–180

Campos CA, Gerschenson LN, Flores SK (2011) Development of edible films and coatings with antimicrobial activity. Food Bioprocess Technol 4:849–875. https://doi.org/10.1007/s11947-010-0434-1

Cao N, Yang X, Fu Y (2009) Effects of various plasticizers on mechanical and water vapor barrier properties of gelatin films. Food Hydrocoll 23:729–735

Carina D, Sharma S, Jaiswal AK, Jaiswal S (2021) seaweeds polysaccharides in active food packaging: a review of recent progress. Trends Food SciTechnol 110(2021):559–572

Cha DS, Choi JH, Chinnan MS, Park HJ (2002) Antimicrobial films based on Na-alginate and κ-carrageenan. LWT Food Sci Technol 35(8):715–719

Changwichan K, Silalertruksa T, Gheewala SH (2018) Eco-efficiency assessment of bioplastics production systems and end-of-life options. Sustainability 10(4):952

Cian RE, Salgado PR, Drago SR, Mauri AN (2014) Effect of glycerol and Ca+2 addition on physicochemical properties of edible carrageenan/porphyran-based films obtained from the red alga, *Pyropia columbina*. J ApplPhycol 27(4):1699–1708

Circuncisao A, Catarino M, Cardoso S, Silva A (2018) Minerals from macroalgae origin: health benefits and risks for consumers. Mar Drugs 16:400

Cunha L, Grenha A (2016) Sulfated seaweed polysaccharides as multifunctional materials in drug delivery applications. Mar Drugs 14(3):42

DeGruson ML (2016) Biobased polymer packaging. In: Smithers G (ed) Reference module in food science. Elsevier, Amsterdam

Desai KGH, Jin Park H (2005) Recent developments in microencapsulation of food ingredients. Dry Technol 23(7):1361–1394

Distantina S, Rochmadi R, Fahrurrozi M, Wiratni W (2013) Synthesis of hydrogel film based on carrageenan extracted from *Kappaphycus alvarezii*. Mod ApplSci 7(8):22

Esmaeili C, Ghasemi M, Heng LY, Hassan SHA, Abdi MM, Daud WRW, Ilbeygi H, Ismail AF (2014) Synthesis and application of polypyrrole/carrageenan nano-bio composite as a cathode catalyst in microbial fuel cells. Carbohydr Polym 114:253–259

Fabra MJ, Talens P, Chiralt A (2008) Effect of alginate and λ-carrageenan on tensile properties and water vapour permeability of sodium caseinate–lipid based films. Carbohydr Polym 74(3): 419–426

Farhan A, Hani NM (2017) Characterization of edible packaging films based on semirefined kappa-carrageenan plasticized with glycerol and sorbitol. Food Hydrocoll 64:48–58

Finkenstadt VL (2005) Natural polysaccharides as electroactive polymers. Appl Microbiol Biotechnol 67(6):735–745

Fouda MMG, El-Aassar MR, El Fawal GF, Hafez EE, Masry SHD, Abdel-Megeed A (2015) K-carrageenan/poly vinyl pyrrolidone/polyethylene glycol/silver nanoparticles film for biomedical application. Int J Biol Macromol 74:179–184

Garcia J, Ghaly E (2001) Evaluation of bioadhesive glipizide spheres and compacts from spheres prepared by extruder/marumerizer technique. Pharm Dev Technol 6(3):407–417

García-González CA, Alnaief M, Smirnova I (2011) Polysaccharide-based aerogels—promising biodegradable carriers for drug delivery systems. Carbohydr Polym 86(4):1425–1438

Goonoo N, Bhaw-Luximon A, Passanha P, Esteves S, Schonherr H, Jhurry D (2017) Biomineralization potential and cellular response of PHB and PHBV blends with natural anionic polysaccharides. Mater Sci Eng 76:13–24

Gu J, Yang S, Ho EA (2015) Biodegradable film for the targeted delivery of siRNA-loaded nanoparticles to vaginal immune cells. Mol Pharm 12(8):2889–2903

Haijin M, Xiaolu J, Huashi G (2003) A κ-carrageenan derived oligosaccharide prepared by enzymatic degradation containing anti-tumor activity. J Appl Phycol 15:297–303

Hamzah HM, Osman A, Tan CP, MohamadGhazali F (2013) Carrageenan as an alternative coating for papaya (*Carica papaya* L. cv. Eksotika). Postharvest Biol Technol 75(supplement C):142–146

Heenan A (1998) Hydrocolloids: frequently asked questions. World Wide Wounds 1:1–17

Hoffman AS (2002) Hydrogels for biomedical applications. Adv Drug Deliv Rev 54(1):3–12

Holdt SL, Kraan S (2011) Bioactive compounds in seaweed: functional food applications and legislation. J Appl Phycol 23(3):543–597

Huq T, Salmieri S, Khan A, Khan RA, Le Tien C, Riedl B, Fraschini C, Bouchard J, Uribe-Calderon J, Kamal MR, Lacroix M (2012) Nanocrystalline cellulose (NCC) reinforced alginate based biodegradable nanocomposite film. Carbohydr Polym 90:1757–1763

Hurtado AQ, Neish IC, Critchley AT (2015) Developments in production technology of Kappaphycus in The Philippines: more than four decades of farming. J ApplPhycol 27(5): 1945–1961

Kampf N, Nussinovitch A (2000) Hydrocolloid coating of cheeses. Food Hydrocoll 14(6):531–537

Kanmani P, Rhim J-W (2014) Development and characterization of carrageenan/grapefruit seed extract composite films for active packaging. Int J Biol Macromol 68:258–266

Konda NVSNM, Singh S, Simmons BA, Klein-Marcuschamer D (2015) An investigation on the economic feasibility of macroalgae as a potential feedstock for biorefineries. Bio Energy Res 8(3):1046–1056

Li L, Ni R, Shao Y, Mao S (2014) Carrageenan and its applications in drug delivery. Carbohydr Polym 103(supplement C):1–11

Martins JT, Bourbon AI, Pinheiro AC, Souza BWS, Cerqueira MA, Vicente AA (2013) Bio-composite films based on κ-carrageenan/locust bean gum blends and clays: physical and antimicrobial properties. Food Bioprocess Technol 6:2081–2092

Martiny TR, Pacheco BS, Pereira CM, Mansilla A, Astorga-España MS, Dotto GL, Moraes CC, Rosa GS (2020) A novel biodegradable film based on κ-carrageenan activated with olive leaves extract. Food Sci Nutr 8(7):3147–3156

McHugh TH, Krochta JM (1994) Sorbitol- vs glycerol-plasticized whey protein edible films: integrated oxygen permeability and tensile property evaluation. J Agric Food Chem 42(4): 841–845

Mekonnen T, Mussone P, Khalil H, Bressler D (2013) Progress in bio-based plastics and plasticizing modifications. J Mater Chem A 1(43):13379–13398

Mohamed SA, El-Sakhawy M, El-Sakhawy MAM (2020) Polysaccharides, protein and lipid-based natural edible films in food packaging: a review. Carbohydr Polym 238:116178

Nagai T (1985) Adhesive topical drug delivery system. J Control Release 2:121–134

Necas J, Bartosikova L (2013) Carrageenan: a review. Vet Med 58(4):187–205

Nussinovitch A (2009) Biopolymer films and composite coatings. In: Norton IT, Ubbink JB (eds) Modern biopolymer science, vol 10. Academic Press, San Diego, pp 295–326

van den Oever M, Molenveld K, van der Zee M, Bos H (2017) Bio-based and biodegradable plastics—facts and figures, Wageningen Food & Biobased Research no.1722. Wageningen Food & Biobased Research, Wageningen

Osorio FA, Molina P, Matiacevich S, Enrione J, Skurtys O (2011) Characteristics of hydroxy propyl methyl cellulose (HPMC) based edible film developed for blueberry coatings. Procedia Food Sci 1(supplement C):287–293

Oun AA, Rhim J-W (2017) Carrageenan-based hydrogels and films: effect of ZnO and CuO nanoparticles on the physical, mechanical, and antimicrobial properties. Food Hydrocoll 67: 45–53

Park SY, Lee BI, Jung ST, Park HJ (2001) Biopolymer composite films based on k-carrageenan and chitosan. Mater Res Bull 36:511–519

Paula GA, Benevides NMB, Cunha AP, de Oliveira AV, Pinto AMB, Morais JPS, Azeredo HMC (2015) Development and characterization of edible films from mixtures of κ-carrageenan, ι-carrageenan, and alginate. Food Hydrocoll 47:140–145

Pawar HV, Tetteh J, Boateng JS (2013) Preparation, optimisation and characterisation of novel wound healing film dressings loaded with streptomycin and diclofenac. Colloids Surf B Biointerfaces 102:102–110

Perumal RK, Perumal S, Thangam R, Gopinath A, Ramadass SK, Madhan B, Sivasubramanian S (2018) Collagen-fucoidan blend film with the potential to induce fibroblast proliferation for regenerative applications. Int J Biol Macromol 106:1032–1040

Picker KM (1999) Matrix tablets of carrageenans. I. A. compaction study. Drug DevInd Pharm 25(3):329–337

Pinheiro AC, Bourbon AI, Quintas MAC, Coimbra MA, Vicente AA (2012) K-carrageenan/ chitosan nanolayered coating for controlled release of a model bioactive compound. Innov Food Sci Emerg Technol 16:227–232

Plotto A, Narciso JA, Rattanapanone N, Baldwin EA (2010) Surface treatments and coatings to maintain fresh-cut mango quality in storage. J Sci Food Agr 90(13):2333–2341

Prasad K, Kadokawa J-I (2010) Preparation of composite materials composed of ι-carrageenan and polymeric ionic liquids. Polym Compos 31(5):799–806

Prata JC (2018) Airborne microplastics: consequences to human health? Environ Pollut 234:115–126
Quinn JF, Caruso F (2004) Facile tailoring of film morphology and release properties using layer-by-layer assembly of thermoresponsive materials. Langmuir 20(1):20–22
Rajendran N, Puppala S, Sneha RM, Angeeleena R, Rajam C (2012) Seaweeds can be a new source for bioplastics. J Pharm Res 5(3):1476–1479
Rhim J-W (2011) Effect of clay contents on mechanical and water vapor barrier properties of agar-based nanocomposite films. Carbohydr Polym 86(2):691–699
Rhim J-W (2012) Physical-mechanical properties of agar/κ-carrageenan blend film and derived clay nanocomposite film. J Food Sci 77(12):N66–N73
Rhim J-W, Wang LF (2013) Mechanical and water barrier properties of agar/κ-carrageenan/konjac glucomannan ternary blend biohydrogel films. Carbohydr Polym 96(1):71–81
Ribeiro C, Vicente AA, Teixeira JA, Miranda C (2007) Optimization of edible coating composition to retard strawberry fruit senescence. Postharvest Biol Technol 44(1):63–70
Rinaudo M (2007) Seaweed polysaccharides. In: Kamerling JP (ed) Comprehensive glycoscience, vol 4. Elsevier, Amsterdam, 2, pp 691–735
Roh YH, Shin CS (2006) Preparation and characterization of alginate–carrageenan complex films. J Appl Polym Sci 99(6):3483–3490
Rozali M, Samsudin A, Isa M (2012) Ion conducting mechanism of carboxy methylcellulose doped with ionic dopant salicylic acid based solid polymer electrolytes. Int J Appl 2(4):113–121
Rudhziah S, Ahmad A, Ahmad I, Mohamed NS (2015) Biopolymer electrolytes based on blend of kappa-carrageenan and cellulose derivatives for potential application in dye sensitized solar cell. Electrochim Acta 175:162–168
Rummel CD, Löder MGJ, Fricke NF, Lang T, Griebeler E-M, Janke M, Gerdts G (2016) Plastic ingestion by pelagic and demersal fish from the North Sea and Baltic Sea. Mar Pollut Bull 102(1):134–141
Sanchez-Garcia MD, Hilliou L, Lagaron JM (2010) Nanobiocomposites of carrageenan, zein, and mica of interest in food packaging and coating applications. J Agric Food Chem 58(11): 6884–6894
Sánchez-González L, Vargas M, González-Martínez C, Chiralt A, Cháfer M (2009) Characterization of edible films based on hydroxypropyl methylcellulose and tea tree essential oil. Food Hydrocoll 23(8):2102–2109
Sankalia MG, Mashru RC, Sankalia JM, Sutariya VB (2006) Physicochemical characterization of papain entrapped in ionotropically cross-linked kappa-carrageenan gel beads for stability improvement using Doehlert shell design. J Pharm Sci 95(9):1994–2013
dos Santos MA, Grenha A (2015) Chapter seven: Polysaccharide nanoparticles for protein and peptide delivery: exploring less-known materials. In: Advances in protein chemistry and structural biology, vol 98, pp 223–261
Sedayu BB, Cran MJ, Bigger SW (2019) A review of property enhancement techniques for carrageenan-based films and coatings. Carbohydr Polym 216(2019):287–302
Sedayu BB, Cran MJ, Bigger SW (2020) Improving the moisture barrier and mechanical properties of semi-refined carrageenan films. J Appl Polym Sci 137:e49238
Seol K-H, Lim D-G, Jang A, Jo C, Lee M (2009) Antimicrobial effect of κ-carrageenan-based edible film containing ovotransferrin in fresh chicken breast stored at 5°C. Meat Sci 83(3): 479–483
Serizawa T, Sakaguchi H, Matsusaki M, Akashi M (2005) Polyelectrolite multilayers prepared on hydrogel surfaces. J Polym Sci 43:1062–1067
Shahbazi M, Rajabzadeh G, Ettelaie R, Rafe A (2016) Kinetic study of kappa-carrageenan degradation and its impact on mechanical and structural properties of chitosan/kappa-carrageenan film. Carbohydr Polym 142:167–176

Shojaee-Aliabadi S, Hosseini H, Mohammadifar MA, Mohammadi A, Ghasemlou M, Ojagh SM, Hosseini SM, Khaksar R (2013) Characterization of antioxidant-antimicrobial kappa-carrageenan films containing *Satureja hortensis* essential oil. Int J BiolMacromol 52:116–124

Shojaee-Aliabadi S, Mohammadifar MA, Hosseini H, Mohammadi A, Ghasemlou Hosseini SM, Haghshenas M, Khaksar R (2014) Characterization of nanobiocomposite kappa-carrageenan film with *Zataria multiflora* essential oil and nanoclay. Int J Biol Macromol 69:282–289

Siah WM, Aminah A, Ishak A (2015) Edible films from seaweed Kappaphycusalvarezii. Int Food Res J 22(6):2230–2236

Simona J, Dani D, Petr S, Marcela N, Jakub T, Bohuslava T (2021) Edible films from carrageenan/orange essential oil/trehalose-structure, optical properties, and antimicrobial activity. Polymers 13:332

Soni A, Kandeepan G, Mendiratta SK, Shukla V, Kumar A (2016) Development and characterization of essential oils incorporated carrageenan based edible film for packaging of chicken patties. Nutr Food Sci 46(1):82–95

Sorrentino A, Gorrasi G, Vittoria V (2007) Potential perspectives of bionanocomposites for food packaging applications. Trend Food Sci Technol 18(2):84–95

Stuchell YM, Krochta JM (1995) Edible coatings on frozen king salmon: effect of whey protein isolate and acetylated monoglycerides on moisture loss and lipid oxidation. J Food Sci 60(1):28–31

The Lancet Planetary Health (2017) Microplastics and human health—an urgent problem. Lancet Planet Health 1(7):e254

Valderrama D, Cai J, Hishamunda N, Ridler N (2013) Social and economic dimensions of carrageenan seaweed farming, Fisheries and aquaculture technical paper no. 580. FAO, Rome

Valero D, Díaz-Mula HM, Zapata PJ, Guillén F, Martínez-Romero D, Castillo S, Serrano M (2013) Effects of alginate edible coating on preserving fruit quality in four plum cultivars during postharvest storage. Postharvest Biol Technol 77(supplement C):1–6

Vieira MGA, da Silva MA, dos Santos LO, Beppu MM (2011) Natural-based plasticizers and biopolymer films: a review. Eur Polym J 47(3):254–263

Wang X, Hu X, Daley A, Rabotyagova O, Cebe P, Kaplan DL (2007) Nanolayer biomaterial coatings of silk fibroin for controlled release. J Control Release 121(3):190–199

Webber V, De Carvalho SM, Barreto PLM (2012) Molecular and rheological characterization of carrageenan solutions extracted from *Kappaphycus alvarezii*. Carbohydr Polym 90:1744–1749

Wibowo AH, Listiyawati O, Purnawan C (2016) The effects of plasticizers and palmitic acid toward the properties of the carrageenan film. IOP Conf Ser Mater Sci Eng 107:012043

Wittaya T (2012) Rice starch-based biodegradable films: properties enhancement. In: Eissa AA (ed) Structure and function of food engineering. IntechOpen, London

Woodall LC, Sanchez-Vidal A, Canals M, Paterson GLJ, Coppock R, Sleight V, Calafat A, Rogers AD, Narayanaswamy BE, Thompson RC (2014) The deep sea is a major sink for microplastic debris. R Soc Open Sci 1(4):140317

Wu Y, Rhim J, Weller C, Hamouz F, Cuppett S, Schnepf M (2000) Moisture loss and lipid oxidation for precooked beef patties stored in edible coatings and films. J Food Sci 65(2):300–304

Yamada T, Ogamo A, Saito T, Uchiyama H, Nakagawa Y (2000) Preparation of o-acylated low-molecular-weight carrageenans with potent anti-HIV activity and low anticoagulant effect. Carbohydr Polym 41:115–120

Yamada K, Kamada N, Odomi M, Okada N, Nabe T, Fujita T, Kohno S, Yamamoto A (2005) Carrageenans can regulate the pulmonary absorption of antiasthmatic drugs and their retention in the rat lung tissues without any membrane damage. Int J Pharm 293(1):63–72

Yang Y, Hu H, Zhou C-H, Xu S, Sebo B, Zhao X-Z (2011) Novel agarose polymer electrolyte for quasi-solid state dye-sensitized solar cell. J Power Sources 196(4):2410–2415

Yu G, Guan H, Ioanoviciu AS, Sikkander SA, Thanawiroon C, Tobacman JK, Toida T, Linhardt RJ (2002) Structural studies on kappa-carrageenan derived oligosaccharides. Carbohydr Res 337(5):433–440

Zarina S, Ahmad I (2015) Biodegradable composite films based on κ-carrageenan reinforced by cellulose nanocrystal from kenaf fibers. Bioresources 10(1):256–271

Chapter 8
Gellan Gum-Based Films

Kanika Agarwal, Rubeka, Swati Jain, Memthoi Devi Heirangkhongjam, and Aparna Agarwal

8.1 Introduction and Sources

Biopolymers are the organic compounds which are produced by microorganisms and have repeating monomer units. These are of two types: homo-polymers which have one type of monomer unit and hetero-polymer which have more than one type of monomer units. These polymer chains are straight or sometimes branched. These are most versatile compounds present naturally. The biologically derived polymers can be further classified into peptides and proteins, polysaccharides, PHAs (polyhydroxyalkanoates) and polynucleotide.

Polysaccharides are polymers made up of many saccharide units; most common monosaccharides are glucose and fructose. Gellan gum is an anionic polysaccharide. It finds application in tissue engineering, encapsulation, drug delivery systems and medical field due to its non-cytotoxicity (Chesterman et al. 2020).

Gellan gum is an extracellular polysaccharide produced by microorganism *Sphingomonas elodea* (ATCC 31461), formerly known as *Pseudomonas elodea*. It was first discovered in 1978 but got its approval in 1992 by FDA. It has application in food industry as thickener and gelling agent and solidifying agent and also contributes as lubricant, flocculant or flavour enhancers.

The process of manufacture of gellan gum involves microbial fermentation which under controlled conditions is relatively reproducible and consistent. This provides the gellan gum a variety of functional properties and makes it a successful biopolymer with varied applications.

K. Agarwal · S. Jain · M. D. Heirangkhongjam · A. Agarwal (✉)
Department of Food and Nutrition, Lady Irwin College, University of Delhi, New Delhi, India
e-mail: aparna.gupta@lic.du.ac.in

Rubeka
Centre for Rural Technology, Indian Institute of Technology, Guwahati, Assam, India

T. Amin et al. (eds.), *Polysaccharide Based Films for Food Packaging: Fundamentals, Properties and Applications*,
https://doi.org/10.1007/978-981-99-4898-7_8

8.2 Structure of Gellan Gum and Gellan Gum-Derived Components

- Gellan gum is a linear, anionic heteropolysaccharide with a molecular weight of 0.5 × 10 6 Da. It is composed of tetra saccharide units 1,3 β-D glucose, 1,4 β-D glucuronic acid, 1,4 β-D glucose and 1,4 α-L-rhamnose (Fig. 8.1). It consists of 60% of glucose, 20% rhamnose and 20% glucuronic acid. Native gellan gum consists of fractions of protein and ash which can be removed by treatments like centrifugation and filtration. The de-esterified product is a polymer with a well-defined, unsubstituted, tetra saccharide repeating unit. In the solid state the molecule forms a parallel, half-staggered intertwined double helix in which each polysaccharide chain is a left-handed, threefold helix. The presence of uronic acid residues in the structure means the polysaccharide can be presented as a variety of salts.

In addition to the native form, gellan gum is available in three modified forms:

- High acyl (HA) gellan—It is partially deacetylated and soluble in hot water. The gel formed is soft, non-brittle and elastic. It can be used for forming films (Xiao et al. 2011).
- Low acyl (LA) gellan—It is highly deacetylated and soluble in hot water. The gels formed are firm/hard and brittle. The low flexibility is due to the double-helix chain structure (Guo et al. 2020). It is usually used in food applications (Giavasis et al. 2000; Sworna and Stouby 2020).
- High clarity gellan—It is highly deacetylated and clarified and is used in confectionery where clarity is crucial (Giavasis et al. 2000; Sworna and Stouby 2020).

8.3 Production of Gellan Gum and Gellan Gum-Derived Compounds

8.3.1 Gellan Gum

During the manufacture of gellan gum, steps may be added depending on which gum is to be prepared. LA gum is prepared by the treatment of the broth with alkali before

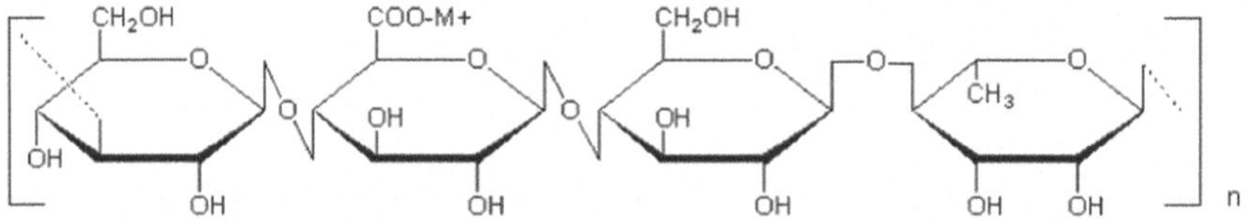

Fig. 8.1 Monomer unit of gellan gum

alcohol precipitation resulting in di-acylation. On the other hand, HA gum is prepared by precipitation immediately after heat treatment.

8.3.2 Gellan Gum-Based Derivatives

The packaging of food needs to keep the food fresh and safe. These films focus on biodegradable, antibacterial, antioxidant properties; thus, the use of biodegradable films made from polysaccharides such as gellan gum is used as edible packaging (Guo et al. 2020).

Studies have also attempted to improve the gelation process resulting in gellan gum derivatives (Agnello et al. 2018; Lee et al. 2004). It has been noted that the flexibility and hardness of the gels can be controlled through blends with either natural or synthetic polymers such as chitosan and carrageenan. These derivatives have good elasticity and break strength in films (Guo et al. 2020). Some of the gellan gum derivatives are discussed below:

(a) Gellan and Guar Gum Blend Films with Nisin

    Nisin is an antimicrobial agent against Gram-positive bacteria and is used as a food additive. It is thermostable, easily digestible and hypoallergenic (Guo et al. 2020). The film was prepared by dissolving gellan and guar gum at 60 °C for 60 min and then stored at 4 °C in the refrigerator. The film-forming solutions of gellan and guar gum were prepared and nisin was added to it. It was observed that the tensile strength, water vapour permeability and elongation at break of the gellan gum film improved due to the addition of guar gum. In addition, nisin added antimicrobial activity against *Bacillus subtilis*. This gellan-based derivative may be beneficial in food packaging and food preservation (Guo et al. 2020).

(b) Gellan Gum and Gelatin Films with Sodium Chloride

    The gellan gum and gelatin films were prepared by mixing 2% total solids of both of them in varying proportions dissolved in distilled water. Sodium chloride and glycerol were added during heating. The increase in gelatin concentration was found to increase the elongation while tensile strength decreased. The addition of sodium chloride was found to enhance the interaction between gellan and gelatin, thereby increasing the tensile strength (Lee et al. 2004).

(c) Methyl Acrylated Gellan Gum Films

    Methyl acrylated gellan gum was formulated by reacting methacrylic anhydride with gellan gum. The hydrogel formed resulted in physical and chemical cross-linking (Coutinho et al. 2010).

(d) Gellan Gum-Based Alkyl Amine Derivative

    A gellan gum derivative was prepared using alkyl amine which activated the hydroxyl group of gellan gum. This resulted in stronger hydrophobic aggregation in water and enhanced ionotropic cross-linking by calcium ions (Agnello et al. 2018).

(e) Miscellaneous

Gellan and xanthan gum were incorporated with zinc oxide nanoparticles which resulted in increased tensile strength, thermal stability and glass transition temperature. These nanocomposites also displayed decreased water vapour permeability and ultraviolet shielding when compared with just gellan and xanthan gum blend films. Thus, these films may be utilized in food packaging applications (Rukmanikrishnan et al. 2020b).

In another study, gellan gum-based derivatives were prepared using 2-hydroxyethyl cellulose and lignin. These blends reportedly had a good dispersion, interfacial adhesion and enhanced mechanical and thermal properties. It was observed that derivatives with a higher lignin content had a higher tensile strength and ultraviolet protection (Rukmanikrishnan et al. 2020a).

Thus, gellan gum and its derivatives can be used as edible films in packaging in the food industry.

8.4 Properties

8.4.1 Thermal Properties

Gellan gum is found to be stable at elevated temperatures and sustains its strength at 90 °C. However, depending on the conditions in gel formation, melting temperature can vary, that is, below or above 100 °C. The factor responsible for such varying melting point is the concentration of cations during gel formation. In this process, divalent and monovalent cations markedly raise the number of junction zones, thereby increasing the resistance to temperature. Other conventional stabilizers or thickeners can be used in lower concentration to modify the melting points (Bajaj et al. 2007; Giavasis et al. 2000).

During the formation of gellan gum edible films, heat treatment is a crucial step. Without proper heat treatment, the edible film surface will crack into small pieces after drying due to lack of intermolecular interactions (Nakamura et al. 1996). Also, the presence of –OH groups has a significant effect as it helps in changing the solution and reaction with other sugars during heating. Hence, an effective drying temperature is essential for a better performance. For instance, edible films dried at 60 °C have highest tensile strength (4.67 ± 0.02 g) and lowest water vapour transmission value (WVP) (4.57 ± 0.02 mg/day cm^2). The main reason is that appropriate temperature during the film formation allows gellan gum to fully stretch, transfer and displace, thus promoting the integration among them. On the other hand, if the drying temperature is elevated, disintegration of the film-forming materials might occur, causing larger cracking of molecules and resulting into lower-quality products (Xiao et al. 2011). Therefore, drying temperature is found to have a significant impact on the tensile strength and mechanical properties of different packaging materials.

Gellan gum-based films have different impacts on the mechanical properties when stored at different temperatures, namely, cold conditions (0 °C), room temperature (25 °C) and high temperature (35 °C). It was observed that at lower temperature (0 °C), it showed a good tensile strength of 4.64 g compared to room temperature (3.26 g) and high temperature (2.85 g). Therefore, this is evident that tensile strength changes with the change in temperature in different storage conditions. Also, considering the membrane performance, room temperature and high temperature showed a greater impact, whereas low temperature had a lesser impact indicating that the membrane performance was relatively stable (Xiao et al. 2011).

Gellan films undergo the aggregation of gellan helices which is inhibited by electrostatic repulsion between the helices. The helix-forming temperature is usually from 30 °C to 40 °C, and the temperature increases with increasing cation concentration, while the aggregate melting temperature changes widely from 40 °C to over 90 °C depending on cation concentrations (Kang et al. 2021). Thermal properties are improved on the addition of gellan into various film-based applications as highlighted in Mohd et al. (2016) and that gellan gum hydrogel improved the thermal stability in tissue engineering applications and was noncytotoxic to human skin fibro-blast cells and the thermal behaviour exhibited improvement in the degradation stability.

8.4.2 Barrier Properties

Gellan-based films are odourless, transparent, tasteless and generally impermeable to oxygen. However, they have certain limitations, for example, hydrophilic nature which affects their tensile strength and barrier properties when compared to conventional synthetic polymers. The water vapour permeability (WVP) of different polymeric films rely on various factors, including the integrity of film matrix, hydrophobicity, thickness, polymeric chain mobility, diffusion rate, solubility coefficient and interactions within the functional groups of the permeant and polymers. Moreover, an effective packaging material should be able to prevent the gaseous and moisture exchanges between the product and the environment. Thus, oxygen and water vapour barrier capacity of the film is the key to determine the efficiency and effectiveness of the film as a packaging material.

Sapper et al. (2019) conducted an experiment where gellan gum blend films were stored at 100–53% RH (relative humidity) gradient at 25 °C for a week. It was observed that WVP of the films significantly increased in the range of 100–75% RH gradient, due to greater or higher degree of plasticization of the material resulting into subsequent rise in molecular mobility and hence promoting the mass transfer rate. In both RH gradients (100–53 and 100–75% RH), gellan blend films exhibited lowest WVP values. This implies that the interactions between the linear amylose chains and negatively charged gellan chains, through hydrogen bonds, are effective enough in reducing the interaction of water molecules with the mixed matrix, thereby reducing their permeation and solubility capacities. Therefore, it is evident that higher plasticization level of polymers at high RH gradients results into differences in the packaging material making it less relevant in determining the barrier capacity of the films.

The oxygen permeability (OP) value is found to be decreased in gellan gum-based films or in polymers where gellan gum has been incorporated. The main reason for the reduction in OP values is due to the interactions between the amylose chain and charged gellan chains as discussed earlier. It helps in creating an extra compact network, thereby hindering the permeation of the gas molecules (Sapper et al. 2019; Forssell et al. 2002).

Gellan gums decrease the water vapour and oxygen permeability of the films which is also affected by the thickness value, moisture content, solubility and the layer in films. It also gets affected by any addition of other gums like xanthan, carrageenan, guar or pullulan in the film and their ratio (Zhu et al. 2013).

8.4.3 Mechanical Properties

Gellan has gained popularity as a gelatinizing agent by replacing gelatin in food items and can be used as thickener, emulsifier and stabilizer in food industry (dos Santos et al. 2020). Films made out of gellan are capable of forming a mechanical layer outside the formulations, matrices (Zhu et al. 2013) and cut fruits and vegetables (Lara Lledó 2016). Gellan gum-structured liquids are used to form these films in which various additives or other polymers such as pullulans and xanthan can be added to it to increase its mechanical strength (Kalia and Roy Choudhury 2019). Various starch-based gellan films exhibit changes in the mechanical strength of the film depending on the filler sources along with the gellan gum. Pliego-Arreaga et al. (2013) developed thermo processed starch-polyester bilayer films by adding gellan and/or xanthan gum and found that the film prepared by adding xanthan gum had lower mechanical strength and storage capability as compared to the gellan one and the bilayer formed with cassava starch with gellan gum appeared as the best option for food packaging purposes taking into account its functional properties and the good layer adhesion of the bilayer. Mohd Azam and Amin (2017) investigated the physical and mechanical properties of gellan gum films incorporated with manuka honey as a wound dressing material and found that increasing the film's composition has shown promising results to be used as wound dressing application, with respect to mechanical, physical and tensile strength. Ismail et al. (2017) tested the mechanical properties of gellan gum thin films and found that the film produced is higher in tensile strength (TS) but lower in strain-at-break (γ). Gellan gum films are found to have satisfying compression and physical strength which was tested by Mohd et al. (2016) for wound dressing applications.

The gelation property of gellan gum-based edible films occurs immediately on heating and cooling of gum solutions with the presence of cations. Such transitions in sol-gel are considered as phase transmission. As discussed earlier, gel formation happens at concentration and temperature and in the presence of monovalent and divalent cations (Yuguchi et al. 1993). Gellan gum forms an ordered double-stranded helix, whereas at high temperature, a single-stranded polysaccharide is formed resulting into the reduction of viscosity of the solution. During gel formation, transition temperature is set approximately at 35 °C, but it can also range between 30 °C and 50 °C. Below this transition temperature, a stiff structure is acquired which results into gel formation.

The gelation mechanism involves the development of double helical junction zones following the aggregation of double helical segments forming a three-dimensional arrangement by coordination complex with cations and hydrogen bond with water (Bajaj et al. 2007). The gelation potential of gellan gum-based films is markedly increased with the addition of monovalent and divalent cations upon cooling, thereby increasing the quantity of salt bridges at junction zones. Several studies have been conducted to study the effect of gel strength. Some of the important factors involved in gel strength are given below.

8.4.3.1 Hydrophilic Ingredients

Addition of hydrophilic ingredients, for example, sucrose (10% mass per volume), helps in decreasing the concentration of ion required to develop optimal strength of gellan gel (Papageorgiou and Kasapis 1995). Similarly, Tang et al. (2001) studied the effects of addition of sucrose and fructose on texture, clarity and gelling temperature. With the incorporation of sucrose and fructose, gel clarity was markedly increased. The gelling temperature also increased with the addition of sucrose. However, no effect was observed with the addition of fructose (up to 35% by mass per volume). Gel strength was dependent on concentration of cation. Sucrose improved the gel strength at low concentration and vice versa.

8.4.3.2 Acetyl Content

Acetyl content is another important factor affecting the strength of gel films. In gelation process, different acetyl contents result in gels with different properties. In general, the presence of large amount of acetyl and glycerol groups prevents the close association between chains of gellan polymer in the formation of bulk helix structure, thereby hindering the compact packing of cross-linked double helix. This in turn is associated with the formation of elastic, soft and thermo-reversible gels, which is the nature of native gellan gum based films. Similarly, deacetylated gellan gum-based films form brittle but firm and thermo-reversible gels in the absence of acetyl and glycerol groups. Deacetylated gellan gum forms firm, brittle and thermo-reversible gel because of the absence of acetyl and glycerol groups (Tang et al. 2001).

8.4.3.3 Concentration and Types of Ions

In the previous section, we have already discussed about the relationship that ions have a great impact on strength of gel and brittleness. Gelation does not occur in deionized water, but these properties can be improved with the addition of salts of sodium, potassium, calcium and magnesium. Notably, divalent cation at very low concentration (0.2%, by mass per volume) is found to be very effective. High gel strength was attained with a maximum of 0.004% calcium (by mass per volume), 0.005% magnesium (by mass per volume), 0.12% potassium (by mass per volume) and 0.16% sodium (by mass per volume). Addition of KCl or NaCl even at high salt concentration (about 1%, by mass per volume) lowered the gel strength (Huang et al. 2004).

8.4.3.4 Gel pH

It is observed that any change in pH does not alter or affect the setting points of the gel, although it affects the melting temperature in few cases. pH range of 3.5–8 (pH range of majority of foods) enhances the gel strength. For instance, gels developed with low levels of monovalent ions tend to melt at about 70 °C at neutral pH, whereas melting point increases slightly at 3.5 pH (Bajaj et al. 2007).

8.4.4 Dielectric Properties

The dielectric constant of gellan gum-based films increases with the increase in temperature while it slightly decreases with frequency. The loss factor increases with temperature while decreases at frequencies lower than 0.9 GHz. This trend shifts to opposite at frequencies higher than 2.7 GHz. The addition of sucrose and salt both affects the dielectric properties of gellan gel. Adding sucrose significantly decreases the dielectric constant, while adding salt is more effective in the adjustment of dielectric loss factor values (Zhang et al. 2015).

8.4.5 Tensile Properties

The tensile properties especially the extensibility properties of the gellan gum-based films are found to be a lot better than other natural polymeric gums such as xanthan (Hernández-García et al. 2021) which even increases with the addition of other natural polymeric sources such as pullulan (Zhu et al. 2013). A study conducted by Mohd Azam and Amin (2017) tested the tensile strength of a film made by gellan gum and manuka honey and found that the addition of higher amount of honey decreased the tensile strength and tensile strain of the film and vice versa. Similar results were reported by Xu et al. (2007) and Ismail et al. (2017) where increased content of gellan in a composite film of konjac-gellan increased the tensile strength of the film.

8.4.6 Heat-Sealing Properties

The heat-sealing characteristics of gellan-based gum films highly depend on the composition of the film. Balasubramanian et al. (2019) studied the effect of titanium oxide on nanocomposite films formed by using a combination of k-carrageenan, xanthan gum and gellan gum and found that the addition of titanium oxide enhances the heat-sealing strength and the antimicrobial activity of the film.

8.4.7 Flexibility Properties

Gellan gum films are found to be flexible enough to be incorporated in in situ preparation of various industrial, pharmaceutical and medical (Mohd Azam and Amin 2017; Patel et al. 2014)(Saha et al. 2018) applications. They provide biocompatibility, adherence properties, pseudo plastic behaviour, good tolerance, optical clarity and suitable tear behaviour. Patel et al. (2014) developed in situ gelling system using the gellan gum films for the treatment of periodontitis and found that the formulated drug was compatible with respect to appearance, pH, viscosity, % drug content, syringe ability, % drug release and effect of sterilization. Various levo floxacin periodontal gel was prepared by different concentrations of gellan gum and poloxamer 407 and was found to be having satisfactory syringe ability, gelling capacity, rheology, flexibility and physical properties (Sapra et al. 2013). Zhang et al. (2021) fabricated a multilayer film by using polyelectrolyte gellan gum-chitosan by layer-by-layer assembly technology with the incorporation of thyme essential oil coarse emulsion or nanoemulsion and found that the film incorporated with nanoemulsion possessed improved mechanical flexibility and UV blocking property as compared to others.

8.4.8 Electrical Properties

The I-V characteristic curve of gellan film conductivity confirms that gellan's conductivity is temperature dependent and an increase in polymer concentration also confirms that counter ion concentration plays an important role in increasing its conductivity. The variation of conductivity as a function of temperature and concentration confirms that gellan conductivity can be controlled. Emphasized that cation mobility and polymer conformation play an important role in gellan conductivity, and the control over conductivity can find its application as a bio-sensor in many electronic devices.

8.4.9 Optical Properties

The optical morphology of the gellan pullulan blend-based films were tested by Zhu et al. (2013) who found that the film was optically smooth, compact and continuous and semitransparent films can be formed.

8.4.10 Water Solubility Properties

Gellan gum solutions tend to form three-dimensional network hydrogels by complication with cations, such as calcium ions, and hydrogen bonding with water (Iglesias et al. 2020). Gellan-based gum films when made with gums like konjac exhibit lower water uptake value and hence possess more water-resistant qualities (Xu et al. 2007).

8.4.11 Viscous Properties

The viscosity of gellan gum-based films decreases with increase in shear rate which exhibits the pseudo plastic behaviour; the overall rheological properties are affected by changes in the acyl substitution (Altan Kamer et al. 2021). The viscosity of gellan-/pullulan-based film is also affected by adding other gums (Zhu et al. 2013) as highlighted in the study conducted by Patel et al. (2014) and Sapra et al. (2013) who developed formulations containing 0.5%w/v of gellan gum and 15%w/v of poloxamer 407 which showed satisfactory results for viscosity, syringe ability, gelation time, gelation temperature and other physical properties. Effect of gellan gum/tuna skin film for guided bone regeneration in artificial bone defect in rabbit calvaria was studied by Jung et al. (2020) who found that the bone formulation by using gellan gum and tuna skin was comparable to collagen-based applications. Based on maximum desirability, the formulation was considered as an optimized formulation. Kamer et al. (2021) highlighted that deacetylated commercial gellan gum showed higher viscosity than the produced native gellan gum at low concentrations of the solution due to concentration dependence of coil-helix transition.

8.4.12 Printing Properties

With gellan gum films, three-dimensional bioprinting is expected to become a strong tool especially for regenerative medicine. Chen et al. (2020) have reported the development of a novel hybrid bio ink by combining rigid gellan gum (GG), flexible sodium alginate (SA) and a bioactive substance thixotropic magnesium phosphate-based gel (TMP-BG) and found that the formulated hybrid bio ink has good print ability, biocompatibility, mechanical support and bioactivity, which is expected to have promising applications in three-dimensional bioprinting. Akkineni et al. (2016) prepared and studied composites of highly concentrated alginate and gellan gum in order to combine the excellent printing properties of alginate with the favourable gelling characteristics of gellan gum and found that the print fidelity of highly concentrated alginate-gellan gum composites was comparable to those of pure alginate; after plotting and cross-linking, the scaffolds possessed improved qualities regarding shape fidelity, mechanical strength and initial cell attachment making them attractive for tissue engineering applications.

8.4.13 Aroma Retention Properties

The retention of aroma compounds can be influenced to a greater or lesser extent by the molecular weight and shape, steric hindrance, chemical functionality and polarity and volatility of the core material of the gellan-based film. For instance, if the volatile compound is small compared with the cavity, only some of the surface of the guest molecule is in contact with the walls of the cavity (Madene et al. 2006).

8.4.14 Antimicrobial Properties

Incorporation of the essential oils increases the antimicrobial activity (Mousavian et al. 2021) of the gellan-based films (Zhang et al. 2021). Xu et al. (2007) prepared and characterized konjac glucomannan-gellan gum blend films and studied their suitability for release of nisin and found that with the increase of the content of gellan gum, the antimicrobial effects were enhanced. Antimicrobial films incorporating nisin were found to have antimicrobial activity against *Staphylococcus aureus* particularly with higher contents of gellan gum, and the final results concluded that the konjac-gellan film could be applied as a potential food package material for releasing the active agent such as nisin. The biocompatibility of gellan gum with desired drug hydrogels in wound dressing material was found to have promising antibacterial and/or bactericidal effect (Mohd et al. 2016).

8.4.15 Moisture Sorption Properties

The moisture sorption capacity of gellan-based films is found to be better than that of films made by xanthan or pullulan wherein the blend of starch with gellan gum was effective in reducing the moisture sorption capacity of starch films while reducing water vapour permeability, enhancing the film strength and resistance to break and preserving films against starch retro gradation throughout the storage time (Sapper et al. 2019). Xanthan on the other hand improvised the tensile strength but did not reduce the water sorption capacity and water vapour permeability. Choudhary et al. (2020) studied the gellan gum cross-linked poly hydrogel for the removal of malachite green dye and methylene blue from wastewater and found that hydrogel can be regarded as an encouraging superabsorbent hydrogel candidate for the remediation of wastewater treatment because of its excellent adsorption capacity.

8.4.16 Release Kinetics of Active Packaging

The formulation of different films with gellan allows a smooth release of bioactive components through the desired packaging material, hence making it suitable for active, smart and intelligent packaging materials. Zhang et al. (2021) found that the multilayer film formed by using gellan gum and chitosan and incorporated with nanoemulsion was a potential antimicrobial film with effective slow-releasing system for food packaging applications. Zhu et al. (2013) found that the SEM (scanning electron microscopy) and FTIR (Fourier transform infrared spectroscopy) analysis of the gellan/pullulan-based film appeared to have a potential for use as a coating and active packaging of food materials. Similar results were reported by Patel et al. (2014) that the film prepared by gellan provided sustained release of the drug over 8 h. Similar facts were highlighted by Iglesias et al. (2020) that incorporation of gellan gums in zinc pectinate matrices increases the viscosity of the polymer matrix. Gellan gum-based films with gellan at 10–20% in the blend were prioritized in the formulation over xanthan and pullulan, and the starch- and gellan-based film was suggested to obtain starch films with more adequate properties for food packaging purposes (Sapper et al. 2019). Gellan with konjac gum as a film is found to have potential for controlled release of nisin and hence can be applied to active packaging of meat and meat products (Xu et al. 2007). Hernández-García et al. (2021) studied release kinetics of nisin from pectin-gellan films intended for active food packaging and found that the release kinetics of nisin was fitted to the analytical solution of the Fick's second law for an infinite plate. The diffusion coefficients of nisin were satisfactory. The mechanical properties were reduced by the nisin incorporation and no significant effect on WVP was observed.

8.4.17 *Safety and Regulations*

The gellan gum and its films, hydrogels and formulations, can be safely used as disintegrating agent, ophthalmic drug delivery, beads, floating in situ gel formulations (Saha et al. 2018). Dental (Sapra et al. 2013) (Patel et al. 2014) wound dressing material (Mohd et al. 2016) and orthopaedic (Jung et al. 2020) applications have been formulated safely. Gellan being a natural excipient has certain advantages over other synthetic and semisynthetic gum applications, but while working with them, the dosage and the regulatory requirement should be looked upon which is specific to each country and continent (Saha et al. 2018). Gellan gum was tested for toxicity in mice, rats and dogs, and 23-day dietary studies in human concluded that there were no adverse effects that could be attributed to chronic exposure to gellan gum, and hence the composite films should be studied for their safety and regulations because of the additional ingredients. The digestibility studies of gellan by the gut micro flora should also be looked into account because it is unlikely that the human gut micro flora have enzymes which can degrade the gellan gum-based films and gellan gum itself (Kang et al. 2021).

8.4.18 *Methods to Improve Properties*

The collaborative researches reached to some conclusions; from optical rotation, light scattering, osmotic pressure and intrinsic viscosity measurements, two disordered chains of deacetylated sodium-type gellan at higher temperatures make a double helix at lower temperatures in dilute solution, and from rheological measurements, gellan forms a gel under conditions favourable for the aggregation of double helices. The aggregation of double helices follows the coil-to-helix transition on further cooling above a certain gellan and cation concentrations (Table 8.1).

8.5 Applications

Hydrocolloids or gums are valuable for industrial use due to their functional properties. These natural polymers form gels when dispersed in water and result in textures of varying consistencies ranging from fluid to hard. They exhibit these properties by acting as thickening, coating, stabilizing and emulsifying agents (Saha and Bhattacharya 2010). Both forms of gellan, that is, acetylated and deacetylated, are popular for various uses among food, pharmacy and biomedicine industries.

Table 8.1 Improving functional properties of gellan gum-based films

Properties	Characteristics
pH	The pH should be from 6.5 to 7; more acidic or more alkaline relatively decreases the growth of cell and gellan production
Dissolved oxygen	High oxygenation rate that promotes optimal gellan synthesis
Temperature	Fermentation involving gellan gum production is done at 30 °C, reaches yield peak at 20 °C, remains relatively high at 25 °C and significantly reduces above 30 °C
Hydrophilic ingredients	Decreases the concentration of ion required in developing optimal strength of gellan gel
Ion concentration	Divalent cation at very low concentration is very effective. Presence of salts of sodium, potassium, calcium and magnesium improves gelation
Moisture sorption	Gellan gum blended with starch is effective in reducing the moisture sorption capacity of starch films, reduces water vapour permeability and enhances the film strength
Antimicrobial properties	Essential oils increase the antimicrobial activity, for example, nisin
Viscosity	Deacetylated gellan gum showed higher viscosity compared to native gellan gum due to its coil-helix transition

8.5.1 *Food Applications*

Typically, and widely, gellan gum is used to enhance the textures in various kinds of products. This polysaccharide is used most often in confectionery, bakery fillings, dessert gels, frostings, icings and glazes, jams and jellies, pet foods, low-fat spreads, sauces, toppings and dairy products (Duxbury 1993). The micelle structure of gellan dispersion helps to create low-fat foods, by acting as a fat substitute ingredient (Iurcius et al. 2016). When used alone or in combination with other hydrocolloids, it is also used in making low-fat (0–20%) meat and poultry products (Milan and Maleki 2012; Laaman and Tye 1991). Gellan gum, when used in batter coatings, also helps to reduce oil absorption by forming a film that acts as a barrier while retaining the desirable qualities of the products like crispiness and juiciness (Chalupa and Sanderson 1994). These biopolymer layers in the form of food films possess the ability to control dehydration and impede the development of textural changes in various food products. In recent years, the popularity and use of gellan have increased for fried foods, resulting in products with remarkably reduced oil absorption (Albert and Mittal 2002; Bajaj and Singhal 2007). These food films can even facilitate adhesion of spices, sweeteners and flavouring agents on products like nuts and crisps. Further, gellan gum when used with poly dextrose is suitable to yield more low-calorie foods like sugar-free jelly sweets, which is usually not attainable with other gums like agar, gelatin and pectin. This is attributable to its unique property of gelling while not adding to the viscosity of the depositing mix (Gibson and Sanderson 1997).

Gellan gum is suitable for producing gelatin desserts and jams, due to its easy gelling and better humidity-controlling properties. It provides greater transparency and heat stability to the desserts at a concentration of 0.3% (Bajaj and Singhal 2007). Gellan can replace agar and provide better textures in foods like hard and soft red bean jelly and certain varieties of noodles (Sanderson 1990). LA gellan gum can modify the properties of traditional gelatin dessert jellies by raising the initial set temperature and the melting point of the gel, which helps to retain and maintain the shape of dessert for longer period, at room temperature. Blends of HA and LA gellan gum can bring about a variety in textures. Also, LA gellan gum when used with carrageenan can produce gelatin-free confectionery. Furthermore, addition of sugar to gellan aids in increasing the strength and clarity of the jelly confectionery, where the concentration and type of sugar would influence the setting and melting temperatures and thereby the gelation process. High concentrations of sugar (70–80%) impede the mixing and deposition process of the confectionery mix, while also ending up making the gels less firm (at >40% sugar) (Sworna and Stouby 2020). Glucose, fructose and corn syrups are known to result in firmer and more brittle gels than sucrose alone (Gibson 1997).

In jams, gellan can be used as a pectin substitute, to make products with lower syneresis while maintaining the desired sensory properties (Duran et al. 1994). While HA gellan gum gives a sheen and spread ability to the product, LA gellan gum gives a more stable and firm jam. However, the type of fruit used with LA gellan gum can considerably affect the product's formulation due to the varying pH and ion contents (Sworna and Stouby 2020). Gellan has an excellent water binding capacity that helps to keep up the shape and facilitates release of fruit flavour in bakery fillings and in brine, cured meat and sea food products. In addition, gellan-made fruit bakery fillings have better texture and mouthfeel in comparison to the traditional modified starch fillings (Sworna and Stouby 2020; Ramalingam et al. 2014; Sutherland 2005).

The food industry utilizes the distinctive rheological properties of gellan. It acts a great stabilizer in reconstituted vegetable juices (Liang et al. 2006) and for water-in-oil dispersion (Norton 1992). Other applications of gellan include making liquid gels with enhanced nutritional value, for example, using carrot pulp (Liang et al. 2006; Banerjee et al. 2013; Manjunatha and Gupta 2006) to prepare dried rehydratable particulates for instant foods like instant gravies and instant sauces and soups.

Gellan gum when used at low concentrations can form a fragile gel network called a fluid gel. These fluid gels have good suspension properties and therefore can be used in beverages ranging from water-based beverages to alcoholic beverages. The role and type of gellan used in each of these preparations are variable. HA gellan can be used to suspend fruit pulp to give a smooth fluid gel and formulate dairy protein beverages as it can gel independently of milk proteins. It is used in soy beverages, whey-substituted beverages, fresh milk, flavoured milk like chocolate milk and fermented milk products like dough (Hasheminya et al. 2011). Yogurts are prepared with LA gellan gum usually at a concentration of 0.04% along with other stabilizers, depending on the desired texture of set or stirred yogurt (Sworna and Stouby 2020). Often, gellan gum is used as fruit toppings in pastries; as stabilizer in ice cream, cream cheese and cottage cheese; and as edible food films for confectionery products.

Gellan gum can be added preceding the fermentation process or as a fining agent in preparation of alcoholic beverages including beers, wines and fortified wines (Dartey 1993; Giese 1995). It has been demonstrated that gellan is also suitable as a support for immobilization of beer yeast because of its ability to form gels in different types of ions and its lack of toxicity and resistance to high differences in pH (Mantaluta et al. 2012). Additionally, being biodegradable and resistant to gastrointestinal enzymes, gellan is known to be a suitable biopolymer for probiotic encapsulations. More recently, HA gellan gum is also being considered appropriate for managing swallowing disorder—dysphagia—where it can be used in powder mixtures or ready-to-eat products to achieve products of desired consistencies and textures.

Owing to its biochemical and biodegradable properties, gellan has applications in other industries as well. Microbiological media, tissue culture, deodorant gels, capsules and photographic emulsions are some other applications of acetylated and deacetylated gellan gum.

8.6 Limitations and Future Trends

Widely, gellan gum is known to be safe for consumption (EFSA Panel on Food Additives and Nutrient Sources Added to Food (ANS) et al. 2018). Studies have demonstrated that even much higher consumption than the usual intake would cause no adverse effects in the human body (Anderson et al. 1988). However, an animal study reported that chronic intake of high doses might lead to some abnormalities in gut lining (Tetsuguchi et al. 1997) and might slow down the digestive processes. In the field of biomedicine, tissue engineering, pharmaceuticals and cosmetic industry, gellan gum is used largely as a culture medium, delivery vehicle and cell encapsulation material since the inception of its discovery. Despite the many desirable properties of gellan gum, these hydrogels are known to possess low mechanical properties like poor toughness and tissue tolerance, which could constrict its usage (Kang et al. 2015; Maia et al. 2019).

With the expanding consumption of convenience foods and rising consumer awareness on nutrient labelling, there is an increased consumer awareness and familiarity to additives. Because of growing application of hydrocolloids in food and beverage industry, consumers now choose and favour ingredients judiciously. Gellan gum offers more versatile functionalities and advantages over other gelling agents, such as being effective even at less concentration. Within the hydrocolloid textural spectrum, the commercial HA and LA gellan gum products can be at opposite ends as they have very different textures. However, more acceptable forms of gellan gum are with medium acyl content, as they can provide more altercations in texture and a single homogeneous setting performance (Baird et al. 1992). A combination of the both types would support intermediate textures while maintaining some properties of both HA and LA gellan gums. Going forward, having a control on the degree of acylation at a commercial scale could result in products with wide spectrum of textural characteristics (Sworna and Stouby 2020).

8.7 Conclusion

This linear polymer has properties that enable applications in food, medicine and pharmaceutical industry. However, its capacity to form gels is the most distinguished attribute which is being utilized well by the food industry. Low acyl and high acyl forms of gellan gum produce biodegradable, edible films for packaging. Further, it is known that when added it aids in improving the thermal behaviour of gels. Gellan has gained popularity as a gelatinizing agent by replacing gelatin in food items and can be used as thickener, emulsifier and stabilizer in food industry. Films made out of gellan are capable of forming a mechanical layer outside the formulations, matrices and cut fruits and vegetables. The tensile properties especially the extensibility properties of the gellan gum-based films are found to be a lot better than other natural polymeric gums. The heat-sealing properties enhance biocompatibility, adherence properties and pseudo plastic behaviour and provide good tolerance, optical clarity and suitable tear behaviour to the products. Apart from this, gellan gum exhibits tremendous abilities in aroma retention, three-dimensional bioprinting and microbial sorption. Despite the many desirable properties of gellan gum, these hydrogels are known to possess low mechanical properties like poor toughness and tissue tolerance, which could constrict its usage. Going forward, having a control on the degree of acylation at a commercial scale could result in products with wide spectrum of textural characteristics.

References

Agnello S, Palumbo FS, Pitarresi G, Fiorica C, Giammona G (2018) Synthesis and evaluation of thermo-rheological behaviour and ionotropic crosslinking of new gellan gum-alkyl derivatives. Carbohydr Polym 185:73–84

Akkineni A, Ahlfeld T, Funk A, Waske A, Lode A, Gelinsky M (2016) Highly concentrated alginate-gellan gum composites for 3D plotting of complex tissue engineering scaffolds. Polymers 8(5):170. https://doi.org/10.3390/polym8050170

Albert S, Mittal GS (2002) Comparative evaluation of edible coatings to reduce fat uptake in a deep-fried cereal product. Food Res Int 35(5):445–448

Altan Kamer DD, Gumus T, Palabiyik I, Demirci AS, Oksuz O (2021) Grape pomace as a promising source for gellan gum production. Food Hydrocoll 114(August 2020):106584. https://doi.org/10.1016/j.foodhyd.2020.106584

Anderson DM, Brydon WG, Eastwood MA (1988) The dietary effects of gellan gum in humans. Food Addit Contam 5(3):237–249

Baird JK, Talashek TA, Chang H (1992) Gellan gum: effect of composition on gel properties. In: Philips GO, Williams PA, Wedlock DJ (eds) Gums and stabilisers for the food industry. Pergamon Press, Oxford, pp 479–487

Bajaj I, Singhal R (2007) Gellan gum in reducing oil uptake in sev, a legume based product during deep fat frying. Food Chem 104(4):1472–1477

Bajaj IB, Survase SA, Saudagar PS, Singhal RS (2007) Gellan gum: fermentative production, downstream processing and applications. Food Technol Biotechnol 45(4):341–354

Balasubramanian R, Kim SS, Lee J, Lee J (2019) Effect of TiO2 on highly elastic, stretchable UV protective nanocomposite films formed by using a combination of k-carrageenan, xanthan gum and gellan gum. Int J Biol Macromol 123:1020–1027. https://doi.org/10.1016/j.ijbiomac.2018.11.151

Banerjee S, Ravi R, Bhattacharya S (2013) Textural characterisation of gellan and agar based fabricated gels with carrot juice. LWT Food Sci Technol 53(1):255–261

Chalupa WF, Sanderson GR (1994) Process for preparing low-fat fried food. US Patent 5,372,829. Assignee: Merck & Co., Inc

Chen Y, Xiong X, Liu X, Cui R, Wang C, Zhao G et al (2020) 3D bioprinting of shear-thinning hybrid bioinks with excellent bioactivity derived from gellan/alginate and thixotropic magnesium phosphate-based gels. J Mater Chem B 8(25):5500–5514. https://doi.org/10.1039/d0tb00060d

Chesterman J, Zhang Z, Ortiz O, Goyal R, Kohn J (2020) Biodegradable polymers. In: Principles of tissue engineering. Elsevier, London, pp 317–342

Choudhary S, Sharma K, Kumar V, Bhatia JK, Sharma S, Sharma V (2020) Microwave-assisted synthesis of gum gellan-cl-poly(acrylic-co-methacrylic acid) hydrogel for cationic dyes removal. Polym Bull 77(9):4917–4935. https://doi.org/10.1007/s00289-019-02998-3

Coutinho DF, Sant SV, Shin H, Oliveira JT, Gomes ME, Neves NM et al (2010) Modified gellan gum hydrogels with tunable physical and mechanical properties. Biomaterials 31(29):7494–7502

Dartey CK (1993) Applications of gellan gum as a fining agent in alcoholic beverages. Res Disclosure 348:256

Duran E, Costell E, Izquierdo L, Duran L (1994) Low sugar bakery jams with gellan gum—guar gum mixtures. Influence of composition on texture. Food Hydrocoll 8(3–4):373–381

Duxbury DO (1993) Fat reduction without adding fat replacers. Food Proc 54(5):68–70

EFSA Panel on Food Additives and Nutrient Sources Added to Food (ANS), Younes M, Aggett P, Aguilar F, Crebelli R, Filipic M, Frutos MJ, Galtier P, Gott D, Gundert-Remy U, Kuhnle GG, Lambré C, Leblanc JC, Lillegaard IT, Moldeus P, Mortensen A, Oskarsson A, Stankovic I, Waalkens-Berendsen I, Woutersen RA, Wright M, Brimer L, Mosesso P, Christodoulidou A, Cascio C, Tard A, Lodi F, Dusemund B (2018) Re-evaluation of gellan gum (E 418) as food additive. EFSA J 16(6):e05296

Forssell P, Lahtinen R, Lahelin M, Myllärinen P (2002) Oxygen permeability of amylose and amylopectin films. Carbohydr Polym 47(2):125–129

Giavasis I, Harvey LM, McNeil B (2000) Gellan gum. Crit Rev Biotechnol 20(3):177–211

Gibson W (1997) Gellan gum. In: Imeson AP (ed) Thickening and gelling agents for food. Springer, Boston, pp 119–143

Gibson W, Sanderson GR (1997) Gellan gum. In: Imeson AP (ed) Thickening and gelling agents for foods, 2nd edn. Blackie, London, pp 119–143

Giese J (1995) Developments in beverage additives. Food Technol 49(9):63–65, 68–70, 72

Guo N, Zhu G, Chen D, Wang D, Zhang F, Zhang Z (2020) Preparation and characterization of gellan gum–guar gum blend films incorporated with nisin. J Food Sci 85(6):1799–1804

Hasheminya SM, Ebrahimzadeh-Mousavi SMA, Ehsani MR, Dehghannya J (2011) Effect of gellan hydrocolloid on rheological properties and stabilization of a fiber-enriched doogh. Food Res 21(2):179–193

Hernández-García E, Vargas M, Chiralt A (2021) Thermoprocessed starch-polyester bilayer films as affected by the addition of gellan or xanthan gum. Food Hydrocoll 113(December 2020):106509. https://doi.org/10.1016/j.foodhyd.2020.106509

Huang Y, Singh PP, Tang J, Swanson BG (2004) Gelling temperatures of high acyl gellan as affected by monovalent and divalent cations with dynamic rheological analysis. Carbohydr Polym 56(1):27–33
Iglesias N, Galbis E, Romero-Azogil L, Benito E, Lucas R, García-Martín MG, De-Paz MV (2020) In-depth study into polymeric materials in low-density gastroretentive formulations. Pharmaceutics 12(7):1–44. https://doi.org/10.3390/pharmaceutics12070636
Ismail NA, Razali MH, Amin KAM (2017) Mechanical and physicochemical properties study on gellan gum thin film prepared using film casting method. AIP Conf Proc 1885. https://doi.org/10.1063/1.5002239
Iurcius CE, Savin A, Lungu C, Martin P, Popa M (2016) Gellan. Food applications. Cellulose Chem Technol 50(1):1–13
Jung S, Oh HK, Kim MS, Lee KY, Park H, Kook MS (2020) Effect of gellan gum/tuna skin film in guided bone regeneration in artificial bone defect in rabbit calvaria. Materials 13(6):1–9. https://doi.org/10.3390/ma13061318
Kalia S, Roy Choudhury A (2019) Synthesis and rheological studies of a novel composite hydrogel of xanthan, gellan and pullulan. Int J Biol Macromol 137:475–482. https://doi.org/10.1016/j.ijbiomac.2019.06.212
Kang D, Zhang HB, Nitta Y, Fang YP, Nishinari K (2015) Gellan. In: Ramawat K, Mérillon JM (eds) Polysaccharides. Springer, Cham. https://doi.org/10.1007/978-3-319-03751-6
Kang D, Zhang H, Nitta Y, Fang Y, Nishinari K (2021) Polysaccharides. Springer, Cham. https://doi.org/10.1007/978-3-319-03751-6
Laaman TR, Tye RI (1991) Application of gellan gum to meat systems. Res Disclosure 323:212
Lara Lledó MI (2016) Antimicrobial packaging system for minimally processed fruit (January). https://riunet.upv.es/handle/10251/61388
Lee KY, Shim J, Lee HG (2004) Mechanical properties of gellan and gelatin composite films. Carbohydr Polym 56(2):251–254
Liang C, Hu X, Ni Y, Wu J, Chen F, Liao X (2006) Effect of hydrocolloids on pulp sediment, white sediment, turbidity and viscosity of reconstituted carrot juice. Food Hydrocoll 20(8):1190–1197
Madene A, Jacquot M, Scher J, Desobry S (2006) Flavour encapsulation and controlled release: a review. Int J Food Sci Technol 41(1):1–21. https://doi.org/10.1111/j.1365-2621.2005.00980.x
Maia FR, Correlo VM, Oliveira JM, Reis RL (2019) Natural origin materials for bone tissue engineering. In: Atala A, Lanza R, Mikos AG, Nerem R (eds) Principles of regenerative medicine, 3rd edn. Academic Press, London, pp 535–558
Manjunatha S, Gupta DKD (2006) Instrumental textural characteristics of restructured carrot cubes. Int J Food Prop 9(3):453–462
Mantaluta A, Cojocaru D, Savin C, Pasa R (2012) The testing of some organic supports for yeasts immobilization technology used in sparkling wine production, annals of the "AlexandruIoanCuza" University. Genet Mol Biol 12(4):81–85
Milan J, Maleki G (2012) Hydrocolloids in food industry. In: Valdez B (ed) Food industrial processes: methods and equipment. InTech Europe, Croatia, pp 17–38
Mohd Azam NAN, Amin KAM (2017) The physical and mechanical properties of gellan gum films incorporated Manuka honey as wound dressing materials. IOP Conf Ser Mater Sci Eng 209(1):012027. https://doi.org/10.1088/1757-899X/209/1/012027
Mohd SS, Abdullah MAA, Mat Amin KA (2016) Gellan gum/clay hydrogels for tissue engineering application: mechanical, thermal behavior, cell viability, and antibacterial properties. J Bioact Compat Polym 31(6):648–666. https://doi.org/10.1177/0883911516643106
Mousavian D, Mohammadi Nafchi A, Nouri L, Abedinia A (2021) Physicomechanical properties, release kinetics, and antimicrobial activity of activated low-density polyethylene and orientated polypropylene films by thyme essential oil active component. J Food Meas Charact 15(1):883–891. https://doi.org/10.1007/s11694-020-00690-z
Nakamura K, Tanaka Y, Sakurai M (1996) Dynamic mechanical properties of aqueous gellan solutions in the sol-gel transition region. Carbohydrpolym 30(2–3):101–108
Norton IT (1992) Water in oil dispersion. European Patent Application 0473854A1

Papageorgiou M, Kasapis S (1995) The effect of added sucrose and corn syrup on the physical properties of gellan—gelatin mixed gels. Food Hydrocoll 9(3):211–220

Patel KS, Vadalia KR, Patel JK (2014) Development and evaluation of in situ gelling system for treatment of periodontitis. Int J Pharm Tech Res 6(7):2102–2112

Pliego-Arreaga R, Regalado C, Amaro-Reyes A, García-Almendárez BE (2013) Revista Mexicana de Ingeniería Química. Rev Mex Ing Quím 12(3):505–511

Ramalingam C, Priya J, Mundra S (2014) Applications of microbial polysaccharides in food industry. Int J Pharm Sci Rev Res 27(1):322–324

Rukmanikrishnan B, Ismail FRM, Manoharan RK, Kim SS, Lee J (2020a) Blends of gellan gum/xanthan gum/zinc oxide based nanocomposites for packaging application: rheological and antimicrobial properties. Int J Biol Macromol 148:1182–1189

Rukmanikrishnan B, Ramalingam S, Rajasekharan SK, Lee J, Lee J (2020b) Binary and ternary sustainable composites of gellan gum, hydroxyethyl cellulose and lignin for food packaging applications: Biocompatibility, antioxidant activity, UV and water barrier properties. Int J Biol Macromol 153:55–62

Saha D, Bhattacharya S (2010) Hydrocolloids as thickening and gelling agents in food: a critical review. J Food Sci Technol 47(6):587–597

Saha T, Masum ZU, Mondal SK, Hossain MS, Md. Abu Jobaer, R. I. S., & Fahad, T. (2018) Application of natural polymers as pharmaceutical excipients. Glob J Life Sci Biol Res Gedik 4(1):1–8. https://doi.org/10.24105/gjlsbr.2018.4.2

Sanderson GR (1990) Gellan gum. In: Harris P (ed) Food gels, 2nd edn. Elsevier Applied Science, London, pp 210–232

dos Santos JWS, de Fátima Silva M, Concha VOC, Yoshida CMP (2020) Diffusion process through biodegradable polymer films. In: Biopolymer membranes and films. Elsevier, Amsterdam, pp 97–118. https://doi.org/10.1016/b978-0-12-818134-8.00004-3

Sapper M, Talens P, Chiralt A (2019) Improving functional properties of cassava starch-based films by incorporating xanthan, gellan, or pullulan gums. Int J Polym Sci 2019(6):1–9. https://doi.org/10.1155/2019/5367164

Sapra P, Patel D, Soniwala M, Chavda J (2013) Development and optimization of in situ periodontal gel containing levofloxacin for the treatment of periodontal diseases. J Sci Innov Res 2(23): 607–626

Sutherland IW (2005) Biotechnology of microbial polysaccharides in food. In: Shetty K, Paliyath G, Pometto A, Levin RE (eds) Food biotechnology, 2nd edn. Taylor & Francis, Boca Raton, pp 193–220

Sworna G, Stouby L (2020) Gellan gum. In: Philips GO, Williams PA (eds) Handbook of hydrocolloids, Food science, technology and nutrition, 3rd edn. Woodhead Publishing, London, pp 855–885. https://doi.org/10.1016/C2018-0-04245-0

Tang J, Mao R, Tung MA, Swanson BG (2001) Gelling temperature, gel clarity and texture of gellan gels containing fructose or sucrose. Carbohydr Polym 44(3):197–209

Tetsuguchi M, Nomura S, Katayama M, Sugawa-Katayama Y (1997) Effects of curdlan and gellan gum on the surface structure of intestinal mucosa in rats. J Nutr Sci Vitaminol (Tokyo) 43(5): 515–527

Xiao G, Zhu Y, Wang L, You Q, Huo P, You Y (2011) Production and storage of edible film using gellan gum. Procedia Environ Sci 8:756–763

Xu X, Li B, Kennedy JF, Xie BJ, Huang M (2007) Characterization of konjac glucomannan-gellan gum blend films and their suitability for release of nisin incorporated therein. Carbohydr Polym 70(2):192–197. https://doi.org/10.1016/j.carbpol.2007.03.017

Yuguchi Y, Mimura M, Kitamura S, Urakawa H, Kajiwara K (1993) Structural characteristics of gellan in aqueous solution. Food Hydrocoll 7(5):373–385

Zhang W, Luan D, Tang J, Sablani SS, Rasco B, Lin H, Liu F (2015) Dielectric properties and other physical properties of low-acyl gellan gel as relevant to microwave assisted pasteurization process. J Food Eng 149:195–203. https://doi.org/10.1016/j.jfoodeng.2014.10.014

Zhang X, Liu D, Jin TZ, Chen W, He Q, Zou Z et al (2021) Preparation and characterization of gellan gum-chitosan polyelectrolyte complex films with the incorporation of thyme essential oil nanoemulsion. Food Hydrocoll 114(September 2020):106570. https://doi.org/10.1016/j.foodhyd.2020.106570

Zhu G, Sheng L, Li J, Tong Q (2013) Preparation and characterisation of gellan/pullulan composite blend films. Int J Food Sci Technol 48(12):2683–2687. https://doi.org/10.1111/ijfs.12235

Chapter 9
Pectin and Pectin Derivative-Based Films

Kamaljit Kaur and Shivani Kaul

9.1 Introduction and Sources

Henri Braconnot coined the term pectin in 1825. It is derived from the Greek word *pektikos* which means congealed and curdled (Naqash et al. 2017). Pectin is a major polysaccharide that serves as a structural component in plant cell walls and growth (Mellinas et al. 2020). Pectin is found not only in plant cells but also in protopectin in the plant cell's middle lamella. Fruit ripening is important in the production of pectin. During ripening, the insoluble protopectin is converted to soluble pectin, and as the ripening proceeds, enzymes such as pectinase and pectin esterase break down pectin resulting into softer fruit (Srivastava and Malviya 2011; Samanta 2019). Naturally occurring pectin has a high molecular weight and high degree of esterification (DE 75–80%) whereas the pectin obtained from processing and modification has low molecular weight (Zhang et al. 2015; Zsivánovits et al. 2005).

Pectin is a hydrophilic heteropolysaccharide derived from fruit and vegetable byproducts, primarily apple pomace and citrus peel. On a dry basis, apple pomace yields 10–15% pectin, and citrus peel yields 20–30% pectin. Pectin can also be obtained in greater quantities from a variety of fruits and byproducts, including passion fruit, mango, banana, and peach (Table 9.1). Pectin naturally contains some functional properties such as thickening, selective gas permeability, gelling, and texturizing and stabilizing properties (da Silva et al. 2017; Mellinas et al. 2020; Vanitha and Khan 2019). It is used in the preparation of edible film and imparts thickening property to the edible coating (Jemilakshmi et al. 2020).

K. Kaur (✉) · S. Kaul
Department of Food Science and Technology, Punjab Agricultural University, Ludhiana, Punjab, India
e-mail: kamalbhella@pau.edu

T. Amin et al. (eds.), *Polysaccharide Based Films for Food Packaging: Fundamentals, Properties and Applications*,
https://doi.org/10.1007/978-981-99-4898-7_9

Table 9.1 Sources of pectin

Source	Percentage (%)	References
Apple pomace	10–15	Srivastava and Malviya (2011)
Pineapple peel	13.78	Karim et al. (2014)
Sugar beet pulp	15–30 (dry weight)	Lv et al. (2013)
Cocoa pod	6–12 (dry weight)	Nazir and Sayuti (2019)
Citrus peel	20–30	Srivastava and Malviya (2011)
Sunflower	15–25	Abbaszadeh (2008)
Passion fruit peel	14.8	Kulkarni and Vijayanand (2010)
Banana peel	5–21	Emaga et al. (2008)
Chick pea husk	8	Urias-Orona et al. (2010)
Mulberry	11.88	Liu et al. (2008)
Dragon fruit peel	18.59	Rahmati et al. (2015)
Potato pulp	4.08–14.34	Yang et al. (2018)
Jackfruit peel	21.5	Xu et al. (2018)

Polysaccharides are widely used in the production of biodegradable edible films today because they have good mechanical resistance, prevent water and oxygen permeation, provide high strength and transparency without developing anaerobic conditions, and can serve as a substitute for petroleum-based polymers. Pectin is a widely used polymer in the preparation of edible films due to its low cost, high-water solubility, edibility, biocompatibility, and biodegradability and the fact that it can have a high or low degree of esterification (Augustia et al. 2020; da Silva et al. 2017; Chambi and Grosso 2011; Dash et al. 2019; Jiménez et al. 2010).

9.2 Structure of Pectin

Pectin belongs to the category of natural polysaccharide containing chains of α-(1–4)-linked D-galacturonic acid and contains substantial amounts of D-arabinose (Ara), L-rhamnose (Rha), and D-galactose (Gal) and 13 diverse monosaccharides along with these (Fissore et al. 2012). Pectin is a good contender in forming a composite film because pectin unit consists of two hydroxyl (OH) groups and one carboxylic (COOH) group which provides various hydrogen bonds for the formation of pliable and durable film (Latif et al. 2013). Pectin has a linear anionic backbone which has two regions (Fig. 9.1): smooth regions and hairy regions. Smooth region contains no side chains whereas hairy region contains nonionic side chains (Naqash et al. 2017). Pectin contains both esterified and nonesterified galacturonic acids, and the ratio between the two determines the nature and activity of pectin in the solution (Nešić et al. 2017). Due to the presence of free carboxyl groups, pectin solutions have acidic pH values (COOH) (Mohnen 2008). Pectin polysaccharides can be classified (Fig. 9.2) into rhamnogalacturonan I, homogalacturonan, and

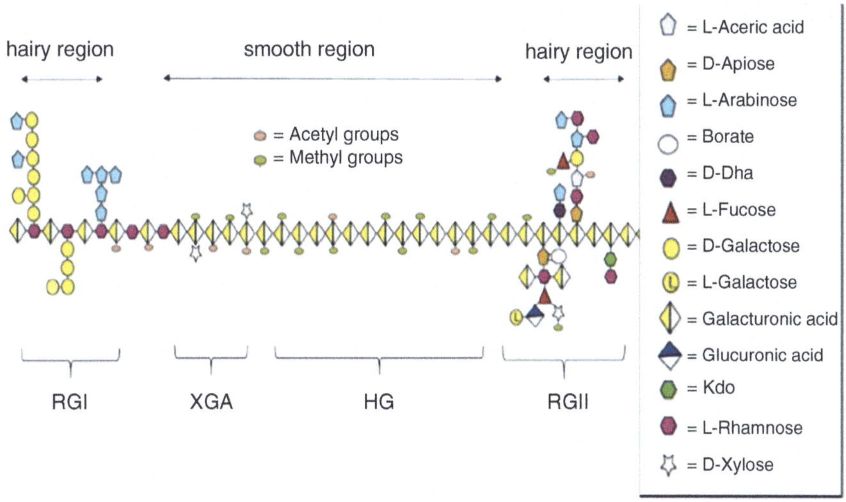

Fig. 9.1 Schematic presentation of pectin structure (Mellinas et al. 2020)

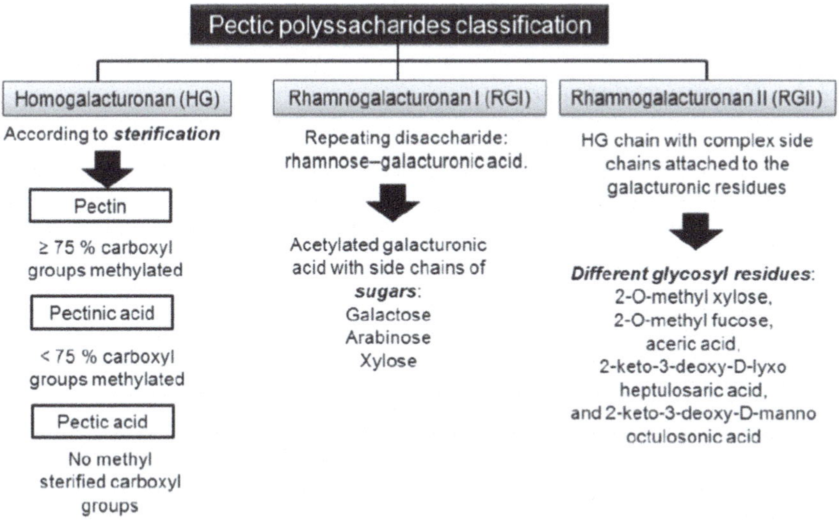

Fig. 9.2 Classification of pectic-based polysaccharides

rhamnogalacturonan II (Valdés et al. 2015). In spite of their complex nature, structural property of the pectin is attributed by the homogalacturonan (Costanza et al. 2019).

9.2.1 *Homogalacturonan (HG)*

HG is a linear homopolymer, which is the most abundant polysaccharide that consists of 60–65% of total pectin and is made up of galacturonic acid residues linked at positions 1 and 4. HG contains approximately 200 galacturonic acid units which makes it about 100 nm long (Samanta 2019). Homogalacturonan contains acetylated oxygen group either at position 2 or 3 and some moderately methyl esterified carboxyl group at position C-6. It may also contain some other cross-linked esters having undefined structures. It has been noticed that shorter areas of homogalacturonan are scattered in between some other pectin polysaccharides whereas the length of stretched homogalacturonan is approximately 100 GalA. Apiogalacturonan, xylogalacturonan, and rhamnogalacturonan II are more complex in structure as compared to homogalacturonan. Homogalacturonan is further classified into three types (Pedrolli et al. 2009):

1. Pectin that contains >75% of methylated carboxyl groups.
2. Pectinic acid containing <75% of methylation.
3. Polygalacturonic acid or pectic acid that contains no methyl-esterified carboxyl groups.

9.2.2 *Rhamnogalacturonan*

Structurally, rhamnogalacturonan I (RG I) is complex than HG and is composed of repeating units of disaccharide rhamnose-galacturonic acid groups that are acetylated at position 2 or 3 and joined to the side chains of various neutral sugars, such as galactose, arabinose, and xylose. RG I forms 20–35% of pectin molecule (Mohnen 2008).

Rhamnogalacturonan II (RGII) forms about 10% of the pectin molecule which is different from RG I on the basis of side groups. In RG II complex, side groups are attached to 12 different types of glycosyl residues, like rare sugar species (2-O-methyl xylose, 2-O-methyl fucose, 2-keto-3-deoxy-D-lyxo heptulosaric acid, acetic acid, and 2-keto-3-deoxy-D-manno octulosonic acid). In the pectin structure, RGI and RGII form hairy regions, whereas HG forms the smooth component of the pectin molecule (Caffall and Mohnen 2009).

9.2.3 *Other Substituted Homogalacturonans*

XGA (xylogalacturonan) and AP (apiogalacturonan) are the two substituted galacturonans, which have restricted expression to some higher extent. When an HG is exchanged at O-3 with a β-linked xylose, the resultant molecule is XGA. Occasionally, the three-linked xylose is further substituted with an extra β-linked xylose at O-4. XGA is predominantly found in reproductive tissues; however some reports suggest detection in stems and leaves of *Arabidopsis*. A recent study reported upregulation of XGA gene in response to pathogen attack. This observation suggests the role of XGA (or some other substituted HGs) in imparting HG resistance to endopolygalacturonase degradation under the event of pathogen attack. In aquatic monocots, for instance, Lemma, another substituted HG apiogalacturonan (AP) is present, where D-apiofuraose substitution takes place at either O-3 or O-2 (Mohnen 2008).

9.3 Manufacturing of Pectin-Based Films

Pectin plays a major role in providing mechanical strength to the cell wall of plants. This natural property of pectin can be used to produce bio-composite materials and food packaging films (Pilla 2011). Apart from mechanical strength, the other natural properties such as gas permeability and gelling make pectin a good contender in preparation of films and coatings from other polysaccharides. Pectin-based films can be produced from various methods (Figs. 9.3 and 9.4) that are explained as below (Lazaridou and Biliaderis 2020; Rodsamran and Sothornvit 2019).

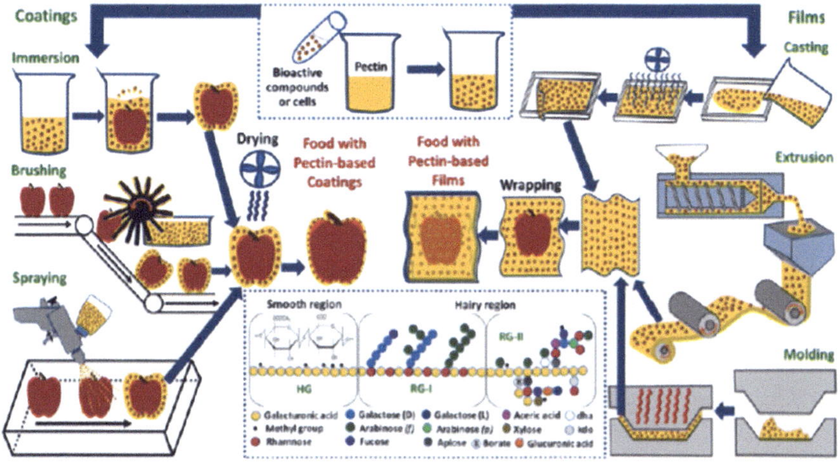

Fig. 9.3 Different methods of coatings and film production from pectin

Fig. 9.4 Pectin film

9.3.1 Casting Method

Casting method for film formation also called as solvent casting method is basically used at lower scale (lab or pilot scale) and basically involves the following three stages (Fig. 9.5) (Suhag et al. 2020):

1. Dispersion of biopolymer in a desired solvent.
2. Casting of the biopolymer solution in an appropriate mold.
3. Drying of ready solution.

Film formation by the casting method entirely depends upon the solubility of the biopolymer in the solvent (Koide et al. 2013). Solvents that are generally used for the solubilization of polymer are ethanol, water, or a combination of both. But in case of pectin, water acts as an effective solvent for the development of films (Campos et al. 2011). Apart from solvent, another important component called as plasticizer is added to the casting solution to modify the mechanical characteristics of the film (Kokoszka et al. 2010). Food-grade plasticizers such as glycerol and sorbitol are used for the film preparation because they are more compatible and stable with the polymeric material (Espitia et al. 2013). The casting stage involves the pouring of homogeneous solution in the desired molds or glass plates coated with Teflon and dried by using various drying techniques such as tray drying, oven drying, vacuum drying, and microwave drying (Cha et al. 2003).

Casting method is further of two types: batch type and continuous type. Batch type casting process is commonly used at laboratory scale where drying time varies from 12 min to 12 h at ambient temperature. On the other hand, continuous process is used at industrial scale where drying is carried out on conveyor belts by using infrared heater. Continuous method is considered to be more advantageous as the product requires less time to dry which further will reduce the chances of microbial contamination in the food product (Espitia et al. 2013; Du et al. 2011).

Advantages of casting method are the following (Chen et al. 2008; Yang et al. 2011; Siemann 2005):

1. Simple process.
2. No special equipment is required.
3. Low cost.

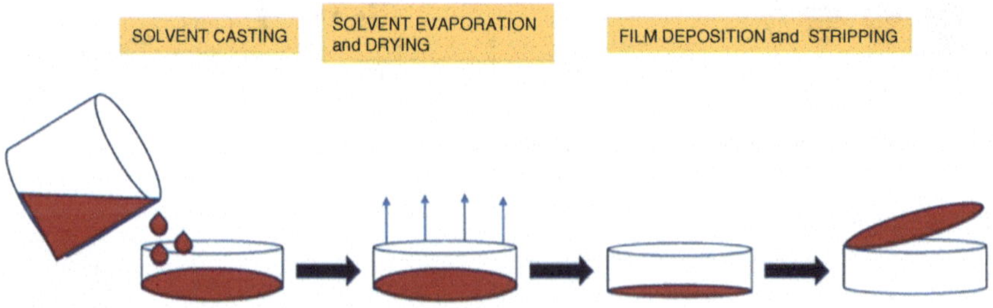

Fig. 9.5 Casting method of film formation at laboratory scale

4. Improved particle to particle interaction.
5. Lower temperature processing does not change conformation of the particles.
6. High optical purity.
7. High transparency.
8. Reduced haziness.

Disadvantages of casting method are the following (Chen et al. 2008; Fakhouri et al. 2013; Jensen et al. 2015):

1. Films are obtained either in the form of simple sheets or tubes.
2. Toxic solvent gets entrapped in the polymer.
3. Addition of solvents causes denaturation of proteins and other particles.
4. A restricted amount of film is produced.
5. Long drying time.

9.3.2 Extrusion Method

Extrusion is a commercial technique that is currently being used in the preparation of conventional packaging films (Fig. 9.6). But now industries have adopted the process of extrusion to produce edible films/packaging materials (Suhag et al. 2020; Anukiruthika et al. 2020). This technique changes the conformation of the molecules, thereby improving the physicochemical properties of the film (Fitch-Vargas et al. 2016). The technique of extrusion is divided into three sections (Calderon-Castro et al. 2018):

1. Feeding section.
2. Kneading section.
3. Heating section.

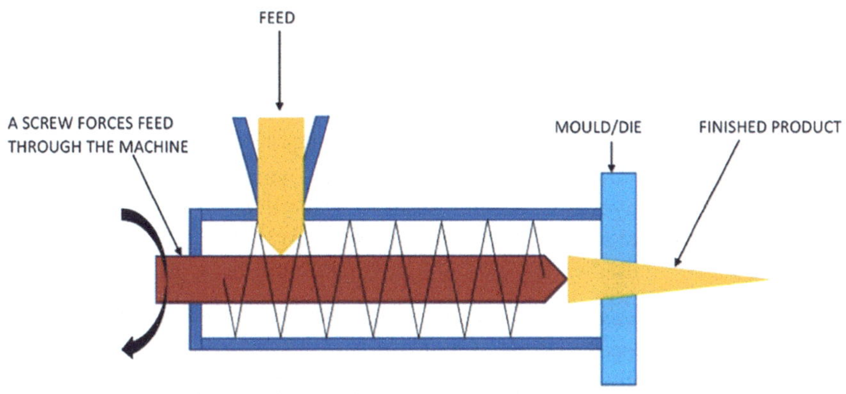

Fig. 9.6 Extrusion process of film formation

In the feeding section, dry mixing of the components is done with compressed air. Plasticizers (approximately 10–60% w/w) and solvents are added, and the solution mix is kneaded in the kneading section (Skurtys et al. 2010). Mechanical and thermal energies are utilized in this process to make extruded films (Fitch-Vargas et al. 2019). For the multilayer film formation, co-extrusion can be used which will help in enhancing the physicochemical properties of the film (Winotapun et al. 2019).

Extrusion technique can prove to be a successful technique only if the process variables such as liquid feed rates, food polymer selection, screw configuration, zone temperatures, screw speed, die configuration and product inlet/outlet pressures are regularly monitored and controlled (Liu et al. 2005). Extrusion has emerged as a promising technique in conventional packaging, but it is still under investigation in the case of pectin-based films. Despite the fact that many studies have been reported for the development of pectin-based films, this area still requires significant research and development. Advantages of extrusion technology are the following (Andreuccetti et al. 2012; Raghav et al. 2016; Liu et al. 2009):

1. Less processing time.
2. Less energy consumption.
3. Better mechanical properties.
4. Improved optical properties.
5. High performance.
6. Low cost.
7. Ease of material handling.
8. No solvents are used.
9. Wide range of processing parameters and conditions.
10. Different forms of product can be obtained.

Disadvantages of extrusion technology are the following (Suhag et al. 2020):

1. Raw material having low moisture can only be processed.
2. Installation cost of equipment is high.
3. High maintenance cost.

9.3.3 Spraying Method

Spraying is one of wet processing techniques in which polymers are solubilized into liquid phase and then dried (Fig. 9.7). In this process, solution is sprayed through nozzle (1 mm) of sprayer on nonsticky surface; hence uniform film is formed by droplets after letting the solution to dry (Espitia et al. 2013). Different types of spray drying that are being used widely are the following (Suhag et al. 2020):

1. *Air spray atomization*: In this technique, a tube is there from which fluid is flowing at a low speed and air of high velocity is sprayed through nozzles in the chamber. A frictional force develops between fluid and air which disrupts the flow of liquid and persuades atomization (Valdés et al. 2017).

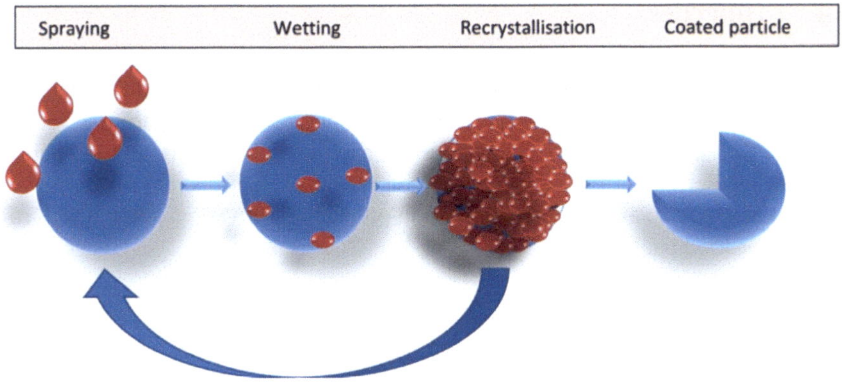

Fig. 9.7 Spraying method of edible coating

2. *Air assisted airless atomization*: In this method, atomization occurs in two steps. In the first step, partial atomization is carried out with a nozzle, and in the second step, small amount of compressed air is passed which completes the atomization process. This type of atomization technique is mainly used in case of highly viscous coating solution (Peretto et al. 2017).
3. *Pressure atomization*: As the name suggests, the atomization is carried out by pressure. This technique is also called as airless atomization because spraying is carried out in the absence of air. In the atomization assembly, nozzles of small size are fitted that generate high energy pressure (max. 3.5 bar) to deliver surface tension in highly viscous polymer solutions (Andrade et al. 2012).

9.3.4 Knife Coating Method

Knife coating is a continuous method of film formation where specific amount of solution is applied in layers with the help of a knife. This method is also known as knife over roll coating. The thickness and the speed of application can be controlled by adjusting the height of the knife (Rossman 2009).

Advantages of knife coating method are the following:

1. Uniform thickness
2. Simpler technique
3. Low maintenance of the device used
4. Reduced drying time

9.4 Properties

Pectin, which is naturally water-loving, can be used in the production of edible films as an additive to provide thickness to the coating solution while preparing the edible coatings. It has good mechanical properties and serves as a gas and oil barrier (Jemilakshmi et al. 2020). Apart from these, some other properties of pectin are mentioned below so as to diversify its use in preparation of various films which can be used for food packaging.

9.4.1 Thermal Properties

Thermal property of a pectin film entirely depends on the chemical nature and state of transition that occurs during processing (Nisar et al. 2018). Studies have disclosed that pectin-based films have the ability to retain nutritional and organoleptic attributes of the food products when subjected to high temperatures such as frying and drying (Vanitha and Khan 2019). Film prepared from pectin and polyvinylpyrrolidone (PVP) not only improves strength and barrier properties but also provides thermal stability up to the temperature of 200 °C (Nešić et al. 2017).

9.4.2 Barrier Properties

Barrier properties are those that provide permeability of water vapors, gases, light, oil, etc. Barrier properties in context to packaging include water vapor transmission rate (WVTR), oxygen transmission rate (OTR), and carbon dioxide transmission rate (CO_2TR) (Siracusa 2012; Jemilakshmi et al. 2020). Films containing pectin and other natural products can prove to be effective packaging materials as they have good carbon dioxide and oxygen barrier properties, but due to poor water vapor transmission rate, they have limited use in food packaging industry and act as sacrificial agents and prevent dehydration (Alves et al. 2011; Chakravartula et al. 2019).

9.4.3 Mechanical Properties

Pectin-based films are fragile in nature that they can be easily broken down by hand; this restricts the use of pectin in packaging. To boost its mechanical properties, stability and adhesion properties of pectin can be compounded with various polymers, plasticizers, emulsifiers, and cross-linking agents (da Silva et al. 2017; Nešić et al. 2017). Plasticizers provide better workability and handling properties to the pectin films (Aider 2010). Glycerol is commonly used as plasticizer for the development of pectin-based films because it has small size and can easily fit into the intermolecular spaces among the pectin chains. This disrupts the hydrogen bonding, reduces the polymer-polymer interaction, and increases the chain mobility, thereby improving the workability of the pectin film. Glycerol can also form covalent ester linkages with the pectin (Cabello et al. 2015). Pectin when combined with starch and glycerol (plasticizer) forms the film having excellent oxygen barrier and mechanical properties (Dash et al. 2019).

9.4.4 Tensile Properties

Pure pectin-based film exhibits a tensile strength of 14.78 MPa, and this can be increased by adding essential oils to polymer solution matrix (Nisar et al. 2018). But in case of glycerol, it should be added in moderation as it lowers the tensile strength of the film but increases the elasticity in the film. This is because high amount of plasticizer hinders cohesiveness of starch, increases mobility of the polymer, and decreases intermolecular interaction (Yahya 2018).

9.4.5 Electrical Properties

As mentioned earlier, plasticizers have a positive impact on mechanical properties, but on the other hand, they have some negative effect on electrical properties too. Glycerol dissolves the polar sites on pectin chains, reduces intermolecular attraction, and increases polymer segmental movement and ion mobility. This results in a film that has lower electrical resistance and lower fracturability (Costanza et al. 2019).

9.4.6 Optical Properties

Optical properties are an essential parameter of the packaging film because it not only affects appealability of the product but also affects consumers' acceptance toward the product. Generally, pectin films have light yellowish color and are translucent in nature. Starch-pectin composite film has white color, but the whiteness of the film can be increased by addition of titanium dioxide nanoparticle to the polymer matrix (Dash et al. 2019). The color of the film produced totally depends on the source from where the pectin has been extracted, for example, pectin obtained from the cocoa pods is brown in color so the film obtained has dull brown color but is transparent in nature (Nazir and Sayuti 2019) whereas films obtained from sugar beets are light red in color (Lv et al. 2013).

9.4.7 Aroma Retention Properties

Pectin films produced by electrospinning when used in combination with PHBV (poly(3-hydroxybutyrate-co-3-hydroxyvalerate) provide retention of volatile compound such as limonene (Akinalan Balik et al. 2019).

9.4.8 Antimicrobial Properties of Pectin-Based Films

Antimicrobial packaging is a valuable preservation technique for preserving food quality and safety. It is a type of active packaging that can prevent the growth of spoilage microorganisms on the surface of food. Pectin is a primary biopolymer derived from plant tissues that has been extensively used in food applications such as a carrier base for various antimicrobial compounds in the design of food packaging films.

Antimicrobial properties of packaging films inhibit the growth of microorganisms on food surfaces by interacting directly with the food surface. Antimicrobial films ensure contact with the food surface, allowing bacteriocins to diffuse to the surface. Effective antimicrobial packaging policy allows the sustainability of food production by imparting good protection to the food. During last decades, an increasing research is on the way to develop antimicrobial systems to be used as biocides in multiple areas (Calce et al. 2014).

Pure pectin films do not possess any antimicrobial property, but natural bioactive compounds can be added to the pectin-based films so as to provide them antimicrobial characteristic against foodborne pathogens. Natamycin, an antimicrobial agent, can be used to create antimicrobial films based on pectin. When natamycin is added to the pectin polymer matrix, it increases water vapor permeability, solubility, and opacity while decreasing the film's tensile strength (Bierhalz et al. 2012).

Aitboulahsen et al. (2020) developed fish Gel/Pec (gelatin-pectin) composite films with the incorporation of *Lavender angustifolia* essential oil (LEO) and *Mentha pulegium* essential oils (MEO) at 0.5% and 1.5% levels and evaluated the antimicrobial properties. Gel/Pec with 1.5% MEO antimicrobial film retarded the total viable count, psychrotrophic bacteria count, lactic acid bacteria, and Enterobacteriaceae throughout 12 days of storage period as compared to control.

Trejo-González et al. (2018) characterized the films of citrus pectin mixed with gellan gum enriched with antimicrobial concentrated supernatant from fermented culture broths of lactic acid bacteria and EDTA (ethylenediaminetetraacetic acid). Films exhibited inhibition against *E. coli*, *L. monocytogenes*, and *S. aureus*.

Ngo et al. (2020) created antimicrobial packaging films with variable ratio of pectin and nanochitosan. These films were tested for thickness, mechanical properties, water vapor permeability, oxygen permeability, and antimicrobial inhibition. It was found that pectin to nanochitosan ratio of 50:50 inhibited the growth of

Colletotrichum gloeosporioides, *E. coli*, *Aspergillus niger*, and *S. cerevisiae*; thus it can be used as an active film to improve the shelf life of food.

To pack the fish, a film was formulated with EDTA, 2% citrus pectin, 0.75% glycerol, 0.4% gellan gum, and 50 AU/mL (arbitrary units) of antimicrobial supernatant from *Streptococcus infantarius*. The films inhibited three bacteria (*E. coli*, *Listeria monocytogenes*, and *Staphylococcus aureus*) in two model fish media (Pérez-Arauz et al. 2021).

Biodegradable alginate-pectin-based bio-composites were developed having possible applications in food packaging. Morphologically surface was rough; mechanical properties were comparable to commercially available films and thermally stable in temperatures below 120 °C. Antibacterial analysis against *E. coli* and *S. aureus* revealed the antibacterial properties of prepared films, thus making them ideal as packaging films (Makaremi et al. 2019). In another study, pectin films activated by copaiba oil nano-emulsions (CP-NE) were prepared and evaluated for morphological, chemical, mechanical, thermal, and antimicrobial properties. The nano-emulsions showed antimicrobial activity against *S. aureus* and *E. coli*; thus this film showed good potential for active food packaging (Norcino et al. 2020).

9.4.9 Moisture Sorption Properties of Pectin-Based Films

Difference in water content among food contents and the environment outside of the package creates a driving force for water transport, but this transport stops when there is zero difference in moisture content. For food systems where equilibrium moisture content is undesirable for one or other components, the food products' shelf life is adjudged by dynamics of water transport across membrane. Thus, kinetics of water transport and equilibrium properties of water through the packaging materials is highly important.

Galus et al. (2013) evaluated pectin films developed by casting method for isotherms and water vapor sorption kinetics. Water vapor sorption kinetics were estimated using a saturated NaCl solution to obtain a relative humidity of 75.3%. At 25 °C, saturated solutions with water activities ranging from 0.113 to 0.901 were used to evaluate water vapor sorption isotherms. Every film's sorption isotherm curve had a sigmoidal shape. As a result of the composition of the film, the microstructure of pectin films exhibited variable internal arrangement.

Pectin films composed of variable sorbitol content were evaluated for moisture sorption behavior and the films with variable equilibrium moisture content were achieved. Addition of sorbitol in pectin films emerged in lowering the equilibrium moisture contents at water activities between low and intermediate levels but more high moisture contents at water activity more than 0.53 (Maftoonazad et al. 2007).

9.4.10 Release Kinetics of Pectin Films in Active Packaging

Utilization of pectin as a source for drug release vehicle has increased tremendously due to its high availability from many sources, as well as biocompatibility and biodegradability. Moreover, its properties such as emulsifying, gelling abilities, less sensitivity to gastric juices, ability to dissolve in alkaline environment, and high resistance to enzymes of upper gastrointestinal tract make the pectin a suitable material for formulation of different drug delivery systems for administration of drug. Antimicrobial packaging is a type of active packaging that slows or prevents the growth of microorganisms that contaminate packaged foods. Biodegradable, bio-based biopolymer films are now used as antimicrobial compound carriers. It is critical to understand the kinetics of antimicrobial compound release from packaging film to food matrix in order to design a suitable antimicrobial film (Rivera-Hernández et al. 2020).

Packaging films were designed by casting 0.5% low acyl gellan, 2% apple pectin, and 2.2% glycerol as plasticizer with an objective to analyze the release kinetics of nisin from film to a food model. The diffusion coefficients of nisin were 7.36 × 10–14 m^2/s and 5.22 × 10–14 for 30 °C and 5 °C, respectively (Rivera-Hernández et al. 2020).

Pectin was extracted by hot water extraction and cold maceration method using two different fruit peels. Pectin was loaded with antipsychotic drug by solvent evaporation method, and its release kinetics were analyzed using in vivo and in vitro models. Pectin obtained from mango peel showed better drug release by both methods and considered as a potential source for design of drug release systems for efficient delivery of drug (Srivastava and Tripathi 2020).

In another study, pectin- and alginate-based single and composite films having natamycin as an active agent were developed, and their diffusion coefficients and release behavior were investigated. Natamycin promoted water vapor permeability, increased soluble matter in water and opacity, and decreases tensile strength as compared to films without added antimicrobial agent (Bierhalz et al. 2012) (Table 9.2).

9.5 Conclusion

Scientists have created a variety of packaging films based on pectin, a natural biopolymer that can be easily extracted from agricultural waste. Due to variability in the structure of pectin, its applications are remarkably diverse and promising. Pectin is a complex macromolecule that exhibits great degree of inter- and intramolecular heterogeneity. Studying its fine structure entails mastering its extraction conditions to create oligomers that can be extracted and thoroughly analyzed. Depending on the treatments employed and origin during extraction, pectin exhibits diverse chemical characteristics and thereby able to link with similar or other

Table 9.2 Properties and application of pectin-based films

Polymer matrix	Properties	Application	References
Pectin + glycerol + polyethylene oxide	• Aroma barrier • Low porosity • High transparency	Multilayer packaging	Akinalan Balik et al. (2019)
Pectin + alginate + natamycin	• Antimicrobial • Lower tensile strength • High water solubility • Water vapor permeability • Opacity	Antimicrobial packaging in cheese wrappings	Bierhalz et al. (2012)
Pectin + alginate + whey protein concentrate	• Hydrophobic nature • Transparent	Food wrapping	Chakravartula et al. (2019)
Starch + pectin + titanium dioxide	• Moisture barrier • Excellent mechanical properties	• Food grade edible films • Substitute of traditional plastic packages	Dash et al. (2019)
Pectin + cellulose nanocrystals + glycerol	• Improved mechanical property • Flexible	• Alternative for polyvinyl chloride film • Packaging of fresh strawberries	da Silva et al. (2017)
Pectin + choline chloride + glycerol	• Yellowish color • Homogenous • Semitransparent • Without pores or cracks • Less rough	• Dry or low moisture foods • Eco-friendly biodegradable plastics	Gouveia et al. (2019)
Pectin + glycerol + fungal biomass	• Reduced water vapor permeability • Improved mechanical properties and tensile strength	Production of bioplastics	Gurram et al. (2018)
Pectin + glycerol + fish gelatin	• High tensile strength • High glass transition temperature • High antioxidant and	• Wrapping cheese • Preservation of beef meat	Jridi et al. (2020) and Bermúdez-Oria et al. (2019)

(continued)

Table 9.2 (continued)

Polymer matrix	Properties	Application	References
	antibacterial property		
Pectin + glycerol + cocoa butter	• High mechanical strength • High transparency • High water absorption • Fragile • Opaque • Thermally stable • Shiny	• Moisture-sensitive food packaging • Food coating or packaging	Mendes et al. (2020)
Pectin + glycerol + Polyvinylpyrrolidone (PVP)	• Improved mechanical resistance • Improved barrier properties • Hydrophobic • Thermally stable	Used as packaging material of nonfood products	Nešić et al. (2017)
Pectin + nanochitosan	• High tensile strength • Lower water solubility • Lower water vapor permeability • Lower oxygen permeability	Used as active packaging films to extend the shelf life of food	Ngo et al. (2020)
Pectin + alginate +zinc oxide nanoparticles	• Increased tensile strength • Antimicrobial properties • Increased elongation at break • Reduced solubility • Reduced water vapor permeability • Low moisture absorption • Low ultraviolet light transmission	Used in active packaging for food preservation	Ngo et al. (2018)

(continued)

Table 9.2 (continued)

Polymer matrix	Properties	Application	References
	• Low oxygen permeability		
Citrus pectin + clove bud essential oil	• Enhanced water barrier properties • Opaque • Resistant to breakage • Flexible • Heat stable	Used as antimicrobial edible films or coatings for various food applications	Nisar et al. (2018)
Pectin + alginate	• Rough surface • Antibacterial • Good mechanical properties • Excellent healing efficiency • Nutrient retention in food product	Used in the food packaging industry	Makaremi et al. (2019)
Pineapple peel pectin	• Opaque • Low water vapor permeability • High nutrient retention • Antioxidant property	Edible films or coatings for various food products	Rodsamran and Sothornvit (2019)
Pectin + carboxymethyl cellulose (CMC)	• Improved mechanical properties • Thermally stable	Used in commercially available food package Materials	Šešlija et al. (2018)
kappa-carrageenan + pectin	• Cohesive • Transparent • High water vapor permeability	–	Alves et al. (2006)
Kappa-carrageenan + pectin + mica flakes	• Enhanced water resistance and barrier properties • Less transparent	–	Alves et al. (2010)
Kappa-carrageenan + pectin+ nanoclay	Low CO_2 and water vapor permeability rate	–	Alves et al. (2011)
	• Smooth surface	–	Augustia et al. (2020)

(continued)

Table 9.2 (continued)

Polymer matrix	Properties	Application	References
Pectin + calcium chloride + cinnamaldehyde oil	• Increased strength		
	• Water resistant		
Pectin + glycerol	• High water vapor permeability	–	Galus et al. (2013)
	• High breakability		
High methoxyl pectin + isomalt	High water vapor barrier properties	–	Matta and Bertola (2020)

biopolymers via ionic, covalent, and nonionic bonds (hydrophilic, van der Waals, H-bonds) forming structures like gels and films. The addition of other active ingredients such as antioxidants, antimicrobials, and so on results in "active" packaging systems that provide additional benefits for controlled release of bioactives as well as an efficient formulation/preservation strategy to improve the shelf life and nutritional quality of processed food products. Future research should focus on deciphering the mechanism of the film structure-composition-processing-properties relationship in order to improve the efficacy and broaden its applicability as an effective packaging ingredient in the food preservation area.

References

Abbaszadeh AH (2008) Pectin and galacturonic acid from citrus wastes. University of Boras, School of Engineering, Boras

Aider M (2010) Chitosan application for active bio-based films production and potential in the food industry. LWT Food Sci Tech 43(6):837–842

Aitboulahsen M, Chairi H, Laglaoui A, Zantar S, Abdelhay A, Bakkali M, Zerrouk MH (2020) Gelatin/pectin-based film incorporated with essential oils: functional characteristics and shelf life extension of tilapia fillets under refrigeration. J Food Saf 40(3):e12774

Akinalan Balik B, Argin S, Lagaron JM, Torres-Giner S (2019) Preparation and characterization of electrospun pectin-based films and their application in sustainable aroma barrier multilayer packaging. Appl Sci 9(23):5136

Alves V, Costa N, Hilliou L, Larotonda F, Gonçalves M, Sereno A, Coelhoso IM (2006) Design of biodegradable composite films for food packaging. Desalination 199:331–333

Alves V, Costa N, Coelhoso IM (2010) Barrier properties of biodegradable composite films based on kappa-carrageenan/pectin blends and mica flakes. Carbohydr Polym 20:269–276

Alves VD, Castelló R, Ferreira AR, Costa N, Fonseca IM, Coelhoso IM (2011) Barrier properties of carrageenan/pectin biodegradable composite films. Procedia Food Sci 1:240–245

Andrade RD, Skurtys O, Osorio FA (2012) Atomizing spray systems for application of edible coatings. Compr Rev Food Sci Food Saf 11(3):323–337

Andreuccetti C, Carvalho RA, Galicia-García T, Martinez-Bustos F, González-Nuñez R, Grosso CRF (2012) Functional properties of gelatin-based films containing Yucca schidigera extract produced via casting, extrusion and blown extrusion processes: a preliminary study. J Food Eng 113(1):33–40

Anukiruthika T, Sethupathy P, Wilson A, Kashampur K, Moses JA, Anandharamakrishnan C (2020) Multilayer packaging: advances in preparation techniques and emerging food applications. Compr Rev Food Sci Food Saf 19(3):1156–1186

Augustia VA, Kusumastuti Y, Wirawan SK (2020) Pectin based edible film for cinnamaldehyde delivery system: effect of calcium chloride content on the physical characteristics of the films. Mater Sci Forum 998:266–270

Bermúdez-Oria A, Rodríguez-Gutiérrez G, Rubio-Senent F, Fernández-Prior Á, Fernández-Bolaños J (2019) Effect of edible pectin-fish gelatin films containing the olive antioxidants hydroxytyrosol and 3, 4-dihydroxyphenylglycol on beef meat during refrigerated storage. Meat Sci 148:213–218

Bierhalz ACK, da Silva MA, Kieckbusch TG (2012) Natamycin release from alginate/pectin films for food packaging applications. J Food Eng 110:18–25

Cabello SP, Takara EA, Marchese J, Ochoa NA (2015) Influence of plasticizers in pectin films: microstructural changes. Mater Chem Phys 162:491–497

Caffall KH, Mohnen D (2009) The structure, function, and biosynthesis of plant cell wall pectic polysaccharides. Carbohydr Res 344:1879–1900

Calce E, Mignogna E, Bugatti V, Galdiero M, Vittoria V, De Luca S (2014) Pectin functionalized with natural fatty acids as antimicrobial agent. Int J Biol Macromol 68:28–32

Calderon-Castro A, Vega-Garcıa M, Zazueta-Morales JD, Fitch-Vargas PR, Carrillo-Lopez A, Gutierrez-Dorado R, Limon-Valenzuela V, Aguilar-Palazuelos E (2018) Effect of extrusion process on the functional properties of high amylose corn starch edible films and its application in mango (Mangifera indica L.) cv. Tommy Atkins. J Food Sci Technol 55(3):905–914

Campos CA, Gerschenson LN, Flores SK (2011) Development of edible films and coatings with antimicrobial activity. Food Bioprocess Technol 4(6):849–875

Cha DS, Cooksey K, Chinna MS, Park HJ (2003) Release of nisin from various heat-pressed and cast films. LWT Food Sci Technol 36(2):209–213

Chakravartula SSN, Soccio M, Lotti N, Balestra F, Dalla Rosa M, Siracusa V (2019) Characterization of composite edible films based on pectin/alginate/whey protein concentrate. Materials 12(15):2454

Chambi HNM, Grosso CRF (2011) Mechanical and water vapor permeability properties of biodegradables films based on methylcellulose, glucomannan, pectin and gelatin. Food Sci Technol 31(3):739–746

Chen Q, Roether J, Boccaccini A (2008) Tissue engineering scaffolds from bioactive glass and composite materials. In: Ashammakhi N, Reis R, Chiellini F (eds) Topics in Tissue Engineering. University of Oulu, Oulu, pp 1–27

Costanza V, Bonanomi L, Moscato G, Wang L, Choi YS, Daraio C (2019) Effect of glycerol on the mechanical and temperature-sensing properties of pectin films. Appl Phys Lett 115(19):193702

Dash KK, Ali NA, Das D, Mohanta D (2019) Thorough evaluation of sweet potato starch and lemon-waste pectin based-edible films with nano-titania inclusions for food packaging applications. Int J Biol Macromol 139:449–458

Du WX, Avena-Bustillos RJ, Hua SST, McHugh TH (2011) Antimicrobial volatile essential oils in edible films for food safety. Sci Again Microb Pathog 2:1124–1134

Emaga TH, Ronkart SN, Robert C, Wathelet B, Paquot M (2008) Characterisation of pectins extracted from banana peels (Musa AAA) under different conditions using an experimental design. Food Chem 108:463–471

Espitia PJP, Du WX, de Jesús A-BR, Soares NDFF, McHugh TH (2013) Edible films from pectin: physical-mechanical and antimicrobial properties-a review. Food Hydrocoll 35:287–296

Fakhouri FM, Costa D, Yamashita F, Martelli SM, Jesus RC, Alganer K, Collares-Queiroz FP, Innocentini-Mei LH (2013) Comparative study of processing methods for starch/gelatin films. Carbohydr Polym 95:681–689

Fissore EN, Rojas AM, Gerschenson LN (2012) Rheological performance of pectin-enriched products isolated from red beet (*Beta vulgaris* L: var. *conditiva*) through alkaline and enzymatic treatments. Food Hydrocoll 26:249–260

Fitch-Vargas PR, Aguilar-Palazuelos E, Zazueta-Morales JDJ, Vega-García MO, Valdez-Morales JE, Fernando Martínez-Bustos F et al (2016) Physicochemical and microstructural characterization of corn starch edible films obtained by a combination of extrusion technology and casting technique. J Food Sci 81(9):2224–2232

Fitch-Vargas PR, Aguilar-Palazuelos E, Vega-García MO, Zazueta-Morales JJ, Calderón-Castro A, Montoya-Rodríguez A et al (2019) Effect of a corn starch coating obtained by the combination of extrusion process and casting technique on the post-harvest quality of tomato. Rev Mex Ing Quim 18(3):789–801

Galus S, Uchanski P, Lenart A (2013) Colour, mechanical properties and water vapour permeability of pectin films. Acta Agrophysica 20(3):375–384

Gouveia TI, Biernacki K, Castro MC, Gonçalves MP, Souza HK (2019) A new approach to develop biodegradable films based on thermoplastic pectin. Food Hydrocoll 97:105175

Gurram R, Souza Filho PF, Taherzadeh MJ, Zaman A (2018) A solvent-free approach for production of films from pectin and fungal biomass. J Polym Environ 26(11):4282–4292

Jemilakshmi TV, Rakshana L, Priya SK, Aishwarya B, Anithaee C (2020) Post harvest quality enhancement of fruits and vegetables using edible coatings: a review. J Crit Rev 7(7):786–790

Jensen A, Lim LT, Barbut S, Marcone M (2015) Development and characterization of soy protein films incorporated with cellulose fibers using a hot surface casting technique. LWT Food Sci Technol 60(1):162–170

Jiménez A, Fabra MJ, Talens P, Chiralt A (2010) Effect of lipid self-association on the microstructure and physical properties of hydroxypropyl-methylcellulose edible films containing fatty acids. Carbohydr Polym 82:585–593

Jridi M, Abdelhedi O, Salem A, Kechaou H, Nasri M, Menchari Y (2020) Physicochemical, antioxidant and antibacterial properties of fish gelatin-based edible films enriched with orange peel pectin: wrapping application. Food Hydrocoll 103:105688

Karim R, Uddin MB, Jubayer Md F (2014) Optimization of pectin isolation method from pineapple (*Ananas comosus* L.) waste Carpath. J Food Sci Technol 6(2):116–122

Koide Y, Ikake H, Muroga Y, Shimizu S (2013) Effect of the cast-solvent on the morphology of cast films formed with a mixture of stereo isomeric poly (lactic acids). Polym J 45:645–650

Kokoszka S, Debeaufort F, Hambleton A, Lenart A, Voilley A (2010) Protein and glycerol contents affect physico-chemical properties of soy protein isolate-based edible films. Innova Food Sci Emerg Technol 11(3):503–510

Kulkarni SG, Vijayanand P (2010) Effect of extraction conditions on the quality characteristics of pectin from passion fruit peel (*Passiflora edulis f. flavicarpa* L.). LWT Food Sci Technol 43(7): 1026–1031

Latif IA, Abbas SM, Kadhum MA (2013) Hydrogen bonds effects on the electrical properties of pectin/PVA graphene nanocomposites. Chem Mater Res 3(1):1–12

Lazaridou A, Biliaderis CG (2020) Edible films and coatings with pectin. In: Kontogiorgos V (ed) Pectin: technological and physiological properties. Springer, Cham, pp 99–123

Liu L, Kerry JF, Kerry JP (2005) Selection of optimum extrusion technology parameters in the manufacture of edible/biodegradable packaging films derived from food-based polymers. J Food Agric Environ 3(3/4):51

Liu H, Xu XM, Guo SD (2008) Comparison of full-fat and low-fat cheese analogues with or without pectin gel through microstructure, texture, rheology, thermal and sensory analysis. Int J Food Sci Technol 43(9):1581–1592

Liu H, Xie F, Yu L, Chen L, Li L (2009) Thermal processing of starch-based polymers. Prog Polym Sci 34(12):1348–1368

Lv C, Wang Y, Wang LJ, Li D, Adhikari B (2013) Optimization of production yield and functional properties of pectin extracted from sugar beet pulp. Carbohydr Polym 95(1):233–240

Maftoonazad N, Ramaswamy HS, Marcotte M (2007) Moisture sorption behavior, and effect of moisture content and sorbitol on thermo-mechanical and barrier properties of pectin based edible films. Int J Food Eng 3(4):10

Makaremi M, Yousefi H, Cavallaro G, Lazzara G, Goh CBS, Lee SM et al (2019) Safely dissolvable and healable active packaging films based on alginate and pectin. Polymers 11(10):1594

Matta E, Bertola N (2020) Development and characterization of high methoxyl pectin film by using isomalt as plasticizer. J Food Process Preserv 44(8):e14568

Mellinas C, Ramos M, Jiménez A, Garrigós MC (2020) Recent trends in the use of pectin from agro-waste residues as a natural-based biopolymer for food packaging applications. Materials 13(3):673

Mendes JF, Norcino LB, Manrich A, Pinheiro ACM, Oliveira JE, Mattoso LHC (2020) Characterization of pectin films integrated with cocoa butter by continuous casting: physical, thermal and barrier properties. J Polym Environ 28:2905–2917

Mohnen D (2008) Pectin structure and biosynthesis. Curr Opin Plant Biol 11:266–277

Naqash F, Masoodi FA, Rather SA, Wani SM, Gani A (2017) Emerging concepts in the nutraceutical and functional properties of pectin—a review. Carbohydr Polym 168:227–239

Nazir N, Sayuti K (2019) Application of pectin extracted from cocoa pod in the production of edible film. IOP Conf Ser Earth Environ Sci 347(1):012060

Nešić A, Ružić J, Gordić M, Ostojić S, Micić D, Onjia A (2017) Pectin-polyvinylpyrrolidone films: a sustainable approach to the development of bio-based packaging materials. Compos Part B 110:56–61

Ngo TMP, Dang TMQ, Tran TX, Rachtanapun P (2018) Effects of zinc oxide nanoparticles on the properties of pectin/alginate edible films. Int J Pol Sci 2018:5645797

Ngo TMP, Nguyen TH, Dang TMQ, Tran TX, Rachtanapun P (2020) Characteristics and antimicrobial properties of active edible films based on pectin and nanochitosan. Int J Mol Sci 21(6): 2224

Nisar T, Wang ZC, Yang X, Tian Y, Iqbal M, Guo Y (2018) Characterization of citrus pectin films integrated with clove bud essential oil: physical, thermal, barrier, antioxidant and antibacterial properties. Int J Biol Macro 106:670–680

Norcino LB, Mendes JF, Natarelli CVL, Manrich A, Oliveira JE, Mattoso LHC (2020) Pectin films loaded with copaiba oil nanoemulsions for potential use as bio-based active packaging. Food Hydrocoll 106:105862

Pedrolli DB, Monteiro AC, Gomes E, Carmona EC (2009) Pectin and pectinases: production, characterization and industrial application of microbial pectinolytic enzymes. Open Biotechnol J 3:9–18

Peretto G, Du WX, Avena-Bustillos RJ, De J, Berrios J, Sambo P et al (2017) Electrostatic and conventional spraying of alginate-based edible coating with natural antimicrobials for preserving fresh strawberry quality. Food Bioprocess Technol 10:165–174

Pérez-Arauz ÁO, Rodríguez-Hernández AI, del Rocío L-CM, Martínez-Juárez VM, Chavarría-Hernández N (2021) Films based on pectin, Gellan, EDTA, and bacteriocin-like compounds produced by Streptococcus infantarius for the bacterial control in fish packaging. J Food Process Preserv 45(1):e15006

Pilla S (2011) Handbook of bioplastics and biocomposites engineering applications, vol 81. John Wiley & Sons, Hoboken

Raghav PK, Agarwal N, Saini M (2016) Edible coating of fruits and vegetables: a review. Int J Sci Res Modern Edu 1(1):2455–5630

Rahmati S, Abdullah A, Momeny E, Kang OL (2015) Optimization studies on microwave assisted extraction of dragon fruit (*Hylocereus polyrhizus*) peel pectin using response surface methodology. Int Food Res J 22(1):233

Rivera-Hernández L, Chavarría-Hernández N, Cuellar MDRL, Martínez-Juárez VM, Rodríguez-Hernández AI (2020) Pectin-gellan films intended for active food packaging: release kinetics of nisin and physico-mechanical characterization. J Food Sci Technol 58(8):2973–2981. https://doi.org/10.1007/s13197-020-04800-z

Rodsamran P, Sothornvit R (2019) Preparation and characterization of pectin fraction from pineapple peel as a natural plasticizer and material for biopolymer film. Food Bio Product Process 18:198–206

Rossman J M (2009) Commercial manufacture of edible films. Edible films and coatings for food applications, 367–390.

Samanta S (2019) Microbial pectinases: a review on molecular and biotechnological perspectives. J Micro Biotech Food 2021:248–266

Šešlija S, Nešić A, Škorić ML, Krušić MK, Santagata G, Malinconico M (2018) Pectin/carboxymethylcellulose films as a potential food packaging material. Macromol Symp 378:1600163

Siemann U (2005) Solvent cast technology—a versatile tool for thin film production. In: Progress in colloid and polymer science. Springer, Berlin

da Silva ISV, Neto WPF, Silvério HA, Pasquini D, Zeni Andrade M, Otaguro H (2017) Mechanical, thermal and barrier properties of pectin/cellulose nanocrystal nanocomposite films and their effect on the storability of strawberries (Fragaria ananassa). Polym Adv Technol 28(8): 1005–1012

Siracusa V (2012) Food packaging permeability behaviour: a report. Int J Polym Sci 2012:302029

Skurtys O, Acevedo C, Pedreschi F, Enrione J, Osorio F, Aguilera JM (2010) Food hydrocolloid edible films and coatings. Food hydrocolloids: characteristics, properties and structures. Nova Science Publishers, Hauppauge

Srivastava P, Malviya R (2011) Sources of pectin, extraction and its applications in pharmaceutical industry: an overview. Indian J Nat Prod Res 2(1):10–18

Srivastava P, Tripathi PK (2020) Polysaccharide-based controlled release systems for antipsychotic therapeutics delivery. Int J Res Pharma Sci 11(3):3496–3505. https://doi.org/10.26452/ijrps.v11i3.2501

Suhag R, Kumar N, Petkoska AT, Upadhyay A (2020) Film formation and deposition methods of edible coating on food products: a review. Food Res Int 136:109582

Trejo-González L, Rodríguez-Hernández AI, del Rocío L-CM, Juárez-Martínez VM, Chavarría-Hernández N (2018) Antimicrobial pectin-gellan films: effects on three foodborne pathogens in a meat medium, and selected physical-mechanical properties. CyTA J Food 16(1):469–476

Urias-Orona V, Rascón-Chu A, Lizardi-Mendoza J, Carvajal-Millán E, Gardea AA, Ramírez-Wong B (2010) A novel pectin material: extraction, characterization and gelling properties. Int J Mol Sci 11(10):3686–3695

Valdés A, Burgos N, Jiménez A, Garrigós MC (2015) Natural pectin polysaccharides as edible coatings. Coatings 5(4):865–886

Valdés A, Ramos M, Beltrán A, Jiménez A, Garrigós MC (2017) State of the art of antimicrobial edible coatings for food packaging applications. Coatings 7(4):56

Vanitha T, Khan M (2019) Role of pectin in food processing and food packaging. In: Masuelli M (ed) Pectins: extraction, purification, characterization and applications. IntechOpen, Norderstedt, pp 85–106

Winotapun C, Phattarateera S, Aontee A, Junsook N, Daud W, Kerddonfag N, Chinsirikul W (2019) Development of multilayer films with improved aroma barrier properties for durian packaging application. Packag Tech Sci 32(8):405–418

Xu SY, Liu JP, Huang X, Du LP, Shi FL, Dong R, Huang XT, Zheng K, Liu Y, Cheong KL (2018) Ultrasonic-microwave assisted extraction, characterization and biological activity of pectin from jackfruit peel. LWT Food Sci Technol 90:577–582

Yahya AM (2018) Edible film from the pectin of papaya skin (the study of cassava starch and glycerol addition). J Phys 953:012248

Yang J, Juanli Y, Yong H (2011) Recent developments in gelcasting of ceramics. J Eur Ceramics Soc 31(14):2569–2591

Yang JS, Mu TH, Ma MM (2018) Extraction, structure, and emulsifying properties of pectin from potato pulp. Food Chem 244:197–205

Zhang W, Xu P, Zhang H (2015) Pectin in cancer therapy: a review. Trends Food Sci Technol 44: 258–271

Zsivánovits G, Marudova M, Ring S (2005) Influence of mechanical properties of pectin films on charge density and charge density distribution in pectin macromolecule. Colloid Polym Sci 284(3):301–308

Chapter 10
Agar-Based Films

Jessica Pandohee and Bo Wang

10.1 Introduction

Food packaging is the final step in the food production system and serves a critical purpose in maintaining food quality and safety and preventing physical damage during transport. For decades, plastic and related products have extensively been used as food packaging. Made from petroleum, plastic is cheap, durable, and malleable; however, nowadays we realize they are causing more damage than good. Due to their slow decomposition rate and low amount of reuse and recycling when they were first introduced, plastic products are causing severe land and water pollution, affecting the lives and well-being of flora and fauna and destroying biodiversity (Jambeck et al. 2015). As, part of the 2030 Agenda for Sustainable Development (United Nations 2015), "Target 12 Responsible Consumption and Production, aims to "substantially reduce waste generation through prevention, reduction, recycling and reuse...promote public procurement practices that are sustainable...ensure that people everywhere have the relevant information and awareness for sustainable development and lifestyles in harmony with nature." Therefore, single-use plastics have been phased out in many countries across the world, and there has been a strong push for the elimination of other plastic products, leaving a need for an alternative material in the food packaging industry (Xanthos and Walker 2017; Vimal et al. 2020).

J. Pandohee (✉)
Centre for Crop and Disease Management, School of Molecular and Life Sciences, Curtin University, Bentley, WA, Australia
e-mail: jessica.pandohee@curtin.edu.au

B. Wang
School of Behavioral and Health Sciences, Australian Catholic University, North Sydney, NSW, Australia

T. Amin et al. (eds.), *Polysaccharide Based Films for Food Packaging: Fundamentals, Properties and Applications*,
https://doi.org/10.1007/978-981-99-4898-7_10

Agar-based films present a potential solution to meet the increased demand for new material needed to replace plastic products in the food packaging industry. Agar is not a new material; it is a pyrocollodion that has been discovered in Japan and used for centuries across the world (Armisén and Gaiatas 2009). Although agar is mainly known as a food ingredient or culinary delicacy, it is widely used in scientific research as a medium to grow microorganism or for seedling germination in plant biology. Agar on its own has little applications due to its brittleness, high moisture permeability, and poor thermal stability (Xiao et al. 2019). Recent research in the development of new agar-based films revealed that the properties of the altered materials can be improved through the addition of essential oil, phytochemicals, metals, nanoparticles, or a combination of those, consequently increasing the applicability of agar-based films as packaging in industries outside the food system but also creating new forms of materials for potential applications in the biomedical, biotechnology, and pharmaceutical industry. This chapter reviews the most recent development in the manufacture, new development, and use of agar-based films. The potential of the enhanced agar-based films to be used as an alternative to plastic packaging is also discussed.

10.2 Sources and Structure of Agar

Agar can be made from several seaweeds found in warm and temperate oceans. The most common species harvest in the production of agar is *Gelidium* cultivated in Spain, Portugal, Morocco, Japan, Korea, Mexico, France, the USA, China, Chile, and South Africa. The *Gracilaria*, *Pterocladia*, and *Gelidiella* species originate from South America, New Zealand, and India, respectively. The use of each seaweed results in a different type of agar being formed and therefore will change the application and use of the agar material. The role of agar in seaweed is to act as a supporting structure in the cellular wall of seaweeds.

Agar is a pyrocollodion, a biopolymer consisting of a mixture of agarose and agaropectin. The structure of agar is shown in Fig. 10.1. Agarose, making up to 70% of the mixture, is a linear polysaccharide consisting of alternating D-galactose and 3,6-anhydro-alpha-L-galactopyranose connected by α-(1 → 3) and β-(1 → 4) link. It is the gelling fraction of agar. The non-gelling fraction of agar is called agaropectin and is made up of alternating D-galactose and L-galactose units that are heavily modified with acidic side groups such as sulfates or pyruvates (Araki 1956). Agaropectin is usually removed during the manufacture process; it is a heterogeneous mixture of smaller molecules.

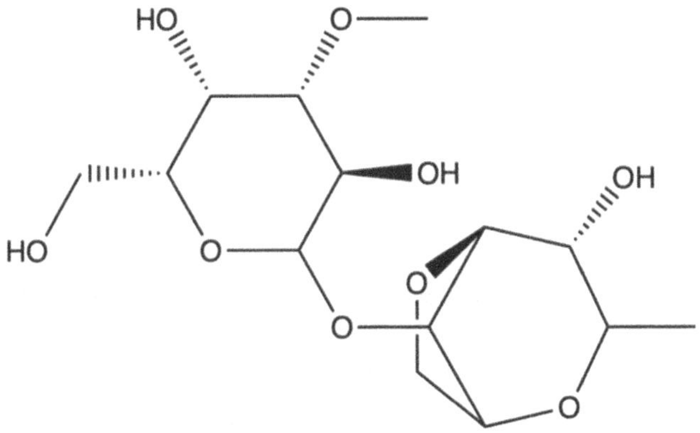

Fig. 10.1 Chemical structure of agar

10.3 Production of Agar and Agar Derivative-Based Films

The manufacturing of agar is a process that involves many steps at several locations (Martínez-Sanz et al. 2019a). Figure 10.2 shows an overview of agar production from its harvest. Raw seaweed material can be obtained through various ways; while farming and culturing seaweed is a common practice in China, Indonesia, and the Philippines, other modes of harvesting seaweed have included cutting or rooting seaweed from the seabed. Once harvested, the seaweeds are stabilized by dehydration to ensure that fermentation does not occur during transport to the processing plant. The *Gracilaria* species has an additional step in its processing; they are usually treatment with a strong sodium hydroxide solution to hydrolyze sulfate groups to increase the strength of the final agar produced. At the processing plant, the dehydrated seaweeds are rehydrated, and any undesired foreign objects and impurities are removed. When soaked in warm water, the agar dissolves out of the seaweed cell walls into the solution which is then filtered and turn into a gel through either freezing or syneresis. The resulting gel is then dried, milled, and screened for quality control and assurance before being distributed as agar.

10.4 Properties of Agar and Agar-Derived-Based Films

Functional properties of agar such as gelling and thickening capability make it a promising material to produce biodegradable packaging material. However, agar-based films with various addictive and modified properties are desired in the applications.

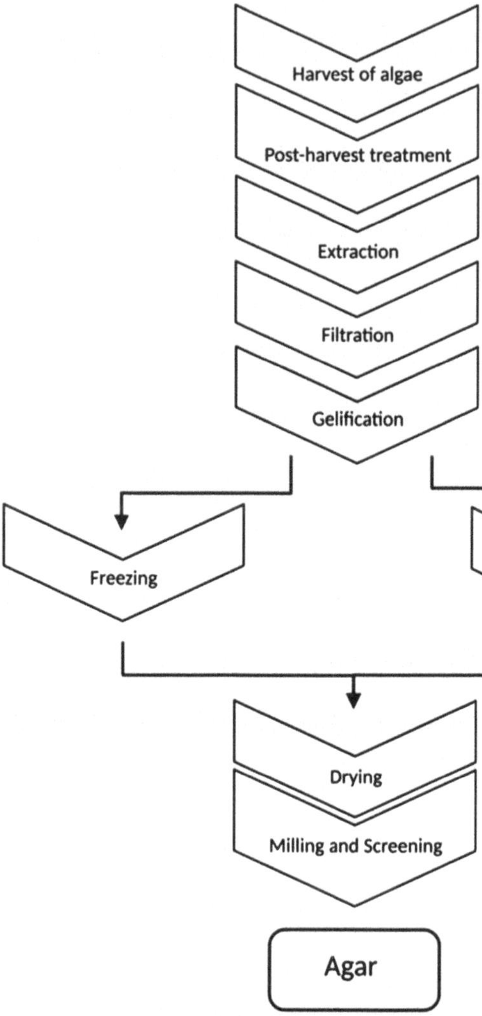

Fig. 10.2 Preparation of agar

10.4.1 Thermal Properties

The thermal property of the packaging material particularly thermal stability is one of the most important qualities which affects their applications. For this purpose, thermogravimetric analysis of the developed films is often performed. Until now, various nanoparticles have been used as fillers into the agar-based film to improve their thermal properties. In the study of Davidović et al. (2019), based on the addition of silver nitrate ($AgNO_3$) and magnesium chloride ($MgCl_2$) at the Ag^+ and Mg^{2+} content of 1.56 and 17.4 mg/g, respectively, the degradation peak temperature of the film was increased from 240 to 300 °C. Similarly, Basumatary et al. (2018) and Malagurski et al. (2017) incorporated silver nanoparticles and zinc minerals to produce agar films, respectively, and these authors observed improved thermal resistance of agar-based film, too.

Several studies also investigated the effect of polysaccharides' incorporation on the agar-based films' thermal property. The incorporation of nano-bacterial cellulose into agar gel at the fiber content of 0–10% (w/w) dramatically improved crystallinity and thermal property of the film, based on the formation of hydrogen bonds between the cellulose and agar (Wang et al. 2018). Similarly, Rukmanikrishnan et al. (2019) observed the more crystalline regions and improved thermal stability of agar film when xanthan gum was incorporated, due to enhanced intermolecular hydrogen bonding interaction among the polymers.

10.4.2 Barrier Properties

The quality of the packaged food products can be deteriorated via multiple mass transfer phenomena including moisture absorption, oil absorption, oxygen invasion, flavor loss, undesirable odor absorption, and the migration of packaging components into the product. Therefore, the packaging materials should act as effective physical barriers against these storage conditions. So far, water vapor and oxygen permeability and UV-blocking properties are the most measured barrier properties of the developed agar-based films, as Table 10.1 shows.

Generally, the incorporation of polymers and/or particles modifies the structure/network of the film to alter their barrier property. For instance, the addition of agar to potato starch at 5% (w/w) significantly decreased water vapor permeability of the film from 6.29 to 4.60 × 10^{-10} g Pa^{-1} s^{-1} m^{-1}, due to the formation of a more compact structure between two polymers, which inhibited the diffusion of water molecules (Wu et al. 2019). In the study carried out by Rhim (2011), when nano clay was incorporated into agar-based films in content of 0–20% (w/w), strong structure was formed between the intercalated silicate layers and agar, creating a tortuous path for water vapor diffusion. Therefore, a decreased water vapor permeability, water vapor uptake ratio, and water solubility were observed. However, at a high incorporation dose, the aggregated nano clay particles negatively affected the water vapor permeability of agar-based films (Lee et al. 2019). Shankar and Rhim (2017) explored the efficacy of silver nanoparticle addition to improve water vapor permeability of agar-based films. The result showed that silver nanoparticles, synthesized using lignin, inhibited water vapor penetration significantly due to the compatibility between lignin and agar. On the other hand, the water vapor and oxygen permeability of agar-based film can be modified by altering the hydrophobicity of the system, too. Zhang et al. (2019) showed that beeswax and liquid paraffin addition into the agar-maltodextrin films resulted in a decreased water vapor and slightly increased oxygen permeability, due to the increase of its hydrophobicity.

Table 10.1 Barrier property of some developed agar-based films

Nature of the agar-based film	Result	References
Agar/maltodextrin/ hydroxypropyl cassava starch films	Higher degree of degrees of substitution of hydroxypropyl cassava starch promoted the interaction between agar and maltodextrin via hydrogen. The water vapor and oxygen permeability were decreased by 25% and 58% with combined agar and maltodextrin	Wongphan and Harnkarnsujarit (2021)
Agar/cassava starch film incorporated with Ag and ZnO nanoparticles	The water solubility and water vapor permeability decreased with the incorporation of Ag and ZnO nanoparticles in the film	Mahuwala et al. (2020)
Agar film made using less purified raw materials	The agar remaining in the less purified fractions had a positive effect on the performance of the films, particularly the improved water vapor barrier property	Martínez-Sanz et al. (2020)
Agar/starch films	Water vapor permeability tended to decrease with increased agar due to the reduced hydrophilic phase of starch and increased hydrophobic aggregates that effectively enhanced water barriers. Meanwhile, interaction between agar and starch macromolecules contribute to rigidity of the film which effectively prevented diffusion of oxygen	Wongphan and Harnkarnsujarit (2020)
Agar/tapioca (cassava) starch with glycerol and span 80	The incorporation of agar in the film created strong inter-molecular links with tapioca starch and formed a dense three-dimensional structure. With the increase of agar content and the presence of glycerol, pores and cracks could be formed which possibly facilitated the permeation of water vapor	Maran et al. (2013)
	Increase in concentration of agar and starch combined with glycerol causes fluctuations in film density and it may be lead to the formation of films with pores and cracks and hence facilitates the permeation of water vapor	
Agar film with curcumin	Addition of a small amount of curcumin (1 wt.%) increased the UV-blocking	Roy and Rhim (2020)
Agar film containing lignin	Compared with agar film, the agar/ligin film exhibited improved UV barrier properties and lower water vapor permeability	Shankar et al. (2015)

The UV-blocking property of the packing film can potentially maintain the quality of the food product by retarding the oxidation. Agar films are usually transparent and known for their poor UV barrier properties; therefore it requires the addition of different additives that would block the passage of light (Shankar et al. 2014). For this purpose, various nanoparticles or polymers are incorporated into the films. Radovanović et al. (2019) further demonstrated that Cu-posnjakite and Cu-phosphate were added into agar films, respectively, and the authors observed the decreased transmittance value of films at 280 nm wavelength, indicating the blocking of UV rays by these particles. Similarly, the titanium dioxide (TiO_2) nanoparticles in the agar/kappa carrageenan film successfully absorbed the UV rays in the wavelength range of 200–300 nm, and it had the potential application in delaying lipid oxidation of photosensitive foods (Hou et al. 2018). Interestingly, when melanin nanoparticles were incorporated into agar-based film at low content ($<1\%$, w/w), the film exhibited UV-blocking effects without compromising transparency comprehensively (Roy and Rhim 2019). Recently, Uranga et al. (2020) observed an improved UV barrier property of the agar/fish gelatin film, in the wavelength range below 300 nm based on the fish gelatin–citric acid nucleophilic substitution and agar–citric acid esterification.

10.4.3 Mechanical Properties

One of the packaging films' aims is to provide effective protection to food products against physical damage caused by mechanical factors. Therefore, typical mechanical characterization methods can be used to evaluate their quality such as tensile strength, elongation-at-break, and elastic modulus (i.e., young's modulus), among others, as Table 10.2 shows.

As a plasticizer, glycerol has been commonly used to improve the flexibility of the packing film and prevent its crack during processing and storage, and the effects of glycerol incorporation into agar-based films on their mechanical properties have been reported (Vieira et al. 2011; Ballesteros-Mártinez et al. 2020). For instance, when glycerol was incorporated in the agar/Algerian *Opuntia ficus-indica* cladodes powder film in the content range of 30–60% (w/w), significantly improved mechanical performance of the film was observed (Makhloufi et al. 2021a) compared with agar film. Moreover, the increase of glycerol in the film resulted in the increase in film's thickness and elongation at break but decreased Young's modulus and tensile strength.

Carbohydrates such as maltodextrin, starch, cellulose, and polysaccharides can be incorporated to agar-based films to additionally alter their mechanical properties. When the potato starch and agar were mixed at different ratios to produce agar-based films using glycerol as plasticizer, the highest value of strength, elastic modulus, and puncture resistance was observed in the film with the agar-to-starch ratio of 7:1–9:1 (w/w) (Tazhibayeva et al. 2020). Moreover, these authors used Mg^{2+} ion to enhance the H-bond grid in the films, and the strength of the film was further improved.

Table 10.2 Effect of incorporated compounds and processing method on mechanical property of agar-based film

Nature of the agar-based film	Result	References
Agar/smooth hound (*Mustelus mustelus*) skin gelatin film	Compared with gelatin-based films, the incorporation of agar at 25% (w/w) improved the film rigidity, by increasing the film tensile strength and decreasing elongation at break	Salem et al. (2021)
Agar/*Opuntia ficus-indica* cladodes powder film containing glycerol as plasticizer	Young's modulus and tensile strength decreased with the increase of glycerol concentration in the film	Makhloufi et al. (2021b)
Rice starch/hydroxypropyl cassava starch/agar film	Higher rice starch ratios significantly increased crystallinity which affected physical, mechanical, and barrier properties during storage for3 months	Phinainitisatra and Harnkarnsujarit (2021)
Agar film containing ZnO or TiO_2 nanoparticles	Incorporation of TiO_2 or ZnO enhanced the mechanical properties of the agar film, and when agar-TiO_2 was used in the combination with κ-carrageenan-*Clitoria ternatea* Linn anthocyanin film, its elongation at break was further improved	Liu et al. (2020)
Agar film containing hydroalcoholic garlic extract with glycerol as a plasticizer	The impregnation of hydroalcoholic garlic extract decreased the strength and elastic modulus of the agar films	Campa-Siqueiros et al. (2020)
Agar film containing maltodextrin and beeswax	The highest tensile strength (20.57 MPa) and young's modulus (640.60 MPa) were observed when the film was prepared by homogenization at 8000 rpm for 1 min	Zhang et al. (2020)
Agar film containing chitosan and halloysite nanotubes	Synergetic effect of chitosan and halloysite on the mechanical properties of the agar film was observed. When 40% chitosan and 8% halloysite are into agar film, the tensile strength and elongation at the break of agar/chitosan/halloysite composite were improved from 35.03 MPa and 14.79% to 48.29 MPa and 22.62%, respectively	Huang et al. (2020)
Agar/gellan gum/montmorillonite based ternary nanocomposite films	The montmorillonite-based ternary nanocomposite nano clay addition significantly enhanced the tensile strength of the film (29.9–44 MPa)	Lee et al. (2019)
Agar/thermoplastic corn starch with high glycerol content	The addition of agar to thermoplastic corn starch resulted in films with promising tensile properties. Stiffness and strength increased considerably by increasing agar content, while deformability of blends was better	Fekete et al. (2019)

(continued)

Table 10.2 (continued)

Nature of the agar-based film	Result	References
	than those of pure thermoplastic corn starch film	
Agar film with impurities	The impurities such as other polysaccharides (mainly floridean starch) and proteins acted as plasticizer in the impurified agars, forming flexible films with no added plasticizer, resulting in an improved mechanical performance, compared with those from the purified agars	Martínez-Sanz et al. (2019b)
Agar/cellulose film	Incorporation of cellulose microfiber into agar film increased tensile strength, young's modulus, and elongation at break	Anand Raj et al. (2019)
Agar film containing SiO_2 nanoparticles	The tensile strength and elongation at break of film containing 10 wt.% nano-SiO_2 increased by 65.29% and 60.38%, respectively, when compared to those of film prepared without nano-SiO_2	Hou et al. (2019)
Agar/kappa carrageenan films containing TiO_2 nanoparticles	As nano-TiO_2 content increased from 0 to 1 g/100 g, tensile strength of the film increases from 30.8 to 60.6 MPa. However, TS decreases gradually as nano-TiO_2 content further raises, possibly due to the formation of intermolecular hydrogen bonding and coordinate bonds between nano-TiO_2 and matrix molecules. Elongation at break decreases gradually as nano-TiO_2 concentration increases	Hou et al. (2019)

Similarly, elephant foot yam starch was combined with xanthan and agar gum to produce edible films, and the film thickness, density, and mechanical strength were improved gradually with the increase of hydrocolloid concentration, and the film with the highest tensile strength was developed when agar was added at 2% (w/w) (Nagar et al. 2020). In Roy et al.'s (2021) study, the incorporation of natural colorant Shikonin into carboxymethyl cellulose/agar film increased film's tensile strength by 37.4%.

10.4.4 Optical Properties

Edible packaging films are applied directly to food surfaces in the wrapping or coating form. Therefore, the color and transparency of these films are significant factors for the visual appearance and consumer acceptance of foods. The optical properties of edible film can be significantly affected by the incorporation of a wide range of compounds including nanoparticles, essential oils, and/or phenolics (Ahmadi et al. 2021; Correa-Pacheco et al. 2019; Cai et al. 2020; Ribeiro et al. 2020). Nagar et al. (2020) reported that when agar was introduced to elephant foot yam starch film, the transparency of the film at 560 nm wavelength was significantly increased due to less wrapping of starch molecules. In da Rocha et al.'s (2018) study, the addition of fish protein hydrolysate and clove essential oil to agar film resulted in an increased *b** (yellowish color) higher opacity of the film, respectively. Similarly, the incorporation of *Larrea nitida* extract containing polyphenols into agar films was responsible for a more saturated orange-brownish color.

10.4.5 Antimicrobial Properties

The antimicrobial and antifungal activity is considered as one of the most promising properties for packaging materials, because it helps extend the shelf life of the food products by inhibiting the growth of pathogenic microorganisms. Generally, the antimicrobial or antifungal films can be prepared by either applying the antimicrobial or antifungal agents on the surface of film or mixing them into the polymer during the preparation. During the storage of the food products, these agents can be controlled-released from the film to ensure product's quality and safety.

Some examples of agar-based films with antimicrobial and/or antifungal properties of agar-based films are shown in Table 10.3. Generally, the agents at the nanoscale are quite effective due to their high surface-to-volume ratio and enhanced surface reactivity, compared with macromolecules. For instance, the agar-based films containing copper (II) carbonate and copper (II) phosphate exhibited distinctive antimicrobial activity against *Escherichia coli* and *Staphylococcus aureus* (Radovanović et al. 2019). In another study, when Cu S nanoparticle, corn starch, and ammonia were used as source of Cu^{2+}/S^{2-}, stabilizing agent, and hydrolyzing agent, respectively, the developed agar-based film showed significant antibacterial activity against *Escherichia coli* and *Listeria monocytogenes* (Roy et al. 2019). Meanwhile, Arfat et al. (2017) reported that the incorporation of silver-copper alloy nanoparticles in the glycerol-plasticized agar solution at 0.5–4% (w/w) contributed to the antibacterial activity against *Listeria monocytogenes* and *Salmonella enterica sv typhimurium*, possibly due to the release of silver and copper ions which penetrate and eliminate the microorganisms, and similarly, the zinc nanoparticles in the agar-based film were found to be responsible for the inhibition of *Staphylococcus aureus* in situ, based on the release of zinc ions to the food at an accepted level (up to 2.5%) (Radovanovic et al. 2019).

The incorporation of plant-based macromolecules such as essential oil and phenolic compounds with antimicrobial into agar-based films has been explored, too. As a natural antimicrobial agent, grapefruit seed extract (GSE) was added into agar to develop agar/GSE films with significant antimicrobial activity against

Table 10.3 Antimicrobial and antifungal property of example agar-based film

Nature of the agar-based film	Result	References
Antimicrobial property		
Agar/carboxymethyl cellulose film with silver-modified montmorillonite	Due to the incorporation of silver particles, this agar-based films exhibited a great antibacterial activity against *Bacillus subtilis* and *Escherichia coli*	Makwana et al. (2020)
Agar film containing protein hydrolysate and clove essential oil	The film incorporated with hydrolysate possessed by Alcalase exhibited the most effective antimicrobial activity against *Staphylococcus aureus*, *Yersinia enterocolitica*, *Aeromonas hydrophila*, *Debaryomyces hansenii* and *Listeria innocua*, due to its lower molecular weight and higher hydrophobic amino acid content	da Rocha et al. (2018)
Agar film with curcumin	The presence of curcumin in the film resulted in strong antioxidant activity in the DPPH and ABTS radical scavenging test, together with some antimicrobial activity against *Escherichia coli* and *Listeria monocytogenes*	Roy and Rhim (2020)
Agar/cellulose film with savory essential oil	Due to the presence of savory essential oil, the developed films containing SEO were quite effective against *Listeria monocytogenes* and *Staphylococcus aureus*	Atef et al. (2015)
Agar film with *Larrea nitida* extract	*Larrea nitida* with antimicrobial activities was used to develop agar films and showed antimicrobial activity against *Listeria innocua* and *Escherichia coli*	Moreno et al. (2020)
Agar film with Zn-carbonate and Zn-phosphate	Treatment of *Staphylococcus aureus* with both nanocomposite formulations resulted in a complete biocidal action for *Staphylococcus aureus* after 24 h. However, *Escherichia coli* was only mildly inhibited	Malagurski et al. (2017)
Antifungal property		
Agar film with Zn-carbonate and Zn-phosphate	Application of both formulations to *Candida albicans* growth media significantly inhibit the growth of fungi, from 8.2 to 2.5 and 2.7 $\log_{10}$ CFU/mL after 24 h incubation	Malagurski et al. (2017)
Agar/gray triggerfish gelatin film with vine leave extract	The formulation of the film inhibited the growth of *Aspergillus flavus*, *Aspergillus niger*, and *Fusarium oxysporum*, although less effective than cycloheximide	Jridi et al. (2018)

Listeria monocytogenes, *Bacillus cereus*, and *Escherichia coli* (Kanmani and Rhim 2014). In Abdollahi et al.'s (2019) study, summer savory essential oil was mixed with carboxymethyl cellulose-agar gel to produce the packaging material, and the developed film showed strong inhibitory activity against *Staphylococcus aureus*, *Bacillus cereus*, and *Listeria monocytogenes*. However, this novel film was less effective against *Escherichia coli*. On the other hand, the incorporation of *Larrea nitida* extract, which is rich in polyphenols into agar film, resulted in antimicrobial activity against *Listeria innocua and Escherichia coli* but no significant antiviral activity against murine norovirus (Moreno et al. 2020).

Compared with antimicrobial property of agar-based film for food applications, the research investigating their antifungal property is less reported. Jridi et al. (2018) explored the antifungal property of agar/gray triggerfish gelatin film with the incorporation of vine leave extract. The result showed that the developed film significantly inhibited the growth of *Aspergillus flavus*, *Aspergillus niger*, and *Fusarium oxysporum*. In Malagurski et al.'s (2017) study, the growth of *Candida albicans* was inhibited by the Zn-carbonate or Zn-phosphate component in the agar film.

10.4.6 Antioxidative Properties

The oxidation of food products during the storage is often a factor limiting their shelf life, particularly when the food is rich in oil and/or fat. Compared with microbial spoilage, oxidation is a relatively slow process. Oxidation can be catalyzed by environmental stress such as heat and light or food component including transition metals and enzymes. Once the oxidation starts, oxidation products with rancid odor and flavor are formed, which significantly affects the quality and acceptability of the food products. Therefore, the packaging material with reduced oxygen permeability and modified opacity can act as an effective physical barrier against oxygen and light to prevent food from oxidation. For this purpose, several recent works investigated the antioxidative property of agar-based film, based on the incorporation of compounds with antioxidative activity such as nanoparticles, phenolic compounds, probiotics, and essential oils, among others, as shown in Table 10.4.

Vine leaf ethanolic extract was incorporated into agar-gray triggerfish skin gelatin films at 5–10 mg/mL, and the authors (Jridi et al. 2018) observed significantly improved antioxidant potential evaluated by iron (Fe^{2+}) chelating activity, ferric (Fe^{3+}) reducing antioxidant power, β-carotene bleaching protection, and 2,2-diphenyl-1-picrylhydrazyl (DPPH) radical scavenging activity. Similarly, Moreno et al. (2020) observed the release of polyphenolic compounds in the *Larrea nitida* extract from the agar film during the 14-day storage, and 2,2′-azino-bis (3-ethylbenzothiazoline-6-sulfonic acid) (ABTS) radical scavenging assays' result indicated a positive correlation between the *Larrea nitida* extract incorporation and oxidative activity of the film. In another study, the pine bark (*Pinus massoniana*) extract in the agar film, with 95% proanthocyanins, resulted in a strong total phenolic content-DPPH radical scavenging activity (Leelaphiwat et al. 2020). The antioxidative capacity of different probiotic bacteria strains have been reported, via chelating the transition metal, secreting antioxidative enzymes, and/or producing antioxidative metabolites by Wang et al. (2017). Therefore, it can be incorporated into the packaging materials. In Yasin et al.'s (2020) study, *Lactobacillus acidophilus* and *Saccharomyces cerevisiae* were added into agar film, and the resultant films exhibited reasonable antioxidative activity.

Table 10.4 Antioxidative property of some agar-based film

Nature of the agar-based film	Result	References
Agar-based film with nanoparticles		
Agar/banana powder film with silver nanoparticles	The incorporation of banana powder increased the 2,2-diphenyl-1-picrylhydrazyl (DPPH) radical scavenging activity of agar film. However, the presence of silver nanoparticles did not contribute to DPPH radical scavenging activity	Orsuwan et al. (2016)
Agar film with melanin nanoparticles	Melanin nanoparticles separated from sepia ink in the agar film contributed to its strong antioxidative property, shown by increased DPPH and ABTS radical scavenging activity	Roy and Rhim (2019)
Agar-based film with phenolic compounds		
Agar/gelatin film with green tea extract	Antioxidant activity was observed in the film. However, the gelatin component hindered the release of total phenolic compounds, catechins and flavanols from the film. Therefore, the antioxidant power released by the films was lower in the case of films containing gelatin	Giménez et al. (2013)
Agar film with curcumin	DPPH and ABTS radical scavenging activity assay result showed the addition of curcumin at 1% (w/w) in the agar film resulted in a strong antioxidative property of the film	Roy and Rhim (2020)
Agar-based film with essential oils		
Agar/alginate bilayer film with cinnamon essential oil	Based on the cinnamon essential oil component, the developed bilayer film exhibited high antioxidative property in ferric reducing antioxidant power (FRAP) and 2,2′-azino-bis (3-ethylbenzothiazoline-6-sulfonic acid) radical (ABTS) scavenging assays	Arancibia et al. (2014)

10.5 Applications

Novel agar-based packing materials with various functional properties are promising systems to improve the quality of the packaged food products. Until now, the quantity of a wide range of food products including seafood, poultry meat, dairy, some vegetables and fruits, etc. during the shelf life has been evaluated when it is packaged using these novel materials.

10.5.1 Animal-Based Products

Among the animal-based products, seafood products are highly prone to microbial deterioration due to its high-water activity and content of nitrogenous substances (Zhai et al. 2017). In De Lacey and López-Caballero's (De Lacey et al. 2014) study, the probiotic component (*Lactobacillus paracasei* L26 and *Bifidobacterium lactis*

B94) in the agar film with green tea extract significantly inhibited the growth of H_2S-producing and total viable bacteria in the hake fillets, finally resulting in a one-week extension of its shelf life. Similarly, in Yasin et al.'s (2020) study, *Lactobacillus acidophilus* and *Saccharomyces cerevisiae* in the agar film successfully extend the shelf life of GIFT (genetically improved farm tilapia) (*Oreochromis mossambicus*), together with the improved color preservation of the fillet during the storage. The effect of the incorporation of essential oils and nanoparticles to agar film on the shelf life of fish products has been investigated by da Rocha et al. (2018). These authors developed agar film incorporated with clove essential oil and fish protein hydrolysate as the packaging materials for flounder (*Paralichthys orbignyanus*) fillet. The result showed that the packaged fillets exhibited lower total volatile bases and pH values. Moreover, the growth of H_2S-producing bacteria was considerably inhibited in this study. Similarly, Arancibia et al. (2014) observed the inhibition of *Listeria monocytogenes* growth in the peeled shrimps using agar/alginate film containing cinnamon essential oil during the chilled storage, without any negative impact on the organoleptic properties of the products. In another study, ZnO nanoparticles, cinnamon essential oil, and nisin were added into agar-based packaging materials to extend the shelf life of minced fish (Abdollahzadeh et al. 2018).

Besides microbial deterioration, the oxidation of polyunsaturated fatty acids in the seafood is another major challenge for the quality deterioration, leading to the formation of oxidative products with odor and flavor. Therefore, agar film with improved oxygen barrier capacity can be potentially used as promising seafood packaging material to retard the lipid oxidation. Vejdan et al. (2017) studied the effects of gelatin/agar bilayer film containing TiO_2 nanoparticles on the oxidative stability of fish oil. Fish oil photooxidation was significantly retarded by increasing TiO_2 concentration in the film. Furthermore, the lowest number of changes in peroxide values, thiobarbituric acid, conjugated diene, and color changes was observed in the samples covered with this novel packaging material.

Compared with seafood products, the application of agar-based film to preserve the quality of meat, poultry, or dairy products has been rarely studied, despite their highly perishable nature. In Jang et al.'s (2010) study, the addition of thymol (1.5%, w/w) into agar-gelatin composite films containing cloisite Na+ inhibited the growth of *Escherichia coli* O157:H7, *Listeria monocytogenes*, *Salmonella Typhimurium*, and *Campylobacter jejuni* during the storage of chicken breast. Malaka et al. (2017) packaged Dangke cheese, a tradition soft cheese in South Sulawesi Indonesia using agar film and monitored the change in its hardness, microstructure, and sensory profile during the storage. The authors observed an extended shelf life, increased hardness, and more compact microstructure of the cheese, as well as a "more milk" smell and smoother texture.

10.5.2 Vegetables and Fruits

Raw vegetables and fruits are also considered as highly perishable food, and they are intended to be consumed "fresh" (Moschopoulou et al. 2019). Nowadays, wax is commonly used as edible coatings for fruits and vegetables, with varying degrees of success in each application (Park 2003). However, the waxy taste is not desired by the consumers; therefore, there is an increasing demand to develop the alternative edible packing materials.

Minh and Van Phuc (2018) investigated the efficacy of agar film in preserving the quality of star apple (*Chrysophyllum cainito*), and the result showed the fruit coated by 2.75% agar and 6.8% glycerol maintained good quality until 30 days of preservation. Recently, due to the superiority in high surface-to-volume ratio and enhanced surface reactivity, the incorporation of various nanoparticles into agar-based film as novel packaging materials to preserve quality of vegetables and fruits has received the most research interest. The presence of silver nanoparticles in the agar/collagen/alginate ternary blend films containing grapefruit seed extract significantly extended the shelf life of fresh potatoes due to its antimicrobial activity (Wang and Rhim 2015). Moreover, these authors observed that this novel film successfully prevented the formation of condensed water on the film surface and the greening of potatoes. In Gudadhe et al.'s (2014) study, silver nanoparticles were biosynthesized using Ocimum sanctum leave extract, and it was added into agar film to improve the quality of *Citrus aurantifolia* (thornless lime) and *Pyrus malus* (apple) during storage. The result showed that the shelf life of both fruits was significantly extended and the penetration of silver nanoparticles inside the fruits was negligible. Similarly, nano clay (Cloisite 30B) and zinc oxide nanoparticles were successfully used in the agar-based films to extend the shelf life of fresh spinach and green grapes, respectively (Kumar et al. 2019; Rhim and Wang 2013). Interestingly, the agar-based film containing Cloisite 30B showed high moisture adsorption capacity, so they can be potentially used as an alternative desiccant to silica gel.

10.6 Limitations and Future Trends

Numerous research studies have explored the development of novel agar-based films with various functional properties based on the incorporation of compounds, and the outcomes seem promising. However, compared with the widely used polyethylene terephthalate (PET) packaging materials, most of these agar-based films still exhibited a relatively high oxygen and water vapor permeability, as well as low mechanic strength particularly tensile strength and elongation at break. These are the challenges to be overcome through the future research. Moreover, most of the reported research focuses on the characteristics of the developed novel agar-based films, rather than their potential toxicity. Particularly, the application of nanoparticles has been well associated with concerns about its potential toxicity and ecotoxicity, and this needs to be addressed before commercial applications (Lead et al. 2018; Wang et al. 2015).

10.7 Conclusion

The novel agar-based films with promising physicochemical properties have been developed by various recent studies. Based on the incorporation of a wide range of compounds, the characteristics of the film such as mechanical, barrier, optical, thermal, antimicrobial, antiviral, and antioxidative properties can be tailored, and the developed novel packaging materials may extend shelf life of a wide range of food products. Although these films still exhibit disadvantages, compared to the commonly used PET ones, they offer a simple, easy-to-produce, and cost-effective solution to maintain the quality of the food products, with potentials for future commercialization.

References

Abdollahi M, Damirchi S, Shafafi M, Rezaei M, Ariaii P (2019) Carboxymethyl cellulose-agar biocomposite film activated with summer savory essential oil as an antimicrobial agent. Int J Biol Macromol 126:561–568

Abdollahzadeh E, Mahmoodzadeh Hosseini H, Imani Fooladi AA (2018) Antibacterial activity of agar-based films containing nisin, cinnamon EO, and ZnO nanoparticles. J Food Saf 38:12440

Ahmadi A, Ahmadi P, Sani MA, Ehsani A, Ghanbarzadeh B (2021) Functional biocompatible nanocomposite films consisting of selenium and zinc oxide nanoparticles embedded in gelatin/cellulose nanofiber matrices. Int J Biol Macromol 175:87–97

Anand Raj LFA, Shanmugapriya R, Jeslin J (2019) Biosynthesis of cellulose microfibre from peanut shell for the preparation of bio-nanocomposite films for food-packaging application. Bull Mater Sci 42:63

Araki C (1956) Structure of Teh Agarose constituent of agar-agar. Short Commun 29(4):543–544

Arancibia M, Giménez B, López-Caballero ME, Gómez-Guillén MC, Montero P (2014) Release of cinnamon essential oil from polysaccharide bilayer films and its use for microbial growth inhibition in chilled shrimps. LWT Food Sci Technol 59:989–995

Arfat YA, Ahmed J, Jacob H (2017) Preparation and characterization of agar-based nanocomposite films reinforced with bimetallic (Ag-cu) alloy nanoparticles. Carbohydr Polym 155:382–390

Armisén R, Gaiatas F (2009) Agar. In: Handbook of hydrocolloids. Woodhead Publishing, Duxford, pp 82–107

Atef M, Rezaei M, Behrooz R (2015) Characterization of physical, mechanical, and antibacterial properties of agar-cellulose bionanocomposite films incorporated with savory essential oil. Food Hydrocoll 45:150–157

Ballesteros-Mártinez L, Pérez-Cervera C, Andrade-Pizarro R (2020) Effect of glycerol and sorbitol concentrations on mechanical, optical, and barrier properties of sweet potato starch film. NFS J 20:1–9

Basumatary K, Daimary P, KumarDas S, Thapa M, Singh M, Mukherjee A, Kumara S (2018) Lagerstroemia speciosa fruit-mediated synthesis of silver nanoparticles and its application as filler in agar based nanocomposite films for antimicrobial food packaging. Food Packag Shelf Life 17:99–106

Cai L, Wang Y, Cao A (2020) The physiochemical and preservation properties of fish sarcoplasmic protein/chitosan composite films containing ginger essential oil emulsions. J Food Process Eng 43:e13495

Campa-Siqueiros PI, Vargas-Arispuro I, Quintana-Owen P, Freile-Pelegrín Y, Azamar-Barrios JA, Madera-Santana TJ (2020) Physicochemical and transport properties of biodegradable agar films impregnated with natural semiochemical based-on hydroalcoholic garlic extract. Int J Biol Macromol 151:27–35

Correa-Pacheco ZN, Bautista-Baños S, Ramos-García MDL, Martínez-González MDC, Hernández-Romano J (2019) Physicochemical characterization and antimicrobial activity of edible propolis-chitosan nanoparticle films. Prog Org Coat 137:105326

Davidović S, Lazić V, Miljković M, Gordić M, Sekulić M, Marinović-Cincović M, Ratnayake IS, Ahrenkiel SP, Nedeljković JM (2019) Antibacterial ability of immobilized silver nanoparticles in agar-agar films co-doped with magnesium ions. Carbohydr Polym 224:115187

De Lacey AL, López-Caballero M, Montero P (2014) Agar films containing green tea extract and probiotic bacteria for extending fish shelf-life. LWT Food Sci Technol 55:559–564

Fekete E, Bella É, Csiszár E, Móczó J (2019) Improving physical properties and retrogradation of thermoplastic starch by incorporating agar. Int J Biol Macromol 136:1026–1033

Giménez B, López de Lacey A, Pérez-Santín E, López-Caballero ME, Montero P (2013) Release of active compounds from agar and agar-gelatin films with green tea extract. Food Hydrocoll 30: 264–271

Gudadhe JA, Yadav A, Gade A, Marcato PD, Durán N, Rai M (2014) Preparation of an agar-silver nanoparticles (A-AgNp) film for increasing the shelf-life of fruits. IET Nanobiotechnol 8:190–195

Hou X, Xue Z, Liu J, Yan M, Xia Y, Ma Z (2018) Characterization and property investigation of novel eco-friendly agar/carrageenan/TiO2 nanocomposite films. J Appl Polym Sci 136:47113

Hou X, Xue Z, Xia Y, Qin Y, Zhang G, Liu H, Li K (2019) Effect of SiO2 nanoparticle on the physical and chemical properties of eco-friendly agar/sodium alginate nanocomposite film. Int J Biol Macromol 125:1289–1298

Huang D, Zhang Z, Zheng Y, Quan Q, Wang W, Wang A (2020) Synergistic effect of chitosan and halloysite nanotubes on improving agar film properties. Food Hydrocoll 101:105471

Jambeck JR, Geyer R, Wilcox C, Siegler TR, Perryman M, Andrady A, Narayan R, Law KL (2015) Marine pollution. Plastic waste inputs from land into the ocean. Science 347(6223):768–771. https://doi.org/10.1126/science.1260879

Jang SA, Lin GO, Song KB (2010) Original article: use of nano-clay (Cloisite Na+) improves tensile strength and vapour permeability in agar rich red algae (Gelidium corneum)–gelatin composite films. Int J Food Sci Technol 45:1883–1888

Jridi M, Abdelhedi O, Nasri M, Fakhfakh N, Nasri M (2018) Development and characterization of grey triggerfish gelatin/agar bilayer and blend films containing vine leaves bioactive compounds. Food Hydrocoll 89:370–378

Kanmani P, Rhim JW (2014) Antimicrobial and physical-mechanical properties of agar-based films incorporated with grapefruit seed extract. Carbohydr Polym 102:708–716

Kumar S, Boro SC, Ray D, Mukherjee A, Dutt J (2019) Bionanocomposite films of agar incorporated with ZnO nanoparticles as an active packaging material for shelf life extension of green grape. Heliyon 5:e01867

Lead JR, Batley GE, Alvarez PJJ, Croteau M-N, Handy RD, McLaughlin MJ, Judy JD, Schirmer K (2018) Nanomaterials in the environment: behavior, fate, bioavailability, and effects—an updated review. Environ Toxicol Chem 37:2029–2063

Lee H, Rukmanikrishnan B, Lee J (2019) Rheological, morphological, mechanical, and water-barrier properties of agar/gellan gum/montmorillonite clay composite films. Int J Biol Macromol 141:538–544

Leelaphiwat P, Chanasinphawatkun N, Prompa K, Harnkarnsujarit N (2020) Properties and release kinetics of pine bark incorporated agar and carrageenan film. Int J Food Sci Technol 55:3392–3402

Liu L, Zhang J, Shi J, Huang X, Zou X, Zhang D, Zhai X, Yang Z, Li Z, Li Y (2020) Preparation and comparison of two functional nanoparticle-based bilayers reinforced with a κ-carrageenan–anthocyanin complex. Int J Biol Macromol 165:758–766

Mahuwala AA, Hemant V, Meharwade SD, Suraj D, Deb A, Chakravorty A, Grace AN, Raghavan V (2020) Synthesis and characterisation of starch/agar nanocomposite films for food packaging application. IET Nanobiotechnol 14:809–814

Makhloufi N, Chougui N, Rezgui F, Benramdane E, Freire CSR, Vilela C, Silvestre AJD (2021a) Bio-based sustainable films from the Algerian Opuntia ficus-indica cladodes powder: effect of plasticizer content. J Appl Polym Sci 138:50450

Makhloufi N, Chougui N, Rezgui F, Benramdane E, Freire CSR, Vilela C, Silvestre AJD (2021b) Bio-based sustainable films from the Algerian Opuntia ficus-indica cladodes powder: effect of plasticizer content. Appl Polym 138:50450

Makwana D, Castaño J, Somani RS, Bajaj H (2020) Characterization of agar-CMC/Ag-MMT nanocomposite and evaluation of antibacterial and mechanical properties for packaging applications. Arab J Chem 13:3092–3099

Malagurski I, Levic S, Nesic A, Mitric M, Pavlovic V, Dimitrijevic-Brankovic S (2017) Mineralized agar-based nanocomposite films: potential food packaging materials with antimicrobial properties. Carbohydr Polym 175:55–62

Malaka R, Hatta W, Baco S (2017) Evaluation of using edible coating and ripening on Dangke, a traditional cheese of Indonesia. Food Res 1:51–56

Maran JP, Sivakumar V, Sridhar R, Thirugnanasambandham K (2013) Development of model for barrier and optical properties of tapioca starch based edible films. Carbohydr Polym 92:1335–1347

Martínez-Sanz M, Martínez-Abad A, López-Rubio A (2019a) Cost-efficient bio-based food packaging films from unpurified agar-based extracts. Food Packag Shelf Life 21:100367

Martínez-Sanz M, Gómez-Mascaraque LG, Ballester AR, Martínez-Abad A, Brodkorb A, López-Rubio A (2019b) Production of unpurified agar-based extracts from red seaweed Gelidium sesquipedale by means of simplified extraction protocols. Algal Res 38:101420. https://doi.org/10.1016/j.algal.2019.101420

Martínez-Sanz M, Cebrián-Lloret V, Mazarro-Ruiz J, López-Rubio A (2020) Improved performance of less purified cellulosic films obtained from agar waste biomass. Carbohydr Polym 233: 115887

Minh NP, Van Phuc N (2018) Application of agar/glycerol as biodegradable coating on storage of star apple (Chrysophyllum cainito). J Glob Pharm Technol 10:191–196

Moreno MA, Bojorges H, Falcó I, Sánchez G, López-Carballo G, López-Rubio A, Zampini IC, Isla MI, Fabra MJ (2020) Active properties of edible marine polysaccharide-based coatings containing Larrea nitida polyphenols enriched extract. Food Hydrocoll 102:105595

Moschopoulou E, Moatsou G, Syrokou MKK, Paramithiotis S, Drosinos EH (2019) Food quality changes during shelf life. In: Caballero B (ed) Food quality and shelf life. Academic Press, Cambridge, MA, pp 1–31

Nagar M, Sharanagat VS, Kumar Y, Singh L (2020) Development and characterization of elephant foot yam starch–hydrocolloids based edible packaging film: physical, optical, thermal and barrier properties. J Food Sci Technol 57:1331–1341

Orsuwan A, Shankar S, Wang L-F, Sothornvit R, Rhim J-W (2016) Preparation of antimicrobial agar/banana powder blend films reinforced with silver nanoparticles. Food Hydrocoll 60:476–485

Park HJ (2003) Edible coatings. In: Zeuthen P, Bøgh-Sørensen L (eds) Food preservation techniques, Food science, technology and nutrition. Woodhead Publishing, Cambridge, pp 90–105

Phinainitisatra T, Harnkarnsujarit N (2021) Development of starch-based peelable coating for edible packaging. Int J Food Sci Technol 56:321–329

Radovanović N, Malagurski I, Lević S, Gordić M, Petrović J, Pavlović V, Mitrić M, Nešić A, Dimitrijević-Branković S (2019) Tailoring the physico-chemical and antimicrobial properties of agar-based films by in situ formation of cu-mineral phase. Eur Polym J 119:352–358

Radovanovic N, Malagurski I, Levic S, Nesic A, Cabrera-Barjas G, Kalusevic A, Nedovic V, Pavlovic V, Dimitrijevic-Brankovic S (2019) Influence of different concentrations of Zn-carbonate phase on physical-chemical properties of antimicrobial agar composite films. Mater Lett 255:126572

Rhim JW (2011) Effect of clay contents on mechanical and water vapor barrier properties of agar-based nanocomposite films. Carbohydr Polym 86:691–699

Rhim J-W, Wang L-F (2013) Mechanical and water barrier properties of agar/κcarrageenan/konjac glucomannan ternary blend biohydrogel films. Carbohydr Polym 96:71–81

Ribeiro AM, Estevinho BN, Rocha F (2020) Edible films prepared with different biopolymers, containing polyphenols extracted from elderberry (Sambucus nigra l.), to protect food products and to improve food functionality. Food Bioprocess Technol 13:1742–1754

da Rocha M, Alemán A, Romani VP, López-Caballero ME, Gómez-Guillén MC, Montero P, Prentice C (2018) Effects of agar films incorporated with fish protein hydrolysate or clove essential oil on flounder (Paralichthys orbignyanus) fillets shelf-life. Food Hydrocoll 81:351–363

Roy S, Rhim JW (2019) Agar-based antioxidant composite films incorporated with melanin nanoparticles. Food Hydrocoll 94:391–398

Roy S, Rhim J-W (2020) Preparation of carbohydrate-based functional composite films incorporated with curcumin. Food Hydrocoll 98:105302

Roy S, Rhim JW, Jaiswal L (2019) Bioactive agar-based functional composite film incorporated with copper sulfide nanoparticles. Food Hydrocoll 93:156–166

Roy S, Kim H-J, Rhim J-W (2021) Synthesis of carboxymethyl cellulose and agar-based multifunctional films reinforced with cellulose nanocrystals and shikonin. ACS Appl Polym Mater 3:1060–1069

Rukmanikrishnan B, Rajasekharan SK, Lee J, Lee J (2019) Biocompatible agar/xanthan gum composite films: thermal, mechanical, UV, and water barrier properties. Polym Adv Technol 30:2750–2758

Salem A, Abdelhedi O, Kchaou H, Nasri M, Jridi M, Zouari N, Debeaufort F (2021) The effects of agar addition and ultrasound treatment on thermomechanical and physical properties of smooth hound (Mustellus mustellus) skin gelatin film. J Food Meas Charact 15:2211–2219. https://doi.org/10.1007/s11694-021-00818-9

Shankar S, Rhim JW (2017) Preparation and characterization of agar/lignin/silver nanoparticles composite films with ultraviolet light barrier and antibacterial properties. Food Hydrocoll 71:76–84

Shankar S, Teng X, Rhim JW (2014) Properties and characterization of agar/CuNP bionanocomposite films prepared with different copper salts and reducing agents. Carbohydr Polym 114:484–492

Shankar S, Reddy JP, Rhim J-W (2015) Effect of lignin on water vapor barrier, mechanical, and structural properties of agar/lignin composite films. Int J Biol Macromol 81:267–273

Tazhibayeva S, Tyussyupova B, Yermagambetova A, Kokanbayev A, Musabekov K (2020) Preparation and regulation of structural-mechanical properties of biodegradable films based on starch and agar. East-Eur J Enterp Technol 5:40–48

United Nations (2015) Transforming our world: the 2030 agenda for sustainable development. https://sdgs.un.org/2030agenda

Uranga J, Nguyen BT, Si TT, Guerrero P, de la Caba K (2020) The effect of cross-linking with citric acid on the properties of agar/fish gelatin films. Polymers 12:291

Vejdan A, Ojagh SM, Abdollahi M (2017) Effect of gelatin/agar bilayer film incorporated with TiO2 nanoparticles as a UV absorbent on fish oil photooxidation. Int J Food Sci Technol 52:1862–1868

Vieira MGA, da Silva MA, dos Santos LO, Beppu MM (2011) Natural-based plasticizers and biopolymer films: a review. Eur Polym J 47:254–263

Vimal KEK, Mathiyazhagan K, Agarwal V, Luthra S, Sivakumar K (2020) Analysis of barriers that impede the elimination of single-use plastic in developing economy context. J Clean Prod 272: 122629. https://doi.org/10.1016/j.jclepro.2020.122629

Wang L-F, Rhim J-W (2015) Preparation and application of agar/alginate/collagen ternary blend functional food packaging films. Int J Biol Macromol 80:460–468

Wang Y, Santos A, Evdokiou A, Losic D (2015) An overview of nanotoxicity and nanomedicine research: principles, progress and implications for cancer therap. J Mater Chem B 3:7153–7172

Wang Y, Wu Y, Wang Y, Xu H, Mei X, Yu D, Wang Y, Li W (2017) Antioxidant properties of probiotic bacteria. Nutrients 9:521

Wang X, Guo C, Hao W, Ullah N, Chen L, Li Z, Feng X (2018) Development and characterization of agar-based edible films reinforced with nano-bacterial cellulose. Int J Biol Macromol 118: 722–730

Wongphan P, Harnkarnsujarit N (2020) Characterization of starch, agar and maltodextrin blends for controlled dissolution of edible films. Int J Biol Macromol 156:80–93

Wongphan P, Harnkarnsujarit N (2021) Edible packaging from hydroxypropyl thermoplastic cassava starch, agar and maltodextrin blends produced by cast extrusion. Int J Food Sci Technol 56:762–772

Wu Y, Geng F, Chang PR, Yu J, Ma X (2019) Carbohydrate polymers effect of agar on the microstructure and performance of potato starch film. Carbohydr Polym 76:299–304

Xanthos D, Walker TR (2017) International policies to reduce plastic marine pollution from single-use plastics (plastic bags and microbeads): a review. Mar Pollut Bull 118(1–2):17–26. https://doi.org/10.1016/j.marpolbul.2017.02.048

Xiao Q, Weng H, Ni H, Hong Q, Lin K, Xiao A (2019) Physicochemical and gel properties of agar extracted by enzyme and enzyme-assisted methods. Food Hydrocoll 87(February):530–540. https://doi.org/10.1016/j.foodhyd.2018.08.041

Yasin R, Samiullah K, Fazal RM, Hussain S, Mahboob S, Al-Ghanim KA, Al-Misned FA, Ahmed Z (2020) Combined effect of probiotics on prolonging the shelf life of GIFT tilapia fillets. Aquac Res 51:5151–5162

Zhai X, Shi J, Zou X, Wang S, Jiang C, Zhang J, Huang X, Zhang W, Holmes M (2017) Novel colorimetric films based on starch/polyvinyl alcohol incorporated with roselle anthocyanins for fish freshness monitoring. Food Hydrocoll 69:308–317

Zhang R, Wang W, Zhang H, Dai Y, Dong H, Hou H (2019) Effects of hydrophobic agents on the physicochemical properties of edible agar/maltodextrin films. Food Hydrocoll 88:283–290

Zhang R, Wang W, Zhang H, Dai Y, Dong H, Kong L, Hou H (2020) Effects of preparation conditions on the properties of agar/maltodextrin-beeswax pseudo-bilayer films. Carbohydr Polym 236:116029

Chapter 11
Testing Methods for Characterisation of Polysaccharide-Based Films

Jahangir A. Rather, Neelofar Manzoor, Sadia Naqash, Qazi Showkat, Darakshan Maji, Hilal A. Makroo, and B. N. Dar

11.1 Introduction

The food packaging is a chief business sector, and frequent novelties in packaging of foods are launched everyday in the market. Owing to the constantly consumer behaviour changes, innovative packaging has been developed. The developed packaging material should meet the market standards (Nešić et al. 2020). Packaging material provides numerous functionalities including protection, containment, convenience and communication to the packaged product (Robertson 2016). The various materials available for packaging purpose of food products are made of plastics, paper, glass, ceramics, biodegradable materials, metals and wood. But nowadays trend is shifting towards the eco-friendly packaging due to their biodegradable nature.

These biodegradable packagings are made from natural and fossil resources and are biodegraded in environment by microorganisms. The polylactic acid and the polyhydroxy butyrate acid biodegradable polymers were used in the food industry for packaging purposes. These bio-based polymers are directly obtained from plant, microorganisms and animal sources and may include polysaccharides and proteins or lipids. Polysaccharides are abundant in nature as they are originated from both plant and marine sources (Ramesh and Tharanathan 2003; Nešić et al. 2020). These polysaccharide materials present an environmentally friendly technical solution, by diminishing dependence on petroleum-based resources (Nešić et al. 2020).

J. A. Rather · N. Manzoor · Q. Showkat · D. Maji · H. A. Makroo · B. N. Dar (✉)
Department of Food Technology, Islamic University of Science and Technology, Awantipora, Jammu and Kashmir, India

S. Naqash
Division of Food Science and Technology, Sher-e-Kashmir University of Agricultural Sciences and Technology of Kashmir, Srinagar, Jammu and Kashmir, India

T. Amin et al. (eds.), *Polysaccharide Based Films for Food Packaging: Fundamentals, Properties and Applications*,
https://doi.org/10.1007/978-981-99-4898-7_11

Polysaccharides are polymeric substances with intrinsic characteristics like biocompatibility, biodegradability and nontoxicity towards living creatures (Chopin et al. 2014). These outstanding properties give them advantage for being used as coatings and films for packaging. The polysaccharides have been recently investigated for active and intelligent packaging purposes. The use of these polysaccharide materials as primary packages partially or wholly replaces conventional packages and reduces the use of synthetic materials (Han 2014; Falguera et al. 2011). Owing to their protective function, incorporation of antimicrobial, antioxidant agents in polysaccharide-based packages may simplify the whole packaging structure (Janjarasskul and Krochta 2010). The film-forming property is owing to the hydrophobic, ionic interactions, electrostatic or other interactive forces such as covalent bonds. For the preparation of these polysaccharide-based packaging materials, appropriate food grade solvents or chemicals should be used (Mahalik and Nambiar 2010).

However, polysaccharide-based films and coatings have poor water barrier and mechanical properties which slow down their industrial applicability (Nešić et al. 2020). Fortunately, their properties can be enhanced by some of the modifications. These modifications enhance their properties and make them suitable for packaging foods. Some opportunities for polysaccharide modification, to achieve more approving characteristics for use in packaging, compromise blending of two or more polymers or combinations with some other constituents (Hussain et al. 2017; Youssef 2013). It is obvious that blending of two or more biopolymers is sometimes obligatory to offer the best packing solution for various foods.

For consumers and manufacturers, packing material should be cost effective. It should provide protection to the food product during storage, transport and distribution. Every packaging material has varying physical, chemical and mechanical properties. The packaging material properties assist as a basis for selection of particular material for packing products. The basic packaging material properties influence the quality and safety (Kim et al. 2014) and are the following:

1. Barrier properties
2. Mechanical properties
3. Chemical reactivity
4. Migration properties

The comprehensive knowledge about the packaging material properties is crucial in the design of packaging materials for particular product. The properties of packaging materials like mechanical strength, gas permeability and sealing ability determine their suitability for intended purposes (Marsh and Bugusu 2007). Hence, it is necessary to determine the packaging material property by suitable testing methods to maintain their safety and quality for packaging purpose.

The packaging material testing generally comprises determination of mechanical strength and permeability as physical properties. But it is too important to study the stability of material for chemicals, principally when used for food applications. The packaging property determination also aids to predict performance of material when the packaging materials are in the actual environmental conditions (Arvanitoyannis and Bosnea 2004). The series of methods are required for confirmation of packaging material for particular purpose. There are numerous crucial aspects to be considered while determination of packaging material properties. The packaging materials should be tested in the similar environment as that of the food product is intended to perform, since practically all the properties change around the environment. It is also important to study the performance of these packaging materials in diverse environmental conditions.

11.2 Need of Testing Methods of Polysaccharide-Based Films

The growth of bio-based films including polysaccharide-based films has increased tremendously since last few years and estimated to increase in share by 70–100% after 2030, with annual growth of 20% and 1–3 MT production in 2020, especially in European market as the consumer's awareness has increased remarkably. Therefore the market potential of the bio-based plastic has huge potential; however their functionality and effectiveness need to be assessed. Generally, the functionalities of the bio-based films including the polysaccharide-based films have been found acceptable as reported by the researchers and have been reconnoitred commercially as well. Although polysaccharide-based films show a low barrier function against water vapour at high relative humidity, they act as good barriers for gaseous transport (O_2, CO_2, etc.) under low or intermediate relative humidity conditions. Furthermore, such films have generated new possibilities of targeting its application in agri-food packaging applications with optimised scientific packaging opportunities as they have better functionality (Briassoulis and Giannoulis 2018). However, to meet the requirements of any food packaging material, either in the form of the legislative framework or as a basic functionality, polysaccharide-based films should be assessed which necessitate the need for testing. Therefore, the polysaccharide-based films represented primarily by cellulose, chitosan, etc. (Cazon and Vázquez 2020) must be inspected through a set of tests such as mechanical and physical properties, besides shelf-life, and quality aspects of packed foods to assess the performance of these films in comparison to the conventionally used films. Some of the aspects to be taken into consideration while understanding the need of testing of polysaccharide-based films and their applicability in food systems and functionality based on the efficiency of the biopolymers are mentioned below:

11.2.1 Assessment of Efficiency of Polysaccharide-Based Films by Employing Testing Methods

This is the established information that the polysaccharide films such as chitosan films have relatively low water barrier properties, likewise also reported by Cazon and Vázquez (2020), owing to their hydrophilic nature in chemical structure. Such properties sometimes limit their utilisation in high moisture or dehydrated foods. Similarly, the oxygen permeability is one of the tools of ascertaining the efficiency of films which is done by standard method ASTM D-3935 (Souza et al. 2015) and ASTM D-1434 (Srinivasa et al. 2004) in chitosan films. Thus, testing of barrier properties of the films is an important criterion in checking the efficacy of polysaccharide-based films in particular and other food packaging materials in general, as it is an important component of enhancement and prediction of shelf life. In order to assess the permeability properties of polysaccharide-based films, the tests are inevitable for its application in food industry which may include the following:

(a) The measurement of molecular weight
(b) Degree of deacetylation done during manufacturing of the film
(c) Conditions of measurement such as temperature, relative humidity, etc.
(d) Time of storage and conditions of storage
(e) Correction of air gap effect

11.2.2 Assessment of Preservation Properties of Polysaccharide-Based Films to Ensure Food Safety

With the new advancements in packaging technology, the aim of developing films is not only to act as protective barrier so that the moisture remains intact but also perform the antimicrobial and antioxidant activities and thereby utilise the films as active packaging system (Pang et al. 2019; Jridi et al. 2014). This way the films operate as carriers for preservation of foods. Therefore, the films used should have the properties of effective functionality and should perform as an active packaging material. Such properties need to be tested broadly to know whether the films are up to the standard mark to perform in the active packaging system both at small scale as well as at commercial level so that it can be used alternatively.

11.2.3 Assessment of Properties of the Polysaccharide-Based Films After Incorporating Natural or Synthetic Ingredients for Various Purposes

The mechanical strength of the films is a very important characteristic of the films required in packaging system which are tested by measuring tensile strength, percentage of elongation at break, deformation, puncture force, etc. Occasionally, the polysaccharide-based films are incorporated with various natural and synthetic components such as plasticisers which may improve its mechanical properties such as elongation. For example, the poor mechanical properties, such as brittleness of starch films acquired due to amorphous structure attributed to amylose, need to be added with plasticisers or other biological and synthetic polymers which enhances the mechanical properties, while maintaining its biodegradability. Consequently, while testing the applicability of such incorporation, the polysaccharide-based films must assure the performance and their relative utilisation with that of other biofilms and petroleum-based films (Bertuzzi et al. 2007).

11.2.4 Assessment of Discrepancies in the Testing Methods

It is found that there is a difference in the actual and reported values of various testing methods which may be due to the testing methods employed by the manufacturers, testing done under different conditions, equipment used for testing or testing procedures employed; for example, the while comparing the testing of water vapour transmission rate, performed under specific environmental conditions such as controlled temperature and relative humidity ranges with the conventional testing conditions, the values were higher. Since such discrepancies in reported and actual values create irregularity and fragmented database of test results, standardisation of the testing methods must be adopted universally for which there is a need to revisit the standard procedures of testing which will help in achieving the targeted testing conditions required for a particular biofilm (Briassoulis and Giannoulis 2018). Unless such standards are not maintained, it becomes impossible to achieve the same results; therefore not only same standard but also testing conditions must be followed by the biofilm-forming manufacturers.

11.2.5 Assessment of Biodegradability

It is the established fact that the bio-based polymers are environmentally friendly due to their biodegradability. However, there are some limitations in terms of their performance such as mechanical, thermal, barrier properties and their economic viability. The research in this filed is required to ensure their commercial applicability which can be done by proper testing of such materials for claiming its biodegradability as well as its viability for application in food packaging (Siracusa et al. 2008). Different techniques are used to assess the biodegradability of biofilms. This may affect the functionality of biofilms by impactful alteration in temperature, water diffusion, moisture, metal impurities from the catalyst particle size and shape, molecular weight, etc. (Kale et al. 2006). The mechanism of biodegradation includes oxo-biodegradation and hydro-biodegradation, while the former is a slower process than later. Hydro-biodegradation process is involved in biodegradation of polysaccharide-based films such as cellulose and starch-based films and oxo-biodegradation mechanism biodegrade natural polymer such as lignin, rubber, etc. and polyolefins (synthetic) (Arvanitoyannis 1999).

Some of the key testing methods and need of testing and the possible concerns or advantages of bio-based/polysaccharide-based films have been enlisted in Table 11.1.

Polysaccharide-based films represent as one of the bio-based films, with promising functionality as it is one of the innovative materials among the natural polymers. Although there are many bio-based poly-films identified by the researchers and the

Table 11.1 Need for testing the functionality of polysaccharide-based films

S. no	Testing method	Need of testing	Possible concerns or advantages of bio-based/ polysaccharide-based films which needs to be tested	References
1.	Tensile properties	To check the tensile strength values of polysaccharide film in comparison to conventional film	Reduced elongation at break in freezing temperatures or water	Weber (2000)
2.	Tear propagation resistance	To evaluate the tearing behaviour for ensuring any accidental damage	Evaluation of tearing behaviour in polysaccharide-based films is very important in thin biofilms including polysaccharide-based films	Briassoulis (2006)
3.	Penetration resistance	Measure the resistance to puncture	Polysaccharide-based films should not lose the package integrity due to gases, odours and contaminates	Briassoulis and Giannoulis (2018)
4.	Seal strength	Facilitation of opening of package	Knowing the minimum and maximum seal strength is important	
5.	Radiation transmittance properties	To check the clarity of film so that it is visible to the customers	The dependence of the total transmittance and haze on the wavelength (390–720 nm) for polysaccharide-based films	
6.	Water vapour transmittance	To avoid infections due to water, contact and excessive humidity	Assessment of interaction of water vapour with the polymeric biomolecules of edible film	Robertson (2012)
7.	Gas transmittance	Controlling the headspace conditions of horticultural produce affected by the respiration and transpiration	Assessment of the interaction of the living fresh produce with the bio-based film	Briassoulis and Giannoulis (2018)
8.	Contact angle	To assess the water condensation behaviour	Interaction of small droplets at the inner package surface polysaccharide film and food material may promote infection and reduce clarity	

continuous research is going on, testing has authenticated their comparable properties with those obtained from other materials in terms of performance and price. There is a need of testing polysaccharide-based films due to the change in the current scenario, while we perceive the market will be difficult as new materials are coming up in the market such as nanocomposites with multifunctional properties. Testing of polysaccharide-based films is therefore an important criterion to assess their appropriate use in packing the foods such as fruits, vegetables, meat, sausages, etc., which should coincide with its resilient mechanical, thermal barrier properties and lastly ensuring the food safety.

11.3 Testing Methods for Characterisation of Polysaccharide-Based Films

Polysaccharide-based films are generally characterised for various properties to achieve optimum packaging for sensitive food products for their safe shelf life extension. In current era, there is a huge demand for the biodegradable films ascribing to their high flexibility, transparency, thermal stability, excellent mechanical properties, biodegradability and barrier properties that result in enhancement of their functionality other than mere packaging. Owing to their close-knit matrix, polysaccharide-based films possess excellent mechanical strength but lack in barrier properties. Reported that the barrier properties of these films can be improved to a larger extent by adding lipids; however, the resulting film does not meet the appealing properties in case of opacity, stability and flexibility and have also been reported to possess and develop waxy taste.

11.3.1 Estimation of Physical Properties of Polysaccharide-Based Films

Polysaccharide-based films are generally opaque and possess an oil-free appearance and have low calorie content. These films can be used for the enhancement of shelf life of perishable food products like fruits, vegetables, shellfish, meat products, etc. and facilitate in reduction of their dehydration, oxidative rancidity and surface darkening (Carneiro-da-Cunha et al. 2009). In addition, these films also possess some other properties that make them more appealing and include barrier properties, high mechanical properties and attractive physical properties. The physical properties of the polysaccharide-based films include film thickness, colour and film solubility that will be discussed in subsequent sections.

11.3.1.1 Estimation of Film Thickness

Thickness of films is an important parameter for estimation of its suitable applications. The precise estimation of thickness is essential for depiction of the set specifications of the films required for storage of products. Thickness of film has a direct impact on the permeability and mechanical strength of the films, thereby effecting the product's shelf life. The thickness values of different polysaccharide-based films are given in Table 11.2. According to the procedure given by ASTM, the thickness of film is generally estimated by using micrometre by placing the film between anvil and spindle (ASTM F2251-13 2013). Similar method has been used by Motedayen et al. (2013); with certain modifications, they estimated the thickness of film (to a precision of 0.001 mm) using a manual digital micrometre at ten different points of the film, followed by averaging the value. The observed thickness values are also used in the estimation of tensile properties and water vapour transmission.

11.3.1.2 Estimation of Gloss Detection of Polysaccharide-Based Films

Surface appearance of the packaging films is an important parameter regarding its overall acceptability. Surface gloss is related to the appearance and shines of the packaging film and is essential for predication of market acceptance. Gloss of film is measured using a gloss metre that involves projection of light at fixed intensity at angles that can be 20°, 60° or 85° over the surface of the film specimen. The amount of reflected light is measured at the opposite angles. As per Singh et al. (2017), gloss values are attained by evaluating the surface reflectance of the film with that of the black standard glass, the gloss value of which is assumed to be 100. Incident angle is directly proportional to the gloss value and depends on the refractive index of the film; however, there is a negative relation between the abrasions created on the film surface during processing with that of the gloss value.

11.3.1.3 Colour Estimation of Polysaccharide-Based Films

Colour is one of the major and fundamental properties of polysaccharide-based film that determines its consumer acceptability. Colour of the films is estimated using CIE (International Commission on Illumination) colorimeter and articulated as L^*, a^* and b^* values determining lightness/brightness, redness/greenness and yellowness/blueness, respectively. The difference in colour is determined using Eq. (11.1)

$$\Delta E = \sqrt{(\Delta L)^2 + (\Delta a)^2 + (\Delta b)^2} \tag{11.1}$$

Table 11.2 Thickness values of different polysaccharide-based films and composites

Sample	Thickness (μm)	References
Pectin: alginate: WPC ratio (1:1:1)	45.00 ± 40.00	Chakravartula et al. (2019)
Chitosan	60.00 ± 0.01	Chan et al. (2020)
Corn starch with 2% papaya polysaccharide	17.1 ± 0.1	
Hyaluronate/chitosan	1.76 ± 0.32	Yamazaki and Iijima (2020)
Pectin: alginate: WPC ratio (0:1.5:1.5)	47.00 ± 1.00	Chakravartula et al. (2019)
Pectin: alginate: WPC ratio (1.5:0:1.5)	64.00 ± 2.00	
Sodium alginate with 2% *Melastoma malabathricum* extract	80.00 ± 0.01	Chan et al. (2020)
Corn starch with 6% papaya polysaccharide	14.9 ± 0.3	
Pectin: alginate: WPC ratio (0.5:2:0.5)	41.00 ± 1.00	Chakravartula et al. (2019)
CMC	40.00 ± 0.01	Chan et al. (2020)
Pectin: chitosan (66:34) with 40% glycerol (unwashed)	55 ± 0.014	Machado et al. (2020)
Sodium alginate	80.00 ± 0.01	Chan et al. (2020)
Corn starch with 10% papaya polysaccharide	13.3 ± 0.	
Pectin: chitosan (66:34) with 15% glycerol (unwashed)	38 ± 0.009	Machado et al. (2020)
Sodium alginate with 4% *Melastoma malabathricum* extract	90.00 ± 0.01	Chan et al. (2020)
Pectin: alginate: WPC ratio (3:0:0)	36.00 ± 1.00	Chakravartula et al. (2019)
Sodium alginate with 0% *Melastoma malabathricum* extract	80.00 ± 0.01	Chan et al. (2020)
Pectin: chitosan (66:34) with 15% glycerol (washed)	28 ± 0.025	Machado et al. (2020)
Pectin: alginate: WPC ratio (2:0.5:0.5)	43.00 ± 2.00	Chakravartula et al. (2019)
Corn starch with 8% papaya polysaccharide	14.7 ± 0.3	
Pectin: alginate: WPC ratio (0.5:0.5:2)	49.00 ± 1.00	Chakravartula et al. (2019)
Sodium alginate with 6% *Melastoma malabathricum* extract	80.00 ± 0.01	Chan et al. (2020)
Corn starch with 0% papaya polysaccharide	18.1 ± 0.2	
Pectin: alginate: WPC ratio (1.5:1.5:0)	43.00 ± 2.00	Chakravartula et al. (2019)
Corn starch with 4% papaya polysaccharide	17.3 ± 0.2	
Pectin: alginate: WPC ratio (0:3:0)	42.00 ± 1.00	Chakravartula et al. (2019)

Colour values of different polysaccharide-based films are displayed in Table 11.3. The colour values of PVA films in comparison of the polysaccharide-based films are given by Feki et al. (2019) who reported that the PVA films possess higher lightness value but lower yellowness values as compared to the polysaccharide-based film, indicating the higher transparency of PVA films. The increased yellowness of the polysaccharide films is ascribed to the presence of various coloured components that are present in polysaccharides.

Table 11.3 Colour values of edible films

Sample	L	A	B	References
Chitosan	13.28 ± 0.13	−1.67 ± 0.25	0.74 ± 0.07	Chan et al. (2020)
Pectin: casein: gelatin (100: 00:00)	43.51 ± 1.30	1.89 ± 0.58	10.18 ± 0.89	Saikia and Badwaik (2018)
Corn starch with 2% papaya polysaccharide	84.76 ± 0.17	2.16 ± 0.19	3.07 ± 0.65	
Sodium alginate with 0% *Melastoma malabathricum* extract	7.36 ± 0.85	−0.28 ± 0.05	−0.94 ± 0.01	Chan et al. (2020)
Pectin: casein: gelatin (50: 00:50)	47.22 ± 1.20	1.10 ± 0.33	9.99 ± 1.56	Saikia and Badwaik (2018)
Corn starch with 0% papaya polysaccharide	86.07 ± 0.16	0.48 ± 0.04	−5.05 ± 0.16	
Sodium alginate	7.36 ± 0.85	−0.28 ± 0.05	−0.94 ± 0.01	Chan et al. (2020)
Sodium alginate with 6% *Melastoma malabathricum* extract	7.36 ± 0.85	−0.28 ± 0.05	−0.94 ± 0.01	
Pectin: casein: gelatin (00: 50:50)	45.48 ± 1.70	1.64 ± 0.33	8.15 ± 1.98	Saikia and Badwaik (2018)
CMC	6.15 ± 0.57	−0.22 ± 0.04	−0.82 ± 0.14	Chan et al. (2020)
Chitosan-glycerol	92.3 ± 0.1	−1.2 ± 0.2	10.4 ± 0.1	
Corn starch with 8% papaya polysaccharide	81.26 ± 0.17	2.35 ± 0.31	4.86 ± 1.07	
Sodium alginate with 2% *Melastoma malabathricum* extract	13.17 ± 0.38	−1.32 ± 0.12	−1.36 ± 0.08	Chan et al. (2020)
Pectin: casein: gelatin (00: 100:00)	38.71 ± 2.56	3.53 ± 0.90	19.93 ± 1.34	Saikia and Badwaik (2018)
Control (chitosan)	92.8 ± 0.2	−0.7 ± 0.0	6.1 ± 0.5	
Pectin: casein: gelatin (00: 00:100)	55.10 ± 1.40	12.06 ± 3.06	34.79 ± 1.89	Saikia and Badwaik (2018)
Corn starch with 10% papaya polysaccharide	81.04 ± 0.15	2.45 ± 0.19	5.05 ± 1.26	
Pectin: casein: gelatin (25: 25:50)	48.50 ± 1.61	1.76 ± 0.66	9.06 ± 1.46	Saikia and Badwaik (2018)
Sodium alginate with 4% *Melastoma malabathricum* extract	17.28 ± 0.32	−2.06 ± 0.10	0.84 ± 0.10	Chan et al. (2020)
Chitosan-glycerol-gelatin	92.6 ± 0.2	−0.8 ± 0.1	6.6 ± 0.1	
Pectin: casein: gelatin (50: 25:25)	44.80 ± 1.72	1.72 ± 0.33	7.90 ± 1.78	Saikia and Badwaik (2018)
Pectin: casein: gelatin (50: 50:00)	33.81 ± 1.76	10.9 ± 1.23	13.00 ± 1.67	
Corn starch with 6% papaya polysaccharide	82.26 ± 0.17	2.31 ± 0.22	4.11 ± 0.52	
Pectin: casein: gelatin (25: 50:25)	48.97 ± 1.67	2.09 ± 0.77	9.93 ± 1.97	Saikia and Badwaik (2018)
Corn starch with 4% papaya polysaccharide	82.81 ± 0.5	2.16 ± 0.19b	4.05 ± 0.65	

11.3.1.4 Estimation of Film Solubility

Solubility of edible/biodegradable films is an important parameter that is associated with the hydrophobicity of the packaging material, therefore predicting its application (Jiang et al. 2010; Mohajer et al. 2017). As reported by Feki et al. (2019), solubility of polysaccharide-based films is generally above 70%; however, with the addition of poly vinyl alcohol, there occurs a significant increase in the film solubility. Solubility of the biodegradable films is estimated according to a method described by Phan et al. (2005). Water solubility index of different polysaccharide-based films is represented in Table 11.4. The determination is carried out at room temperature (25 °C). The film specimen is placed in a desiccator containing silica gel, until a constant weight is obtained. Later approximately 500 mg of specimen is immersed in a beaker filled with 50 mL distilled water and stirred continuously at 500 rpm for 24 h. After that the mixture from the beaker is filtered using glass microfibre filter. Oven method is used for determination of insoluble matter, and the percentage of soluble matter is estimated and considered as the initial dry sample weight. Film water solubility is estimated using Eq. (11.2):

$$\text{Film water solubility}\% = \frac{W_i - W_f}{W_i} \times 100 \tag{11.2}$$

where W_i is the initial weight expressed as dry matter and W_f is the weight of undissolved film materials.

11.3.1.5 Estimation of Swelling Index

Swelling index or swelling capacity of polysaccharide-based films plays a crucial role in lesion curing and biomedical usage. Considering the high hydrophobicity of the polysaccharide-based films, these films possess low swelling capacity as compared to the pure PVA films (Feki et al. 2019). Estimation of swelling index is performed to determine the water uptake capacity of the films. The pre-weighed dried film specimen are immersed in distilled water for 12 h and kept at room temperature with varying pH. The experiment is carried out in replicates and is estimated using Eq. (11.3):

$$\text{Swellingindex}\% = \frac{W_s - W_d}{W_d} \times 100 \tag{11.3}$$

where W_s is the weight of swollen film and W_d is the weight of dry film.

The decline in swelling capacity of the polysaccharide-based films is ascribed to their polymeric binding and a shift from hydrophobic to hydrophilic interactions in the polymeric structure.

Table 11.4 Water solubility and water activity of edible films

Sample	Water solubility (%)	Water activity	Moisture absorption (%)	References
Chitosan	25.60 ± 0.90	0.571 ± 0.025	NR	Chan et al. (2020)
Pectin: casein: gelatin (00:50:50)	57.19 ± 1.70	NR	32.17 ± 1.20	Saikia and Badwaik (2018)
Sodium alginate with 2% *Melastoma malabathricum* extract	100.00 ± 0.00	0.424 ± 0.012	NR	Chan et al. (2020)
Soybean polysaccharide: gelatin (40:60)	38.1 ± 0.6	NR	NR	Liu et al. (2020)
Corn starch with 10% papaya polysaccharide	10.17 ± 0.67	NR	NR	
Pectin: casein: gelatin (00:00:100)	59.24 ± 1.30	NR	30.75 ± 0.88	Saikia and Badwaik (2018)
Soybean polysaccharide: gelatin (20:80)	33.8 ± 0.2	NR	NR	Liu et al. (2020)
CMC	100.00 ± 0.00	0.545 ± 0.031	NR	Chan et al. (2020)
Sodium alginate	100.00 ± 0.00	0.524 ± 0.036	NR	
Pectin: casein: gelatin (100:00:00)	75.64 ± 2.56	NR	48.55 ± 1.98	Saikia and Badwaik (2018)
Sodium alginate with 0% *Melastoma malabathricum* extract	100.00 ± 0.00	0.524 ± 0.03	NR	Chan et al. (2020)
Soybean polysaccharide: gelatin (60:40)	55.2 ± 0.2	NR	NR	Liu et al. (2020)
Sodium alginate with 4% *Melastoma malabathricum* extract	100.00 ± 0.00	0.417 ± 0.006	NR	Chan et al. (2020)
Pectin: casein: gelatin (50:25:25)	40.59 ± 1.7	NR	17.75 ± 1.30	Saikia and Badwaik (2018)
Soybean polysaccharide: gelatin (80:20)	100.00 ± 0.00	NR	NR	Liu et al. (2020)
Pectin: casein: gelatin (00:100:00)	67.51 ± 1.43	NR	36.30 ± 1.40	Saikia and Badwaik (2018)
Sodium alginate with 6% *Melastoma malabathricum* extract	100.00 ± 0.00	0.524 ± 0.036	NR	Chan et al. (2020)
Pectin: casein: gelatin (50:00:50)	51.69 ± 1.50	NR	28. 94 ± 1.60	Saikia and Badwaik (2018)
Soybean polysaccharide: gelatin (100:0)	100.00 ± 0.00	NR	NR	Liu et al. (2020)

(continued)

Table 11.4 (continued)

Sample	Water solubility (%)	Water activity	Moisture absorption (%)	References
Pectin: casein: gelatin (50:50:00)	67.91 ± 2.02	NR	43.18 ± 1.01	Saikia and Badwaik (2018)
Soybean polysaccharide: gelatin (0:100)	31.5 ± 0.2	NR	NR	Liu et al. (2020)
Corn starch	69.86 ± 1.75	NR	NR	
Pectin: casein: gelatin (25:25:50)	46.45 ± 1.20	NR	24.87 ± 1.59	Saikia and Badwaik (2018)
Pectin: casein: gelatin (25:50:25)	47.50 ± 2.50	NR	27.07 ± 1.33	

11.3.1.6 Estimation of Optical Properties of Polysaccharide-Based Films

Optical properties of packaging materials tend to have a critical importance related to their application and acceptability. Low relative opacity of the films indicates their high transparency (Mali et al. 2004). For determination of optical properties of polysaccharide-based films, the film specimen are cut into small rectangular pieces and positioned in the test cell of UV-vis spectrophotometer, and observations were made at the wavelengths ranging between 200 nm and 800 nm (Feki et al. 2019). Utilisation of UV-vis spectroscopy is majorly estimated to determine the capacity of films to protect the photosensitivity of food products. Transmission rate of the films in visible areas of light varied between 3% and 70%, and the transmission is found to be lesser in case of PVA films. Higher transmission of polysaccharide-based films renders them excellent barriers to food products against oxidative damage caused and induced by ultraviolet radiations (Iwata et al. 2000).

11.3.2 Mechanical Properties of Polysaccharide-Based Films

Mechanical properties of packaging films are correlated to distribution and density of interactions (intermolecular and intramolecular) in the networks formed (Leceta et al. 2013). The tensile strength and elongation at break tests are used to show the mechanical properties of the polysaccharide-based packaging films (Mali et al. 2006; Pan et al. 2014). The tensile strength is to be determined at the specimen break point under tensile stress as the capacity of resistance to rupture, and elongation at break (EAB) is the protraction in the original length to the break point length of specimen, which is related to the elasticity of a polymeric film (Silva and Dujovne 2013). The TS and EAB of polysaccharide-based films and composites are shown in Table 11.5.

Table 11.5 Tensile properties and elongation at break of several polysaccharide-based films

Film composition	Tensile strength (Mpa)	Elongation (%)	Elastic modulus (MPa)	References
Pectin: alginate: WPC ratio (2:0.5: 0.5)	27.00 ± 5.00	26.00 ± 4.00	555 ± 129	Chakravartula et al. (2019)
Pectin: alginate: WPC ratio (0:3:0)	39.00 ± 16.00	27.00 ± 14.00	607 ± 243	
Sodium alginate with 2% *Melastoma malabathricum* extract	5.92 ± 0.38	16.03 ± 0.64	NR	Chan et al. (2020)
Polysaccharide mixtures with methylcellulose, glucomannan and pectin ratio (0: 1:0)	58.5 ± 0.00	5.6 ± 0.00	NR	Chambi and Grosso (2011)
Sodium alginate with 4% *Melastoma malabathricum* extract	10.24 ± 0.31	22.91 ± 1.27	NR	Chan et al. (2020)
Pectin: casein: gelatin (00:00:100)	0.489 ± 0.45	31.05 ± 1.56	NR	Saikia and Badwaik (2018)
Pectin: alginate: WPC ratio (0.5: 2:0.5)	31.00 ± 10.00	24.00 ± 12.00	746 ± 127	Chakravartula et al. (2019)
Methylcellulose: glucomannan: pectin (0.33:0.33:0.33)	47.1 ± 0.00	3.00 ± 0.00	NR	Chambi and Grosso (2011)
Chitosan mixtures with different solvents and Ph	0.56–19.2	22–494.8	NR	Kim et al. (2006)
Methylcellulose	33.39 ± 1.60	59.84 ± 5.20	819.60 ± 32.39	Łupina et al. (2019)
Gelatin: methyl cellulose (25:75)	30.90 ± 1.56	38.97 ± 4.55	772.31 ± 35.10	
Polysaccharide mixtures with methylcellulose, glucomannan and pectin ratio (0.17:0.17:0.67)	42.7 ± 0.00	2.3 ± 0.00	NR	Chambi and Grosso (2011)
Chitosan	15.45 ± 1.25	7.83 ± 0.76	NR	Chan et al. (2020)
Pullulan: ethyl cellulose (1:3)	1.83 ± 0.29	1.95 ± 0.26	99.11 ± 13.79	Yang et al. (2020)
Gelatin: octenylsuccinic anhydride modified starch (75:25)	33.10 ± 2.55	17.44 ± 3.24	561.28 ± 31.07	Łupina et al. (2019)
Bran gum with glycerol and with/without beeswax mixtures	1.64–7.58	8.22–4.85	NR	Spotti et al. (2016)

(continued)

Table 11.5 (continued)

Film composition	Tensile strength (Mpa)	Elongation (%)	Elastic modulus (MPa)	References
Pectin mixtures with/without chitosan nanoparticles	58.5–26.07	2.91–0.94	NR	Lorevice et al. (2016)
Gelatin: methyl cellulose (75:25)	38.12 ± 2.05	19.13 ± 2.17	952.47 ± 58.93	Łupina et al. (2019)
Sodium alginate with 6% *Melastoma malabathricum* extract	NR	17.17 ± 0.16	NR	Chan et al. (2020)
Polysaccharide mixtures with methylcellulose, glucomannan and pectin ratio (0.5: 0.5:0)	41.2 ± 0.00	3.7 ± 0.00	NR	Chambi and Grosso (2011)
Methylcellulose: glucomannan: pectin (0.67:0.67:0.17)	37.3 ± 0.00	3.9 ± 0.00	NR	
Pectin: casein: gelatin (00:100:00)	0.336 ± 0.61	17.12 ± 1.50	NR	Saikia and Badwaik (2018)
Soybean polysaccharide: gelatin (0:100)	6.08 ± 1.01	317.70 ± 58.87	6.4 ± 0.5	Liu et al. (2020)
Sodium alginate with 0% *Melastoma malabathricum* extract	NR	17.17 ± 0.16	NR	Chan et al. (2020)
Pectin: casein: gelatin (00:50:50)	0.632 ± 1.03	23.95 ± 0.50	NR	Saikia and Badwaik (2018)
Gelatin: octenylsuccinic anhydride modified starch (25:75)	4.50 ± 1.05	93.46 ± 29.04	26.84 ± 3.14	Łupina et al. (2019)
Soybean polysaccharide: gelatin (40:60)	4.75 ± 0.99	253.20 ± 43.80	7.5 ± 0.6	Liu et al. (2020)
Corn starch with 2% papaya polysaccharide	3.66 ± 0.35	58.60 ± 3.48	NR	Ma et al. (2021)
Methylcellulose: glucomannan: pectin (0.17:0.67:0.17)	72.6 ± 0.00	9.9 ± 0.00	NR	Chambi and Grosso (2011)
Heparin/chitosan	11.8 ± 0.4	NR	NR	Yamazaki and Iijima (2020)
Corn starch with 4% papaya polysaccharide	4.28 ± 0.26	59.89 ± 3.44	NR	Ma et al. (2021)
Gum arabic	4.60 ± 1.53	23.47 ± 6.71	196.43 ± 38.39	Łupina et al. (2019)
Pectin: casein: gelatin (25:25:50)	0.500 ± 0.27	29.22 ± 0.86	NR	Saikia and Badwaik (2018)

(continued)

Table 11.5 (continued)

Film composition	Tensile strength (Mpa)	Elongation (%)	Elastic modulus (MPa)	References
Sodium alginate	NR	3.5 ± 0.00	NR	Alves et al. (2011)
Soybean polysaccharide: gelatin (100:0)	2.57 ± 0.39	41.20 ± 8.25	11.7 ± 0.7	Liu et al. (2020)
Pectin: casein: gelatin (50:25:25)	0.863 ± 0.30	21.23 ± 1.62	NR	Saikia and Badwaik (2018)
Soybean polysaccharide: gelatin (20:80)	5.20 ± 0.50	252.70 ± 34.24	6.9 ± 0.5	Liu et al. (2020)
Chondroitin sulphate/chitosan	121.6 ± 56	NR	NR	Yamazaki and Iijima (2020)
Gelatin	65.31 ± 1.75	8.55 ± 1.73	1335.32 ± 74.19	Łupina et al. (2019)
Chitosan-glycerol	38.2 ± 2.7	31.8 ± 2.2	NR	
Gelatin: gum arabic (25:75)	23.71 ± 1.46	11.64 ± 3.95	659.26 ± 44.58	Łupina et al. (2019)
Pullulan: ethyl cellulose (1:1)	2.49 ± 0.37	2.79 ± 0.39	53.31 ± 8.29	Yang et al. (2020)
Arabinoxylan +15% glycerin	22.30 ± 2.97	5.46 ± 1.89	NR	Phan et al. (2005)
Pectin: casein: gelatin (50:00:50)	0.892 ± 1.12	18.83 ± 1.12	NR	Saikia and Badwaik (2018)
Gelatin: methyl cellulose (50:50)	24.34 ± 2.51	6.70 ± 2.44	706.94 ± 94.59	Łupina et al. (2019)
Chitosan-glycerol-gelatin	51.7 ± 3.1	38.3 ± 2.7	NR	
Hyaluronate/chitosan	296.5 ± 69	NR	NR	Yamazaki and Iijima (2020)
Potato starch with/without glycerol	30–68	3.0–5.0	NR	Sessini et al. (2016)
Tara gum mixtures with sorbitol and glycerol as plasticisers and with/without oleic acid	26.8–57.4	8.5–2.7	NR	Ma et al. (2016)
Pullulan: ethyl cellulose (3:1)	1.66 ± 0.21	8.35 ± 1.09	17.95 ± 3.36	Yang et al. (2020)
Quinoa-chitosan mixtures	2.3–8.3	273–117.4	NR	Abugoch et al. (2011)
Starch mixtures with glycerol and xylitol as plasticisers	5.06–44.3	2.4–70.7	NR	Muscat et al. (2012)
Pullulan: ethyl cellulose (1:2)	1.77 ± 0.26	2.37 ± 0.32	64.63 ± 7.47	Yang et al. (2020)
Corn starch with 8% papaya polysaccharide	4.86 ± 0.46	64.17 ± 4.31	NR	

(continued)

Table 11.5 (continued)

Film composition	Tensile strength (Mpa)	Elongation (%)	Elastic modulus (MPa)	References
Tara gum mixtures with glycerol and with/without chitosan	58.44–22.71	44–46	NR	Antoniou et al. (2015)
Hydroxypropyl methyl cellulose mixtures with/without silver nanoparticles	51–28	NR	NR	de Moura et al. (2012)
Chitosan	22.2–29.6	13–73.6	NR	Shen and Kamdem (2015)
Pectin: alginate: WPC ratio (0.5:0.5: 2)	11.00 ± 3.00	19.00 ± 2.00	288 ± 59	Chakravartula et al. (2019)
Polysaccharide mixtures with methylcellulose, glucomannan and pectin ratio (1: 0:0)	38.6 ± 0.00	4.4 ± 0.00	NR	Chambi and Grosso (2011)
Sodium alginate	17.17 ± 0.16	42.70 ± 1.26	NR	Chan et al. (2020)
Pectin: casein: gelatin (100:00:00)	1.353 ± 2.11	11.25 ± 1.24	NR	Saikia and Badwaik (2018)
Chitosan	NR	16.00 ± 0.00	NR	Alves et al. (2011)
Gelatin: octenylsuccinic anhydride modified starch (50:50)	15.69 ± 1.71	27.13 ± 5.42	187.67 ± 13.32	Łupina et al. (2019)
Agar with/without silver nanoparticles	51.5–46.38	33.02–33.64	NR	Rhim et al. (2013)
Cellulose acetate	1.43 ± 0.00	NR	NR	Goetz et al. (2016)
Control (chitosan)	60.9 ± 2.8	15.6 ± 1.8	NR	
Gelatin: water-soluble soy polysaccharides (75:25)	48.18 ± 1.08	4.97 ± 0.84	992.81 ± 56.77	Łupina et al. (2019)
Pectin: alginate: WPC ratio (1:1:1)	23 ± 6	22.00 ± 9.00	444 ± 107	Chakravartula et al. (2019)
Pullulan: ethyl cellulose (2:1)	1.75 ± 0.17	5.83 ± 0.88	27.15 ± 3.75	Yang et al. (2020)
Carrageenan	NR	6.8 ± 0.00	NR	Alves et al. (2011)
Soybean polysaccharide: gelatin (80:20)	2.59 ± 0.48	86.40 ± 20.28	10.3 ± 0.6	Liu et al. (2020)
Soybean polysaccharide: gelatin (60:40)	2.90 ± 0.39	139.30 ± 14.81	7.5 ± 0.6	
Corn starch with 0% papaya polysaccharide	3.03 ± 0.13	43.25 ± 2.71	NR	

(continued)

Table 11.5 (continued)

Film composition	Tensile strength (Mpa)	Elongation (%)	Elastic modulus (MPa)	References
Gelatin: water-soluble soy polysaccharides (25:75)	26.88 ± 1.72	11.82 ± 1.25	533.55 ± 48.31	Łupina et al. (2019)
CMC/chitosan	14.1 ± 4			Yamazaki and Iijima (2020)
Pullulan: ethyl cellulose (1:0)	1.56 ± 0.17	17.30 ± 4.34	9.01 ± 1.27	Yang et al. (2020)
Corn starch with 6% papaya polysaccharide	4.52 ± 0.38	63.10 ± 4.83	NR	
Pectin: alginate: WPC ratio (3:0:0)	62.00 ± 20.00	20.00 ± 8.00	1467 ± 303	Chakravartula et al. (2019)
Water-soluble soy polysaccharides	12.67 ± 1.55	23.78 ± 3.77	332.86 ± 45.61	Łupina et al. (2019)
Corn starch with 10% papaya polysaccharide	3.57 ± 0.17	68.38 ± 4.75	NR	
Agar +15% glycerine	42.11 ± 3.27	6.51 ± 0.96	NR	Phan et al. (2005)
Pectin: casein: gelatin (25:50:25)	0.347 ± 0.33	12.93 ± 1.76	NR	Saikia and Badwaik (2018)
Methyl cellulose and ethanol mixtures with/without polyethylene glycol	25–33	29–14	NR	Nazan Turhan and Sahbaz (2004)
Gelatin: gum arabic	45.81 ± 3.02	5.68 ± 1.0	1185 ± 124.89	Łupina et al. (2019)
Pectin: alginate: WPC ratio (1.5: 0:1.5)	13.00 ± 4.00	16.00 ± 7.00	352 ± 140	Chakravartula et al. (2019)
Methylcellulose: glucomannan: pectin (0.5:0:0.5)	42.7 ± 0.00	3.5 ± 0.00	NR	Chambi and Grosso (2011)
Methylcellulose: glucomannan: pectin (0:0.5:0.5)	64.7 ± 0.00	3.3 ± 0.00	NR	
Pectin: casein: gelatin (50:50:00)	0.700 ± 0.96	21.78 ± 1.52	NR	Saikia and Badwaik (2018)
CMC	12.61 ± 1.46	23.84 ± 3.09	NR	Chan et al. (2020)
Pectin: alginate: WPC ratio (0:1.5: 1.5)	22.00 ± 9.00	40.00 ± 23.00	406 ± 145	Chakravartula et al. (2019)
Gelatin: um arabic (50:50)	35.35 ± 2.95	8.62 ± 5.46	101.77 ± 65	Łupina et al. (2019)
Pectin: alginate: WPC ratio (1.5:1.5:0	61.00 ± 11.00	23.00 ± 10.00	958 ± 240	Chakravartula et al. (2019)

(continued)

Table 11.5 (continued)

Film composition	Tensile strength (Mpa)	Elongation (%)	Elastic modulus (MPa)	References
Polysaccharide mixtures with methylcellulose, glucomannan and pectin ratio (0: 0:1)	51.1 ± 0.00	2.6 ± 0.00	NR	Chambi and Grosso (2011)
Gelatin: water-soluble soy polysaccharides (50:50)	37.42 ± 2.34	7.17 ± 1.21	882.07 ± 70.72	Łupina et al. (2019)

11.3.2.1 Estimation of Tensile Strength and Elongation at Break

The two typical tests for determination of mechanical properties of material are tensile strength and the elongation at break and are determined by using texture analyser. The film specimens are equilibrated at 25 °C and 55% RH for three days and then cut into 10 cm × 3 cm rectangular sections. Film specimens are fixed with grips, and measurements are carried out with 50 mm grip separation initially with 1 mm/s constant speed at 25 °C until sample ruptures. The tests are done at least eight times for each kind of polysaccharide film. The TS (MPa) and EAB (%) are calculated by Eqs. (11.2) and (11.3), respectively (Du et al. 2016):

$$\text{Tensile strength (TS)} = \frac{F_{\max}}{A}.$$

where $F_{\max}$ (N) is the maximum load and A (m^2) is the strip cross-sectional area

$$\text{Elongation at break (EAB)} = \frac{\Delta L}{L0} \times 100$$

where $L0$ (cm) is the initial film strips height and ΔL (cm) is the increased height before film strip breaks.

11.3.3 Barrier Properties

Determination of barrier properties of the polysaccharide-based films is vital for calculating and predicting the shelf life of any food product. In case of packaging materials, water vapour transmission and oxygen transmission are two important properties, owing to their considerable impact on the food quality. Oxygen and water vapour tend to pass through the walls of polymeric film that result in incessant changes in the quality of the product leading to decline in their shelf life (Siracusa et al. 2008).

11.3.3.1 Estimation of Water Vapour Barrier Properties

Water vapour permeability (WVP) can be used for the quantification of water vapour barrier properties of film, indicating the amount of moisture that seeps in or out of the film per unit area and time (kg/ms Pa). WVP takes into account various parameters including the thickness of the film (packaging material) and the differential pressure. This property is essential in a number of food products like fruits and vegetables where dehydration of product needs to be avoided and also in case of dry products like bread or other dried foods where seeping in of water from the surroundings needs to be averted. Water vapour barrier properties of different polysaccharide-based films are given in Table 11.6. Major information regarding WVP of polysaccharide-based films has been attained gravimetrically following ASTM standard test method known as cup method (E96/E96M). According to this method, an open-mouthed cup of known area is filled with distilled water or any other desiccant or salt solution. The film of interest is stuck completely on the cup and the whole assembly is weighed. The weighed cup along with the film is placed under controlled temperature and low relative humidity conditions. Permeation of water vapour is facilitated by the partial pressure gradient between the inside and the outside film walls. Change in weight of the cup is continuously observed (Rezvani et al. 2013).The data attained from the observations is used to calculate WVP according to the gas diffusion laws of Fick and Henry, using Eq. (11.4):

$$\text{WVP} = \frac{\Delta w \cdot X}{A \cdot \Delta P} \tag{11.4}$$

where Δw is the weight gain of the cup after 24 h, X is the thickness of the film, A is the exposed film area and Δp represents differential vapour pressure of water across the film (Khan et al. 2010). Water vapour transmission rate can also be used to express water movement and can be calculated as per Eq. (11.5):

$$\text{WVP} = \text{WVTR} \cdot \frac{X}{\Delta P} \tag{11.5}$$

where X is the film thickness and ΔP represents the partial pressure difference across the film.

As reported by Cazón et al. (2017), water transfer barrier properties of polysaccharide-based films are higher as compared to the non-biodegradable and partially degradable synthetic films.

11.3.3.2 Estimation of Gas Barrier Properties

Most of the food materials are sensitive to oxygen that has huge impact on their shelf life; therefore, optimum gas permeability of the packaging material is an important parameter in retaining the safe shelf life of the food products. Permeation of gasses

Table 11.6 Water vapour permeability (WVP) of polysaccharide-based and composite films

Film composition	WVP (g/ms Pa)	References
Methyl cellulose	52.29 ± 3.08	Łupina et al. (2019)
Pectin: alginate: WPC ratio (1:1:1)	$7.9 \pm 0.4 \times 10^{10}$	Chakravartula et al. (2019)
Pectin: casein: gelatin (100:00:00)	1.75×10^{-12}	Saikia and Badwaik (2018)
Soybean polysaccharide: gelatin (100:0)	2.478 ± 0.170	Liu et al. (2020)
Gelatin	50.79 ± 0.79	Łupina et al. (2019)
Pectin: chitosan (66:34) with 40% glycerol	0.004	Machado et al. (2020)
Pectin: alginate: WPC ratio (1.5:0:1.5)	$9.6 \pm 0.5 \times 10^{10}$	Chakravartula et al. (2019)
Pectin: casein: gelatin (100:00:00)	1.75×10^{-12}	Saikia and Badwaik (2018)
Tara gum mixtures with glycerol and with/without chitosan	9.91×10^{-10} – 1.28×10^{-10}	Antoniou et al. (2015)
Gelatin: octenylsuccinic anhydride modified starch (50:50)	51.35 ± 0.56	Łupina et al. (2019)
Pectin: casein: gelatin (00:00:100)	1.46×10^{-12}	Saikia and Badwaik (2018)
Soybean polysaccharide: gelatin (80:20)	2.509 ± 0.156	Liu et al. (2020)
Pectin: chitosan (66:34) with 40% glycerol	0.072	Machado et al. (2020)
Pectin: alginate: WPC ratio (2:0.5:0.5)	$7.9 \pm 0.2 \times 10^{10}$	Chakravartula et al. (2019)
Gelatin: gum arabic (50:50)	48.76 ± 0.27	Łupina et al. (2019)
Pectin: alginate: WPC ratio (3:0:0)	$7.3 \pm 0.4 \times 10^{10}$	Chakravartula et al. (2019)
Soybean polysaccharide: gelatin (20:80)	1.926 ± 0.048	Liu et al. (2020)
Soybean polysaccharide: gelatin (60:40)	2.386 ± 0.324	
Gelatin: methyl cellulose (75:25)	47.28 ± 1.78	Łupina et al. (2019)
Pectin: casein: gelatin (50:50:00)	2.05×10^{-12}	Saikia and Badwaik (2018)
Methyl cellulose mixtures with ethanol	0.51×10^{-10} – 1.08×10^{-10}	Nazan Turhan and Sahbaz (2004)
Pectin: casein: gelatin (50:25:25)	1.07×10^{-12}	Saikia and Badwaik (2018)
Bran gum with glycerol and beeswax mixtures	2.5×10^{-10} – 4.75×10^{-11}	Spotti et al. (2016)
Gelatin: octenylsuccinic anhydride modified starch (25:75)	46.57 ± 0.73	Łupina et al. (2019)
Pectin: alginate: WPC (0.5:2:0.5)	$7.5 \pm 0.4 \times 10^{10}$	Chakravartula et al. (2019)
Pectin: alginate: WPC (1.5:1.5:0	$7.4 \pm 0.3 \times 10^{10}$	
Gelatin: methyl cellulose (50:50)	45.42 ± 1.35	Łupina et al. (2019)
Gelatin: methyl cellulose (25:75)	48.86 ± 1.97	
Pectin: alginate: WPC (0:3:0)	$7.2 \pm 0.1 \times 10^{10}$	Chakravartula et al. (2019)

(continued)

Table 11.6 (continued)

Film composition	WVP (g/ms Pa)	References
Gelatin: water-soluble soy polysaccharides (75:25)	46.98 ± 0.36	Łupina et al. (2019)
Pectin: casein: gelatin (00:100:00)	2.15×10^{-12}	Saikia and Badwaik (2018)
Gelatin: water-soluble soy polysaccharides (25:75)	51.97 ± 0.98	Łupina et al. (2019)
Gelatin: gum arabic (25:75)	51.76 ± 1.42	
Pectin: casein: gelatin (00:50:50)	1.67×10^{-12}	Saikia and Badwaik (2018)
Tapioca starch mixtures	12.1×10^{-10}	Vasconez et al. (2009)
Pectin: casein: gelatin (25:25:50)	1.18×10^{-12}	Saikia and Badwaik (2018)
Water-soluble soy polysaccharides	65.04 ± 1.69	Łupina et al. (2019)
Chitosan and starch mixtures	1.18×10^{-10} – 1.55×10^{-10}	
Soybean polysaccharide: gelatin (40:60)	2.034 ± 0.057	Liu et al. (2020)
Pectin: casein: gelatin (25:50:25)	1.44×10^{-12}	Saikia and Badwaik (2018)
Gelatin: gum arabic	48.66 ± 018 g	Łupina et al. (2019)
Pectin: alginate: WPC ratio (0.5:0.5:2)	$9.0 \pm 0.1 \times 10^{10}$	Chakravartula et al. (2019)
Pectin: alginate: WPC ratio (0:1.5:1.5)	$8.7 \pm 1.1 \times 10^{10}$	
Gum arabic	60.15 ± 1.99	Łupina et al. (2019)
Pectin: casein: gelatin (50:00:50)	1.81×10^{-12}	Saikia and Badwaik (2018)
Tara gum mixtures with sorbitol and glycerol	0.52×10^{-10} – 0.69×10^{-10}	Ma et al. (2016)
Soybean polysaccharide: gelatin (0:100)	2.278 ± 0.138	Liu et al. (2020)
Gelatin: octenylsuccinic anhydride modified starch (75:25)	49.62 ± 1.67	Łupina et al. (2019)
Cassava starch mixtures with glycerol	1.8×10^{-10} – 2.8×10^{-10}	
Gelatin: water-soluble soy polysaccharides (50:50)	51.74 ± 0.62	Łupina et al. (2019)

through films is majorly dependent on the solubility of gas in the polymer material and gas diffusivity through them. Estimation of gas barrier properties of polymeric films is vital to design food packaging materials. As reported by Ferreira et al. (2016), at low humidity, polysaccharide-based films pose excellent gas barrier properties owing to their densely packed matrix with extensively packed hydrogen bonding. Gas barrier properties of different polysaccharide films are represented in Table 11.7. Oxygen barrier property of films is estimated using the ASTM D1434 methods that provide two procedures for determination of gas transmission rate and permeability of the polymeric matrix. In this method, the sample film acts as a semipermeable membrane because of being sealed between two chambers. One

Table 11.7 Oxygen and carbon dioxide permeability of polysaccharide-based films and composite edible films

Sample	O2 permeability (cm^3/m^2 day bar)	CO2 permeability (cm^3/m^2 day bar)	References
Pectin: alginate: WPC ratio (1:1:1)	16 ± 21	68 ± 2	Chakravartula et al. (2019)
Pectin: alginate: WPC ratio (3:0:0)	10 ± 2	119 ± 2	
Pectin: alginate: WPC ratio (1.5:1.5:0	49 ± 3	123 ± 2	
Pectin: alginate: WPC ratio (0:1.5:1.5)	83 ± 1	87 ± 2	
Pectin: alginate: WPC ratio (1.5:0:1.5)	1023 ± 3	115 ± 4	
Pectin: alginate: WPC ratio (2:0.5:0.5)	166 ± 1	303 ± 6	
Pectin: alginate: WPC ratio (0.5:2:0.5)	17 ± 3	75 ± 4	
Pectin: alginate: WPC ratio (0.5:0.5:2)	149 ± 4	152 ± 4	
Pectin: alginate: WPC ratio (0:3:0)	438 ± 16	1043 ± 59	
Pectin: chitosan (66:34) with 40% glycerol	1.8 ± 0.00 (cm^3/m^2 day)	NR	Machado et al. (2020)
Pectin: chitosan (66:34) with 40% glycerol	0.7 ± 0.00 (cm^3/m^2 day)	NR	

chamber is filled with gas kept at high pressure while the other chamber is held at low pressure to create gradient between the two chambers. Gases are allowed to pass through the membrane film and gets build up in the evacuated chamber. Depending on the means of gas measurement, methods for estimation of barrier properties vary. In manometric method, the chamber kept at lower pressure is evacuated first, and the increase in pressure is employed in measuring the gas concentration, while in volumetric method, gas concentration is estimated by the increase in volume and the lower pressure chamber is kept around the atmospheric pressure. In addition, ASTM (F1307, D3985, F1927) provides coulometric sensor-based method for estimation of oxygen transmission through the packaging films. This method employs a setup similar to that of water vapour transmission rate (ASTM F1249) where one chamber is packed with oxygen and another one is filled with nitrogen gas that acts as a carrier for oxygen towards the coulometric sensor. An electric current is produced by oxygen that is directly proportional to the concentration of oxygen at the sensor per unit time. With an increase in time, there is also an increase in the generation of electric current, and the values are observed till they become constant. Transmission of oxygen is generally determined by using Eq. (11.6) given by Singh et al. (2017):

$$\mathrm{OTR} = \frac{E_e - E_0}{A * RL} \tag{11.6}$$

where E_e is the steady-state voltage level, E_0 is the zero voltage level, A is the specimen area, Q is the calibration constant and RL is the value of load resistance.

11.3.4 Thermal Analysis of Polysaccharide-Based Films

The kinetic and thermophysical properties of the polymeric material are determined by using thermal analysis. These properties are measured as function of temperature/time over varying temperature ranges. The thermal analysis is usually used to determine the char content, thermal stability and breakdown temperatures of polymeric materials. Numerous thermal analytical techniques are usually used to examine thermal properties of materials including plyboards, polymers, composites and films. The thermal analytical techniques are thermo gravimetric analysis, differential scanning calorimetry, thermomechanical analysis, dynamic mechanical analysis, differential thermal analysis, etc.

Among these techniques, thermomechanical analysis (TGA) and differential scanning calorimetry (DSC) have been widely used in polymer industries. TGA is one of the most sensitive and widely used thermal technique for monitoring the composition and structure dependence of natural polymers or composites on the thermal degradation behaviour (Izani et al. 2013).

11.3.4.1 Differential Scanning Calorimetry (DSC)

The DSC finds applicability in characterising materials like biopolymeric films. For crystalline materials, the quantitative applicability includes determination of heat of fusion and the degree of crystallisation. The glass transition temperature (T_{g}) and melting points are valuable for qualitative material classification, though thermal methods cannot be used alone for identification. The determination is one of the most important applicability of DSC. There are dramatic changes in physical properties of a polymeric substance at T_g, where polymer changes from glassy to the rubbery state. At T_g, the polymeric substance undergoes volume and expansion, heat capacity and heat flow changes. The DSC instrument measures the changes in heat capacity of materials (Schick et al. 2012).

The degree of crystallinity is an essential factor for determining polymer properties of crystalline materials. The IR spectroscopy, density measurements, X-ray diffraction and thermal methods are used to determine the crystallinity degree. In maximum cases, DSC is one of the easy methods for determining the crystallinity level and is also employed for determining amorphous content levels in polymeric materials (Koshy et al. 2017).

11.3.4.2 Thermogravimetric Analysis (TGA)

TGA is the thermal analysis technique in which the sample weight is continuously being measured under controlled temperature (Yahiaoui et al. 2015). The TGA analyser has the advantage of easy operation, precise control and record of temperature and weight loss of samples. TGA has also been used for performing the biomass proximate analysis (Saldarriaga et al. 2015).

TGA is usually done for studying the thermal stability of materials. The TGA curve obtained is an effective tool for observing the polymer thermal degradation under inert and oxidative environments (Barneto et al. 2009). The method comprises the measurement of sample weight loss being progressively heated. The temperature is increased constantly or in a series of steps, and the curves of weight loss versus temperature are obtained. There is thermal stability of material when there is negligible change in weight loss and slope of the material in particular range of temperature (Singh et al. 2017). The TGA is carried out under of 40 mL/min nitrogen atmosphere, gas flow and a heating scan of 10 °C/min, over 40–800 °C temperature range. Initially, the film samples are held for 1 min at 40 °C. The 10 mg sample mass is used for the TGA experiments (Chakravartula et al. 2019).

11.3.4.3 Differential Scanning Calorimetry (DSC)

DSC is used to calculate the characteristic temperature and the corresponding enthalpies of phase changes using DSC analysis, equipped with a liquid sub-ambient accessory and standardised with standards of high purity. In DSC analysis, films are cut into 2 mm^2 small pieces and 10–12 mg in weight, placed in sealed aluminium pan. After 5 min isothermal −20 °C, the film samples are heated at 10 °C/min, from −20 to 280 °C (first scan) and then after a further 2 min isotherm at 280 °C were cooled to −20°C at 80 °C/min rate. Finally, after 3 min isotherm, samples are reheated at 10 °C/min from −20 to 280 °C (second scan). All the experiments are being performed under (20 cm^3/min) nitrogen flow. The melting temperature (T_m) is the peak value of DSC curve in the endothermic phenomena. The melting enthalpy (ΔH_m) is obtained from the area of the DSC endothermic peak. From the first scan, T_m and ΔH_m values are collected.

11.3.5 FTIR Analysis of Polysaccharide-Based Films

The FTIR analysis is conducted to get the transmission or absorption infrared spectrum of film samples. In FTIR analysis, IR rays are being absorbed by the bonds of molecules. The IR ray energy can be correlated to the vibrational energy of various bonds present in functional groups of compound. The FTIR analytical technique is used to detect the chemical structure of the developed films and the possible interactions arising between components of films. The FTIR spectra are collected at the average of 50 scans recorded at a resolution of 4 cm^{-1} in the range from 4000 to 500 cm^{-1} (Pérez-Vergara et al. 2020).

A broad stretching in 3423 cm^{-1} regions in polysaccharide films signifies the stretching vibration of the OH groups of carbohydrates. This is the typical absorption band of carbohydrates and is responsible for the water solubility of polysaccharides (Karbowiak et al. 2011). The CH stretching of methyl and methylene groups is represented at absorption bands of 2922 and 2853 cm^{-1} (Wang et al. 2010). The enol and amide stretching vibrations are represented at 1650–1540 cm^{-1} region. The CO group stretching is indicated at 1628 cm^{-1} absorption band. Similarly, 1415–1382 cm^{-1} peaks represent >CO stretch of (COO) groups and CO bond from (COO) groups (Wang et al. 2010). At 1044 cm^{-1}, the sharp absorption band represents the (CO) stretching vibration (Sajna et al. 2013). The 1200–800 cm^{-1} wave number region is the fingerprint region and is used to describe different polysaccharides (Copikova et al. 2006). The absorption bands in 983–1200 cm^{-1} region showed presence of sugar monomers like glucose, galactose and mannose. The absorption band in the anomeric region at 796 cm^{-1} showed possible existence of alpha glycosidic linkages. The absorption band at 870–890 cm^{-1} represents glycosidic linkages. The absorption band at 734 cm^{-1} is attributed due to COC bending vibrations (Sajna et al. 2013).

11.3.6 SEM Analysis of Polysaccharide-Based Films

The determination of microscopic structure is a principal index for polysaccharide-based films and composites and shows the surface morphology and internal organisation structures of these films. It tells about the smoothness or roughness of the polysaccharide-based films. For conducting SEM analysis, these films are mounted onto aluminium specimen stubs by using double-faced adhesive tape, and samples are coated with a gold layer by using sputter coater. The samples are observed using (15.0 kV) accelerating voltage with magnification of 4000× and 5000× (Du et al. 2016)

11.4 Conclusion

Food packaging is necessary for protecting the foods from external environment. Food packaging material mainly used was derived from petroleum products. However, this packaging material has harmful impact on human health and is also non-biodegradable and causes pollution problems. This leads to the shifting of packing material development from plant- and animal-based materials. Edible packaging material development is now gaining importance as people are health conscious now. The various substances used for development of edible packages are proteins, carbohydrates, lipids or composites. Among these, polysaccharides act as important packaging materials due to their abundance in nature and good functional properties. The packaging material developed from these substances should have certain properties to maintain the standards and act as better packaging material. This leads to the characterisation of these polysaccharide-based films for evaluating whether they can act as efficient packaging material.

References

Abugoch LE, Tapia C, Villaman MC, Yazdani-Pedram M, Díaz-Dosque M (2011) Characterization of quinoa protein chitosan blend edible films. Food Hydrocoll 25(5):879–886

Alves VD, Ferreira AR, Costa N, Freitas F, Reis MA, Coelhoso IM (2011) Characterization of biodegradable films from the extracellular polysaccharide produced by Pseudomonas oleovorans grown on glycerol by-product. Carbohydr Polym 83(4):1582–1590

Antoniou J, Liu F, Majeed H, Zhong F (2015) Characterization of tara gum edible films incorporated with bulk chitosan and chitosan nanoparticles: a comparative study. Food Hydrocoll 44: 309–319

Arvanitoyannis IS (1999) Totally and partially biodegradable polymer blends based on natural synthetic macromolecules: preparation, physical properties, and potential as food packaging materials. J Macromol Sci Rev Macromol Chem Phys C39(2):205–271

Arvanitoyannis IS, Bosnea L (2004) Migration of substances from food packaging materials to foods. Crit Rev Food Sci Nutr 44(2):63–76

Barneto AG, Ariza J, Martín JE, Jiménez L (2009) Use of autocatalytic kinetics to obtain composition of lignocellulosic materials. Bioresour Technol 100:3963–3973

Bertuzzi MA, Castro Vidaurre EF, Armada M, Gottifredi JC (2007) Water vapour permeability of edible starch based films. J Food Eng 80(3):972–978

Briassoulis D (2006) Mechanical behaviour of biodegradable agricultural films under real field conditions. Polym Degrad Stabil 91(2006):1256–1272

Briassoulis D, Giannoulis A (2018) Evaluation of the functionality of bio-based food packaging films. Polym Test 69:39–51

Carneiro-da-Cunha MG, Cerqueira MA, Souza BW, Souza MP, Teixeira JA, Vicente AA (2009) Physical properties of edible coatings and films made with a polysaccharide from Anacardium occidentale L. J Food Eng 95(3):379–385

Cazon P, Vázquez M (2020) Mechanical and barrier properties of chitosan combined with other components as food packaging film. Environ Chem Lett 18(2):257–267

Cazón P, Velazquez G, Ramírez JA, Vázquez M (2017) Polysaccharide-based films and coatings for food packaging: a review. Food Hydrocoll 68:136–148

Chakravartula SSN, Soccio M, Lotti N, Balestra F, Dalla Rosa M, Siracusa V (2019) Characterization of composite edible films based on pectin/alginate/whey protein concentrate. Materials 12(15):2454

Chambi HNM, Grosso CRF (2011) Mechanical and water vapor permeability properties of biodegradables films based on methylcellulose, glucomannan, pectin and gelatin. Food Sci Technol 31(3):739–746

Chan HM, Nyam KL, Yusof YA, Pui LP (2020) Investigation of properties of polysaccharide-based edible film incorporated with functional melastoma malabathricum extract. Carpathian J Food Sci Technol 12(1):120–134

Chopin N, Guillory X, Weiss P, Bideau J, Colliec-Jouault S (2014) Design polysaccharides of marine origin: chemical modifications to reach advanced versatile compounds. Curr Org Chem 18:867–895

Copikova AS, Barros I, Smidova M, Cerna DH, Teixeira I, Delgadillo A, Syystya A, Coimbra MA (2006) Influence of hydration of food additive polysaccharides on FT-IR spectra distinction. Carbohydr Polym 63:355–359

Du H, Hu Q, Yang W, Pei F, Kimatu BM, Ma N, Zhao L (2016) Development, physiochemical characterization and forming mechanism of flammulina velutipes polysaccharide-based edible films. Carbohydr Polym 152:214–221

Falguera V, Quintero JP, Jiménez A, Muñoz JA, Ibarz A (2011) Edible films and coatings: structures, active functions, and trends in their use. Trends Food Sci Technol 22:292–303

Feki A, Ben Amara I, Bardaa S, Hajji S, Chabchoub N, Kallel R, Ktari N (2019) Preparation and characterization of polysaccharide-based films and evaluation of their healing effects on dermal laser burns in rats. Eur Polym J 115:147–156. https://doi.org/10.1016/j.eurpolymj.2019.02.043

Ferreira AR, Alves VD, Coelhoso IM (2016) Polysaccharide-based membranes in food packaging applications. Membranes 6(2):22

Goetz LA, Jalvo B, Rosal R, Mathew AP (2016) Superhydrophilic anti-fouling electrospun cellulose acetate membranes coated with chitin nanocrystals for water filtration. J Membr Sci 510:238–248

Han JH (2014) Edible films and coatings. In: Innovations in food packaging. Elsevier, Amsterdam, pp 213–255

Hussain A, Zia KM, Tabasum S, Noreen A, Ali M, Iqbal R, Zuber M (2017) Blends and composites of exopolysaccharides; properties and applications: a review. Int J Biol Macromol 94:10–27

Iwata KI, Ishizaki SH, Handa AK, Tanaka MU (2000) Preparation and characterization of edible films from fish water-soluble proteins. Fish Sci 66:372–378

Izani MAN, Paridah MT, Anwar UMK, Nor MYM, Hng PS (2013) Effects of fiber treatment on morphology, tensile and thermogravimetric analysis of oil palm empty fruit bunches fibers. Compos Part B 45:1251–1257

Janjarasskul T, Krochta JM (2010) Edible packaging materials. Annu Rev Food Sci Technol 1:415–448

Jiang Y, Li Y, Chai Z, Leng X (2010) Study of the physical properties of whey protein isolate and gelatin composite films. J Agric Food Chem 58(8):5100–5108

Jridi M, Hajji S, Ayed HB, Lassoued I, Mbarek A, Kammoun M, Nasri M (2014) Physical, structural, antioxidant and antimicrobial properties of gelatin–chitosan composite edible films. Int J Biol Macromol 67:373–379

Kale G, Auras R, Singh SP (2006) Degradation of commercial biodegradable packages under real composting and ambient exposure conditions. J Polym Environ 14:317–334

Karbowiak T, Ferret E, Debeaufort F, Voilley A, Cayot C (2011) Investigation of water transfer across thin layer biopolymer films by infrared spectroscopy. J Membr Sci 370(1–2):82–90

Khan RA, Salmieri S, Dussault D, Uribe-Calderon J, Kamal MR, Safrany A et al (2010) Production and properties of nanocellulose-reinforced methylcellulose-based biodegradable films. J Agric Food Chem 58(13):7878–7885

Kim KM, Son JH, Kim S-K, Weller CL, Hanna MA (2006) Properties of chitosan films as a function of pH and solvent type. J Food Sci 71(3):E119–E124

Kim YT, Min B, Kim KW (2014) General characteristics of packaging materials for food system. In: Han JH (ed) Innovations in food packaging, 2nd edn. Academic Press, London, pp 13–35

Koshy O, Subramanian L, Thomas S (2017) Differential scanning calorimetry in nanoscience and nanotechnology. In: Thermal and rheological measurement techniques for nanomaterials characterization. Elsevier, Oxford, pp 109–122

Leceta I, Guerrero P, de la Caba K (2013) Functional properties of chitosan-based films. Carbohydr Polym 93(1):339–346

Liu C, Huang J, Zheng X, Liu S, Lu K, Tang K, Liu J (2020) Heat sealable soluble soybean polysaccharide/gelatin blend edible films for food packaging applications. Food Packag Shelf Life 24:100485

Lorevice MV, Otoni CG, Moura MRD, Mattoso LHC (2016) Chitosan nanoparticles on the improvement of thermal, barrier, and mechanical properties of high- and low-methyl pectin films. Food Hydrocoll 52:732–740
Łupina K, Kowalczyk D, Zięba E, Kazimierczak W, Mężyńska M, Basiura-Cembala M, Wiącek AE (2019) Edible films made from blends of gelatin and polysaccharide-based emulsifiers-a comparative study. Food Hydrocoll 96:555–567
Ma Q, Hu D, Wang L (2016) Preparation and physical properties of tara gum film reinforced with cellulose nanocrystals. Int J Biol Macromol 86:606–612
Ma Y, Zhao Y, Xie J, Sameen DE, Ahmed S, Dai J, Liu Y (2021) Optimization, characterization and evaluation of papaya polysaccharide-corn starch film for fresh cut apples. Int J Biol Macromol 166:1057–1071
Machado BR, Facchi SP, de Oliveira AC, Nunes CS, Souza PR, Vilsinski BH, Martins AF (2020) Bactericidal pectin/chitosan/glycerol films for food pack coatings: a critical viewpoint. Int J Mol Sci 21(22):8663
Mahalik NP, Nambiar AN (2010) Trends in food packaging and manufacturing systems and technology. Trends Food Sci Technol 21:117–128
Mali S, Grossmann MVE, Garcıa MA, Martino MN, Zaritzky NE (2004) Barrier, mechanical and optical properties of plasticized yam starch films. Carbohydr Polym 56(2):129–135
Mali S, Grossmann MVE, García MA, Martino MN, Zaritzky NE (2006) Effects of controlled storage on thermal, mechanical and barrier properties of plasticized films from different starch sources. J Food Eng 75(4):453–460
Marsh K, Bugusu B (2007) Food packaging—roles, materials, and environmental issues. J Food Sci 72(3):R39–R55
Mohajer S, Rezaei M, Hosseini SF (2017) Physico-chemical and microstructural properties of fish gelatin/agar bio-based blend films. Carbohydr Polym 157:784–793
Motedayen AA, Khodaiyan F, Salehi EA (2013) Development and characterisation of composite films made of kefiran and starch. Food Chem 136(3–4):1231–1238
de Moura MR, Mattoso LHC, Zucolotto V (2012) Development of cellulose based bactericidal nanocomposites containing silver nanoparticles and their use as active food packaging. J Food Eng 109(3):520–524
Muscat D, Adhikari B, Adhikari R, Chaudhary DS (2012) Comparative study of film forming behaviour of low and high amylose starches using glycerol and xylitol as plasticizers. J Food Eng 109(2):189–201
Nazan Turhan K, Sahbaz F (2004) Water vapor permeability, tensile properties and solubility of methylcellulose-based edible films. J Food Eng 61(3):459–466
Nešić A, Cabrera-Barjas G, Dimitrijević-Branković S, Davidović S, Radovanović N, Delattre C (2020) Prospect of polysaccharide-based materials as advanced food packaging. Molecules 25(1):135
Pan H, Jiang B, Chen J, Jin Z (2014) Blend-modification of soy protein/lauric acid edible films using polysaccharides. Food Chem 151:1–6
Pang M, Cao L, Cao L, Sheb Y, Wang H (2019) Properties of nisin incorporated ZrO_2/poly (vinyl alcohol)—wheat gluten antimicrobial barrier films. CyTA J Food 17(1):400–407
Pérez-Vergara LD, Cifuentes MT, Franco AP, Pérez-Cervera CE, Andrade-Pizarro RD (2020) Development and characterization of edible films based on native cassava starch, beeswax, and propolis. NFS J 21:39–49
Phan TD, Debeaufort F, Luu D, Voilley A (2005) Functional properties of edible agar-based and starch-based films for food quality preservation. J Agric Food Chem 53(4):973–981
Ramesh HP, Tharanathan RN (2003) Carbohydrates—the renewable raw materials of high biotechnological value. Crit Rev Biotechnol 23:149–173
Rezvani E, Schleining G, Sümen G, Taherian AR (2013) Assessment of physical and mechanical properties of sodium caseinate and stearic acid-based film-forming emulsions and edible films. J Food Eng 116(2):598–605

Rhim JW, Wang LF, Hong SI (2013) Preparation and characterization of agar/silver nanoparticles composite films with antimicrobial activity. Food Hydrocoll 33(2):327–335
Robertson GL (2012) Food packaging: principles and practice, 3rd edn. CRC Press, New York
Robertson GL (2016) Food packaging: principles and practice. CRC Press, New York
Saikia M, Badwaik LS (2018) Characterization and antimicrobial property of casein, gelatin and pectin based active composite films. J Packag Technol Res 2(3):233–242
Sajna VK, Sukumaran RK, Gottumukkala LD, Jayamurthy H, Dhar KS, Pandey A (2013) Studies on structural and physical characteristics of a novel exopolysaccharide from Pseudozyma sp. Int J Biol Macromol 59(2013):84–89
Saldarriaga JF, Aguado R, Pablos A, Amutio M, Olazar M, Bilbao J (2015) Fast characterization of biomass fuels by thermogravimetric analysis (TGA). Fuel 140:744–751
Schick C, Lexa D, Leibowitz L (2012) Differential scanning calorimetry and differential thermal analysis. In: Kaufmann EN (ed) Characterization of materials. John Wiley & Sons, New York, pp 483–495
Sessini V, Arrieta MP, Kenny JM, Peponi L (2016) Processing of edible films based on nanoreinforced gelatinized starch. Polym Degrad Stab 132:157–168
Shen Z, Kamdem DP (2015) Development and characterization of biodegradable chitosan films containing two essential oils. Int J Biol Macromol 74:289–296
Silva F, Dujovne D (2013) Predictive mobility applied to content centric networks. Communications (LATINCOM), 2013 IEEE Latin-America conference on (pp. 1–6). IEEE
Singh P, Wani AA, Langowski HC (2017) Food packaging materials: testing & quality assurance. CRC Press, Boa Raton
Siracusa V, Rocculi P, Romani S, Dalla Rosa M (2008) Biodegradable polymers for food packaging: a review. Trends Food Sci Technol 19(12):634–643
Souza MP, Vaz AFM, Silva HD, Cerqueira MA, Vicente AA, Carneiro-da-Cunha MG (2015) Development and characterization of an active chitosan-based film containing quercetin. Food Bioprocess Technol 8(11):2183–2191
Spotti ML, Cecchini JP, Spotti MJ, Carrara CR (2016) Brea gum (from Cercidium praecox) as a structural support for emulsion-based edible films. LWT Food Sci Technol 68:127–134
Srinivasa PC, Ramesh MN, Kumar KR, Tharanathan RN (2004) Properties and sorption studies of chitosan–polyvinyl alcohol blend films. Carbohydr Polym 53(4):431–438
Vasconez MB, Flores SK, Campos CA, Alvarado J, Gerschenson LN (2009) Antimicrobial activity and physical properties of chitosan tapioca starch based edible films and coatings. Food Res Int 42(7):762–769
Wang Y, Li C, Liu P, Ahmed Z, Xiao P, Bai X (2010) Physical characterization of exopolysaccharide produced by *Lactobacillus plantarum* KF5 isolated from Tibet Kefir. Carbohydr Polym 82:895–903
Weber CJ (2000) Bio-based packaging: materials for the food industry—status and perspectives, a European concerted action report. KVL, Frederiksberg
Yahiaoui M, Hadoun H, Toumert I, Hassani A (2015) Determination of kinetic parameters of Phlomis bovei de Noé using thermogravimetric analysis. Bioresour Technol 196:441–447
Yamazaki M, Iijima K (2020) Fabrication and characterization of polysaccharide composite films from polyion complex particles. Polymers 12(2):435
Yang Y, Xie B, Liu Q, Kong B, Wang H (2020) Fabrication and characterization of a novel polysaccharide based composite nanofiber films with tunable physical properties. Carbohydr Polym 236:116054
Youssef AM (2013) Polymer nanocomposites as a new trend for packaging applications. Polym-Plast Technol Eng 52:635–660

Chapter 12
Modeling of Moisture Transfer in Food Systems Packed in Polysaccharide-Based Films

Preetinder Kaur, Gurjeet Kaur, Kirandeep, and Gagandeep Kaur

12.1 Introduction

Packaging plays a vital role in containment of foods and conveys nutritional information to consumers. Moreover, it helps in prolonging the shelf life of foods by providing barrier against moisture, oxygen, microbial and insect attack, and environmental factors such as dust, pollution, etc. In food packaging sector, the primary choice of manufacturers is the plastic owing to its less cost and weight, excellent mechanical strength, and barrier properties. However, these petrochemical synthetic films are resistant to degradation and are posing great threat to environment in the present era (Muscat et al. 2012). Due to the worldwide awareness among people regarding pollution from packaging material, alternate options along with waste management programs are a need of the hour. One solution could be the adoption of edible and bio-based polymers due to their compatible properties to avoid moisture, aroma, and oxygen penetration (Aider 2010). These bio-based polymers can be manufactured from biomaterials that are less costly, and most of them are discarded as waste or by-products. These can be considered as novel food packaging systems with no negative impact from food packaging waste on ecosystem.

The most abundant polymers such as polysaccharides can be extracted from plants or marine organisms' biomass directly and microorganisms through fermentation (Ramesh and Tharanathan 2003). These include starch, cellulose,

P. Kaur (✉) · G. Kaur · G. Kaur
Department of Processing and Food Engineering, Punjab Agricultural University, Ludhiana, Punjab, India
e-mail: preetinder72@pau.edu

Kirandeep
KVK Langroya, Nawanshahr, Punjab Agricultural University, Ludhiana, Punjab, India

T. Amin et al. (eds.), *Polysaccharide Based Films for Food Packaging: Fundamentals, Properties and Applications*,
https://doi.org/10.1007/978-981-99-4898-7_12

hemicelluloses, agar-agar, alginate, dextran, kefiran, pullulan, etc. These polymeric carbohydrates are biocompatible, biodegradable, and nontoxic toward living organisms (Chopin et al. 2014). For these properties, polysaccharides are most suitable to use in food packaging as edible coatings and films. Nonetheless, characteristics of polysaccharide-based packaging, namely, high water vapor permeability and poor mechanical strength, hinder their applications at industrial scale (Debeaufort et al. 1998). According to the final use, these limitations can be eliminated by modification of structure of bio-based films by combining two or more films or combining with other constituents (Hussain et al. 2017). Moreover, for hygroscopic films like alginate and carrageenan, these can be applied in thick form on the food to avoid further moisture absorption from the food. Thus, the product retains moisture until the coating becomes dehydrated (Bourtoom 2008; Embuscado and Huber 2009).

12.2 Need for Moisture Control

Moisture is a critical quality parameter as it directly influences the physical, chemical, and sensory attributes of foods. Fresh agricultural produce loses water due to transpiration and respiration. Thus, it is quite challenging for producers, processors, and retailers to control these losses (Mahajan et al. 2014). It also contributes to economic losses as reduction in weight because moisture loss affects the sale of food (Caleb et al. 2013). A food product from a processing unit is developed with defined moisture content at which product remains safe for long time period. The most important factors considered by food manufacturers are physical characteristics, chemical properties, sensory attributes, and microbial stability. All these factors are related to moisture content and water activity (a_w) of components of foods in a food system. a_w is a ratio of partial pressure of water in the food to partial pressure of pure water and is related to equilibrium relative humidity. These food components continue to adsorb or desorb water to attain a state of equilibrium. Moisture transfer is a physical phenomenon which occurs during storage and transportation of foods. This transfer could be due to diffusion of water inside the food or moisture migration within a multicomponent food or between food and surroundings. Variations in moisture content of food system affect stability and shelf life of foods due to variation in physical (crystallization, stickiness, and texture) and chemical properties. Similar to food systems, bio-based films from polysaccharides are hydrophilic and can be used for short-term applications or in foods having very low moisture content. The water barrier property of films is significant to prevent the dehydration of fresh products or sogginess in dry products. In this case, the water vapor permeability of films remains low when a_w is low. However, a rise in a_w accompanies increase in moisture content of film causing plasticization and decreasing water vapor barrier (Gontard et al. 1993; Cuq et al. 1997). Moreover, increasing temperature negatively affects the moisture barrier.

These changes can be understood from glass transition. Foods or films exist in thermodynamically stable crystalline or unstable amorphous state. The amorphous region can exist in a rubbery or glassy state and exhibit a property called the glass transition temperature (T_g). Whenever there is a shift from the glassy state to the rubbery state or vice versa in a material, thermodynamic properties, molecular mobility, dielectric constant, and mechanical properties also change (Sperling 1986). The factors influencing this transition are plasticization by water, weight of ingredients, and the amount and type of bonding interactions (Levine and Slade 1988). Plasticization of films due to water results in decline of glass transition temperature of amorphous hydrocolloids. The solubility and mobility of water lead to inconsistent diffusion behavior in glassy polymers due to structural changes (Crank 1975).

The physical property, that is, rate of crystallization, increases as moisture increases and reduces shelf life of foods. It also increases along with increasing temperature above the glass transition temperature until the aqueous system remains saturated with the diffusing molecules (Crank 1975). Below the T_g, reactions occur very slowly as the food is in the glassy state and has a very high viscosity of about 10^{12} Pa s (White and Cakebread 1966). Chemical reactions such as lipid oxidation and nonenzymatic browning are also affected by moisture transfer in food systems and depend upon ingredients, viscosity of the system, and the mobility of the reactants. The reaction rate of non-enzymatic browning becomes maximum when a_w falls within 0.6 ± 0.8 range. Similarly, for lipid oxidation, the rate increases both above and below the monolayer moisture content; thus prevention of moisture exchange is critical unless oxygen is excluded. Furthermore, as a_w goes above the 0.60 for salt-tolerant microorganisms, 0.80 for molds, about 0.87 for yeasts, and 0.91 for most pathogens, microbial activities take place (Beuchat 1981). Thus, to maintain the quality and desirable textural attributes of food products, moisture as well as a_w of foods needs to be controlled in food systems.

12.3 Modeling of Moisture Transfer

Fresh produce remains metabolically active during the storage. The ongoing metabolic processes release heat of respiration which is transferred directly or through the packaging surroundings to film surface and then to atmosphere. The bio-based films have been considered as thin layers which provide barrier to mass transfer (moisture, oxygen, and solute movement) within the food or between the food and environment (Gutiérrez and Álvarez 2017).

For thin layer geometry, that is, a polysaccharide-based film, diffusion is responsible for mass exchange. On the basis of the property governing the movement of components, diffusion can be classified into two types—pressure diffusion due to difference in pressure and thermal diffusion due to the available thermal gradient (Ibarz and Barbosa-Cánovas 2002). Diffusion is influenced by various factors that are inherent to the food product and its surroundings and are governed by different heat transfer mechanisms, namely, conduction, convection, and radiation. Based upon the assumption that out of the two components on both sides of film, component A has higher concentration than component B, that is, $C_A > C_B$ or the temperature $T_A > T_B$. The rate of transfer of mass or heat (χ), as shown in Eq. (12.1), is dependent upon concentration or temperature difference across the thin film:

$$\chi = K \cdot \Delta\omega \tag{12.1}$$

where K is proportionality constant and $\Delta\omega$ is the driving force (concentration or temperature)

Various researchers have developed models for moisture transfer across permeable films based upon different theories of heat and mass transfer. A variation of assumptions is also seen which depends upon the type of packaged product or the packaging material. Diffusion is a complex phenomenon as it involves transfer of species within a food system naturally or forcefully. It follows unsteady state as pressure difference across the film surface decreases with time. Moreover, it is highly dependent upon moisture content and geometric configurations and takes place from higher to lower concentration (Labuza and Hyman 1998). As per Fick's law (Eq. 12.4), the molar flux ($\bar{J}*_A$ expressed as mol m^{-1} s^{-1}), due to diffusion is directly related to molar concentration (C expressed as mol m^{-3}), effective diffusion coefficient (D_{AB} expressed as m^2 s^{-1}), and concentration gradient ($\overline{\nabla} X_A$):

$$\bar{J}*_A = -C \cdot D_{AB} \cdot \overline{\nabla} X_A \tag{12.2}$$

X defines the species molar fraction divided by the mixture total molar density: $X_i = C_i/C$.

D_{AB} changes to an effective diffusion coefficient, that is, D_{eff}, when material has porosity and expressed as

$$D_{\text{eff}} = D_{AB} \cdot \frac{\varepsilon}{\tau} \tag{12.3}$$

where ε is the porosity and τ is the tortuosity.

The mass diffusion through a thin slab is explained by Crank (1975). The moisture transfer to surroundings can be modeled as

$$W = \frac{X - X_e}{X_0 - X_e} = -\frac{8}{\pi^2} \sum_{n=0}^{\infty} \frac{1}{(1 + 2n)^2} \exp\left[-\frac{\left((1 + 2n)^2 \cdot D_{\text{eff}} \cdot t\right)}{4L^2}\right] \tag{12.4}$$

where X_0 is the initial moisture content (g water/g dry solids), X_e is the equilibrium moisture content, t is the time (s), L is the thickness of thin slab (m), D_{eff} is the effective diffusion coefficient.

Polysaccharide-based films become hydrated when they come in contact with liquid. As a result, the relaxation of film matrix causes expansion or swelling of film, which affects the transportation process of active agents in the film matrix. The active agent diffuses easily due to increased diffusion coefficient in the swollen matrix. Mass transfer in such conditions can be categorized as ideal Fickian diffusion, anomalous behavior, and Case II transport (Mastromatteo et al. (2010). In case of hydrophilic packaging films, active compounds diffuse through Fickian diffusion and relaxation process. Various studies have been documented which considered a linear superimpose approach to model the diffusion mechanism (Mastromatteo et al. 2010; Pinheiro et al. 2013). These models explained the moisture transfer phenomenon by addition of molecule movement through Brownian motion and by polymer relaxation:

$$M_t = M_{t,F} + M_{t,R} \tag{12.5}$$

where $M_{t,f}$ describes the diffusion through a thin sheet at any time t using Fickian approach as follows assuming the boundary conditions are constant:

$$M_{t,F} = M_F^{\mathrm{eq}} \left\{ 1 - \frac{8}{\pi^2} \sum_{n=0}^{n=\infty} \frac{1}{(2 \cdot n + 1)^2} \exp\left[-D \cdot (2 \cdot n + 1)^2 \cdot \pi^2 \cdot \frac{t}{l^2} \right] \right\} \tag{12.6}$$

where M_F^{eq} is the amount of compound with low molecular mass released at equilibrium as a consequence of stochastic phenomenon.

D = diffusion coefficient through the swollen polymer matrix;
l = film thickness.
$M_{t,\,R}$ denotes the relaxation caused by stress dissipation as penetrant enters.

The relaxation process can be expressed using a first-order kinetics equation:

$$M_{R,t} = M_F^{\mathrm{eq}} \left[1 - \exp\left(-\frac{t}{\tau} \right) \right] \tag{12.7}$$

M_F^{eq} = amount of compound with low molecular weight released at equilibrium as a consequence of polymer relaxation;
τ = relaxation time associated with polymer relaxation.

12.3.1 Conduction

According to Fourier's law of heat conduction (Eq. 12.2), the heat flux or heat flow density is a function of negative temperature gradient along one direction, that is, x (Ibarz and Barbosa-Cánovas 2002), as other directions do not contribute significantly to cause temperature difference:

$$q = -kA\frac{dT}{dx} \tag{12.8}$$

Here k is isotropic and represents the thermal conductivity of material (W m^{-1} $°C^{-1}$), A is the area, and dT/dx is the thermal gradient along x direction (°C m^{-1}).

12.3.2 Convection

In this phenomenon, the thermal energy transfers by the movement of molecules from one location to another within a material. It occurs naturally or by mechanical means known as natural convection and forced convection, respectively. If we assume transfer of heat from the surface of food having temperature T_f to the surrounding air with temperature T_s, it will be proportional to the surface area of food (A) and the temperature difference, that is, $T_f - T_s$ as follows:

$$q = hA(T_f - T_s) \tag{12.9}$$

where h represents the convective heat transfer coefficient (W m^{-2} $°C^{-1}$).

12.3.3 Radiation

The heat transfer by radiation takes place by emission and then absorption of electromagnetic waves. Any material has absolute temperature above 0 release radiations. This radiation is proportional to the fourth power of absolute temperature as shown in equation:

$$q = \sigma\varepsilon AT^4 \tag{12.10}$$

where A represents surface area of food, σ is Stefan-Boltzmann constant, and ε is emissivity.

12.4 Factors Influencing Diffusion of Moisture Through Polysaccharide Films

Foods are hygroscopic in nature. During storage, moisture movement within food materials and between foods and surroundings is dependent upon various factors. The food characteristics as well as external factors affects the diffusion process and is discussed below:

12.4.1 Moisture Content

Foods can be categorized into three categories on the basis of moisture content as perishable, semi-perishable, and nonperishable foods. Perishable foods like fruits, vegetables, and dairy and meat products have short shelf life, whereas semi-perishable foods such as nuts, butter, oils, and cheese have limited storage life. Nonperishable foods represent the most stable class of fruits as these can be stored for months or years without the need of refrigeration, for example cereals, sugar, coffee, etc. Another classification based upon the water activity defines low moisture foods with water activity <0.6, intermediate moisture foods having water activity 0.6–0.9, and high moisture foods with water activity >0.9. These foods with varied moisture levels have different packaging requirement to ensure the best quality and meet consumer expectations. Moreover, the fresh produce keeps on losing weight owing to moisture loss even after the harvesting due to transpiration and respiration. The moisture starts accumulating underneath the package surface. This leads to condensation, alteration of permeance of package, and further quality loss of fresh produce due to microbial contamination (Linke and Geyer 2013).

Generally, as moisture content increases, effective moisture diffusivity also increases. However, certain exceptions were recorded by scientists based upon the structure of food material. Karathanos et al. (1990) observed an increase in effective diffusivity as moisture increased to a value of 10%. Further increase in moisture led to decrease and finally a constant value of effective diffusivity at high moisture. This was attributed due to structural changes related to moisture content. Considering this, food products can be categorized as follows:

- **Hydrophilic materials**: In case of materials such as starch or wheat gluten, alterations in mobility of polymers resulted owing the plasticizing effect of water as effective diffusivity remained low as the water activity was below 0.4, increased up to 0.8 water activities, and then decreased above 0.8 water activities. According to Gouanvé et al. (2007), the underlying fact for such observations was the water cluster formation in the rubbery state of gluten which led to poor movement of water molecules and thus lowers effective diffusivity.
- **Porous materials:** The porous nature of dried products affects the moisture migration phenomenon. Liquids diffuse slowly as compared to gases in porous materials. Porosity is a measure of the free space occupied in the food material. When porous materials such as sponge cake absorb water, swelling of matrix occurs. This further leads to lowering the porosity of food material and hence the effective diffusivity (Guillard et al. 2003).
- **Hydrophobic materials:** Contrary to hydrophilic materials, Bourlieu et al. (2008) observed a quasi-constant effective diffusivity when a lipid-based material absorbed a very small amount of water. It was due to partial crystalline nature of the material which was unaffected by increased moisture content.

12.4.2 Composition

Foods are composed of a variety of constituents such as carbohydrates, fats, vitamins, minerals, protein, fiber, and water. Packaging requirements vary based upon the composition of foods. An increase in moisture leads to plasticization of gluten network and less resistance to deformation of texture. However, decreased moisture content gives rise to recrystallization of starch along with redistribution of moisture within starch from gluten, which brings an antiplasticizing outcome on the protein arrangement (Schiraldi and Fessas 2001). Moisture transfer modeling is less complex in single component foods as compared to mixture of different foods. The most common example of multicomponent foods is breakfast cereals. Any deviations in the food properties such as moisture content lead to sogginess, clump formation, etc. owing to concentration gradient. Gulati et al. (2015) conducted a study on storage of sandwich system containing bread and chicken. It was observed that bread gained the moisture and chicken loosed the moisture to attain equilibrium conditions of water activity. Another study by O'Connor et al. (2017) demonstrated that a cheese sandwiched in breads led to decrease firmness of breads due to moisture migration.

12.4.3 Thermal Properties

Thermal properties influence the heat transfer mechanisms in foods, for example, thermal conductivity influences the diffusion of thermal energy regions at high temperature to region at low temperature. For high moisture foods, thermal conductivity approaches to that of pure water, whereas in dried and porous foods, thermal conductivity depends upon the presence of air. Specific heat is highly dependent upon the composition of foods. Other properties which have significant impact on moisture migration process are as follows:

- **Specific heat:** It is a measure of quantity of heat gained or lost to cause one unit change in the temperature of food material with negligible change in state and is calculated as follows:

$$C_p = \frac{q}{m\Delta T} \tag{12.11}$$

where q is the quantity of heat gained or lost (kJ), mass of product (kg), ΔT is the change in temperature (°C), and C_p is the specific heat (kJ kg^{-1} °C^{-1}).

- **Thermal conductivity**: This term represents the ability of food to transfer heat by measuring quantity of heat transferred through a material of unit thickness with unit temperature gradient and per unit time as described in the equation.
- **Thermal diffusivity:** It represents how quickly material transfers heat in relation to its ability to store it. Thermal diffusivity can be expressed as

$$\alpha = \frac{k}{\rho \cdot C_p} \tag{12.12}$$

where k is the thermal conductivity of food material, ρ is the density, C_p is the specific heat, and α is the thermal diffusivity (m^2s^{-1}).

12.4.4 Surrounding Environment

- **Temperature:** Temperature of packaged product, its surroundings within the package as well as outside the package, controls the moisture migration. For every 10 °C rise in temperature, biological reaction rate increases to 2–3-folds. At higher temperatures, enzymatic denaturation results in reduced respiration rates. However, at low temperatures, physiological injuries cause respiration rates to rise. Rise in temperature alters the barrier properties of films as it leads to volume expansion (Lim et al. 2001). Permeability of films also changes following Eq. (12.20) due to dependence on temperature. This expanded volume affects the gas composition over the food product and hence the biochemical, physical, and microbial food quality. As documented by Bertuzzi et al. (2007), diffusion and solubility phenomenon are dependent upon temperature and obey Arrhenius law and Van't Hoff law, respectively. When temperature varies over a narrow range, Arrhenius law can explain the effect of temperature on solubility and diffusivity using following equations:

$$S = S_0 \exp\left(\frac{\Delta H_s}{RT}\right) \tag{12.13}$$

$$D = D_0 \exp\left(\frac{-E_D}{RT}\right) \tag{12.14}$$

where E_D is the apparent activation energies for diffusion (kJ/mol), ΔH_s is the enthalpy change of sorption (kJ/mol), R is the gas constant (8.314 kJ/mol K), T is the the absolute temperature (K).

- **Relative humidity:** An increase in relative humidity results in an increase in effective diffusivity as reported by Guillard et al. (2013) and Turan (2019). Relative humidity represents the amount of water vapors in air in percentage of total needed to attain saturated conditions of air. As the concentration of water vapors increases in surroundings, a gradient results in diffusion of moisture to the region of lower concentration.

12.5 Model Components and Parameters

The basic components to be considered for modeling of moisture through polysaccharide films are presented in Fig. 12.1, where a food material packaged into edible films is exposed to surroundings during storage. Models can be developed on the basis of factors such as foods and surrounding environment affecting the diffusion of moisture. While establishing a model, after defining the model components, the next important consideration is the parameters. Model equations include parameters related to food, surroundings (as discussed in previous section), packaging material, and package properties as discussed below:

12.5.1 Packaging Material

- **Oxygen transmission rate:** The storability of packaged foods such as fruits and vegetables depends upon the efficiency of polysaccharide-based films to avoid transfer of oxygen through these films. Oxygen permeation through a material occurs when a gas molecule adsorbs at interface 1, diffuses through the material, and desorbs at interface 2 as shown in Fig. 12.2 (Wang et al. 2017).
  Oxygen is detrimental to the quality of food products containing lipids as these compounds get easily oxidized in the presence of oxygen. Oxygen permeability coefficient (OPC) gives measure of oxygen transmission rate by measuring amount of oxygen permeation through film per unit area and time (kg.m/(m^2.s. Pa)). Oxygen transmission rate (OTR) is measured by an instrument following the ASTM D1434-82 method at atmospheric pressure and 23 °C. Using this OTR, OPC can be calculated using the following equation:

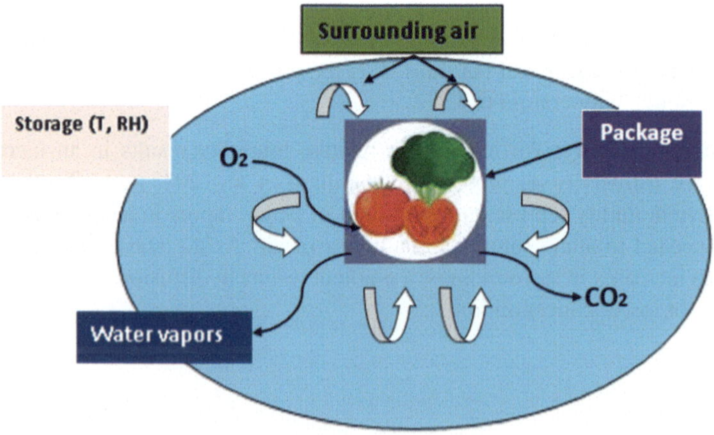

Fig. 12.1 Conceptual model for moisture transfer

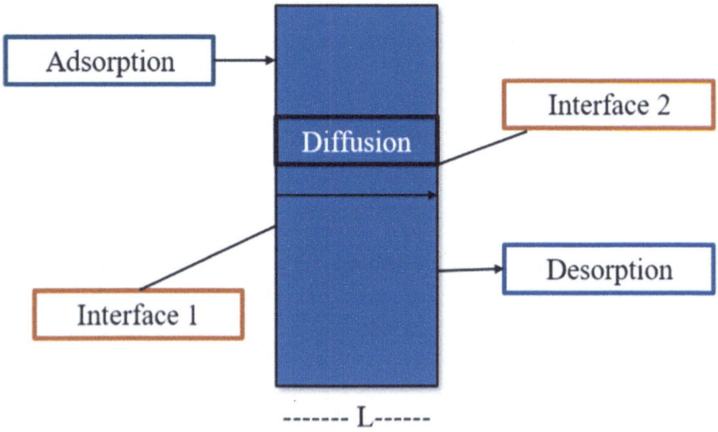

Fig. 12.2 Oxygen permeation through film

$$\mathrm{OPC} = \frac{\mathrm{OTR} \times \mathrm{thickness}}{\Delta P} \tag{12.15}$$

ΔP defines the difference in partial pressure of oxygen. Lower values of OPC indicate smaller concentrations of oxygen inside the package, hence extending the shelf life of foods prone to oxidation.

- **Water vapor permeability:** Moisture permeating through packaging materials during storage results in changing food texture, color, and sensory attributes of a food (Maria Rubino et al. 2001). Therefore, measurement of water vapor transmission rate through films is a matter of concern (Hu et al. 2001). Water vapor permeability (WVP) defines the barrier of film against permeability of water vapor through film. The most common method to determine WVP is ASTM E96/E96M gravimetric method, also known as wet cup method. As per this method, a test film is restrained and sealed above distilled water in an acrylic cup in a container, exposing the film to high humidity. The whole setup is weighed and placed in desiccators with desiccant to provide low relative humidity. Air movement is provided by fan to ensure uniform conditions over the film. The ambient temperature is maintained at 23 °C. The water transmission occurs due to difference in partial pressure on both sides of film. Weight change with time is measured periodically to obtain water vapor transmission rate (WVTR) of the film using equation:

$$W_t = (P_t - P_i) - (C_t - C_i) \tag{12.16}$$

where $W_t$is the net weight change of desiccant, P_t is the weight of pouch + desiccant at time t, P_i is the initial weight of pouch + desiccant, C_t is the weight of control pouch at time t, C_i initial weight of control pouch.

A graph obtained is plotted between net weight change of desiccant and time. From the slope of this graph, the transmission rate can be calculated as follows:

$$\text{Transmission rate}\left(\text{g}\cdot\text{m}^{-2}\cdot\text{day}^{-1}\right)=\frac{\text{Slope}}{\text{Area}}=\frac{\frac{W_t}{t}}{A} \quad (12.17)$$

$$\text{Permeance}\left(\text{g}\cdot\text{m}^{-2}\cdot\text{day}^{-1}\cdot\text{mmHg}^{-1}\right)=\frac{\text{Transmission rate}}{\text{pressure gradient}} \quad (12.18)$$

$$\text{Permeability}\left(\text{g}\cdot\mu\text{m}\cdot\text{m}^{-2}\cdot\text{day}^{-1}\cdot\text{mmHg}^{-1}\right)=\text{Thickness}\times\text{Permeance} \quad (12.19)$$

$$\text{WVP}=\text{WVP}_0\exp\left(\frac{-E_p}{RT}\right) \quad (12.20)$$

where E_P = apparent activation energy for permeation phenomena (kJ/mol).

As per permeability definition, energies are related as

$$E_p=E_D+\Delta H_s \quad (12.21)$$

Polysaccharide-based films have poor water vapor permeability. Therefore, a lot of research is being conducted on the modification of these films by incorporating mineral fillers as documented by Spence et al. (2011) in layered microfibrillated cellulose films. The addition of fillers resulted in lowering of WVP of films as compared to low-density polyethylene packaging films. Another study by Parra et al. (2004) demonstrated that cassava starch films blended with plasticizers such as glycerol, polyethylene glycol, and cross-linking agent glutaraldehyde exhibited better elongation with 0.3 g of polyethylene glycol and less WVP with 0.5 g of glutaraldehyde owing to cross-linking effect. Similarly, other researchers have conducted studies on film barriers against water vapor which are tabulated in Table 12.1.

- **Thickness of Films**

  The barrier properties such as oxygen transmission rate and water vapor transmission rate are highly influenced by thickness of films as shown in Eq. (12.19). Effects of film thickness on the barrier properties vary depending upon the composition (flavors, antimicrobial agents, antioxidants, emulsifiers, surfactants, plasticizers) of film. Bravin et al. (2006) studied the effect of thickness (30–90 μm) their functional properties of polysaccharide-lipid based edible films. No changes were observed in WVTR of films. However, WVP increased along with increasing film thickness (Fig. 12.3). WVP of starch-methylcellulose film without oil had higher dependence on thickness. Williams and Mittal (1999) found that the increase in thickness of polysaccharide films led to increase the diffusivity of water and fat. This variation may be attributed due to structural changes and water-film interactions leading to swelling of film due to hydrophilic nature of polysaccharide films. Apart from this, the plasticizing effect of water on the film matrix increases the matrix density and hence fast-moving chain extremities for films having 25% chitosan and 75% gelatin (Benbettaïeb et al. 2014).

Table 12.1 Models for determination of water vapor permeability of polysaccharide-based films

Polysaccharide	WVP	Salient findings	References
Banana starch	Starch: 8.09×10^{-11} g/m s Pa Starch with nanoparticles: 2.24 g/m s Pa	0.5 g of nanoparticles from plantain rachis in 11.5 g of banana starch gave good barrier against water vapor	García-Ramon et al. (2021)
Cellulose	Cellulose: 0.223×10^{-7} g mm^{-2} s^{-1} Pa^{-1} Cellulose with nanocapsules 0.09–0.55 × 10–7 g mm^{-2} s^{-1} Pa^{-1}	α-tocopherol nanocapsules in cellulose films lowered water vapor permeability	Mirzaei-Mohkam et al. (2020)
Chitosan	Chitosan: 19.50 g mm/kPa h m^2 Chitosan + carrot fibers: 13.39–18.14 g mm/kPa h m^2	Higher concentrations of fiber cellulose reduced the water vapor permeability of film	Sogut and Cakmak (2020)
Cassava starch	0.19–0.48 g mm m^{-2} h^{-1} kPa^{-1}	Lower wettability and good mechanical properties; suitable for packaging of fruits and vegetables	Luchese et al. (2018)
Cassava starch	Starch: 1.15 g mm Kpa^{-1} h^{-1} m^{-2} Starch with nanocomposite: 0.41–0.66 g mm Kpa^{-1} h^{-1} m^{-2}	Starch and cellulose nanoparticles improved the barrier against moisture transfer	Santana et al. (2018)
Regenerated cellulose	Regenerated cellulose: 2.28×10^{-9}g m/m^2 Pa s Regenerated cellulose with nanocomposite films: 1.62–1.9428 × 10^{-9} g m/m^2 Pa s	2% Nano-SiO_2 by weight resulted in improved tensile and thermal properties as well as water vapor barrier properties	Reddy et al. (2018)
Alginate	Alginate: 2.0×10^{-4} g m/m^2 Pa s Alginate + plasticizers: 1.79–2.11 × 10^{-4} g m/m^2 Pa s	Alginate+diethylene glycol film provided minimum water vapor permeability	El Miri et al. (2018)
Potato starch	Potato starch: 5.0 g mm/kPa day m^2 Potato starch + cellulose, nanoclay, and their combination 1.19–7.1 g mm/kPa day m^2	Nanoclay and cellulose fibers reduced the water vapor permeability	El Halal et al. (2017)

12.5.2 Package Properties

Package properties such as size, food package void volume, and perforations contribute to great extent in establishing moisture transfer models for food systems. The dimensions of packages such as length, breadth, height, and exposed area influence the moisture migration. A package with more surface area will provide

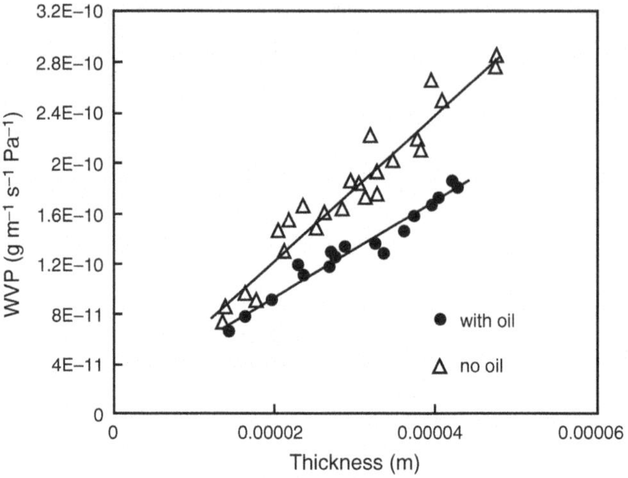

Fig. 12.3 Effect of thickness on WVP of starch-methylcellulose film with oil and without oil (Bravin et al. 2006)

more space for diffusion phenomenon. The unoccupied volume inside the package serves as a medium for gas or water vapor migration. Also composition of gases within this volume has huge impact on storability of food products. More is the void volume; more time is needed to attain steady state by diffusion. Food packages which are having equal volume to foods packed inside them such as vacuum packaging provide negligible medium to transfer moisture through packages. There is another technique of using perforated films in modified atmosphere technology for prolonging shelf life of fresh produce. This controls the permeation of gases and water vapors through the films at a rate which is in balance with the outer environment (Paul and Clarke 2002).

12.6 Modeling Approaches

Models explaining the moisture transfer can be categorized by considering the type of foods as sorption studies demonstrate adsorption and desorption of moisture in intermediate moisture or processed foods. Equations used to determine the transpiration and respiration rates are suitable for fresh produce as discussed in the following section.

12.6.1 Sorption Models

As polysaccharide films are hydrophilic in nature, it is necessary to study their sorption behavior to examine effectiveness for food packaging applications. These films are susceptible to absorption or desorption of moisture depending upon the temperature and relative humidity of surrounding air.

There are various techniques to describe the relationship between the water content and the surroundings. One method is through sorption phenomenon conducted at a constant temperature and under equilibrium conditions (Perdomo et al. 2009) using desiccators. In this traditional method, samples placed in desiccators equilibrate to a known relative humidity at a particular temperature. This relative humidity is maintained by using saturated salt solutions, acids, humidifiers, or air mixtures. Sample weight is examined until weight change stops. The resulting moisture of samples is known as equilibrium moisture content at which sample does not gain or lose moisture. Moisture sorption isotherm is obtained by plotting the moisture content and a_w values at a constant temperature. As this method is time-consuming, other automated fast systems such as dynamic vapor sorption are being used. An internal programming automatically changes a_w of sample progressively once equilibrium is achieved. Dynamic dew point isotherm method is independent of equilibrium state as adsorption is achieved using wet air and desorption occurs as desiccated air passes over sample.

There are around 200 models that have to describe sorption isotherm, and not even single equation can describe the sorption behavior accurately over whole range of water activity and temperature for the hydrophilic polysaccharide-based materials (Zhang et al. 2014). This is due to the dependency of sorption isotherm on the structural arrangements of the polysaccharide and the sorption system under different environmental conditions (Mali et al. 2005). The common models used for describing sorption behavior of biological materials are given in Table 12.2. The Guggenheim, Anderson, and de Boer (GAB) model is considered best as it covers the whole range of water activity. Brunauer, Emmett, and Teller (BET) model is also one of the popular models. However, it is of limited use in describing sorption phenomenon of polysaccharide-based films as it is applicable in 0.11–0.55 a_w range. These films dissolve once a_w crosses 0.7, and within the applicability range of BET, there is no dissolution of film structure (Mathlouthi 2001). Using BET model, the monolayer moisture content can be determined, which is considered as the most stable moisture content. At this level, the chemical reaction is not effective to deteriorate film quality (Zimeri 2002). The specific surface area, which describes the surface characteristics and water-binding properties of the film, can be determined from Eq. (12.22) using monolayer moisture content (Cassini et al. 2006; Rosa et al. 2010):

Table 12.2 Common mathematical models for sorption studies

S. No.	Name	Expression	a_w range
1.	Guggenheim, Anderson, and de Boer (GAB)	$\text{EMC} = \frac{M_O C K a_w}{(1 - K a_w)(1 - K a_w + C K a_w)}$	<0.95
2.	Brunauer, Emmett, and Teller (BET)	$\frac{\text{EMC}}{M_0} = \frac{C a_w}{(1 - a_w)(1 - a_w + C a_w)}$	<0.50
3.	Modified Halsey	$\text{EMC} = \left[\frac{\ln(a_w)}{-\exp(A+BT)}\right]^{\left(\frac{1}{-C}\right)}$	0.10–0.80
4.	Modified Oswin	$\text{EMC} = (A + BT)\left[\frac{a_w}{1 - a_w}\right]^{\left(\frac{1}{C}\right)}$	NA
5.	Modified-Henderson	$\text{EMC} = \left[\frac{\ln(1 - a_w)}{-A(T+C)}\right]^{\left(\frac{1}{B}\right)}$	0.50–0.95
6.	Smith	$\text{EMC} = A + B \times \ln(1 - a_w)$	0.50–0.95

EMC is equilibrium moisture content (g water/g dry weight), a_w is water activity, M_O is monolayer moisture content, and A, B, and C are model constants

$$S_O = M_o \times \frac{N_0 \times A_{H_2O}}{M_{H_2O}} = 3.5 \times 10^3 \times M_o \tag{12.22}$$

where S_O is the specific surface area ($m^2\ g^{-1}$), M_O is the monolayer moisture (dry base), M_{H2O} is the molecular weight of water (18 g mol^{-1}), N_O is the Avogadro's constant (6×10^{23} mol^{-1}), and A_{H2O} is the area of one water molecular (1.06×10^{-19} m^2).

- **Model fitting:** The goodness of fitted models can be evaluated on the basis of residual sum of squares (RSS), root mean square error (RMSE), and correlation coefficient (R^2) using Eqs. (12.23, 12.24, and 12.25). A model is considered best when the root mean square value is minimum and correlation coefficient approaches to 1. Moreover, plots of residuals are also indicator of fitness of model. A random residual plot provides best fit for the sorption studies:

$$\text{RSS} = \sum (\text{EMC} - \text{PMC})^2 \tag{12.23}$$

$$\text{RMSE} = \frac{\sum (\text{EMC} - \text{PMC})^2}{n} \tag{12.24}$$

$$R^2 = 1 - \sqrt{\frac{\sum (\text{EMC} - \text{PMC})^2}{\sum (\text{EMC} - \overline{\text{EMC}})^2}} \tag{12.25}$$

where PMC is a predicted moisture content.

The impact of temperature variation on sorption behavior is evaluated using Clausiuse-Clapeyron equation given as follows:

$$\frac{-q_{st}}{R} = \left[\frac{dln(a_w)}{d\left(\frac{1}{T}\right)}\right] \tag{12.26}$$

where q_{st} is the net isosteric heat of sorption ($Jmol^{-1}$), R is universal gas constant ($8.314\ J\ mol^{-1}\ K^{-1}$), and T is absolute temperature (K).

Moisture content and water activity of permeable films is an important characteristic that defines its use and suitability for packaging of both fresh produce and processed food products. Fresh foods packaging requirements vary in comparison to those for processed foods. Fresh foods continue to respire after harvesting. Transpiration along with respiration causes condensation problems inside the package. It can alter the functioning of packaging films due to exposure to excess moisture inside the package from the fresh foods. Thus, in such case, film selection should be made considering these factors. In case of processed foods, final products from processing units are manufactured in such a way to have maximum stability on the basis of set water activity and moisture content. Packaging requirements demand those bio-based films which can avoid disturbance of these food properties by providing barrier against external environmental conditions. Owing to these, film properties need to be examined prior to its application on foods.

Various studies on sorption behavior of films are documented and some of them are shown in Table 12.3. Dias et al. (2010) reported the sorption behavior of films made from rice flour and rice starch, and sorption data was fitted to GAB model. Another study on hydroxypropyl methylcellulose-based composite (HPMC) films by Klangmuang and Sothornvit (2016) revealed that for same a_w, HPMC composite film with beeswax showed lower moisture content than those with nanoclay. This may be attributed due to less water binding ability of beeswax in contrast to nanoclay.

12.6.2 Respiration

Respiration is a process by which plants acquire energy. During this process, plants produce oxygen by breakdown of carbon dioxide as a part of photosynthesis process. The energy produced during respiration is utilized for complete oxidation of a mole of hexose (Taiz and Zeiger 1998). In some research studies (Kaur et al. 2010; Singh et al. 2012; Devgan et al. 2019; Singla et al. 2020), respiration rate was expressed using partial pressure of gases in the headspace of package at particular temperatures as follows:

$$R_{O_2} = \frac{\left(p^0_{O_2} - p^t_{O_2}\right) \times V_v}{100 \times W \times (t_t - t_0)} \tag{12.30}$$

Table 12.3 Models for determination of moisture characteristics of polysaccharide films based upon sorption

Film	Model used	Environmental conditions	Salient findings	References
Potato starch blended with polyvinyl alcohol	GAB	11.28–90.10 a_w at 30 °C	Moisture adsorption as well as M_O of blended films increased as the proportion of polyvinyl alcohol increased	Gómez-Aldapa et al. (2020)
Carboxymethyl cellulose films enriched with spent coffee grounds	GAB	0.11, 0.33, 0.53, 0.75, and 0.97 a_w at 25 °C	Lower M_O value was obtained for the film with 0.20% of polysaccharides extracted through hydrolysis from spent coffee, sorption behavior typical of hydrophilic films	Ballesteros et al. (2018)
Mango kernel starch with plasticizers	GAB	0.11–0.90 a_w at 25 °C	Moisture contents of both unplasticized and plasticized films were found to be similar till 0.6 a_w	Nawab et al. (2016)
			The order of sensitivity of the films for a_w was found to be glycerol > glycerol þ sorbitol > sorbitol > control	
			Lowest M_O was found in sorbitol (0.128)	
Pea starch	BET, GAB, Peleg, Oswin, Ferro-Fontan, Henderson, Lewicki, Iglesias-Chirife, and Flory-Huggins	0.11–0.96 a_w at 5, 15, 25, and 40 °C	Sigmoidal shape curve, M_O varied between 0.061 and 0.036 g water/g dry solids for BET model and from 0.105 to 0.050 g water/g dry solids for the GAB model	Saberi et al. (2015)
Starch blended with polyethylene oxide	BET, Smith, Halsey, Oswin, Henderson, Kuhn, Iglesias and Chirife, and Freundlich	0.11–0.92 a_w at 27 + 1 °C	Lowest M_O, that is, 4.98% found in polyethylene oxide/starch films (%wt) 80/20 with R^2 0.99	Jagadish and Raj (2013)
Sago starch	GAB, Halsey, Oswin, and Henderson	0.23–0.90 a_w at 23, 31, and 37 °C	Sigmoidal shape curve, fitness order of models was Halsey>Oswin>GAB>Henderson	Bajpai et al. (2011)
Corn starch	GAB, Peleg, and modified Oswin			

		0.1–0.9 and at three temperatures, namely, 25, 35, and 45 °C	Edible starch films with and without antimicrobial agents type II BET isotherm; modified Oswin model fitted best	Chowdhury and Das (2010)
Pea starch with plasticizers	BET, GAB, Smith, and Flory-Huggins	0.12–0.85a_w at 25 °C	GAB model fitted all kinds of films well, except for glucose and sorbitol-plasticized films	Zhang and Han (2008)

$$R_{CO_2} = \frac{\left(p^0_{CO_2} - p^t_{CO_2}\right) \times V_v}{100 \times W \times (t_t - t_0)} \quad (12.31)$$

where R_{O_2} and R_{CO_2} represent respiration rates (mL kg^{-1} h^{-1}), $p^0_{O_2}$ and $p^0_{CO_2}$ partial pressures of O_2 and CO_2, respectively, at the initial stage (kPa), $p^t_{O_2}$ and $p^t_{CO_2}$ partial pressures of O_2 and CO_2, respectively, at the final stage (kPa), V_v void volume (ml), W weight of food (kg), t_t time at the final stage (h), t_0 time at the initial stage (h).

The respiratory quotient which is the ratio of evolution of CO_2 and consumption of O_2 can be expressed (Torrieri et al. 2010; Devgan et al. 2019) as

$$RQ = \frac{R_{CO_2}}{R_{O_2}} \quad (12.32)$$

Alternate models based on Michaelis-Menten equations provide good insight about role of concentrations of O_2 and CO_2 in respiration (Helena Gomes et al. 2010; Ho et al. 2013). The following equation gives the simplest method of estimation of rate of oxygen consumption on the basis of a limiting enzymatic reaction using O_2 as substrate while assuming no inhibition of rate of oxygen consumption by evolved CO_2:

$$R_{O_2} = \frac{R^{max}_{O_2} m_{O_2}}{k_m + m_{O_2}} \quad (12.33)$$

where R_{O_2} is the consumption rate of O_2 (cm^3/kg day), $R^{max}_{O_2}$ is the maximum rate of O_2 consumption (cm^3/kg day), K_m is the enzyme-substrate complex dissociation constant.

On the basis of inhibition of rate of oxygen consumption on account of generated CO_2, these equations can be modified as

Competitive inhibition:

$$R_{O_2} = \frac{R^{max}_{O_2} p^{in}_{O_2}}{k_{mO_2}\left(1 + \frac{p^{in}_{CO_2}}{k_{mcCO_2}}\right) + p^{in}_{O_2}} \quad (12.34)$$

Uncompetitive inhibition:

$$R_{O_2} = \frac{R^{max}_{O_2} p^{in}_{O_2}}{k_{mO_2} + p^{in}_{O_2}\left(1 + \frac{p^{in}_{CO_2}}{k_{muCO_2}}\right)} \quad (12.35)$$

Noncompetitive inhibition:

$$R_{O_2} = \frac{R_{O_2}^{\max} p_{O_2}^{\text{in}}}{k_{mO_2} + p_{O_2}^{\text{in}} \left(1 + \frac{p_{CO_2}^{\text{in}}}{k_{mnCO_2}}\right)} \tag{12.36}$$

where R_{O_2} is the maximum consumption rate of oxygen (mL kg^{-1} h^{-1}), k_{mO_2} is the constant for oxygen consumption, k_{mcCO_2} is the constant for competitive enzyme inhibition, k_{muCO_2} is the constant for uncompetitive enzyme inhibition, k_{mnCO_2} is the constant for noncompetitive inhibition.

These constants can be calculated using nonlinear regression of respiration data. Inhibition constant defines a value to which respiration rate is inhibited by carbon dioxide. Singh et al. (2012) and Kaur et al. (2010) conducted studies on baby corn and spinach, respectively, in which they analyzed the respiration behavior by enzyme kinetics on the assumption of mixed inhibition using the following equation given by Peppelenbos and Van't Leven (2017):

$$R_{O_2} = \frac{R_{O_2}^{\max} \times p_{O_2}^{\text{in}}}{k_{mO_2}\left(1 + \frac{p_{CO_2}^{\text{in}}}{k_{mcCO_2}}\right) + p_{O_2}^{\text{in}} \left(1 + \frac{p_{CO_2}^{\text{in}}}{k_{muCO_2}}\right)} \tag{12.37}$$

Respiration rates of fresh produce are influenced by both internal and external factors. Internal factors have lot of complexity on the basis of different types of food and are difficult to analyze. However, external factor such as temperature controls the respiration rate to a large extent. This dependence of respiration rate on temperature can be described using Arrhenius law for fresh foods (Fonseca et al. 2002). Respiration rates on the basis of gaseous composition and parameters used in enzyme kinetic model can be calculated as follows (Kaur et al. 2010; Singh et al. 2012; Singla et al. 2020):

$$R_{O_2} = R_{O_2}^{*} \times e^{\frac{Ea_1}{R}\left(\frac{1}{T} - \frac{1}{T_{\text{avg}}}\right)} \tag{12.38}$$

$$R_{co_2} = R_{CO_2}^{*} \times e^{-\frac{Ea2}{R}\left(\frac{1}{T} - \frac{1}{T_{\text{avg}}}\right)} \tag{12.39}$$

$$RQ = RQ^{*} \times e^{-\frac{Ea_3}{R}\left(\frac{1}{T} - \frac{1}{T_{\text{avg}}}\right)} \tag{12.40}$$

$$R_{O_2}^{\max} = R_{O_2}^{\max *} \times e^{-\frac{Ea_4}{R}\left(\frac{1}{T} - \frac{1}{T_{\text{avg}}}\right)} \tag{12.41}$$

$$k_{mO_2} = k_{mO_2}^{*} \times e^{-\frac{Ea_5}{R}\left(\frac{1}{T} - \frac{1}{T_{\text{avg}}}\right)} \tag{12.42}$$

$$k_{mcCO_2} = k^*_{mCO_2} \times e^{\frac{-Ea_6}{R}\left(\frac{1}{T} - \frac{1}{T_{avg}}\right)} \tag{12.43}$$

$$k_{muCO_2} = k^*_{muCO_2} \times e^{\frac{-Ea_7}{R}\left(\frac{1}{T} - \frac{1}{T_{avg}}\right)} \tag{12.44}$$

where $k^*_{mO_2}, k^*_{mCO_2}, k^*_{muCO_2}$ are pre-exponential enzyme kinetics parameters, $-Ea_1, -Ea_2, -Ea_3, -Ea_4, -Ea_5, -Ea_6$, and $-Ea_7$ are activation energies (kJg^{-1} mol^{-1}), R is universal gas constant (8.314 kJ g^{-1}mol^{-1} K^{-1}), T is storage temperature (K), T_{avg} is average of experimental temperatures (K).

12.6.3 Transpiration

Transpiration is a process of water movement in the plants in which water transfers through intracellular spaces and skin of fruits and vegetables. This moisture evaporates from the skin surface to the surroundings (Veraverbeke et al. 2003). Thus, commonly it is referred as moisture loss or transpiration rate (Maguire et al. 2001). This water loss from the food leads to condensation problems, shriveling, wiling, and reduced glossiness. Transpiration process depends upon concentration difference in terms of water activity or water vapor pressure between food product and outer environment. Considering this, at 100% relative humidity, transpiration rate must be zero. However, this is not true in actual conditions as transpiration takes place as respiration process releases heat (Tano et al. 2005). Thus, this complex phenomenon of transpiration at saturated conditions includes various types of heats as follows:

1. Heat of respiration produced by food product
2. Heat of evaporation due to cooling at surface of food
3. Heat transfer due to convection between food and surroundings

12.6.4 Two Approaches Have Been Usually Adopted to Model the Transpiration Rates

12.6.5 Theoretical

The transpiration rate on the basis of mass is a function of vapor pressure gradient and resistance as given in following equation (Leonardi et al. 2000; Sastry 1985):

$$TR_M = k_T(P_{FS} - P_A) \tag{12.45}$$

where TR_M is the transpiration rate the basis of mass (mg/kg.s), k_T is the transpiration coefficient (mg/kg.s.MPa), P_{FS} is the water vapor pressure at the surface of food (MPa), P_A is the ambient vapor pressure (MPa).

The inverse of transpiration coefficient, that is, $1/k_T$, represents resistance, and it can be expressed as

$$\frac{1}{k_T} = \frac{1}{k_{FS}} + \frac{1}{k_A} \tag{12.46}$$

where k_{FS} is the food skin mass transfer coefficient (mg/kg.s.MPa), k_A is the air mass transfer coefficient or convective mass transfer (mg/kg.s.MPa).

Combining these two equations,

$$TR_M = \frac{(P_{FS} - P_A)}{\frac{1}{k_{FS}} + \frac{1}{k_A}} \tag{12.47}$$

Mahajan et al. (2016) modified Eq. (12.45) as when P_A or water activity of surroundings becomes greater than 0.984 (which is generally considered as water activity of fresh fruits and vegetables), transpiration rate will be zero. Therefore, a modified version of this equation includes the term describing transpiration rate due to generation of respiratory heat.

$$TR_M = k_T(a_{wF} - a_{wA}) \cdot \left(1 - e^{-aT}\right) \tag{12.48}$$

where a_{wF} and a_{wA} represent water activities of food and surroundings, k_T and T constants depend upon product.

12.6.6 Gravimetric

This is the commonly used technique to find out mass loss by weighing the foods after defined interval under given conditions of temperature and relative humidity. This mass loss or transpiration rate can be expressed in terms of surface area or initial mass of the food:

$$TR_{SA} = \frac{(M_0 - M_t)}{t \cdot A_s} \tag{12.49}$$

$$TR_M = \frac{(M_0 - M_t)}{t \cdot M_0} \tag{12.50}$$

where TR_{SA} expressed as mg/cm^2 h or mg/cm^2 s and TR_M as g/kg.h or mg/kg.h or mg/kg.s, M_0 is the initial mass of food, M_t is the mass of food after defined interval, t is the time interval, $A_s$surface area of food initially.

Xanthopoulos et al. (2014) modeled the transpiration process of grape tomatoes at three temperatures, that is, 10, 15, and 20 °C, and relative humidity levels 70%, 80%, and 92% using the above equations. These models provided satisfactory fit to observations recorded from experiments. Caleb et al. (2013) examined the transpiration rate of pomegranate (*Punica granatum L.*) using conditions such as temperature at 5, 10, and 15 °C and relative humidity of 76%, 86%, and 96%. Maximum loss in weight was found for the combination 15 °C and 76% RH. Bovi et al. (2018) evaluated transpiration phenomenon of both packed and unpacked strawberries stored at 4, 12, and 20 °C temperature and 76%, 86%, 96%, and 100% relative humidity. For unpackaged strawberries, a combination of Arrhenius equation and Fick's first law of diffusion-based model was used, whereas for packaged material, a model was developed considering the total mass loss owing to transfer of volatiles and degree of filling using linear regression.

12.7 Moisture Transfer in Food Systems: Case Studies

Foods carrying various components such as biscuits with cream filling, hard candies with liquid at the center, etc. are highly in demand. Such compositions having varied proportions of different ingredients with different a_w are challenging to preserve for longer periods. Foods packed in permeable films keep on exchanging water with outside environment, which can be monitored by controlling the package permeability, moisture diffusion within components, and sorption characteristics of foods packed inside these films. The mathematical models to explain the moisture transfer process have been mostly applied for uncoated or foods packed in plastic films (Rumsey and Krochta 1994). Modeling data on moisture transfer in foods packed in edible films is scarce.

Based upon the model proposed by Chu et al. (2012) for two components, composite foods containing biscuits as dry and pork slice and a piece of hawthorn as wet components, unidirectional moisture transfer was considered firstly, within the food components and then between external and internal package environment. Moisture migration occurs due to moisture concentration gradient and thermodynamic parameters. The setup to study moisture transfer is shown in Fig. 12.4a, b.

12.7.1 Moisture Diffusion Between Food Components and Surrounding Air

According to Fick's law, considering dry component as thin sheet, diffusion between air and surface of thin dry food is calculated as

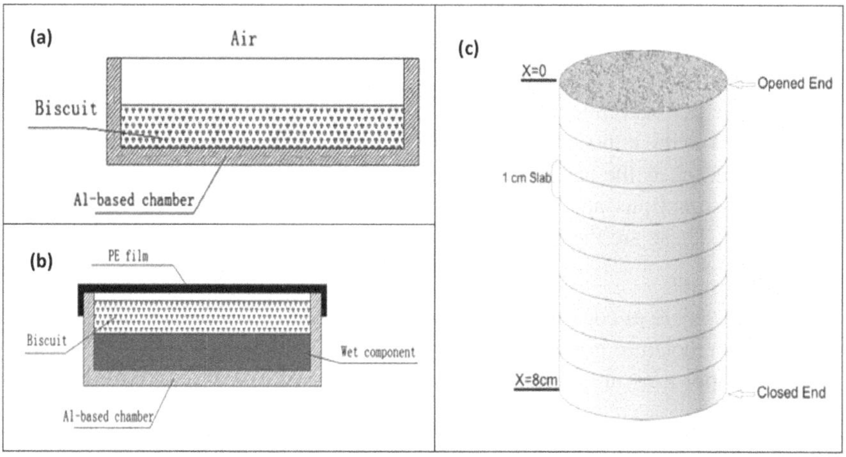

Fig. 12.4 Moisture transfer setup for (**a**) food component and surrounding air and (**b**) packaged multicomponent food and surrounding air (Chu et al. 2012) and (**c**) moisture transfer setup for packaged yerba mate (Ramallo and Albani 2007)

$$\frac{\partial X}{\partial t} = D_{\text{eff}} \frac{\partial^2 X}{\partial x^2} \tag{12.51}$$

where X is the moisture content (db), D_{eff} is the effective moisture diffusivity at a given RH, x is the distance from interface (m).

Equation (12.6) given by Crank (1975) is used to find solution of the above equation.

12.7.2 *Moisture Transfer Between Food and External Environment Through Packaging Film*

Assuming that the moisture from outside is absorbed by dry component only, a pseudo-steady-state equation derived from Henry and Fick's law can be used to estimate moisture migration through packaging film as follows:

$$\frac{W \cdot \mathrm{d}x}{\mathrm{d}t} = \frac{P \cdot P_0 \cdot S}{w} \left(RH_p - RH_s \right) \tag{12.52}$$

where W is the dry weight of food, dx/dt is the rate of moisture gain, P is the permeability of packaging film, P_0 is the saturated vapor pressure, S is the effective area for diffusion, RH_p is the relative humidity inside the package, RH_s is the relative humidity of surrounding environment, w thickness of film.

Robertson (1993) gave another equation to calculate moisture transfer through packaging material as

$$P = \frac{Q \cdot w}{A \cdot t(p_{\text{out}} - p_{\text{in}})} \tag{12.53}$$

where P is the permeability coefficient, Q is the permeant passing through w thickness of film, A is the exposed area, t is the time, p_{out} is the partial pressure on outer surface of film, p_{in} is the partial pressure on inner surface of film.

Certain assumptions are made to determine the moisture transfer in packaged food systems. These are the following:

- P/w is constant.
- Surroundings are at constant temperature and RH.
- There is no moisture gradient inside the package.
- Internal vapor pressure, that is, p_{in}, is constant.

p_{in} is a function of moisture and is calculated from moisture sorption data. As it changes with time, Ramallo and Albani (2007) made the calculation using Eq. (12.51) for diffusion of moisture between yerba mate and open environment. To study the water uptake by packaged yerba mate, a system made up of eight compartments was used (Fig. 12.4c). However, for moisture transfer through packaging films as described above in Eq. (12.53), they introduced an empirical equation describing relationship between product moisture content and internal RH and made further calculations using finite difference methods for eight compartments based upon diffusion, permeability, and sorption isotherms. Two membranes of known permeability were used and food was placed side by side in each section with the origin of x-axis in the membrane.

For Section 1

$$\Delta W_P = \frac{P}{w} A\left(p_{\text{out}} - p_{\text{in}}^t\right)\Delta t \tag{12.54}$$

$$\Delta W_D = D_{\text{eff}}\left(X_e^t - X_1^t\right) A\rho \frac{\Delta t}{\Delta l} \tag{12.55}$$

$$X_i^{t+\Delta t} = \left[X_1^t W(\Delta W_P - \Delta W_D)\right] \frac{1}{W} \tag{12.56}$$

For sections 2–15

$$X_i^{t+\Delta t} = D_{\text{eff}}\left(X_{i+1}^t - 2X_i^t + X_{i-1}^t\right) \frac{\Delta t^2}{\Delta l^2} + X_i^t \tag{12.57}$$

For Section 16

$$X_{16}^{t+\Delta t} = D_{eff}\left(2X_{15}^t - 2X_{16}^t\right) \frac{\Delta t}{\Delta l^2} + X_{16}^t \tag{12.58}$$

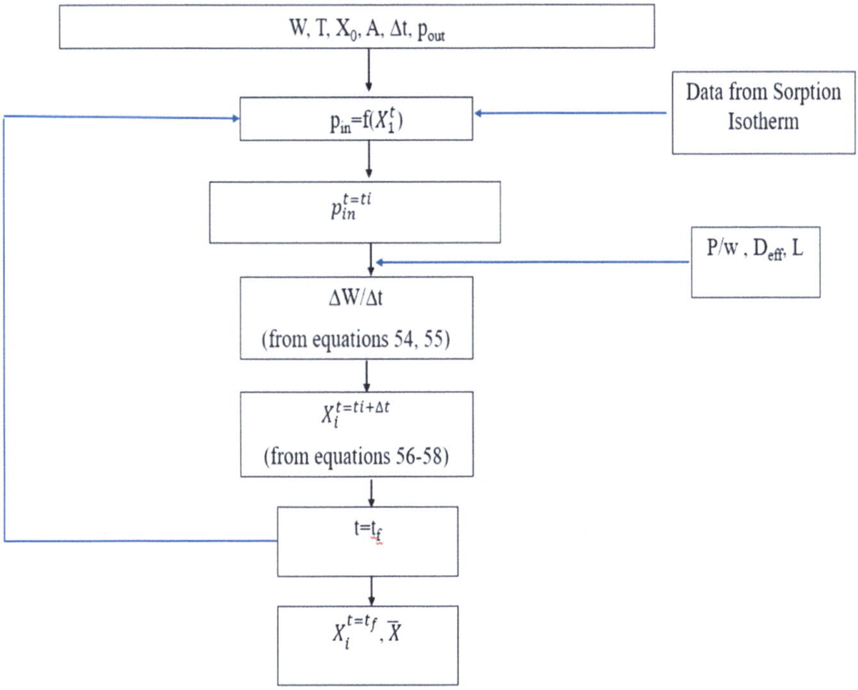

Fig. 12.5 Procedure to calculate moisture gain in yerba mate bed (Ramallo and Albani 2007)

where ΔW_P and ΔW_D are change in mass of food in Section 1 due to permeability and diffusion processes, respectively, and i and t represent sections and time, respectively. W is the weight of dry solids.

At T absolute temperature and X_0 initial moisture content, numerical procedures are carried out to calculate water uptake using arrangement shown in Fig. 12.5 for all sections with $\Delta l = 0.5$ cm.

This new moisture content ($X_i^{t=ti+\Delta t}$) postulates a new equilibrium vapor pressure achieved from the balance of water gain. The whole calculation process is repeated until the final time of the test (t_f) is reached.

12.7.3 Moisture Transfer Between Food Components

Modeling moisture transfer between different components of foods can be well understood by sorption isotherms of these components. For two component-based food systems, a moisture transfer will occur until they attain equilibrium conditions, that is, foods having same a_w level, as shown in Fig. 12.6 given below:

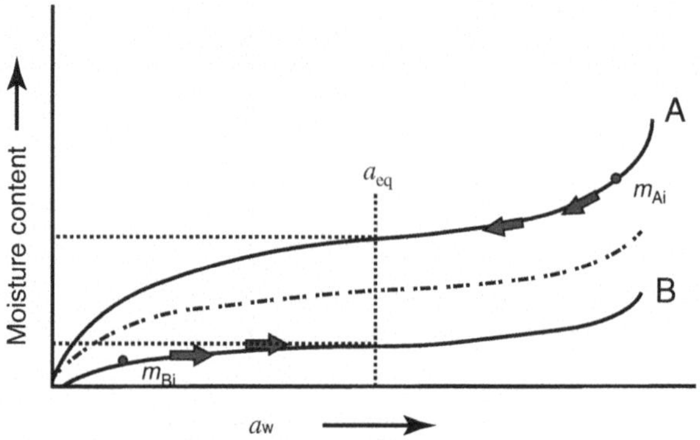

Fig. 12.6 Moisture sorption isotherm for a two-component system (A and B) (Chu et al. 2012)

Thus, from the sorption curve, a point having same a_w although different moisture contents represents the most stable condition for storage. For these foods with different components, mixture isotherm is given as (Xiong 2002)

$$X = f(A)X_{Ai} + f(B)X_{Bi} \tag{12.59}$$

where X is the total moisture content of multicomponent food, X_{Ai} is the initial moisture content of component A is the, X_{Bi} = initial moisture content of component B, $f(A)$ is the solids of A/total solids, $f(B)$ is the solids of B/total solids.

12.8 Shelf Life Prediction

On the basis of equations described above, shelf life of foods can be determined considering diffusion within the food components and moisture transfer from environment to food product in a package over the time using the following equation:

$$X(t) = X_w(t) + X_{\text{out}}(t) + X_d \tag{12.60}$$

where $X_w(t)$ is themoisture contribution by wet component of food, $X_{\text{out}}(t)$ is themoisture contribution due to environment, X_d is initial moisture content of dry component

Heiss and Eichner (1971) developed a model for shelf life prediction of packaged foods. According to this model, shelf life (S) can be calculated as follows:

$$S = \frac{\ln(X_e - X_i)/(X_e - X_c)}{P\left(\frac{A}{w}\right)\left(\frac{p_{\text{out}}}{\text{slope}}\right)} \tag{12.61}$$

Apart from this, on the basis of packaging film properties, shelf life simulation can be done as proposed by Hernandez and Giacin (1997):

$$\int_{t=0}^{t=t} dt = \frac{wW}{PA} \int_{X_i}^{X_c} \frac{dx}{(p_{\text{out}} - p_{\text{in}})} \tag{12.62}$$

where X_e is the equilibrium moisture content of foods, X_c is the critical moisture content of food, X_i is the initial moisture content of food, P is the permeability constant, S is the slope of sorption isotherm curve, A is the exposed area, p is the pressure.

12.9 Application of Polysaccharide-Based Films on Foods

12.9.1 Fresh Produce

Perishable foods such as raw fruits and vegetables remain metabolically active after harvesting. Water loss is a major concern in such foods due to respiration and transpiration (Bovi et al. 2016). Bio-based packaging films are playing a pivotal role in maintaining the quality of foods by slowing down the decay processes. Additionally, these films are being welcomed by the consumers due to their biodegradable, edible, and compatible properties in comparison to synthetic polymers. Polysaccharide films or coatings have been used by various researchers to access their applicability on the raw products.

A guar gum- and pea starch-based coating was applied to oranges by Saberi et al. (2018). The quality of coated oranges was evaluated during storage to examine the effectiveness of edible coatings. Weight loss, firmness, respiration rate, ethylene production, color sensory analysis, and overall acceptability results demonstrated that respiration rates were lower, along with less weight loss, firmness loss, and flavor loss in coated fruits. Thus, it resulted in increasing shelf life of citrus fruits. Zhou et al. (2021) observed that film based on carboxymethyl chitosan (CMCS) and pullulan (Pul) with galangal essential oil (GEO) was highly effective in preserving quality of mangoes stored at 25 ± 1 °C in terms of titratable acidity, weight loss, firmness, and soluble solids for 15 days. A glance at Table 12.4 indicates the functionality of polysaccharide-based materials to maintain fresh-like quality of foods. Nawab et al. (2017) aimed to analyze the effect of mango kernel starch coatings with plasticizers on the shelf life of tomatoes at room temperature (20 °C). Results for titratable acidity, firmness, ascorbic acid content, soluble solid concentration, and decay percentage of coated tomatoes were significantly different from control as ripening process was delayed by coatings. Among different plasticizers, maximum effectiveness was observed with sorbitol followed by combination of glycerol: sorbitol and then glycerol. Sensory evaluation for color, texture, and aroma also proved that edible coating with sorbitol prolonged the ripening up to 20 days when stored at 20 °C. Li et al. (2017) and Sucheta et al. (2019) have reported on preservation effect of coatings on tomatoes. Guar gum edible coating was used by

Table 12.4 Polysaccharide-based film application on fresh foods

Food	Formulation of film	Salient findings	References
Guava	Gum arabic + cinnamon essential oil	10% gum arabic +1% cinnamon essential oil maintain carotenoids, chlorophylls, soluble solid color, firmness, content, Ph, and flavor index at 10 ± 1 °C, 90–95% RH for 28 days	Etemadipoor et al. (2019)
Fresh-cut apples	Fish gelatin and chitosan	Efficient for controlling microbial activity, maintaining firmness for four days of storage at 22 °C and 12 days at 5 °C	Shyu et al. (2019)
Strawberry	Chitosan+ thymus capitatus essential oil	Shelf life enhancement of the strawberries 15 days at 5 °C on the basis of Ph, titratable acidity, weight loss, decay index, and microbiological quality	Martínez et al. (2018)
Green bell pepper	Chitosan	Delayed a loss in firmness, weight loss, and vitamin C content and inhibited the growth of heterotrophic bacteria, mesophilic bacteria, yeast, and mold during the five weeks of storage	Adetunji et al. (2019)
Apricot	Chitosan and soybean protein isolate	Coating significantly reduced weight loss and prevented decrease in firmness and textural properties	Zhang et al. (2018)
Tomatoes	Chitosan	0.25% chitosan better retained Ph, acidity, ascorbic acid, weight loss, moisture, and reduced microbial growth for 30 days	Sucharitha et al. (2018)
Fresh-cut melons	Chitosan from mushroom waste	Enhance fruit firmness, inhibit off-flavors, and reduce the microbial counts	Poverenov et al. (2018)

Saha et al. (2016) to investigate its effect on shelf life of cucumber at 25 ± 2 °C and 70 ± 5%. The coating reduced the weight loss, decay loss, acidity, total phenolics, relatively maintained the antioxidant activity, decreased the microbial infection, and thereby increased the postharvest storage life of cucumbers at ambient conditions.

12.9.2 Meat and Poultry Foods

Meat is a perishable, nutritious food and a prime source of protein. Its high moisture attributes to the bacterial contamination. Furthermore, lipid oxidation is also a deteriorating phenomenon which affects the stability of meat and results in social health issues. Owing to this, edible coatings are widely used to extend the shelf life of meat.

Zimoch-Korzycka and Jarmoluk (2017) used coatings with different concentrations of chitosan and hydroxypropyl methylcellulose on pork meat. Meat samples were analyzed for pH, water activity, and total number of microorganisms, psychrotrophs, yeast and molds, color, and thiobarbituric acid-reactive substances. The pH and a_w values ranged 5.42–5.54 and 0.919–0.926, respectively. Total number of microorganisms, psychrotrophs, yeast, and molds reduced to 2.71, 1.46, and 0.78 log cycles, respectively, after two weeks of storage. Thiobarbituric acid-reactive substances significantly reduced with the inclusion of cellulase in the coating material. Alginate coatings along with inclusions of thyme oil and garlic, when applied to lamb meat and stored under frozen conditions, significantly reduced exudative losses and increased yellowness and chroma (Guerrero et al. 2019). Šuput et al. (2019) studied the effect of starch-based coating containing oregano oil on the osmotically dried meat. All physicochemical quality parameter as well as overall meat quality was improved due to less lipid oxidation in samples packed under modified atmosphere.

The chitosan-coated pastirma were stored by Abdallah et al. (2017) for four weeks and analyzed for sensory, physicochemical, and microbiological quality. The results indicated an improvement in sensory characteristics and antioxidant effect, less moisture loss, and approximately 1 log cfu/g reduction in aerobic plate count in chitosan-coated samples. Xiong et al. (2019) conducted a study to develop chitosan-gelatine edible coating with grape seed extract and/or nisin and investigated its effectiveness to maintain quality of fresh pork stored at 4 °C for 20 days. The 1% chitosan (3%) and gelatin (0.5%) grape seed extract formulation was extremely effective in pork preservation. Beef samples coated with alginate-based coating incorporated with basil extracts were examined for polyphenols, antioxidant activity, and other meat characteristics (Alexandre et al. 2021). The incorporated basil resulted in high levels of polyphenols and antioxidant activity and reduced lipid oxidation of meat.

Mukhtar et al. (2018) analyzed chitosan and clove oil-coated fresh poultry meat for quality attributes, microbial analysis (TPC), and sensory characteristics. 1% chitosan solution +2% starch solution +1.5 mL clove oil gave best outcomes by exhibiting synergistic barrier properties against aerobic microbes. Another study by Abdel-Naeem et al. (2021) on bacteriological and sensory quality of chicken meat coated with chitosan and lauric arginate demonstrated that all coated chicken drumsticks had a significant reduced bacterial counts, pH, and thiobarbituric acid values along with a significant enhancement in sensory attributes. Farhan and Hani (2020) conducted an experiment to produce films from semi-refined κ-carrageenan with extract of germinated fenugreek seeds and used for preservation of fresh chicken breast. Results obtained showed that phenolic content and antioxidant activity increased and semi-refined κ-carrageenan films containing 10%, 15%, and 20% germinated fenugreek seeds showed 2.20, 2.35, and 2.39 log reductions.

12.10 Conclusion

Food products contain substantial amounts of water, which poses great threat to the long-term storage of foods. Packaging provides a mean of protection, containment, and handling of foods. Bio-based packaging materials are gaining importance due to their environment-friendly properties. Polysaccharide-based films are being used in various food systems, and transfer of moisture through these films is mediated by diffusion process. This mainly occurs owing to difference in concentration and thermal energy across the barrier films. Moisture follows a path from the surface of food to the inner package environment, from there to package inside surface, through the package material, and finally from package surface to the storage environment. To understand this, models are defined on the basis of type of food (hygroscopic, hydrophilic, porous, perishable, nonperishable, stable, etc.); thermal properties of foods; package and packaging material characteristics, namely, size, void volume, surface area, water vapor permeability, water vapor transmission rate, and thickness and external temperature; and relative humidity of surrounding environment. A rise in temperature is accompanied by change in glass transition temperature which further alters the thermodynamic, dielectric, mechanical properties of foods. Also, high relative humidity mediates the path of moisture from outer environment toward the inner of food materials. A gain in mass of food materials by adsorption of moisture and lose of mass as moisture leaves the food surface (desorption) during sorption phenomenon provides information about most stable state of moisture known as monolayer moisture. Thus, food materials possessing such moisture level are less prone to degradation due to physicochemical, microbial, and enzymatic changes. In case of fresh produce, the concentration of gases such as oxygen and carbon dioxide gives a measure of respiration rates based upon their concentration at different times, void volume, and weight of the produce. The respiration behavior can be modeled using the enzyme kinetic approach. Transpiration in fresh foods leads to moisture loss as moisture evaporates from the food surface to surroundings. Its rate can be expressed on the basis of mass and surface area of food product. Generally respiration is assumed negligible while calculating transpiration rate, but in high humidity conditions, modifications are made to the basic equations to incorporate term defining transpiration rate influenced by respiration heat. Different modeling approaches can be adopted on the basis of processes involved in transfer of moisture that vary with characteristics of food under consideration. These models and their solutions can be used to provide a sound knowledge about suitability of polysaccharide-based films for packaging of a variety of foods and maintaining their structural integrity and nutritional quality during storage.

References

Abdallah MRS, Mohamed MA, Mohamed HMH, Emara MMT (2017) Improving the sensory, physicochemical and microbiological quality of pastirma (a traditional dry cured meat product) using chitosan coating. LWT 86:247–253. https://doi.org/10.1016/j.lwt.2017.08.006

Abdel-Naeem HHS, Zayed NER, Mansour HA (2021) Effect of chitosan and lauric arginate edible coating on bacteriological quality, deterioration criteria, and sensory attributes of frozen stored chicken meat. LWT 150:111928. https://doi.org/10.1016/j.lwt.2021.111928

Adetunji CO, Ojediran JO, Adetunji JB, Owa SO (2019) Influence of chitosan edible coating on postharvest qualities of Capsicum annum L. during storage in evaporative cooling system. Croatian J Food Sci Technol 11(1):59–66. https://doi.org/10.17508/cjfst.2019.11.1

Aider M (2010) Chitosan application for active bio-based films production and potential in the food industry: review. LWT Food Sci Technol 43(6):837–842. https://doi.org/10.1016/j.lwt.2010.01.021

Alexandre S, Vital ACP, Mottin C, do Prado RM, Ornaghi MG, Ramos TR, do Prado IN (2021) Use of alginate edible coating and basil (Ocimum spp) extracts on beef characteristics during storage. J Food Sci Technol 58(10):3835–3843. https://doi.org/10.1007/s13197-020-04844-1

Bajpai SK, Navin C, Ruchi L (2011) Water vapor permeation and antimicrobial properties of sago starch based films formed via microwave irradiation. Int Food Res J 18:417–426

Ballesteros LF, Cerqueira MA, Teixeira JA, Mussatto SI (2018) Production and physicochemical properties of carboxymethyl cellulose films enriched with spent coffee grounds polysaccharides. Int J Biol Macromol 106:647–655. https://doi.org/10.1016/j.ijbiomac.2017.08.060

Benbettaïeb N, Kurek M, Bornaz S, Debeaufort F (2014) Barrier, structural and mechanical properties of bovine gelatin-chitosan blend films related to biopolymer interactions. J Sci Food Agric 94(12):2409–2419. https://doi.org/10.1002/jsfa.6570

Bertuzzi MA, Armada M, Gottifredi JC (2007) Physicochemical characterization of starch based films. J Food Eng 82(1):17–25. https://doi.org/10.1016/j.jfoodeng.2006.12.016

Beuchat LR (1981) Microbial stability as affected by water activity. Cereal Foods World 26:345–349

Bourlieu C, Guillard V, Powell H, Vallès-Pàmies B, Guilbert S, Gontard N (2008) Modelling and control of moisture transfers in high, intermediate and low a_w composite food. Food Chem 106(4):1350–1358. https://doi.org/10.1016/j.foodchem.2007.03.079

Bourtoom T (2008) Edible films and coatings: characteristics and properties. Int Food Res J 15(3): 237–248

Bovi GG, Caleb OJ, Linke M, Rauh C, Mahajan PV (2016) Transpiration and moisture evolution in packaged fresh horticultural produce and the role of integrated mathematical models: a review. Biosyst Eng 150:24–39. https://doi.org/10.1016/j.biosystemseng.2016.07.013

Bovi GG, Rux G, Caleb OJ, Herppich WB, Linke M, Rauh C, Mahajan PV (2018) Measurement and modelling of transpiration losses in packaged and unpackaged strawberries. Biosyst Eng 174:1–9. https://doi.org/10.1016/j.biosystemseng.2018.06.012

Bravin B, Peressini D, Sensidoni A (2006) Development and application of polysaccharide–lipid edible coating to extend shelf-life of dry bakery products. J Food Eng 76(3):280–290. https://doi.org/10.1016/j.jfoodeng.2005.05.021

Caleb OJ, Mahajan PV, Al-Said FA, Opara UL (2013) Transpiration rate and quality of pomegranate arils as affected by storage conditions. CyTA J Food 11(3):199–207

Cassini AS, Marczak LDF, Noreña CPZ (2006) Water adsorption isotherms of texturized soy protein. J Food Eng 77(1):194–199. https://doi.org/10.1016/j.jfoodeng.2005.05.059

Chopin N, Guillory X, Weiss P, Bideau J, Colliec-Jouault S (2014) Design polysaccharides of marine origin: chemical modifications to reach advanced versatile compounds. Curr Org Chem 18:867–895

Chowdhury T, Das M (2010) Moisture sorption isotherm and isosteric heat of sorption characteristics of starch based edible films containing antimicrobial preservative. Int Food Res J 17:601–614

Chu Z, Lu L, Wang J (2012) Mathematical model for water transfer in multidomain food packed in permeable packaging. Packag Technol Sci 26:11–22. https://doi.org/10.1002/pts.1992

Crank J (1975) The mathematics of diffusion, 2nd edn. Oxford University Press, London

Cuq B, Gontard N, Aymard C, Guilbert S (1997) Relative humidity and temperature effects on mechanical and water vapor barrier properties of myofibrillar protein-based films. Polym Gels Netw 5:1–15

Debeaufort F, Quezada-Gallo J-A, Voilley A (1998) Edible films and coatings: tomorrow's packagings: a review. Crit Rev Food Sci Nutr 38:299–313

Devgan K, Kaur P, Kumar N, Kaur A (2019) Active modified atmosphere packaging of yellow bell pepper for retention of physico-chemical quality attributes. J Food Sci Technol 56:878–888. https://doi.org/10.1007/s13197-018-3548-5

Dias AB, Müller CMO, Larotonda FDS, Laurindo JB (2010) Biodegradable films based on rice starch and rice flour. J Cereal Sci 51(2):213–219. https://doi.org/10.1016/j.jcs.2009.11.014

El Halal SLM, Bruni GP, do Evangelho JA, Biduski B, Silva FT, Dias ARG et al (2017) The properties of potato and cassava starch films combined with cellulose fibers and/or nanoclay. Starch Stärke 70(1–2):1700115. https://doi.org/10.1002/star.201700115

El Miri N, Aziz F, Aboulkas A, El Bouchti M, Ben Youcef H, El Achaby M (2018) Effect of plasticizers on physicochemical properties of cellulose nanocrystals filled alginate bionanocomposite films. Adv Polym Technol 37:3171–3185. https://doi.org/10.1002/adv.22087

Embuscado ME, Huber KC (eds) (2009) Edible films and coatings for food applications. Springer, New York

Etemadipoor R, Ramezanian A, Dastjerdi AM, Shamili M (2019) The potential of gum arabic enriched with cinnamon essential oil for improving the qualitative characteristics and storability of guava (Psidium guajava L.) fruit. Sci Hortic 251:101–107. https://doi.org/10.1016/j.scienta.2019.03.021

Farhan A, Hani NM (2020) Active edible films based on semi-refined κ-carrageenan: antioxidant and color properties and application in chicken breast packaging. Food Packag Shelf Life 24: 100476. https://doi.org/10.1016/j.fpsl.2020.100476

Fonseca SC, Oliveira FA, Brecht JK (2002) Modelling respiration rate of fresh fruits and vegetables for modified atmosphere packages: a review. J Food Eng 52(2):99–119. https://doi.org/10.1016/s0260-8774(01)00106-6

García-Ramon JA, Carmona-García R, Valera-Zaragoza M, Aparicio-Saguilan A, Bello-P'erez LA, Aguirre-Cruz A, Alvarez-Ramirez J (2021) Morphological, barrier, and mechanical properties of banana starch films reinforced with cellulose nanoparticles from plantain rachis. Int J Biol Macromol 187:35–42. https://doi.org/10.1016/j.ijbiomac.2021.07.112

Gómez-Aldapa CA, Velazquez G, Gutierrez MC, Rangel-Vargas E, Castro-Rosas J, Aguirre-Loredo RY (2020) Effect of polyvinyl alcohol on the physicochemical properties of biodegradable starch films. Mater Chem Phys 239:122027. https://doi.org/10.1016/j.matchemphys.201

Gontard N, Guilbert S, Cuq JL (1993) Water and glycerol as plasticizers affect mechanical and water vapor barrier properties of an edible wheat gluten film. J Food Sci 58:206–211

Gouanvé F, Marais S, Bessadok A, Langevin D, Métayer M (2007) Kinetics of water sorption in flax and PET fibers. Eur Polym J 43(2):586–598. https://doi.org/10.1016/j.eurpolymj.2006.10.023

Guerrero A, Ferrero S, Barahona M, Boito B, Lisbinski E, Maggi F, Sañudo C (2019) Effects of active edible coating based on thyme and garlic essential oils on lamb meat shelf life after long-term frozen storage. J Sci Food Agric 100:656–664. https://doi.org/10.1002/jsfa.10061

Guillard V, Broyart B, Bonazzi C, Guilbert S, Gontard N (2003) Moisture diffusivity in sponge cake as related to porous structure evaluation and moisture content. J Food Sci 68(2):555–562. https://doi.org/10.1111/j.1365-2621.2003.tb05711.x

Guillard V, Bourlieu C, Gontard N (2013) Food structure and moisture transfer. Springer, New York. https://doi.org/10.1007/978-1-4614-6342-9

Gulati T, Datta AK, Doona CJ, Ruan RR, Feeherry FE (2015) Modeling moisture migration in a multi-domain food system: application to storage of a sandwich system. Food Res Int 76:427–438. https://doi.org/10.1016/j.foodres.2015.06.022

Gutiérrez TJ, Álvarez K (2017) Transport phenomena in biodegradable and edible films. In: Masuelli MA (ed) Biopackaging. CRC Press, Boca Raton, pp 58–88. https://doi.org/10.1201/9781315152349-4

Heiss R, Eichner E (1971) Moisture content and shelf-life. Part 2. Food Manuf 46:37–42

Helena Gomes M, Beaudry RM, Almeida DPF, Xavier Malcata F (2010) Modelling respiration of packaged fresh-cut "Rocha" pear as affected by oxygen concentration and temperature. J Food Eng 96(1):74–79. https://doi.org/10.1016/j.jfoodeng.2009.06.043

Hernandez RJ, Giacin JR (1997) Factor affecting permeation, sorption and migration processes in packaged-product systems. In: Taub IA, Singh RP (eds) Food Storage Stability. CRC Press, New York, pp 269–330
Ho QT, Verboven P, Verlinden BE, Schenk A, Nicolaï BM (2013) Controlled atmosphere storage may lead to local ATP deficiency in apple. Postharvest Biol Technol 78:103–112. https://doi.org/10.1016/j.postharvbio.2012.12.014
Hu Y, Topolkaraev V, Hiltner A, Baer E (2001) Measurement of water vapor transmission rate in highly permeable films. J Appl Polym Sci 81(7):1624–1633. https://doi.org/10.1002/app.1593
Hussain A, Zia KM, Tabasum S, Noreen A, Ali M, Iqbal R, Zuber M (2017) Blends and composites of exopolysaccharides; properties and applications: a review. Int J Biol Macromol 94:10–27
Ibarz A, Barbosa-Cánovas GV (2002) Unit operations in food engineering. CRC Press, Boca Raton
Jagadish RS, Raj B (2013) Moisture sorption studies of PEO/starch blended films for food packaging applications. Int J Agric Food Sci Technol 4(9):923–932
Karathanos VT, Villalobos G, Saravacos GD (1990) Comparison of two methods of estimation of the effective moisture diffusivity from drying data. J Food Sci 55(1):218–223. https://doi.org/10.1111/j.1365-2621.1990.tb06056.x
Kaur P, Rai DR, Paul S (2010) Nonlinear estimation of respiratory dynamics of fresh-cut spinach (Spinacia Oleracea) based on enzyme kinetics. J Food Process Eng 34(6):2137–2155. https://doi.org/10.1111/j.1745-4530.2009.00508.x
Klangmuang P, Sothornvit R (2016) Combination of beeswax and nanoclay on barriers, sorption isotherm and mechanical properties of hydroxypropyl methylcellulose-based composite films. LWT Food Sci Technol 65:222–227. https://doi.org/10.1016/j.lwt.2015.08.003
Labuza TP, Hyman CR (1998) Moisture migration and control in multi-domain foods. Trends Food Sci Technol 9(2):47–55. https://doi.org/10.1016/S0924-2244(98)00005-3
Leonardi C, Baille A, Guichard S (2000) Predicting transpiration of shaded and non-shaded tomato fruits under greenhouse environments. Sci Hortic 84(3–4):297–307. https://doi.org/10.1016/s0304-4238(99)00130-2
Levine H, Slade L (1988) Water as a plasticizer: physico-chemical aspects of low-moisture polymeric systems. In: Franks F (ed) Water science reviews 3: water dynamics, Water science review. Cambridge University Press, Cambridge, pp 79–185. https://doi.org/10.1017/CBO9780511552083.002
Li C, Tao J, Zhang H (2017) Peach gum polysaccharides-based edible coatings extend shelf life of cherry tomatoes. 3 Biotech 7(3):168. https://doi.org/10.1007/s13205-017-0845-z
Lim J, Park ES, Cheigh HS, Lee DS (2001) Effect of package size and temperature on the volume expansion of flexible permeable packages of Kimchi, a Korean fermented vegetable. Packaging Technol Sci: An Int J 14(2):41–48. https://doi.org/10.1002/pts.536
Linke M, Geyer M (2013) Condensation dynamics in plastic film packaging of fruit and vegetables. J Food Eng 116(1):144–154. https://doi.org/10.1016/j.jfoodeng.2012.11.026
Luchese CL, Benelli P, Spada JC, Tessaro IC (2018) Impact of the starch source on the physico-chemical properties and biodegradability of different starch-based films. J Appl Polym Sci 135(33):46564. https://doi.org/10.1002/app.46564
Maguire KM, Banks NH, Opara LU (2001) Factors affecting weight loss of apples. In: Janick J (ed) Horticultural reviews. John Wiley & Sons, Hoboken, pp 197–234
Mahajan PV, Caleb OJ, Singh Z, Watkins CB, Geyer M (2014) Postharvest treatments of fresh produce. Phil Trans R Soc A 372:1–16
Mahajan P, Rux G, Caleb O, Linke M, Herppich W, Geyer M (2016) Mathematical model for transpiration rate at 100% humidity for designing modified humidity packaging. Acta Hortic 1141:269–274. https://doi.org/10.17660/actahortic.2016.1141.33
Mali S, Sakanaka LS, Yamashita F, Grossmann MVE (2005) Water sorption and mechanical properties of cassava starch films and their relation to plasticizing effect. Carbohydr Polym 60(3):283–289. https://doi.org/10.1016/j.carbpol.2005.01.003
Martínez K, Ortiz M, Albis A, Gilma Gutiérrez Castañeda C, Valencia M, Grande Tovar C (2018) The effect of edible chitosan coatings incorporated with Thymus capitatus essential oil on the

shelf-life of strawberry (Fragaria x ananassa) during cold storage. Biomol Ther 8(4):155. https://doi.org/10.3390/biom8040155

Mastromatteo M, Mastromatteo M, Conte A, Del Nobile MA (2010) Advances in controlled release devices for food packaging applications. Trends Food Sci Technol 21(12):591–598. https://doi.org/10.1016/j.tifs.2010.07.010

Mathlouthi M (2001) Water content, water activity, water structure and the stability of foodstuffs. Food Control 12(7):409–417. https://doi.org/10.1016/s0956-7135(01)00032-9

Mirzaei-Mohkam A, Garavand F, Dehnad D, Keramat J, Nasirpour A (2020) Physical, mechanical, thermal and structural characteristics of nanoencapsulated vitamin E loaded carboxymethyl cellulose films. Prog Org Coat 138:105383. https://doi.org/10.1016/j.porgcoat.2019.105383

Mukhtar S, Zahoor T, Randhawa MA, Iqbal R, Shabbir A, Liaqat A, Ahsan S (2018) Synergistic effect of chitosan and clove oil on raw poultry meat. J Food Process Technol 9(6):1–5. https://doi.org/10.4172/2157-7110.1000737

Muscat D, Adhikari B, Adhikari R, Chaudhary DS (2012) Comparative study of film forming behaviour of low and high amylose starches using glycerol and xylitol as plasticizers. J Food Eng 109(2):189–201

Nawab A, Alam F, Haq MA, Hasnain A (2016) Biodegradable film from mango kernel starch: effect of plasticizers on physical, barrier, and mechanical properties. Starch Stärke 68(9–10): 919–928. https://doi.org/10.1002/star.201500349

Nawab A, Alam F, Hasnain A (2017) Mango kernel starch as a novel edible coating for enhancing shelf- life of tomato (Solanum lycopersicum) fruit. Int J Biol Macromol 103:581–586. https://doi.org/10.1016/j.ijbiomac.2017.057

O'Connor LJ, Favreau-Farhadi N, Barrett AH (2017) Use of edible barriers in intermediate moisture food systems to inhibit moisture migration. J Food Process Preserv 42(2):13512. https://doi.org/10.1111/jfpp.13512

Parra D, Tadini C, Ponce P, Lugao A (2004) Mechanical properties and water vapor transmission in some blends of cassava starch edible films. Carbohydr Polym 58(4):475–481. https://doi.org/10.1016/j.carbpol.2004.08.021

Paul DR, Clarke R (2002) Modeling of modified atmosphere packaging based on designs with a membrane and perforations. J Membr Sci 208(1–2):269–283. https://doi.org/10.1016/s0376-7388(02)00303-4

Peppelenbos HW, Van't Leven J (2017) Evaluation of four types of inhibition for modelling the influence of carbon dioxide on oxygen consumption of fruits and vegetables. Postharvest Biol Technol 7(1-2):27–40. https://doi.org/10.1016/0925-5214(96)80995-1

Perdomo J, Cova A, Sandoval AJ, García L, Laredo E, Müller AJ (2009) Glass transition temperatures and water sorption isotherms of cassava starch. Carbohydr Polym 76(2): 305–313. https://doi.org/10.1016/j.carbpol.2008.10.023

Pinheiro AC, Bourbon AI, Vicente AA, Quintas MAC (2013) Transport mechanism of macromolecules on hydrophilic bio-polymeric matrices—diffusion of protein-based compounds from chitosan films. J Food Eng 116(3):633–638. https://doi.org/10.1016/j.jfoodeng.2012.12.038

Poverenov E, Arnon-Rips H, Zaitsev Y, Bar V, Danay O, Horev B, Rodov V (2018) Potential of chitosan from mushroom waste to enhance quality and storability of fresh-cut melons. Food Chem 268:233–241. https://doi.org/10.1016/j.foodchem.2018.045

Ramallo LA, Albani OA (2007) Water diffusion coefficient and modeling of water uptake in packaged yerba mate. J Food Process Preserv 31(4):406–419. https://doi.org/10.1111/j.1745-4549.2007.00137.x

Ramesh HP, Tharanathan RN (2003) The renewable raw materials of high biotechnological value. Crit Rev Biotechnol 23:149–173

Reddy JP, Varada Rajulu A, Rhim J-W, Seo J (2018) Mechanical, thermal, and water vapor barrier properties of regenerated cellulose/nano-SiO_2 composite films. Cellulose 25:7153–7165. https://doi.org/10.1007/s10570-018-2059-x

Robertson G (1993) Food packaging. Marcel Dekker, New York, NY

Rosa GS, Moraes MA, Pinto LAA (2010) Moisture sorption properties of chitosan. LWT Food Sci Technol 43(3):415–420. https://doi.org/10.1016/j.lwt.2009.09.003

Rubino M, Tung MA, Yada S, Britt IJ (2001) Permeation of oxygen, water vapor, and limonene through printed and unprinted biaxially oriented polypropylene films. J Agric Food Chem 49(6): 3041–3045. https://doi.org/10.1021/jf001427s

Rumsey T, Krochta J (1994) Mathematical modeling of moisture transfer in food systems with edible coatings. In: Krochta J et al (eds) Edible coatings and films to improve food quality. Technomic Publishing, Lancaster, PA, pp 337–356

Saberi B, Vuong Q, Chockchaisawasdee S, Golding J, Scarlett C, Stathopoulos C (2015) Water sorption isotherm of pea starch edible films and prediction models. Foods 5(4):1. https://doi.org/10.3390/foods5010001

Saberi B, Golding JB, Marques JR, Pristijono P, Chockchaisawasdee S, Scarlett CJ, Stathopoulos CE (2018) Application of biocomposite edible coatings based on pea starch and guar gum on quality, storability and shelf life of "valencia" oranges. Postharvest Biol Technol 137:9–20. https://doi.org/10.1016/j.postharvbio.2017.11.003

Saha A, Tyagi S, Gupta RK, Tyagi YK (2016) Guar gum based edible coating on cucumber (cucumis sativus l.). Eur J Pharm Med Res 3(9):558–570

Santana JS, de Carvalho Costa ÉK, Rodrigues PR, Correia PRC, Cruz RS, Druzian JI (2018) Morphological, barrier, and mechanical properties of cassava starch films reinforced with cellulose and starch nanoparticles. J Appl Polym Sci 47001:1–10. https://doi.org/10.1002/app.47001

Sastry S (1985) Moisture losses from perishable commodities: recent research and developments. Int J Refrig 8(6):343–346. https://doi.org/10.1016/0140-7007(85)90029-5

Schiraldi A, Fessas D (2001) Mechanism of staling: an overview. In: Chinachoti P, Vodovotz Y (eds) Bread staling. CRC Press, Boca Raton, pp 1–17

Shyu YS, Chen GW, Chiang SC, Sung WC (2019) Effect of chitosan and fish gelatin coatings on preventing the deterioration and preserving the quality of fresh-cut apples. Molecules 24(10): 2008. https://doi.org/10.3390/molecules24102008

Singh M, Kumar A, Kaur P (2012) Respiratory dynamics of fresh baby corn (Zea mays L.) under modified atmospheres based on enzyme kinetics. J Food Sci Technol 51(9):1911–1919. https://doi.org/10.1007/s13197-012-0735-7

Singla M, Kumar A, Kaur P, Goraya RK (2020) Respiratory properties of fresh black carrot (Dacus carota L.) based upon non-linear enzyme kinetics approach. J Food Sci Technol 57(10): 3903–3912. https://doi.org/10.1007/s13197-020-04422-5

Sogut E, Cakmak H (2020) Utilization of carrot (Daucus carota L) Fiber as a filler for chitosan-based films. Food Hydrocoll 106:105861. https://doi.org/10.1016/j.foodhyd.2020.105861

Spence K, Venditti R, Rojas O, Pawlak J, Hubbe M (2011) Water vapor barrier properties of coated and filled microfibrillated cellulose composite films. BioResources 6(4):4370–4388. https://doi.org/10.15376/biores.6.4.4370-4388

Sperling LH (1986) Introduction to physical polymer science. John Wiley & Sons, New York

Sucharitha KV, Beulah AM, Ravikiran K (2018) Effect of chitosan coating on storage stability of tomatoes (Lycopersicon esculentum Mill). Int Food Res J 25(1):93–99

Sucheta, Chaturvedi K, Sharma N, Yadav SK (2019) Composite edible coatings from commercial pectin, corn flour and beetroot powder minimize post-harvest decay, reduces ripening and improves sensory liking of tomatoes. Int J Biol Macromol 133:283–294. https://doi.org/10.1016/j.ijbiomac.2019.04.13

Šuput D, Lazić V, Pezo L, Gubić J, Šojić B, Plavšić D, Lončar B, Nićetin M, Filipović V, Knežević V (2019) Shelf life and quality of dehydrated meat packed in edible coating under modified atmosphere. Roman Biotechnol Lett 24(3):545–553. https://doi.org/10.25083/rbl/24.3/545.553

Taiz L, Zeiger E (1998) Plant physiology, 2nd edn. Sinauer Associates, Inc., Sunderland, MA

Tano K, Kamenan A, Arul J (2005) Respiration and transpiration characteristics of selected fresh fruits and vegetables. Agron Afr 17(2):103–115. https://doi.org/10.4314/aga.v17i2.1662

Torrieri E, Perone N, Cavella S, Masi P (2010) Modelling the respiration rate of minimally processed broccoli (Brassica rapa var. sylvestris) for modified atmosphere package design. Int J Food Sci Technol 45(10):2186–2193. https://doi.org/10.1111/j.1365-2621.2010.02387.x

Turan D (2019) Water vapor transport properties of polyurethane films for packaging of respiring foods. Food Eng Rev 13:54–65. https://doi.org/10.1007/s12393-019-09205-z

Veraverbeke EA, Verboven P, Van Oostveldt P, Nicolaï BM (2003) Prediction of moisture loss across the cuticle of apple (Malus sylvestris subsp. mitis (Wallr.)) during storage. Postharvest Biol Technol 30(1):75–88. https://doi.org/10.1016/s0925-5214(03)00083-8

Wang J, Gardner DJ, Stark NM, Bousfield DW, Tajvidi M, Cai Z (2017) Moisture and oxygen barrier properties of cellulose nanomaterial-based films. ACS Sustain Chem Eng 6(1):49–70. https://doi.org/10.1021/acssuschemeng.7b03523

White GW, Cakebread SH (1966) The glassy state in certain sugar-containing food products. Int J Food Sci Technol 1(1):73–82. https://doi.org/10.1111/j.1365-2621.1966.tb01031.x

Williams R, Mittal G (1999) Water and fat transfer properties of polysaccharide films on fried pastry mix. LWT Food Sci Technol 32(7):440–445. https://doi.org/10.1006/fstl.1999.0573

Xanthopoulos GT, Athanasiou AA, Lentzou DI, Boudouvis AG, Lambrinos GP (2014) Modelling of transpiration rate of grape tomatoes. Semi-empirical and analytical approach. Biosyst Eng 124:16–23. https://doi.org/10.1016/j.biosystemseng.2014.06.005

Xiong Y, Chen M, Warner RD, Fang Z (2019) Incorporating nisin and grape seed extract in chitosan-gelatine edible coating and its effect on cold storage of fresh pork. Food Contr 110: 107018. https://doi.org/10.1016/j.foodcont.2019.107018

Xiong L (2002) Determination and prediction of shelf life of moisture sensitive post shredded wheat cereal. School of Packaging, Michigan State University 12:77–94

Zhang Y, Han JH (2008) Sorption isotherm and plasticization effect of moisture and plasticizers in pea starch film. J Food Sci 73(7):E313–E324. https://doi.org/10.1111/j.1750-3841.2008.00867.x

Zhang Y, Rempel C, Liu Q (2014) Thermoplastic starch processing and characteristics—a review. Crit Rev Food Sci Nutr 54(10):1353–1370. https://doi.org/10.1080/10408398.2011.636156

Zhang L, Chen F, Lai S, Wang H, Yang H (2018) Impact of soybean protein isolate-chitosan edible coating on the softening of apricot fruit during storage. LWT 96:604–611. https://doi.org/10.1016/j.lwt.2018.06.011

Zhou W, He Y, Liu F, Liao L, Huang X, Li R, Zou Y, Zhou L, Zou L, Liu Y, Ruan R, Li J (2021) Carboxymethyl chitosan-pullulan edible films enriched with galangal essential oil: characterization and application in mango preservation. Carbohydr Polym 256:117579. https://doi.org/10.1016/j.carbpol.2020.117579

Zimeri J (2002) The effect of moisture content on the crystallinity and glass transition temperature of inulin. Carbohydr Polym 48(3):299–304. https://doi.org/10.1016/s0144-8617(01)00260-0

Zimoch-Korzycka A, Jarmoluk A (2017) Polysaccharide-based edible coatings containing cellulase for improved preservation of meat quality during storage. Molecules 22(3):390. https://doi.org/10.3390/molecules22030390

Chapter 13
Modelling of Mass and Gas Transfer Through Polysaccharide-Based Films

Vikash Chandra Verma, Rifat Javaid, Saamira, Omar Bashir, and Tawheed Amin

13.1 Introduction

Integrated food product packaging provides protection of food products from external influences and harm and food content and provides ingredient and nutritional information to the consumers (Coles et al. 2003), and other information about food being offered for better marketing significantly contribute to a business profit. After the process is completed, packing maintains the benefits of food processing, enables food to travel long distances safely from where they originated and is still important during consumption. We should be careful towards packaging technology development with other issues. Packaging technology needs, however, maintain a balance between the security of the food supply and other issues including rising energy and commodity prices, heightened social and environmental awareness and severe municipal waste and solid waste disposal legislation. Food consumption is an essential daily requirement and belongs to the unique category of product that is usually consumed three times a day by everyone. As a result, food packaging includes about two-thirds of volume waste (Hunt and Auster 1990), as food packaging is around 50% (by weight) of total packages. It is well known that plastics are durable materials, but the major disadvantages belong to their nonrenewable nature.

V. C. Verma (✉)
Department of Agricultural Engineering, V K S College of Agriculture, Dumraon (Buxar), Bihar Agricultural University, Bhagalpur, Bihar, India

R. Javaid · Saamira · T. Amin
Sher-e-Kashmir University of Agricultural Sciences and Technology of Kashmir, Shalimar, Srinagar, Jammu and Kashmir, India

O. Bashir
Department of Food Technology and Nutrition, Lovely Professional University, Phagwara, Punjab, India

T. Amin et al. (eds.), *Polysaccharide Based Films for Food Packaging: Fundamentals, Properties and Applications*,
https://doi.org/10.1007/978-981-99-4898-7_13

Table 13.1 Trends in the evolution of food packaging

Period	Function and issues
1960s	Convenience and marketing
1970s	Lightweight, source reduction and energy saving
1980s	Safety and temper evidence
1990s	Environmental impact
2000s	Safety and security

Actually these are made of petrochemicals, a nonrenewable source of energy. Therefore, they are challenging to dispose of once used, resulting in a large amount of waste. There is no doubt that food packaging shares major portion of total major packaging material. Food packaging has changed over time from simplified storage methods to simplification, marketing, inventory reduction, safety, proof of disruption and environmental issues (Stilwell et al. 1991; see Table 13.1).

The accumulation of food packaging material (nonrenewable) in nature due to evolution of above said different prospects during the industrialised era is causing serious environmental problems. These days, consumers prefer to minimise the environmental issues caused by food packaging and search for materials that can be degraded quickly. Therefore, it is essential to concentrate research efforts on the development of substitutes for synthetic packaging materials. Due to the ability to stop moisture loss, odour loss, solute transport and water absorption in food or oxygen penetration, bio-films can be a better choice to synthetic materials as a food packaging system. They also allow for better transport to longer distances (Aider 2010; Dutta et al. 2009). Food scientists and engineers are putting a lot of efforts into developing new biodegradable film materials that are abundant, renewable and reasonably priced. Many of these items are considered as agricultural waste or byproducts and are often inexpensive (Kim et al. 2006). Disposable crockery, drinking cups, feeding spoons, bottles, plates, wraps and laminating films, straws, stirrers, food containers purchased at grocery shops and fast food restaurants or food delivered to your home all are degradable materials that are currently used in the food industry (Siracusa et al. 2008). The use of films made of polysaccharide in food products could create new possibilities for the production of new food packaging systems. In order to make edible films and coatings that may be utilised as packaging for food preservation, polysaccharides have reportedly been employed as a raw material. These polysaccharides include cellulose derivatives, chitosan, starches and gums (Aider 2010; Elsabee and Abdou 2013; Jimenez et al. 2012; Klemm et al. 2005; Psomiadou et al. 1996; Vargas and Gonzalez-Martinez 2010; Zhang et al. 2008). Due to their simple availability from agricultural waste, polysaccharide-based packaging materials are inexpensive and also possess certain functional or specific characteristics. There have been claims that the gas barrier properties of the polysaccharide-based films can prevent solvent from penetrating meals and causing toxicity or quality loss by delaying the organic vapour loss (aroma compounds) during storage (Dutta et al. 2009; Vieira et al. 2011). The availability of oxygen in packaging systems is also important because it allows you to control fruit ripening or

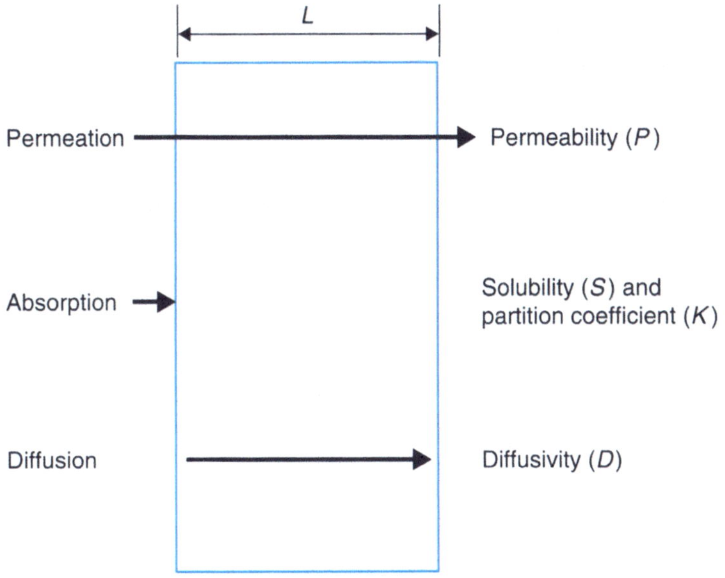

Fig. 13.1 Mass transfer phenomena and their characteristic coefficients

reduce the oxidation in some foods, for example, polyunsaturated fats result into better shelf life (Dutta et al. 2009; Vieira et al. 2011). The components must first be dissolved or dispersed using a solvent, such as water, alcohol, a mixture of the two or another solvent, in order to create the film. To attain a variety of packaging objectives, including greater bonding and strength and shelf stability, plasticisers, antibacterial agents, colouring agents or flavouring compounds can be added during this procedure. PH adjustment and/or heating the solution is essentially required for the solubility of some biopolymers. In order to make free-standing films, the film-forming solution is casted and dried at the desired temperature and relative humidity conditions. Food may be directly coated with film-forming solutions using a variety of techniques, including dipping, spraying, brushing and panning followed by drying as packaging material (Bourtoom 2008). The characteristics of the food and packaging materials have a direct impact on the quality of packaged food.

As shown below, common mass and gas transfer phenomena in food packaging systems include permeation, absorption and diffusion (Fig. 13.1).

Permeation is the permeants' capacity to enter an object directly in response to inadequate pressure changes. An effective indicator of permeation is how quickly gases and liquids pass through packing materials; this attribute of the packaging material, which only depends on the thickness of the film, may be referred to as the permeance (ASTM 1993; Segall and Scanlon 1996). The permeability (P), an important characteristic of the film, is created by multiplying the permeance by the thickness of the film. As a result, permeability is the mass transfer coefficient or permeation (P).

Diffusion refers to the movement of a diffusant inside a packaging system as a result of a concentration difference, which acts as a driving force. The term "diffusivity" (D) is used to describe in what way a substance diffuses in a medium. A substance's affinity for or through the two separate mediums with which it comes into contact is measured by its absorption and desorption. The solubility (S) or partition (K) coefficients can be used to represent a substance's affinity for a material. A component that is migrating through a specific packaging system has characteristic values like its permeability, solubility and diffusivity. Therefore, these variables are important for simulating the system's mass transfer profile.

The circumstance of massive exchange of gas like oxygen and water vapour, flavour loss, unpleasant odour, permeability and diffusion of packaging components towards food are the main causes of quality degradation in food products (Kester and Fennema 1986; Debeaufort et al. 1998). These occurrences might have set between food and packaging materials and within the food substance or through the complex constituents included inside the food product (Krochta 1997). For this reason, designing food packaging systems requires a thorough understanding of mass transfer experiments mostly on package element migration and the components of food; this includes flavour profiles, moisture and the uptake and release of volatile substances, migration of gases and the reaction rate of oxidation and raw material breakdown. The quality of packages, including their seals and closures, as well as the permeability of the packing materials themselves, determines how well foods are protected from gas and vapour exchange with the environment. Gases and vapours can flow through polymeric materials via one of the two methods:

1. When gases and vapours pass through micropores, point defects and fractures in building supplies, this is known as the "pore effect".
2. The emulsification action, in which gases and vapours dissolve on the surface of the polymer, disperses within the polymer because of a difference in concentration and evaporates somewhere inside the polymer. True permeability, also known as activated diffusion, is the term used to characterise this "solution-diffusion" phenomenon.

Characterising the transfer rate of molecular mass through a packaging substance involves irreversible mechanisms. The chemical potential gradient is a driving force necessary for the movement of vapours and soluble compounds. They frequently change between greater and lower chemical potentials ((M)fx), modifying the particle number, which itself is conjugate to chemical potential. The foundation of this hypothesis relies on kinetic and random molecular motion. Furthermore, the process may simply explain in terms of chemical potential. The molecule exhibits strong chemical potential at high concentrations and low chemical potential at the lowest concentrations at a given temperature. There is emission of free energy when the molecules migrate from high charge density to low charge density. In other terms, it may be claimed that the concentration gradient of migratory species is connected to the gradient in chemical potential.

A single dimensional flow (or flux) consideration is necessary because the migration of gases and solutes in packing material is considerable through the

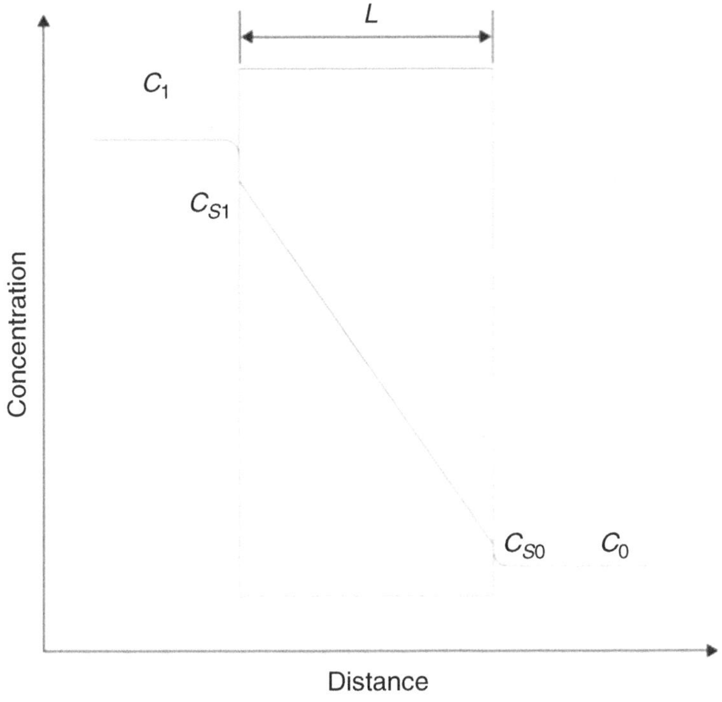

Fig. 13.2 Mass transfer through packaging material from C_1 to C_0

surface area related to the thickness. According to the linear flux (flow per unit cross-section in unit time) towards driving force may be interpreted as the material's resistance to the flow of specified molecules:

$$\text{Driving force} = \frac{\text{Flux}}{\text{Material resistance}}$$

A concentration profile is a graphic that represents the intensity of a dose as it varies with position. This drawing could be placed on a process to highlight where certain changes take place. Figure 13.2 illustrates the transfer of mass via packing material. A molecule begins to permeate from the area where its concentration is maximum, C_1, to the area where it is minimum, C_0.

Diffusion affects migration within the packaging, and the diffusion action reflects the molecule migration towards intensity C_{S1} to C_{S0}. Absorbance on the surface indicates the transfer from concentration C_1 to C_{S1}. There can be evaporation from the surface and hence desorption makes a distinction in concentration across C_{S0} and

C_0. The material susceptibility to gases or solutes underneath the effect of thermodynamic force is treated independently.

13.2 Diffusion Laws

When a constant pressure difference is retained throughout the polymer, diffusion of a gas or vapour occurs at a consistent rate during steady-state circumstances. The partial diffusive flux, J, in a polymer may be described as the quantity of material moving over a surface of a certain area per unit time, normal towards the flow direction, and is given as

$$J = \frac{Q}{At} \tag{13.1}$$

where Q is the net quantity of permeant passing over the area A during time t.

Fick's first law expresses the direct proportionality correlation between the diffusion rate and the concentration gradient. Fick's law governs the process of diffusion (Crank 1975). Diffusion is a significant element within the material. Fick's first law may be stated as follows:

$$J_d = -D\frac{\partial C}{\partial X} \tag{13.2}$$

Here J_d, D, C and x represent the flow rate for each cross-section, diffusivity (the rate at which the permeant disperses across the film occurs), diffusant concentration and the length from where the diffusant needs to flow, respectively.

The discharge of diffusant at the constant rate is given by Eqs. (13.2) and (13.3).

Equation (13.3) yields the discharge of diffusant at the constant rate following Eq. (13.2) for both cases, when the quantities C_{s1} and C_{s0} remain unaffected (and the D is constant):

$$\text{Flux}, J_d = \frac{Q}{At} = D\frac{(C_{s_1} - C_{s_0})}{L} \tag{13.3}$$

where L denotes the membrane thickness, Q represents the amount of dispersed moving material (in mol or kg) and A denotes the area of cross-sectional diffusion. The diffusivity D is measured in $m^2\ s^{-1}$, while the flux is measured in mol $m^{-2}\ s^{-1}$ or kg $m^{-2}\ s^{-1}$. The following is the formulation of Fick's second law of diffusion, which holds true when D is independent of concentration and diffusion is restricted to the x-direction only. One-dimensional diffusion in an unsteady or transient condition with respect to time "t" can be studied using this.

$$\frac{\partial C}{\partial t} = -D\frac{\partial^2 C}{\partial^2 X} \tag{13.4}$$

By using different starting boundary conditions, Crank (1975) proposed several results for Eq. (13.3) with various material geometries (such as infinite slabs, spheres, etc.).

13.3 Solubility

Gas first should absorb in the substance for diffusing via packing material from C_{s1} to C_{s0}. Henry's law states that, provided the gas concentration below its saturation point or maximum solubility, the sorption (on the packing material's surface) and the relative pressure of the gas constituents are linearly correlated.

$$P = \sigma X_s \tag{13.5}$$

where X_s is the molar proportion of gas in the packing material, pi represents the partial pressure of the gas in the atmosphere and Pa is Henry's law constant (Moore 1972).

Equation (13.6) also illustrates the linear connection between the partial pressure of gas and the concentration at the surface of the packing material:

$$C_s = H^{-1} P_1 \tag{13.6}$$

P_1 denotes the partial pressure of gas on the high concentration (C_1). When discussing diluted solutions, like the permeation condition, Eq. (13.5) is consistent with Eq. (13.5). Therefore, $C_{s1} \propto X_s$. Equation (13.7) may be used to represent the constant H^{-1}:

$$H^{-1} = \frac{C_{s1}}{p_1} = \frac{n}{V_{\text{material}} \cdot p_1} = \frac{n}{V_{\text{STP}}} \frac{V_{\text{STP}}}{V_{\text{material}} \cdot p_1} = \frac{S}{V_{\text{STP}}} \tag{13.7}$$

where "n" signifies the dissipation of gas molecules in the packaging material in terms of moles, within packing material, the gas molecule dissipates via volume V at standard temperature and pressure (STP) conditions and the volume of one mole of ideal gas has a value of $22{,}414 \times 10^{-6} \mathrm{m}^3 \mathrm{mol}^{-1}$ and is represented as VSTP. Equation (13.8) may be used to represent the solubility "S" of the gas in Eq. (13.7):

$$S = \frac{n V_{\text{STP}}}{V_{\text{material}} \cdot p_1} \tag{13.8}$$

where "S" denotes its solubility in mol m^3 gas mol^{-1}m^{-3} material Pa^{-1} and Pa^{-1} denotes the volume fraction of the absorbed constituent in a substance at normal gas requirements. This is because the solubility is uniform and unaffected by the concentration absorbed at a particular temperature (Robertson 1993). At the margins of package material, the gas concentrations may be stated using Eqs. (13.7) and (13.8) as follows:

$$C_{s1} = \frac{S.p_1}{V_{\mathrm{STP}}}, C_{s0} = \frac{S.p_0}{V_{\mathrm{STP}}}. \tag{13.9}$$

There is a logarithmic nonlinear correlation of sorption phenomena, which is conveyed as a Langmuir-type sorption, if a gas molecule and the packaging material have an affiliation towards each other or if a gas molecule is entrapped in the micro-void's matrix of the packaging material at a relatively low pressure (Robertson 1993).

13.4 Gaseous Permeability and Diffusivity

Gas flux Jgo among both absorption and dispersion may utilise the partial pressure concept rather than the concentration gradient, since the factor that motivates the permeability of gas inside the packaging material is the change in partial pressure or gas concentration within the opposite extremes of packaging films. This is significant because measuring the vapour pressure in equilibrium with the polymer is more practical than determining the actual concentration the process of mass transfer, the P which represents the partial pressure and the S as solubility from Eqs. (13.4) and (13.9) may be used to replace the concentration:

$$\text{Flux gas}, J_g = \frac{Q}{At} = D\frac{(C_{s1} - C_{s0})}{L} = D\frac{S\,(p_1 - p_0)}{V_{\mathrm{STP}}\,L} = P.\frac{(p_1 - p_0)}{V_{\mathrm{STP}}\,L} \tag{13.13}$$

where J_g is the flux, on the left side, that is, C_1, and on the right side, that is, C_0 of Fig. 2.2, the partial pressures of the gas are p_1 and p_0, respectively. At ambient conditions, the partial pressure p_1 would typically remain constant, but p_0 depends on the gas reactivity with food or its ability to diffuse into the food substance.

Equation (13.14) illustrates a link between a gas permeability, diffusivity and solubility, which is implied by Eq. (13.13).

$$P = D \times S \tag{13.14}$$

The kinetic term (D), which represents the dynamics of the penetrant-film system, and the thermodynamic term (S), which depends on the interactions between the penetrant and the film, together make up the permeability coefficient.

Consequently, the SI units of S, P and D are Pa^{-1}, $m^2s^{-1}Pa^{-1}$ and m^2s^{-1}, respectively. Alternately, Eq. (13.13) can be rewritten to keep all critical dimensions by utilising the flow (for gas) formulation in Eq. (13.3):

$$P = \frac{Q_{\text{gas}}}{A_{\text{material}}} \frac{V_{\text{STP}}\, L}{t(p_1 - p_0)} \tag{13.15}$$

$$\frac{Q_{\text{gas}}}{t} = \frac{P}{L} \frac{A_{\text{material}}(p_1 - p_0)}{V_{\text{STP}}}$$

The units of P are $m^3_{\text{gas}} m_{\text{thickness}} m_{\text{area}}^{-2} s^{-1} Pa^{-1}$, consistent with those used in various packaging texts for permeability: mL (=0.001 inch) of package thickness times cubic centimetres (or mL) of gas (under STP or other prescribed circumstances) per 100 square inches per day per atmosphere (ASTM 1993). P is a measure of how quickly a gas may pass through a film under a pressure gradient. The symbol R stands for the ratio P/L, which is also known as the permeance. Permeance is an indicator of performance and not a measure of material quality. For the purpose of studying permeation, the following four assumptions are taken into consideration:

1. There is steady-state diffusion.
2. The polymer exhibits a linear relationship between concentration and distance.
3. There is only one mode of diffusion, that is, via the film, no net diffusion occurs along with it or across it.
4. The concentration has no effect on D or S.

The already defined permeability coefficient is independent of thickness because P has already taken thickness into account. However, the permeance P/L—the overall level of protection offered by a barrier material per unit area—only approaches zero as an approximation. Due to this, it is no longer feasible to increase polymer thickness (X) beyond a certain point in order to achieve decreased permeability.

13.5 Convective Mass Transfer

The concentration difference between the two regions is the only dynamic force causing the mass transfer of solute molecules in a specific region or content of material. The dynamic pressures necessary to separate the solute between the solution and the material are determined by the concentration difference in the film layer that forms the border between the solution and the packing material. Convective mass transfer coefficient (h_m) may need to be taken into account in this case because it depends on the agitation of the solution and the solute's affinity for both the material and the solution. The mass transfer of a solute to the packing material is depicted in Fig. 13.3. The parameter that represents the resistance to mass transfer at the material's surface is the convective mass transfer coefficient (h_m) and velocity

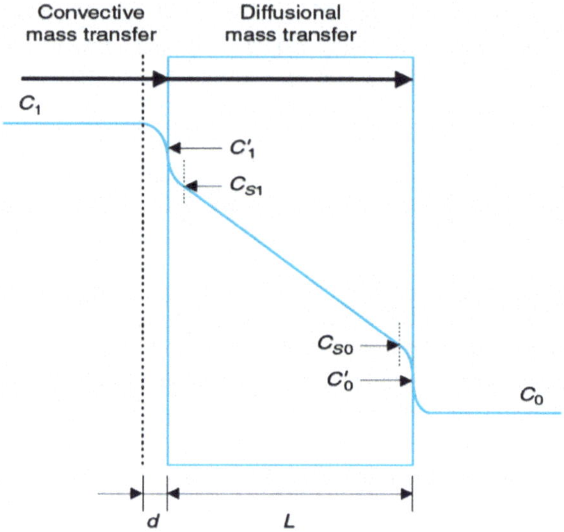

Fig. 13.3 Mass transfer of a solute through a membrane

(ms^{-1}). This parameter is also known as "surface mass transfer coefficient". The flux on the solution side of the boundary film layer is represented by

$$J_h = h_{m1}\left(C_1 - C_1'\right) \tag{13.10}$$

On the solution side, which is represented in Fig. 13.3, mass transfer resistance might not be the only barrier. The additional resistance to the mass transfer at the package's interface is expressed as

$$J_h = h_{ms}\left(C_{s1}' - C_{s1}\right) \tag{13.11}$$

The partition coefficient "K", defined as the concentration ratio for the solute across the two separate phases, can be used to indicate the absorption affinity of the migrating solute over the interface itself, that is, solution and packaging material. The partition coefficient of the solute at the interface between the solution and the substance is written as follows:

$$K = \frac{C_{s1}'}{C_{s1}} \tag{13.12}$$

Desorption uses expressions that are similar. The next transfer phenomenon that occurs after the solute molecule is absorbed into the material layer is diffusion, which is governed by the aforementioned Fick's equations.

13.6 Overall Mass Transfer Coefficient (U)

The total resistance (R_t) takes into consideration the resistance inside the interface region of packaging material owing to diffusion phenomena as well as resistance on the surface due to absorption and desorption phenomena. The overall mass transfer coefficient (U) takes into account solute diffusivity through the membrane as well as mass transfer resistances at the surfaces. The sum of all previously discussed resistances to mass transfer is used to express the total resistance (R_t):

$$R_t = R_{\text{surface}} + R_{\text{diffusion}} = \frac{1}{U} = \frac{1}{h_{ma}} + \frac{L}{D} + \frac{1}{h_{md}} \tag{13.16}$$

where the convective mass transfer coefficients for absorption and desorption phenomena are h_{ma} and h_{md}, respectively.

The properties of the solutes, solvents and package, as well as the level of solvent agitation on both sides of the packing material, all play a role in the resistance to solute absorption and desorption into the package and on the low concentration side, respectively (Fig. 2.3). The partition coefficient is the only factor that influences the rate of mass transfer over the interface when resistances are minimal, such as when agitation on the solution side yields identical diffusion in the interface region and that across the membrane material as a whole, $C'_1 \approx C_1$ and $C'_{S1} \approx C_{S1}$ in this instance.

The flux of the overall mass transfer (J_T), which is determined by the overall mass transfer coefficient (U) and the concentration difference driving the mass transfer, is as follows:

$$J_T = U\Delta C = U(C_1 - C_0) \tag{13.17}$$

Equation (13.17) depicts the overall mass transfer via a packing material from the side with C_1, that is, high concentration to the side with C_0, that is, low concentration. The total flux (J_T) in such a mass transfer phenomena is equal to the fluxes at the interfaces (J_h) and the flux of diffusional mass transfer (J_d), therefore taking into account that the concentration difference is a driving force rather than a difference in partial pressure. The permeation coefficient (the permeability P) and the flux of permeation in the transfer of gas through a package described above are often equivalent to the total flux (J_T) and the overall mass transfer coefficient (U).

$$\begin{aligned} J_T &= J_{ha} = J_d = J_{hd} = U(C_1 - C_0) = h_{ma}(C_1 - C_{s1}) = \frac{D(C_{s1} - C_{s0})}{L} \\ &= h_{md}(C_{s0} - C_0) \end{aligned} \tag{13.18}$$

The partition coefficient dominates surface resistance, according to Eqs. (13.12) and (13.18). Considering that the partition coefficient is same for both absorption and desorption, the diffusion flux (J_d) can be written by replacing C_{s1} with $K\,C_1$.

When the dimensionless partition coefficient (K) is introduced, as stated in Eq. (13.19), empirically dealing with the surface mass transfer coefficient (h_m) and convective mass transfer is not difficult:

$$J_T = \frac{D(C_{s1} - C_{s0})}{L} = \frac{DK(C_1 - C_0)}{L} \tag{13.19}$$

13.7 Modelling

Mathematical models are used to describe physical processes of practical interest. As an alternative to the experimental examination of the actual process, these are quite helpful. Therefore, for both industry professionals and regulators, models that describe the mass and gas transfer via packaging materials to foodstuffs are extremely useful tools. These models can, at least in part, replace time- and cost-consuming migration studies. The models may be used to evaluate and demonstrate compliance with certain regulatory mobility limitations and to represent changes in migratory species concentration over time, either in polymer or in food-dependent diets.

Recently, there has been a lot of research done on modelling the transfer of additives, particularly antioxidants and stabilisers, as well as monomers and other compounds utilised in the manufacturing or conversion of other substances. Recently, mathematical modelling has been applied to improve the efficacy and safety of virgin polymer coatings (also known as functional barriers) in preventing any potential food contamination due to recycled plastic use in adhesive or composite structures (Feigenbaum et al. 2005; Dole et al. 2006a, 2006b).

The most significant physicochemical interactions inside the food-packaging-environment system are migration, sorption and penetration across numerous categories of chemicals, and these interactions have a significant impact on the quality of food (Fig. 13.4). Substances that come from the environment (SE) or food (SF) may be appropriately captured by food or packaging. Migration is shown as MF for chemical substances coming from the packing material to its enclosing food volume and ME for the reverse direction of migration in the same figure.

Fig. 13.4 Food-packaging-environment possible mass transfer processes

All of the aforementioned phenomena could possibly have an impact on the general product quality. These processes evolve according to nature, composition and physical qualities of the food as well as the physicochemical properties of the polymer. Consumers must feel safe using any existing packaging materials that contain process aids, colourants, active chemicals or functional elements. Thus, it is a food with ingredients that either interact or not with the packaging material. As a result, this could alter the materials' initial mechanical and barrier qualities in addition to the product's safety when consumed. Migration is by far the most important process in terms of public health because chemical contaminants that migrate into foods may pose a risk to people's health (Sanches-Silva et al. 2008). Understanding and regulating the mechanism by which a potentially harmful ingredient migrates from a packaging material is crucial to the selection and use of food packaging materials for the possible influence on human health. The literature contains a variety of references pertaining to migration rates, their response items and the function of supplements (Gilbert et al. 1980; Tehrany and Desobry 2004). There is a need for additional research from both an experimental and theoretical perspectives since the mass transport phenomena associated with food packing can have a considerable impact on many technological sectors (Vitrac and Hayert 2006).

The term "migration", as it is frequently used, refers to the chemical compounds' diffusion from a zone of greater concentration (the food-contact layer) to a zone of lower concentration (typically the food surface), as a result of the concentration gradient.

The three steps that can be utilised to monitor the migration process are typically diffusion of chemical compounds through the matrix, desorption of the diffused molecules from the polymer surface in the food volume and sorption of compounds at the polymer-food interface. Diffusion is primarily responsible for the mass transfer mechanism that takes place when chemical components migrate from packing material to food. Even though solubility is not a transport phenomenon, it has an impact on other mass transport phenomena wherever it occurs in a system because of the partition coefficient, which reflects the compatibility of the polymer with its solvent (food) (Mangaraj 1963; Tehrany and Desobry 2004). Solubility factors were used by Halek and Luttmann (1991), Nielsen (1994) and Paik and Writer (1995) to predict migration rates from polymeric packaging to food. The mass movement caused by sorption or desorption phenomena may be assessed according to studies that made it possible to distinguish between the dominant internal, interfacial and thermodynamic systemic phenomena.

According to Piringer and Baner (2008) and Pennarun et al. (2004), the following two hypotheses were represented:

1. The migrant has a high solubility in food.
2. The diffusion coefficient of the migrant has "upper bound" value, D*, which really reflects the most unpleasant case scenario, with a major emphasis on security rather than quality control. Within a specified level of statistical confidence, D* is the value of the inflated additive diffusion coefficient (DP), which is

higher than any actual additive diffusion coefficient (DP) for the particular migrant.

While the primary hypothesis results in a direct relationship, $K_{P,F} \leq 1$ (where $K_{P,F}$ is the polymer-food partition coefficient), the second hypothesis is far more challenging to enumerate. This is due to the fact that a realistic D_P might be anywhere between 10^{-7} cm^2 s^{-1} and 10^{-18} cm^2 s^{-1}. Together with the diffusion equations and $K_{P,F} = 1$, this can produce an approximate estimate of the migration value based on calculations that is equivalent to or larger than the actual migration value obtainable in the similar situation.

The evolution of the migrant concentration in the feeding or packing phase is depicted in a novel approximation space known as the "kinetic phase diagram", the result of a revolutionary solution to the general dimensionless mass transport issue. The interior and exterior mass transfer resistances, which could be easily differentiated, might be used to easily explain the migrant concentration at equilibrium. The three characteristics (especially diffusion, partition coefficient and interfacial mass transfer coefficient) governing the sorption kinetics were simultaneously identified using the estimate approach as suggested.

The assumption that the migratory process was controlled by the migrant's diffusion, as illustrated by Fick's law in Eq. (13.1), was made by the majority of the migration modelling studies that used a deterministic method (Crank 1975; Limm and Hollifield 1996; Reynier et al. 1999; Poças et al. 2008). Equation (13.1) shows that the gradient of a substance's concentration is inversely proportional to the transient mass flow of that substance through a control volume. It should also be noted that this method was relied on particular assumptions, that is, homogeneous media, constant packaging thickness, no chemical processes, no boundary effects, etc. Along with diffusion, the following modifications are made to the chemical processes that are considered:

$$\frac{\partial C}{\partial t} = -D\frac{\partial^2 C}{\partial^2 X} + kC^n \tag{13.20}$$

where "n" denotes the order of the reaction and k the reaction rate constant.

The chemical process and Fickian diffusion are both explained by Eq. (13.20). A few assumptions must be made so as to obtain a critical solution for the second-order partial differential equation. Initially, it was assumed that the diffusion coefficient was constant for both the food and the packaging (Brandsch et al. 2006). The basic diffusion (Eq. (13.1)) can be resolved, and it can be inferred that for the equation to be most effectively used, the migrant's diffusion and partition coefficient must be identified. It is commonly believed that migrant has a very high solubility in the polymer. As a result, the hypothesis that the partition coefficient $K_{P/F} = 1$ is taken, avoiding any further problems in determining the partition coefficient for a particular migrant-packaging-food system (Brandsch et al. 2002). In order to estimate the diffusion coefficient for a specific migrant, many models are prepared and stated over time (Piringer 1994; Limm and Hollifield 1996; Han et al. 2003; Pennarun et al. 2004; Begley et al. 2005).

The most widely used model now is the one developed by "Piringer". Piringer's approach is used to mathematically characterise the diffusion coefficients of gases and condensed phases, notably polymeric products, which is written as follows (Piringer 1994; Brandsch et al. 2002):

$$D_P \leq D_P{}^* = 10000e^{\left(A - aM_i - \frac{b}{T}\right)} \tag{13.21}$$

The molecular weight and temperature of additives are denoted by M and T, respectively; also D_p, D_p^* and A represent the diffusion coefficient of additive, overestimated diffusion coefficient of additive and polymer-specific constant, respectively, where a and b are an additive molecular weight-specific constant and temperature-specific constant, respectively.

The diffusion coefficients have been estimated utilising experimental investigation in Formula (13.21), placing the study-derived coefficients against the migrant's molecular mass, well with precise component A_p, and for the actual temperature (T) in K (Brandsch et al. 2002). By merely altering the model's unique polymer properties, new types of polymers may be accommodated (Pennarun et al. 2004). According to Brandsch et al. (2002), the model's conclusions overstated experimental values by more than 95%, with high temperatures showing the lowest discrepancies between "worst-case" estimates and experimental data. Without utilising experimental data, Brandsch et al. (2002) calculated the diffusion coefficients using the formula below:

$$D_P = D_0 \exp\left(A_P - 0.1351M_i^{\frac{2}{3}} + 0.003M_i - \frac{10454}{T}\right) \tag{13.22}$$

The equation linked the molecular masses to the quantity, A_P, which acted as the polymer matrix's "conductance" to migratory diffusion (Brandsch et al. 2002). That is feasible to relate "conductance" in Formula (13.4) for providing "worst case" migration prediction. You may then estimate the migration rate that fall inside the margin of safety of the permissible level. The temperature (T) that is truly an input from the matrix material towards the diffusion activation energy, the polymer-specific parameter's upper limit (A_P) and the polymer-specific parameter (t) may all be used to achieve this:

$$A_P = A'_P - \frac{\tau}{T} \tag{13.23}$$

As demonstrated by Brandsch et al. (2002), under these circumstances, the migrate frequency of a migrant out from the material to the food-in-contact may be determined using the calculation presented below, which is created after Fick's law calculation that is quantitatively addressed using planar geometry:

$$\frac{M_{F,t}}{A} = C_{P,0}\rho_P d_P\left(\frac{a}{1+a}\right)\left[1-\sum_{n=1}^{\infty}\frac{2\alpha(1+\alpha)}{1+\alpha+\alpha^2 q_n{}^2}\,exp\left(-D_{Pt}\frac{q_n{}^2}{d_P{}^2}\right)\right] \tag{13.24}$$

For which $M\ (F,t)/A$ represents the value of migration at time t, also A, is the contact surface area, C_{P0} is the initial concentration of migrant inside the polymer, while as the density of polymer and food is represented by P and F, respectively, d_P is the polymer thickness, and $=((VF/VP))/K(P,F)$ where VF and VP are the food volume and polymer volumes, respectively, and $K(P,F)$ partition coefficient of food-polymer. The equation's positive roots are the parameter's q_n, which are as follows: $T_{an}\ (q_n) = D = q_n$.

A model based on part of experimental results and approximations was developed by Limm and Hollifield (1996). The model of additive distribution anticipation approach is a first step in methodically planning the migration of the additives. Their strategy made use of the correlations among activation energies and molecular diameters that were found for relatively light molecules (100 Daltons or less in weight). It was believed that an Arrhenius-type equation of the same kind might appropriately explain the molecular diffusion of additives across a polymer matrix and is also reported in Helmroth et al. (2002) and Poças et al. (2008):

$$D = D' e^{\frac{-E_D}{RT}} \tag{13.25}$$

The Piringer's model, which has been the subject of prior research, established a clear connection within the diffusion coefficient and the migrant's molar concentration. The Arrhenius behaviour is used in Limm and Hollifield's model to estimate the temperature dependence of the dispersion of different additives. However, this model, unlike Piringer's model, needed very little data, and its predictive method relied upon established physical diffusion theories. The Limm and Hollified's model appears to have a close relationship with experimental data since it relies on empirical constants acquired via practical migrant data. The models developed by Piringer, Limm and Hollified also seem to have a number of restrictions that must be considered before applying the model. Because a correct calculation of the diffusion coefficient is required, it is vital to emphasise that these two models may only be accurately applied to polyolefins. It must be remembered that polyolefins have become the most often utilised material in the food packaging business right now.

As mentioned below, numerous models were discovered in the literature that aim to forecast the migration pace from the opposing viewpoint. A model utilising the finite element approach was developed and implemented by Han et al. in 2003 for quantifying migration in multilayer frames (FME), which is a numerical treatment technique. Numerous presumptions were formed in this manner, such as the absence of polymer swelling, the coefficient of independent separation, a huge weight transfer, the dependence of diffusion coefficients on temperature, the existence of complete packing, etc. (Poças et al. 2008), assuming minimum strength and a constant diffusion coefficient of the properties of food packaging and mass transportation. The Fick's laws analytically condensed simplified analytical solution yielded the diffusion coefficient employed in the simulations, which is written as

$$\frac{M_{F,t}}{A} = 2C_{P,0}\left(\frac{D_t}{\pi}\right)^{0.5} \tag{13.26}$$

where A represents the portion of the polymer's surface that is interacting with food, $C_{P,0}$ represents the starting percentage of the migrants' polymer volume at time t D_0, D represents the migrant's diffusivity within the polymer and $M_{F,t}$ represents the mass amount which has moved towards the food after time "t" from the polymer.

The coefficient of diffusion was computed by considering that the entire process is exclusively diffusion driven since multilayer polymer packaging was the prime focus of our modelling process, and the present research value was restricted in the same manner.

13.8 Significance

As it influences food manufacturing, preservation, distribution, marketing, advertising and even meal preparation, food packaging relations have grown to be a relatively important field of study, although colourants, active chemicals or functional processes that serve as process aids in the currently used packaging materials should be safe for consumers.

As a result, it is a food which may possibly interact with the packaging. Thus, basic mechanical and barrier characteristics of the materials as well as the product's safety during consumption may be changed. The selection and usage of food packaging materials must take into account the possible influence on human health by identifying and controlling the migration of a potentially hazardous component (toxins) from a packaging material. The early usage of two very basic models, that is, the Piringer's model (Piringer 1994) or the Limm and Hollified's model, has been the most popular method for practical migration modelling for quality assurance (Limm and Hollifield 1996).

In Piringer's model, the absolute temperature and a specific parameter are employed to relate the relative molecular mass of the migrant with the diffusion coefficients. For the purpose of migration modelling, this method is typically the simplest and most popular. The Limm and Holifield's model, which works effectively with only polyolefins, seems to be restricting the model's potential to be used to migration modelling for other materials. The Limm and Holifield's model, on the other hand, proposed a similar approach for migration modelling. However, the study by O'Brien et al. (1999) demonstrates that the Limma and Hollifield's model was more accurate in most situations when a direct comparison of the two models for the migration of a variety of compounds from HDPE in olive oil was done. All approaches meet with the core principle that the mass migrant transfer is diffusion-driven and is thus suitably described by Fick's laws. As a result, while the models are otherwise essentially equal in their insight, the diffusion coefficient has actually been estimated by efficiently categorising the variations among the most common migration predictive models.

The following particular assumptions serve as the foundation for the Fick's law's valid application:

1. The distribution of migrants was uniform, whether it was during the packaging process or the food phase.
2. In the migrant's transition between the packaging and food stages, there was no evidence of boundary resistance.
3. Packaging and food do not interact in any manner.
4. Within the food packaging system, there was no swelling phenomenon.
5. It is possible to assume that food and polymer have a specific partition coefficient.
6. No external provision for migrants during the migration process was seen, that is, the total amount of migration occurs in the food packaging system remains same.

Although the statements stated above may be true for the great majority of migration applications, some more complex phenomena, including the sorption of specific food pollutants by the packing material, partitioning at the food packaging interface, chemical interactions of the packaging material or some of its byproducts with the food content, etc., appear to be less frequent and will be taken into account even though they all have a great potential for occurring inside real food packaging systems. It is safe to presume that none of the above assumptions can always be true.

In general, it appears to be rather simple to detect an incidence when at least one of the abovementioned assumptions is false. For example, we might report on the assumption in a more specific situation of pragmatic hesitations (1) which, according to the assumption, packaging materials except plastic polymers could not be used and (2) that might not applicable to sticky migrants and so on. By summarising the research on various materials in interaction with food simulants used to validate the abovementioned models and the findings of that research, it is possible to come to the conclusion that migrants exhibit an apparent non-Fickian behaviour. Each model made extensive use of oversimplified assumptions, which increased inaccuracies but was still suitable for resolving food safety issues. When a more in-depth, comprehensive, highly descriptive and scientifically accurate insight may be needed, Eq.'s (13.1) output results seem to be quite lacking. Equation (13.1) is sufficient for defining standards and laws and for estimating mass transportation for industrial applications. It is unquestionable that Eq. (13.20) is a considerable advance, despite the fact that its apparent applicability has been quite limited, likely primarily due to its inevitable complexity.

The following system consideration is the latter disadvantage: rather than being an essential part of the system itself, the environment is frequently thought as an outside force that creates the circumstances (boundary/initial conditions and parameter values employed in the equations). As a result, only certain variables and parameters rather than the environmental impact directly contributed to the transport issue. Therefore, such involvement can be regarded as being somewhat arbitrary. Since the unique method now suggested is "environment-packaging food", the old "packaging food" system has to be revised, according to the authors, and may even need to be restructured entirely. We want to promote the concept that taking such a position enables a more in-depth and deeply integrated study of the whole spectrum of all transport phenomena that take place inside an entire system.

13.9 Conclusion

The main models reviewed are predicting the movement of known and already characterised migrants. Because the material could include a number of fully unidentified compounds, these models cannot exactly predict how a chemical will migrate from one contact medium to another to the amount of food. Evidently, these models have a tendency to overestimate the migration value, making it impossible for them to be employed in the food manufacturing process where an additive's migration may be crucial and/or even required to maintain the product's optimum quality, the same as in case of active packaging. Calculation of the accurate value of migration procedure is necessary in order to achieve optimum quality. As a result, there is a need of a new study that identifies an appropriate model which is required, and this seems to be a top priority. Therefore, when doing research, it is important to include all phenomena of mass transfer that occur inside the "food-packaging-environment" system. This will be particularly significant for the food industry when ensuring food safety through cost-effective checks of fulfilment with current regulations in addition to when focusing on additional quality-based food packaging and system design. For industrial operations, a method for analysing the relation among the environment, packaging and food is becoming necessary to optimise how well the three components of the system work together (food-packaging-environment). Therefore, future research and modelling efforts should presumably focus more on developing a model that can produce accurate results under the complex assumptions and situations indicated above.

Additionally, polysaccharide films must meet the same standards as traditional food packaging materials. These are correlated majorly to their barrier properties, optical properties and strength and moulding properties. They should also have migration/scalping resistance for food quality purposes, as food quality duration compatibility with their shelf life is a major challenge with bio-packaging. Legislations are similar to plastics and bio-plastics so that these must compete with legislation. In addition, they are biodegradable or edible and put constraints for food application purposes.

References

Aider M (2010) Chitosan application for active bio-based films production and potential in the food industry. LWT Food Sci Technol 43(6):837–842

ASTM Subcommittee E35. 15 on Antimicrobial Agents (1993) ASTM standards on materials and environmental microbiology. ASTM International

Begley T, Castle L, Feigenbaum A, Franz R, Hinrichs K, Lickly T, Piringer O (2005) Evaluation of migration models that might be used in support of regulations for food-contact plastics. Food Addit Contam 22(1):73–90

Bourtoom T (2008) Edible films and coatings: characteristics and properties. Int Food Res J 15(3): 237–248

Brandsch M (2006) Transport of L-proline, L-proline-containing peptides and related drugs at mammalian epithelial cell membranes. Amino Acids 31:119–136

Brandsch J, Mercea P, Rüter M, Tosa V, Piringer O (2002) Migration modelling as a tool for quality assurance of food packaging. Food Addit Contam 19(S1):29–41

Coles R, McDowell D, Kirwan MJ (eds) (2003) Food packaging technology, vol 5. CRC Press, Boca Raton

Crank J (1975) The mathematics of diffusion. Oxford University Press, New York

Debeaufort F, Quezada-Gallo JA, Voilley A (1998) Edible films and coatings: tomorrow's packagings: a review. Crit Rev Food Sci 38(4):299–313

Dole P, Feigenbaum AE, Cruz CDL, Pastorelli S, Paseiro P, Hankemeier T, Papaspyrides C (2006a) Typical diffusion behaviour in packaging polymers–application to functional barriers. Food Addit Contam 23(2):202–211

Dole P, Voulzatis Y, Vitrac O, Reynier A, Hankemeier T, Aucejo S, Feigenbaum A (2006b) Modelling of migration from multi-layers and functional barriers: estimation of parameters. Food Addit Contam 23(10):1038–1052

Dutta PK, Tripathi S, Mehrotra GK, Dutta J (2009) Perspectives for chitosan based antimicrobial films in food applications. Food Chem 114(4):1173–1182

Elsabee MZ, Abdou ES (2013) Chitosan based edible films and coatings: A review. Mater Sci Eng C 33(4):1819–1841

Feigenbaum A, Dole P, Aucejo S, Dainelli D, Garcia CDLC, Hankemeier T, Voulzatis Y (2005) Functional barriers: properties and evaluation. Food Addit Contam 22(10):956–967

Gilbert SG, Miltz J, Giacin JR (1980) Transport considerations of potential migrants from food packaging materials. J Food Process Preserv 4(1–2):27–49

Halek GW, Luttmann JP (1991) Sorption behaviour of citrus-flavour compounds in polyethylenes and polypropylenes: effects of permeant functional groups and polymer structure. In: Hotchkiss JH (ed) Food and packaging interaction II. Symposium Series American Chemical Society, Washington, DC, pp 212–226

Han JK, Selke SE, Downes TW, Harte BR (2003) Application of a computer model to evaluate the ability of plastics to act as functional barriers. Packag Technol Sci Int J 16(3):107–118

Helmroth E, Rijk R, Dekker M, Jongen W (2002) Predictive modelling of migration from packaging materials into food products for regulatory purposes. Trends Food Sci Technol 13(3):102–109

Hunt CB, Auster ER (1990) Proactive environmental management: avoiding the toxic trap. MIT Sloan Manag Rev 31(2):7

Jimenez A, Fabra MJ, Talens P, Chiralt A (2012) Edible and biodegradable starch films: a review. Food Bioprocess Technol 5:2058–2076

Kester JJ, Fennema OR (1986) Edible films and coatings: a review. Food Technol 8(12):47–59

Kim KM, Son JH, Kim SK, Weller CL, Hanna MA (2006) Properties of chitosan films as a function of pH and solvent type. J Food Sci 71(3):E119–E124

Klemm D, Heublein B, Fink HP, Bohn A (2005) Cellulose: fascinating biopolymer and sustainable raw material. Angew Chem Int Ed 44(22):3358–3393

Krochta JM (1997) Edible protein films and coating. In: Damodaran S, Paraf A (eds) Food proteins and their applications. Marcel Dekker, New York, pp 529–549

Limm W, Hollifield HC (1996) Modelling of additive diffusion in polyolefins. Food Addit Contam 13(8):949–967

Mangaraj D (1963) Cohesive energy densities of high polymers. Part I determination of cohesive energy density from swelling measurements. Macromol Chem Phys 65(1):29–38
Moore WJ (1972) Physical chemistry, 5th edn. Longman Group Limited, London
Nielsen TJ (1994) Limonene and myrcene sorption into refillable polyethylene terephthalate bottles, and washing effects on removal of sorbed compounds. J Food Sci 59(1):227–230
O'brien DF, Baldo MA, Thompson ME, Forrest SR (1999) Improved energy transfer in electrophosphorescent devices. Appl Phys Lett 74(3):442–444
Paik JS, Writer MS (1995) Prediction of flavor sorption using the Flory-Huggins equation. J Agric Food Chem 43(1):175–178
Pennarun PY, Dole P, Feigenbaum A (2004) Overestimated diffusion coefficients for the prediction of worst case migration from PET: application to recycled PET and to functional barriers assessment. Packag Technol Sci Int J 17(6):307–320
Piringer OG (1994) Evaluation of plastics for food packaging. Food Addit Contam 11(2):221–230
Piringer OG, Baner AL (2008) Plastic packaging: interactions with food and pharmaceuticals. John Wiley & Sons, Weinheim
Poças MF, Oliveira JC, Oliveira FA, Hogg T (2008) A critical survey of predictive mathematical models for migration from packaging. Crit Rev Food Sci Nutr 48(10):913–928
Psomiadou E, Arvanitoyannis I, Yamamoto N (1996) Edible films made from natural resources; microcrystalline cellulose (MCC), methylcellulose (MC) and corn starch and polyols—Part 2. Carbohydr Polym 31(4):193–204
Reynier A, Dole P, Feigenbaum A, Feigenbaum A (1999) Prediction of worst case migration: presentation of a rigorous methodology. Food Addit Contam 16(4):137–152
Robertson GL (1993) Permeability of thermo plastic polymer. In: Food packaging, principles and practice. Marcel Dekker, New York, pp 73–110
Segall KI, Scanlon MG (1996) Design and analysis of a modified-atmosphere package for minimally processed romaine lettuce. J Am Soc Hortic Sci 121(4):722–729
Silva AS, Freire JC, Franz R, Losada PP (2008) Mass transport studies of model migrants within dry foodstuffs. J Cereal Sci 48(3):662–669
Siracusa V, Rocculi P, Romani S, Dalla Rosa M (2008) Biodegradable polymers for food packaging: a review. Trends Food Sci Technol 19(12):634–643
Stilwell EJ, Canty RC, Kopf PW, Montrone AM, Arthur DL, Inc (1991) Packaging for the environment, a partnership for progress. AMACOM, American Management Association, New York
Tehrany EA, Desobry S (2004) Partition coefficients in food/packaging systems: a review. Food Addit Contam 21(12):1186–1202
Vargas M, González-Martínez C (2010) Recent patents on food applications of chitosan. Recent Pat Food Nutr Agric 2(2):121–128
Vieira MGA, Da Silva MA, Dos Santos LO, Beppu MM (2011) Natural-based plasticizers and biopolymer films: A review. Eur Polym J 47(3):254–263
Vitrac O, Hayert M (2006) Identification of diffusion transport properties from desorption/sorption kinetics: an analysis based on a new approximation of fick equation during solid–liquid contact. Ind Eng Chem Res 45(23):7941–7956
Zhang L, Gu FX, Chan JM, Wang AZ, Langer RS, Farokhzad OC (2008) Nanoparticles in medicine: therapeutic applications and developments. Clin Pharmacol Ther 83(5):761–769

Chapter 14
Modelling and Dimensioning of Polysaccharide-Based Edible Films

Mohammad U. Makeri

14.1 Introduction

Food packaging is employed not only to serve the protection and provision of nutritional information of the content but also to impart safety advice, extend shelf life and improve textural characteristics in addition to primary role of containing food product. The use of petroleum-based plastic packages such as polyethylene and polystyrene constitutes non-biodegradable waste that poses serious environmental challenges in addition to hazards to aquatic life. This led to the quest for packaging materials that have close or superior aesthetic properties and that can be degraded biologically (biodegradable) and/or can even be ingested by humans (edible) along with the food being contained and protected. The term "biodegradable material" describes materials that when disposed of, break down within a short time and return to their natural state.

Edible films are pre-formed, thin layer of material (usually transparent) wrapped directly on the surface of food product as a primary packaging to prevent against physical and mechanical damage, preserve nutrients, protect against microbes and act as a fluid barrier (Pirozzi et al. 2020). They function based on their property to act as variable barriers against water and/or gas or vapour transmission between foods and packaging (Cuq et al. 1995).

The use of biodegradable polymers for edible film packaging offers a good alternative to traditional food packaging and a potential, viable solution to the environmental waste generated from non-biodegradable and other synthetic polymers (Cazón et al. 2018a; Silva et al. 2008). Consequently, renewable raw materials

M. U. Makeri (✉)
Food Technology Department, NAERLS, Ahmadu Bello University, Zaria, Nigeria

Analytical Laboratory Unit, Product Development Research Programme, Institute for Agricultural Research, Ahmadu Bello University, Samaru, Zaria, Nigeria

T. Amin et al. (eds.), *Polysaccharide Based Films for Food Packaging: Fundamentals, Properties and Applications*,
https://doi.org/10.1007/978-981-99-4898-7_14

such as proteins and polysaccharides are the focus of most research efforts aimed at developing biodegradable polymers due in part to their film forming capability, versatility, flexibility and edibility.

Polysaccharide-based films for use as edible packaging must be nontoxic and can be made from biopolymer including starch, carrageenan, chitosan, etc. or their composites (Cerqueira et al. 2013). Polysaccharides such as cellulose, pectin, chitosan, carrageenan, starch, pullulan, gellan gum and xanthan gum alginate have been employed as edible food films to serve as a barrier layer against fluid diffusion, water movement, aroma degradation and exchange of solutes (Sabbah et al. 2019; Kõrge et al. 2020). Any contemplated edible food films should be able to offer protection to the food against physical and microbial incidences, maintain sensory characteristics, offer appropriate barrier against moisture and gases of fresh produce during storage, retard enzymatic oxidation and provide protection of the food from discoloration and texture softening, be safe, offer biochemical and physicochemical stability to the food, delay food deterioration and be produced by simple technology using low-cost biopolymers (Nešić et al. 2019; Alvarado-González et al. 2012). Cerqueira et al. (2013) reported that the functionality and performance of edible films mainly depend on their wetability and barrier and mechanical properties, which on the other hand are governed by film composition, method of film formation and the technique of application of the film to the food product. Thus, the design and selection of polysaccharide material for use in edible films are carried out based on their physical, mechanical, optical, structural, permeability properties and microstructural properties that dictate their structural functionality and protective properties (Alvarado-González et al. 2012; Mohammadi et al. 2018; Taghizadeh et al. 2018; Chen et al. 2020). However, polysaccharide-based films together with plasticizer yield better film compared to plain films because polysaccharides (including cellulose, starch and starch derivatives, pectin derivatives, sea weed extracts, gums, microbial gums and chitosan) are generally very hydrophilic and result in poor water vapour and permeability properties (Bourtoom 2008).

This chapter therefore presents an overview of film formulation and dimensioning along with model polysaccharide edible films and some important considerations in the design and evaluation of contemplated edible films.

14.2 Rationale for Dimensioning

Composite polymer offers a simple and relatively inexpensive source of film forming material for developing biodegradable edible packaging. Developing an edible material requires the use of three key ingredients to produce a standard edible film: film forming substance, plasticizer and additive (Han 2014). A suitable solvent usually water is also required for the preparation of a film-forming dispersion for edible films. Generally, aqueous alcohols, water, or alcohol with a natural solubility are utilized as solvents for compounds that form films. The main components that can produce films include proteins, polysaccharides, lipids, and combinations or mixtures of these (Han 2014). Additionally, to enhance the protective qualities of edible films, functional additives such plasticizers, antioxidants, vitamins, antibacterial agents, essential oils, pigments, and chemical preservatives are utilized. For example, mixing a brittle polymer with a rubbery one may result in a material with good rigidity and toughness (Morgado et al. 2011). Polyvinyl alcohol (PVA) is a polymer suitable for mixing with cellulose to improve the mechanical properties of the composite film. Besides, it is easily soluble in water, nontoxic and biodegradable (Frone et al. 2015; Wahyuningsih et al. 2016). Understandably though, the function of biodegradable films is not to completely replace other forms of food packaging, but to provide value-added packaging that may have less negative impacts on the environment as a result of non-biodegradable solid waste in addition to extending the useful life of foods. To meet the desired objectives, the design of an edible food packaging must take into consideration following important physicochemical properties (water vapour and oxygen permeability, opacity, light transparency, solubility in water, hydrophilic-hydrophobic interactions), mechanical properties and structural properties (tensile strength, percent elongation, elongation at break, elastic modulus and glass transition temperature) and nutrition and safety concerns (non-toxic, non-allergic, fully digestible).

14.2.1 Mechanical Properties

Food items can benefit from the mechanical protection provided by edible films by serving as an obstacle between the food and the outside world, adding extra security while being stored. Edible films' mechanical attributes such as tension and elasticity enable them to endure physical force and shield the food from harm. Mechanical properties of films are important to ensure integrity during transportation, handling and storage of packaged foods (Rouhi et al. 2017). Mechanical properties including toughness, burst strength and distance to burst (maximum deformation) depend on film composition and component properties (Vieira et al. 2011). The mechanical properties such as elastic modulus are critical in the design of edible films (Romero-Bastida et al. 2011; Kechichian et al. 2010; Yener et al. 2009). These properties can also alter how heat is transferred while food is being dried or fried, significantly lowering the danger of mechanical harm. Generally, however, the mechanical damage that foods may sustain depends on the type of edible film used. While potato starch films are less strong mechanically than wheat and maize starch films, they are more effective in blocking oxygen and water vapour than the latter (Basiak et al. 2017). Beeswax, candelilla wax, and carnauba wax are examples of natural waxes that can be added to starch/gelatin films to enhance water resistance and surface hydrophobicity while lowering tensile strength (Cheng et al 2023). Measurement of these properties allows predicting how the material will behave under different food processing conditions and comparing performance with that of other polymers.

14.2.2 Barrier Properties

One of the major functions of edible films in food packaging is to minimize transfer of molecules between the food and the environment with the aim of preserving the food quality. Thus, the barrier properties of a polysaccharide-based food package films are critical for prediction and/or estimation of product shelf life. Measuring these properties allows to ascertain the permeability and transfer mechanisms of the film of molecules such as O_2, CO_2, water vapour, organic vapour or liquids. Among these fluids, water vapour permeability that measures the amount of water permeating a unit area and time (g/s m Pa) taking into consideration the differential pressure and thickness of the material, is the most extensively studied property of biodegradable, edible films principally because of the important role of water activity in enzymatic, microbial growth and other biochemical reactions that determine food safety and edibility. Film materials with low WVP values are desired in application such as wrapping and packaging in addition to flexibility and high elastic modulus, tensile strength and elongation at break values (Liu et al. 2012). Optical and micro-structural properties, on the other hands, are critical in the selection and functional analysis of properties of contemplated material for use as edible film base or composite (Liu et al. 2012; Abugoch et al. 2011).

Water adsorption is important because packaging films should maintain moisture levels in a packaged product, depending on the type and state of the foods. This property implies swelling capacity of the film in contact with food and liquids or during storage at high relative humidity and can affect mechanical properties of films. The presence of micropores in structure of some biopolymers such as cellulose and their combination with plasticizers and other hydrophilic polymers is relevant to the formation of a suitable packaging film. Although commercially available synthetic polymers effectively control water vapour and gas permeability, edible films are preferably employed for packaging processed foods to prevent movement of gases and water because of their low WVP (Mohammadi et al. 2018). In addition to their hydrophilic nature and variable water solubility in hydrophilic compounds, polysaccharide-based edible film tends to reduce the lipid uptake in fried and fatty food products, and when used as food film, they minimize moisture loss of products during frying, while maintaining acceptable organoleptic and aesthetic characteristics of food products (Albert and Mittal 2002).

Consequently, the packaging properties of edible film can be deduced from characterization and evaluation of the functionality, and performance of edible films relates to their physicochemical and mechanical properties that can be modelled using different techniques.

14.2.3 Colour and Optical Properties

The colour of a film is an important factor in terms of consumer acceptance. The overall optical and micro-structural properties denote the transparency and ease of transmission of light through the film. Thus, the determination of these properties is critical for the final analysis of the functional characteristics of edible films and applicability to specific food products (Liu et al. 2012; Abugoch et al. 2011). The barrier properties to ultraviolet (UV) radiation of edible films are important characteristic because UV radiation causes degradation of lipids and fatty foods via oxidation reaction and consequently affects the appearance and flavour characteristics of film-packaged food products and decreased shelf life (Cazón et al. 2018b). The range of wavelength between 200 and 315 nm is the most susceptible region for lipid oxidation; hence, an appropriate formulation of biodegradable films with barrier properties against the UV light can help to slow the oxidation process of the lipids (Cazón et al. 2018b; Guo et al. 2014). Transparency is an important optical property of edible films that has a direct impact on the consumer acceptability because transparent packaging allows the consumer to have a good view of the filmed product content to judge its quality and make decision to buy or not.

14.2.4 Physicochemical Properties

Biopolymers are employed as alternatives to synthetic polymers in the food industry for use as edible packaging or films for foods because the non-biodegradable nature of the synthetic polymers used in containing most food products presents numerous environmental and health hazards ranging from waste accumulation and threats to aquatic lives. Edible, biodegradable polymers such as polysaccharides, however, presents a number of advantages when employed as food film. Polysaccharide-based edible films have been shown to regulate ethylene generation and delay senescence by decreasing the climacteric process of respiration and ethylene production in fruits, leading to a delay in ripening and senescence. They can also reduce the rates of harmful chemical reactions and stop moisture loss, maintaining the integrity and tenderness of food products (Pellá et al. 2020). Edible films also restricts how much oxygen is absorbed by foods by serving as a barrier to oxygen by regulating the passage of oxygen across a food item and the environment thus dictating the stability of foods especially produce (Sahraee et al. 2019; Tapia-Blácido et al. 2018). Film design and formation, on the other hand, are mechanisms whereby interaction within polymer(s) and between co-polymers plus additives leads to creation of cross-linkages and inter- and intra-molecular networks (Cazón et al. 2018a). Thus, the multitude of functional and physicochemical properties played by polysaccharide-based edible films ranged mainly from cohesiveness, swelling, water retention capacity, water vapour permeability, fat adsorption capacity and gelation capacity, in addition to other structural and mechanical properties. However, no single, plain polysaccharide or biopolymer possesses all these characteristics but may require sort of cooperation between more than one polysaccharide type to yield film with the desired mechanical and physicochemical properties good to withstand food handling and transportation stresses. Thus, the molecular association and interaction between polymers leads to reinforcement of film networks and stability during application on food products and also during handling of the filmed foods. For example, aloe gel films have high water vapour permeability but lack some barrier properties, whereas gellan gum, a polysaccharide, is hard and transparent and possesses low water vapour permeability and smooth surfaces; thus, when mixed as blend with polyvinyl alcohol and/or glycerol, the resulting film could have better water permeability, toughness, transparency and water and gas diffusion properties (Saibuatong and Phisalaphong 2010; Cazón et al. 2017). Furthermore, the design and development of functional edible films from pure components and blends necessitate the evaluation of optical, physical, mechanical and microstructural properties as they relate to functionality and structure of edible films. Additionally, the water sorption capacity, water vapour permeability, toughness and elastic modulus of *aloe vera-gellan* gum composite edible film synergistically improved compared to plain *aloe vera* or plain *gellan* gum film alone (Alvarado-González et al. 2012).

Cellulose, for example, is a linear chain with two anhydroglucose rings covalently bonded in a β-1-4 glycosidic bond isolated from plants and other plant-based materials with over ten billion tons produced each year globally and considered as one of the largest polysaccharide reservoirs with inexhaustible source of raw materials for edible food packaging (Wang et al. 2016; Xu et al. 2016). However, due to its crystallinity, cellulose is insoluble in water, limiting its potential food applications as base material for biodegradable films, but instead is used as a reinforcing agent in other polymer-based films (Fernandes et al. 2010; Khan et al. 2012; Roohani et al. 2008). Recently, though, studies have shown that cellulose or cellulose derivatives along with appropriate solvent can be used to develop biodegradable films or as composite along with plasticizers (Cazón et al. 2017; Park et al. 1993; Shivakumara and Demappa 2016). Table 14.1 shows the effects of incorporating levels of

Table 14.1 Effect of plasticizer type and concentration on mechanical properties, solubility and moisture content of *tara* gum film

Film type	Plasticizer content (gg^{-1})	Tensile strength (MPa)	Elongation (%)	Solubility (%)	Moisture content (%)
TG	0.0	97.24 ± 7.60	11.20 ± 4.66	12.39 ± 4.81	12.82 ± 0.54
TG + glycerol	0.075	90.60 ± 11.33	15.90 ± 5.40	14.37 ± 6.06	12.83 ± 0.19
TG + glycerol	0.150	78.26 ± 10.91	23.05 ± 2.69	17.98 ± 4.41	13.91 ± 0.12
TG + glycerol	0.225	54.00 ± 2.74	33.50 ± 2.12	25.37 ± 3.05	16.86 ± 0.17
TG + glycerol	0.300	45.36 ± 2.26	44.07 ± 3.81	32.40 ± 1.45	18.89 ± 0.76
TG + sorbitol	0.075	81.95 ± 6.91	12.00 ± 2.96	13.62 ± 1.03	12.93 ± 0.07
TG + sorbitol	0.150	62.87 ± 8.77	16.47 ± 3.74	18.67 ± 3.43	12.29 ± 0.18
TG + sorbitol	0.225	50.11 ± 11.70	20.87 ± 1.50	25.09 ± 2.96	12.44 ± 0.73
TG + sorbitol	0.300	43.64 ± 1.18	26.30 ± 0.42	34.16 ± 1.50	12.62 ± 0.47
TG + PEG400	0.075	69.52 ± 11.29	4.60 ± 1.13	29.72 ± 4.65	12.89 ± 0.83
TG + PEG400	0.150	46.75 ± 3.00	6.10 ± 0.42	32.83 ± 5.84	12.66 ± 0.11
TG + PEG400	0.225	29.29 ± 3.79	8.47 ± 0.76	33.92 ± 4.52	13.34 ± 0

TG tara gum; *PEG-400* polyethylene glycol 400 (Antoniou et al. 2014)

plasticizers on some mechanical properties of a polysaccharide (*tara* gum)-based film.

Chitosan, another important polysaccharide, has been used extensively in edible films and coatings among many other polysaccharides due to its abundance, versatility and excellent physicochemical, structural and antimicrobial properties (Cazón et al. 2018a, 2019a). The recent discoveries about the potential of chitosan as a biodegradable edible film highlight other important food packaging applications. Several studies placed chitosan-based composite as a good replacement to synthetic polymers due to its availability, biodegradability and amenability (Aşik and Candoğan 2014). Chitosan has good antimicrobial activity against most spoilage and pathogenic microorganisms because of its poly-cationic property and exhibits anti-oxidative activity owing to its ability to chelate free metal ions (Jridi et al. 2014; Leceta et al. 2013). Chitosan-based material also forms semipermeable films that are transparent, tough and flexible and have good oxygen barrier properties across the internal atmosphere for delayed ripening and decrease transpiration rates in fruits and vegetables (Li and Yu 2001). Thus, combination of polysaccharide with other biodegradable polymers (as composite) or the addition of plasticizers (as additive) is an effective strategy that can be employed to improve the physicochemical and mechanical properties and performance of polysaccharide-based films. Additionally, film blend including chitosan-gelatin edible film has fundamental properties of compact structure and enhanced physical, mechanical and transport properties superior to those of films based on single components (Benbettaïeb et al. 2014).

Consequently, many polysaccharide-based edible films are built upon chitosan (Van den Broek et al. 2015; Jridi et al. 2014; Elsabee and Abdou 2013), chitosan and polyvinyl alcohol and PVA (Cazón et al. 2018a, 2019a; Liu et al. 2018) as base material mainly due to its high-polar character, good mechanical properties, high thermal stability and amenability (Cazón et al. 2018a), tensile strength, barrier and water vapour permeability characteristics, water adsorption and mechanical properties, light barrier properties and transparency (Cazón et al. 2018a). Polyvinyl alcohol (PVA) is also a polymer suitable for mixing with cellulose or chitosan to modify the mechanical properties of the resulting composite film due to its solubility in water, biodegradability and nontoxicity (Frone et al. 2015; Wahyuningsih et al. 2016).

14.2.5 Effects of Plasticizers

Plasticizers are non-volatile compounds usually of low molecular weight that makes them fill the intermolecular spaces within and between polymeric chains, thus changing the three-dimensional molecular organization of polymers, reducing the energy required for molecular motion and the formation of hydrogen bonding between chains improving their molecular mobility (Cazón et al. 2018b).

Most plasticizers are polymers rich in hydroxyl groups that increase the plasticization of polymer network in edible films and prevent cracking during handling, transportation and storage of filmed food products (Cazón et al. 2018b). They are added as additives in film formulation mainly to increase the molecular mobility of biopolymers and decrease intermolecular attractions between adjacent polymeric chains via reduced hydrogen bonding between polymer chains. They also confer flexibility, reduce brittleness and improve tear-impact resistance of the packaging material, ultimately extending the product's shelf life (Barreto et al. 2003; Myllärinen et al. 2002; Rojas-Graü et al. 2008).

Addition of plasticizers such as glycerol and sorbitol also significantly improved extensibility, plasticity, toughness and workability properties of edible films (Bangyekan et al. 2006; Razmjoo et al. 2021). Thus, as shown in Table 14.1, the physical, mechanical, structural and permeability properties of polysaccharide-based edible films improve when developed as composite of chitosan-polyvinyl alcohol (PVA), gelatin-PVA, cellolose-PVA, carrageenan-glycerol films, etc. compared to plain films (Mohammadi et al. 2018; Taghizadeh et al. 2018; Chen et al. 2020), because polysaccharides (including cellulose, starch or their derivatives, pectin derivatives, gums and chitosan) alone are generally very hydrophilic and result in poor water vapour permeability properties (Bourtoom 2008). Polyvinyl alcohol enhances and strengthens the toughness of films, increases film thickness and bursts strength than other polymer plasticizers in polysaccharide-based films (Cazón et al. 2018b). The cellulose-glycerol-polyvinyl alcohol composite film showed better mechanical, structural and water absorption properties with improved burst strength; however, increasing the glycerol level above 5% (w/w) decreases the toughness (Cazón et al. 2018b).

14.3 Modelling and Dimensioning

14.3.1 Film Formulation and Preparation

To illustrate a model technique, an experimental design of a cellulose, glycerol and polyvinyl alcohol composite films conducted at different levels is shown in Table 14.2. Following the Box-Behnken design (Box and Behnken 1960), a second-order factorial design with polysaccharides (cellulose, chitosan, carrageenan, etc.) and polyvinyl alcohol plus a plasticizer (e.g. glycerol) as operational variable, and the effects of composite on the select dependent variable (T, BS, DB and %W) were considered as an example of film solution preparation. The dependent variables including the cellulose (range 3–4%, w/w), glycerol (0–5%, w/w) and PVA (2% w/w) were optimized to prevent rupture of films during drying and storage (Cazón et al. 2018b). The relationship between independent (cellulose, glycerol and polyvinyl alcohol) and dependent variables (T, BS, DB and %W) of the developed films was modelled using polynomial equations. The response of the dependent variables, toughness (T), burst strength (BS) and distance to burst DB and moisture content (% W) are also shown in Table 14.2. Samples were examined by UV-Vis-NIR spectrophotometer, FTIR, SEM and a_w, to analyse how the variation of glycerol and PVA

Table 14.2 Formulations assayed for the study of cellulose-based films reinforced with glycerol and polyvinyl alcohol and experimental results for the properties of the films (Cazón et al. 2018b)

Exp.	Cellulose (%w/w)	Glycerol (%w/w)	PVA (%w/w)	Toughness (MJ/m^3)	Burst strength (g)	Distance to burst (mm)	Water adsorption (%w/w)
1	3.505	–	2.00	0.30	1770.035	1.60	137.03
2	3.500	–	2.00	0.71	2356.91	1.74	166.49
3	3.502	–	5.00	2.60	5403.29	2.98	175.57
4	4.000	–	5.00	2.40	4805.65	2.75	199.36
5	3.502	–	5.00	0.43	1953.32	2.22	193.47
6	3.505	–	10.00	10.55	6400.57	3.22	202.16
7	3.002	–	10.00	28.45	10781.55	6.04	184.59
8	4.002	–	2.00	1.66	2303.49	1.67	141.72
9	4.005	–	5.00	0.52	9575.09	4.40	196.72
10	3.00	5.00	5.00	44.30	5322.73	2.72	197.48
11	3.50	2.50	5.00	0.95	3495.12	2.72	173.57
12	3.00	2.50	2.00	0.25	1395.09	4.76	136.35
13	3.50	0.00	10.00	23.41	12014.24	4.92	222.09
14	4.00	2.50	10.00	10.04	10291.13	5.11	219.89
15	4.00	2.50	10.00	10.04	10291.13	5.11	219.89

content could affect the water vapour permeability and optical, structural, morphological and water/O_2 barrier properties of the films, respectively.

14.3.2 Thermal and Mechanical Properties

Mechanical properties such as toughness, burst strength and distance to burst (maximum deformation) of polysaccharide-based edible films are governed by film composition, component properties and environmental conditions (Vieira et al. 2011). Mechanical properties of edible films are important as they guarantee the integrity of the packaged food products during transportation, handling and storage (Rouhi et al. 2017). Measurement of these properties allows prediction of how the film will behave under different food processing and handling conditions in comparison with the performance of other polymers. Studies showed that composite films of bacterial cellulose with polyvinyl alcohol and chitosan had improved mechanical and puncture properties and increased resistance and deformation at break.

Additionally, the porosity and three-dimensional structure of the cellulose facilitated the penetration of chitosan and PVA into the cellulose matrix, thus reinforcing the films (Cazón et al. 2019b). Further, the presence of chitosan and polyvinyl alcohol in cellulose films increased the thermal stability of the films and caused higher puncture properties than other polysaccharide films due to the interactions among the co-polymers (Cazón et al. 2019b).

Toughness (or tensile energy to break), also termed as the ability of a material to absorb energy during deformation up to the point of fracture (Zhang and Zhang 2012), is an important mechanical property in polymeric materials intended as packaging material that can withstand handling stress. Accordingly, the toughness (TS) encompasses the energy required both to create the crack and to propagate the crack until fracture (Zhang et al. 2005).

Another important and useful parameter employed to assess the mechanical strength of packaging films is the puncture strength. For a food film to be able to maintain its structural integrity during handling, transportation and storage of packed product without protuberances, it should have good uni-axial mechanical properties because films that have much strength will be expected to have better mechanical properties (Zhang et al. 2005). Puncture strength is another important parameter that determines the burst strength (BS) and distance to burst (DB) properties of the film. The toughness property of film can be measured using an Instron Universal Testing Machine following the procedures of the ASTM Standard Method D 882–91 (Cerqueira et al. 2011) (Fig. 14.1).

The test for this mechanical property is also performed using a texturometer such as TA-XT Plus at room temperature attached with appropriate software, whereas toughness (MPa) is measured employing the standard method D-882 of the American Society for Testing Materials (ASTM) (Cazón et al. 2018a). To do this, an initial grip separation of 40 mm and crosshead speed of 4 mm/s is set up, and ten samples of 15 mm × 100 mm are cut from each film formulation and stored in desiccators at 51% RH for five days fixed on a film supporting (HDP/FSR) and submitted to perforation by a cylindrical probe ($d = 3$ mm), moving at 60 mm/min until rupture (to stabilize the moisture content). Curves of force (g) as a function of deformation (mm) are obtained, and the maximum force of the peak is recorded as burst strength (BS) and distance to the peak recorded as distance to burst (DB) (Cazón et al. 2018b). The films are then fixed at both ends and the strength measured while in the tension mode and the toughness (T) calculated from the stress-strain plot as the area under the curve per specimen volume, following Eq. (14.1). TS and EB tests are

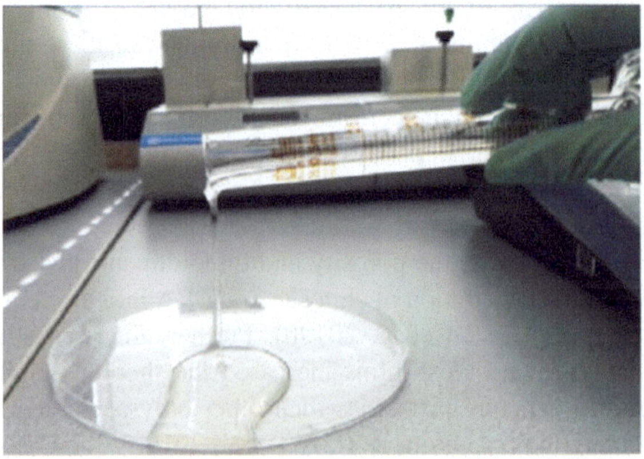

Fig. 14.1 Plain cellulose film solution prepared by NaOH/urea method (Cazón et al. 2018a)

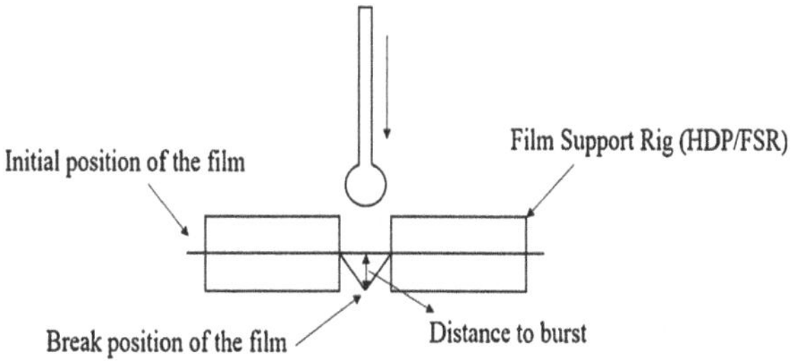

Fig. 14.2 Schematic diagram of method of puncture test for edible film

replicated at least five times (using five different film samples obtained from five different films) for each type of film to ensure precision and accuracy of the data:

$$T = \frac{\text{Energy}}{\text{Volume}} = \int_0^{\varepsilon f} \mho d\varepsilon \tag{14.1}$$

where εf is the maximum strain, $\mho$ is the maximum stress before rupturing and $d\varepsilon$ is the derivative as a function of the strain.

Figure 14.2 illustrates the procedure employed in the determination of TS, BS and DB of edible films.

The film thickness is another important parameter because it affects the opacity, adsorption and light transmittance properties of the film. The film thickness (in mm) is obtained using a thickness metre with measurement taken at about five random locations on the film surface, and mean values reported of the films were measured (Cazón et al. 2019b). The results of a formulation and the corresponding results obtained for dependent variables (thickness, T, BS, DB, %W and %S) are shown in Table 14.3. Thus, the effect of the PVA and chitosan on the dependent variables as modelled employing a second-order polynomial equation shows how the film thickness is affected by the formulation. As the level of PVA increases, the thickness increases with the chitosan impacting negatively on the thickness with increase in level (Table 14.3). Thus, the formulation of the films affected the thickness of the composite films with values ranged from 2.08 to 5.71×10^{-2} mm. Data fitted well to a linear model where the F value of the model was 15.03 at significant level of $p < 0.05$. According to the p values of the model terms, only the PVA concentration showed a significant ($p < 0.05$) effect on the film thickness (Fig. 14.3). Thus, the chitosan concentration was not significant as reported in the study by Cazón et al. (2018b). Table 14.3 shows the experimental design of a regenerated cellulose-chitosan-PVA composite films where the maximum concentration of chitosan and PVA was set at 1 and 4% (w/w), respectively (Cazón et al. 2018b).

As shown in the response surface curve, Fig. 14.3, PVA increased the thickness of the BC films as indicated also in Table 14.2; however, chitosan did not show significant effect on film thickness of the chitosan-cellulose-PVA model composite film (Cazón et al. 2019b).

Table 14.3 Formulations of composite bacterial cellulose-based films along with chitosan and polyvinyl alcohol and the experimental results for the thickness, mechanical and puncture properties, water retention and percentage film solubility (Cazón et al. 2019b)

Exp.	Chitosan % (w/w)	PVA (% w/w)	Thickness (mm)	*T* (MJ/m^3)	BS (g)	DB (mm)	%W (%)	%S (%)
1	0	0	2.08×10^{-2}	0.26	58.88	0.39	364.78	9.37
2	0	2.0	5.07×10^{-2}	1.16	543.02	0.97	224.52	46.91
3	0	4.0	6.62×10^{-2}	2.48	1054.04	1.37	164.49	41.96
4	0.5	0	3.98×10^{-2}	1.12	489.98	1.31	157.79	4.53
5	0.5	2.0	3.98×10^{-2}	1.29	1312.33	1.60	345.99	32.08
6	0.5	4.0	5.04×10^{-2}	5.94	2878.19	2.75	368.35	36.82
7	1.0	0	2.99×10^{-2}	1.92	1067.35	1.71	96.16	8.14
8	1.0	2.0	5.17×10^{-2}	3.41	2640.13	2.36	478.73	29.71
9	1.0	4.0	5.71×10^{-2}	7.18	3234.62	3.24	397.09	33.13

T toughness, *BS* bursts strength, *DB* distance to burst, *%W* percent water retention or swelling index, *%S* percent solubility

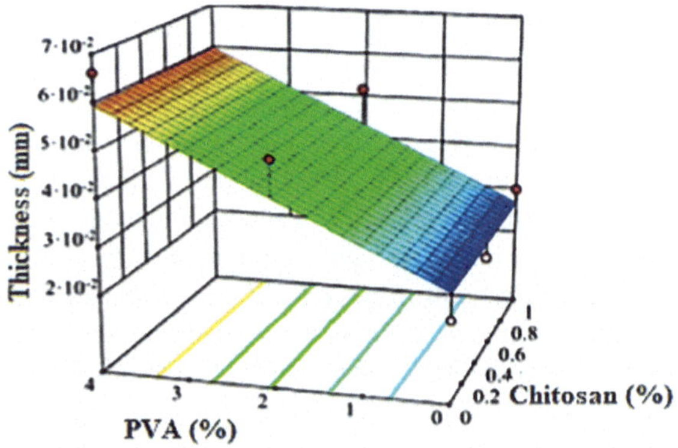

Fig. 14.3 Prediction of the model for the effect of the polyvinyl alcohol (PVA) and chitosan on thickness

In the bacterial cellulose-PVA-chitosan composite film (Fig. 14.3), the thickness values ranged from 2.08 to 5.71 × 10^{-2} mm, with the chitosan impacting less than PVA on the thickness. The chitosan and PVA exhibit similar effects as can be observed in the other mechanical properties of the edible films.

Equation (14.2) is a model that forecasts the values of thickness of a cellulose-PVA-chitosan composite films as a function of chitosan and PVA concentrations:

$$\text{Thickness (mm)} = 2.66 \times 10^{-2} + 3.37 \times 10^{-4}\ \text{Chitosan (\%)} + 8.29 \times 10^{-3}\ \text{PVA(\%)} \tag{14.2}$$

Similarly, Eq. (14.3) predicts the values of toughness of a bacterial cellulose films with chitosan and PVA concentrations (Cazón et al. 2019b):

$$\text{Toughness}\left(\text{MJ/m}^3\right) = -0.74 + 2.87\ \text{Chitosan(\%)} + 1.03\ \text{PVA(\%)} \tag{14.3}$$

Or Eq. (14.4), as proposed by Cazón et al. (2018b).

$$\begin{aligned} \text{T} = {} & 65.18 - 43.18\ \text{Cellulose (\%)} - 1.36\ \text{Glycerol (\%)} - 7.99\ \text{PVA (\%)} \\ & + 0.52\ \text{Cellulose (\%) Glycerol} - 2.53\ \text{Cellulose (\%) PVA(\%)} \\ & - 0.32\ \text{Glycerol (\%) PVA (\%)} + 7.12\ \text{Cellulose (\%) 2} \\ & + 0.07\ \text{Glycerol (\%) 2} + 0.32\ \text{PVA (\%) 2} \end{aligned} \tag{14.4}$$

Similarly, Eq. (14.5) forecasts the values of the DB as a function of the chitosan and PVA concentrations:

$$\text{DB (mm)} = 0.32 + 1.52\ \text{Chitosan (\%)} + 0.33\ \text{PVA (\%)} \tag{14.5}$$

Thus, the model equation shows that the DB values increase with increase in the concentrations of chitosan and PVA and that the increases will be greater in the presence of both the polymers with their plasticizing effect observed on bacterial cellulose films because of the good extensibility of pure PVA and chitosan films (Cazón et al. 2018b). Additionally, PVA and chitosan are miscible, and this property presumably accelerated strong intermolecular hydrogen bonds and increased van der Waals forces between the co-polymers resulting in increased T values, as were also reported in other composite edible film formulations such as kefiran-chitosan films (Sabaghi et al. 2015), wheat starch-chitosan films (Bonilla et al. 2013) and PVA-chitosan films (Hajji et al. 2016; Bonilla et al. 2014).

The response surface describing the effect of the levels of cellulose, glycerol and polyvinyl alcohol (PVOH) on toughness is shown in Fig. 14.4. Thus, the response surface showed that chitosan and PVOH increased the toughness of the composite films. Cazón et al. (2019b) reported that increasing polyvinyl alcohol concentration increases toughness values, reaching higher toughness values at lower concentrations of cellulose (from 0.30 to 10.55 MJ/m^3 with the increase of PVA concentration) (Cazón et al. 2019b).

Model Eq. (14.6) forecasts the values of BS as a function of graded levels of chitosan and PVA.

$$\sqrt{BS}\ \text{(g)} = 8.82 + 25.83\ \text{Chitosan} + 6.71\ \text{PVA (\%)} \tag{14.6}$$

where data fit well to a linear model, the Box-Cox plot recommends transforming the data to the square root of BS to get a better fit of the equation, and the response of the BS is obtained as a function of the concentration of chitosan and PVA on the polysaccharide-based films (Eq. 14.6).

Thus, according to the equation, chitosan and PVA cause increase in the value of BS, the increase of which becomes more noticeable when both the polymers are added at the same time. The synergistic effects of the two polymers could be due to the molecular interaction of the cellulose-chitosan-PVA in the film matrix. Figure 14.4 shows that plain chitosan films had a BS value ranging from 1234 to 1892 g as a function of the molecular weight.

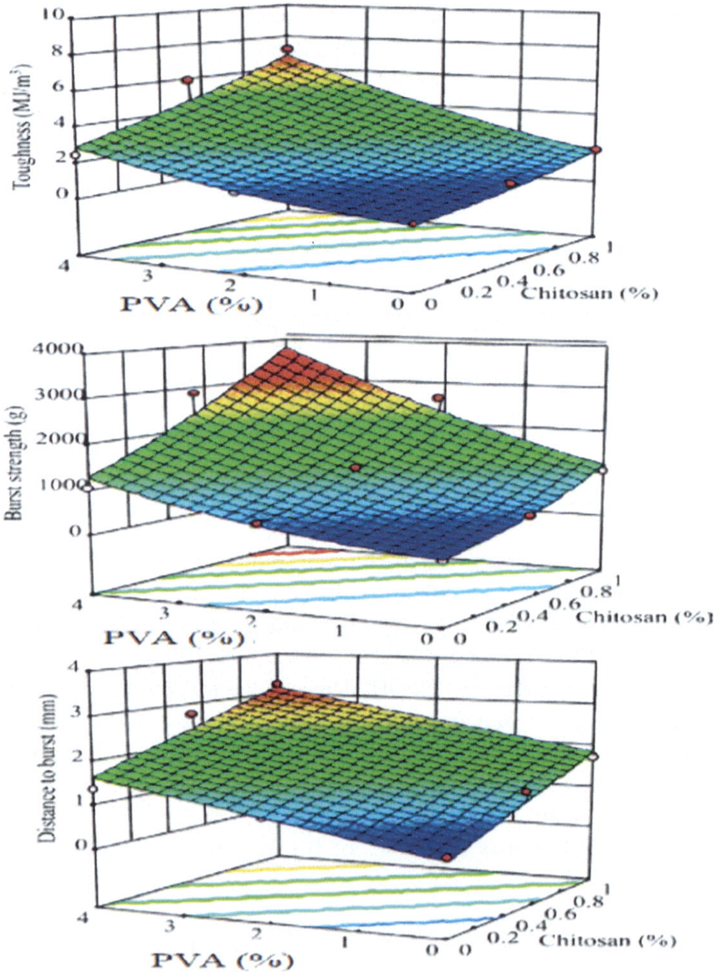

Fig. 14.4 Prediction of the model for the effect of the polyvinyl alcohol and chitosan on toughness, burst strength and distance to burst (Cazón et al. 2019b)

14.3.3 Water Adsorption, Solubility and Swelling Properties

Water adsorption property denotes the capacity of a film to swell when in contact with food (solid, semisolid or liquid) during handling and storage at high relative humidity (Cazón et al. 2018b). Water adsorption capacity is important because it affect the mechanical properties of packaging films that should acts to maintain acceptable moisture levels in a packaged product. Most biodegradable, polysaccharide films are sensitive to moisture, and thus their properties change with change in relative humidity (Cazón et al. 2020). The water adsorption and solubility property is defined by the content of dry matter solubilized after 24 h immersion in water. Thus, the presence of microspores in the structure of some biopolymers and also the resultant film formed from mixtures with plasticizers and other hydrophilic polymers greatly affect film swelling capacity (El Seoud et al. 2008; Han 2014; Leceta et al. 2013; Zhang and Han 2008).

Water adsorption property can be determined gravimetrically by placing film samples (3 or 4) of defined size (3 × 3 or 4 × 4 cm^2) in desiccators with silica gel for about five days to stabilize moisture and then weighed. Thereafter, each sample is then immersed in a capped container with 100 mL of distilled water and stored at 20 °C for 24 h to allow samples to equilibrate before the excess surface water from the swollen film is removed with the aid of a filter paper. The wet films are weighed and their moisture content (%W) calculated according to Eq. (14.7):

$$\%W = \frac{\left(W_{\text{wet}} - W_{\text{dry}}\right)}{W_{\text{dry}}} \times 100 \tag{14.7}$$

where W_{wet} is the wet weight and W_{dry} is the dried weight.

The water solubility of the films is increased when PVA is added to an edible film formulation. Pure polysaccharide film and polysaccharide-chitosan composite film showed similar lower solubility values. Plain cellulose exhibited a water retention capacity of about five times its dry weight, giving a swelling degree of 364.78% (Cazón et al. 2019b). Plain polysaccharides such as cellulose showed a water retention capacity much higher than that of other hydrocolloid-based films. Chitosan conferred a significant decrease in the swelling degree of polysaccharide films; however, the addition of combined reinforcement polymers such as PVA and chitosan improved the degree of swelling as did plain cellulose. Table 14.2 shows that 1% (w/w) chitosan in the film formulation conferred lower T, low BS, low DB and lower swelling index.

The water solubility (%S) of edible films is similarly determined gravimetrically as above using film of size 3 × 3 cm^2 previously dried at 105 °C for 24 h in vacuum oven and weighed to the nearest 0.0001 g as initial dry weight (W_0). The dry sample is then placed in a glass beaker containing 100 mL of distilled water with gentle shaking at 25 °C for 24 h; thereafter, remaining undissolved film is dried at 105 °C for 24 h until constant weight to obtain the final dry weight (W_f) (Gómez-Estaca et al. 2010; Rhim 2004). The solubility in water (%S) is calculated as follows:

$$\%S = \frac{W_0 - W_f}{W_0} \times 100 \tag{14.8}$$

$$\begin{aligned} \text{Solubility}\%S\ (\%) = {} & 10.80 - 27.97\ \text{Chitosan}\ (\%) + 23.0\ \text{PVA}\ (\%) \\ & - 5.80\ \text{Chitosan}\ (\%)\ \text{PVA}\ (\%) + 25.30\ \text{Chitosan}\ (\%)\ 2 \\ & - 2.83\ \text{PVA}\ (\%) \end{aligned} \tag{14.9}$$

Similarly, Eq. (14.10) can be used to estimate the values of the swelling property ($\%W$) as a function of chitosan and PVA:

$$\begin{aligned} \%W = {} & 335.28 - 235.20\ \text{Chitosan}\ (\%) - 51.05\ \text{PVA}\ (\%) \\ & + 211.30\ \text{Chitosan PVA}\ (\%) \end{aligned} \tag{14.10}$$

As shown by the response surface curve above (Fig. 14.5), the film solubility is mainly influenced by the PVA content, and thus, because of its hydrophilic nature and affinity to water, part of the PVA becomes dissolved in water. Cellulose, however, is insoluble in water and showed a very low solubility value of about 9% (Cazón et al. 2017). Previous study reported a 65.5% solubility for plain PVA films (Hajji et al. 2016). Thus the value obtained for pure bacterial cellulose was probably due to the presence of soluble solids remaining from the culture medium in the dry samples. Bacterial cellulose-chitosan films do not show significant changes in the solubility values compared with plain bacterial films (Cazón et al. 2017). Meanwhile, chitosan is retained in the film because it is soluble at slightly acidic pH; however, bacterial cellulose-chitosan-PVA composite films showed a light decrease in the percentage of soluble matter compared with the bacterial cellulose-PVA films. It is thus probable that well-packed structures between chitosan and PVA caused by cross-linking between the amine and hydroxyl moieties limit the interactions with water, rendering the PVA partially soluble (Cazón et al. 2017).

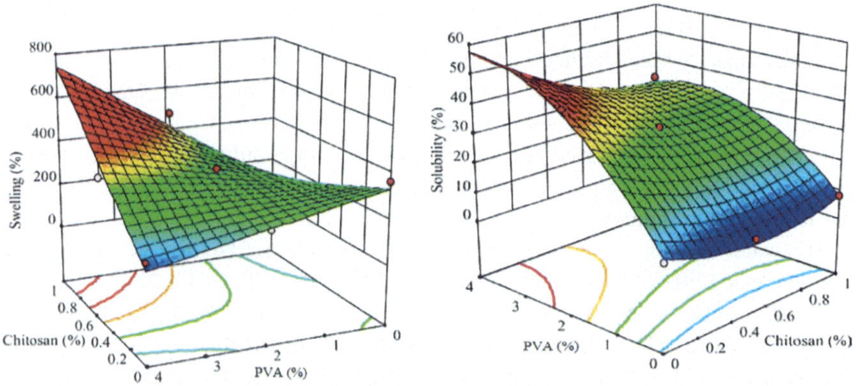

Fig. 14.5 Prediction model for the effect of the polyvinyl alcohol (PVA) and chitosan on the solubility and swelling of the films

The swelling behaviour of edible films is governed by several factors including the network density of polymers (Pavaloiu et al. 2014). The swelling ratio of plain bacterial cellulose edible film is about five times its dry weight, yielding a water retention capacity of about 364.78%, much higher than that of other polysaccharide films, because the porous and three-dimensional structure of the cellulose permits greater contact surface, easing the physical entrapment of water molecules inside the cellulose matrix (Cazón et al. 2019b). In a well-fitted, two-factor interaction mathematical model, Cazón et al. (2019b) found that the swelling or water retention values ranged from 96.16% to 478.73%, as shown in Fig. 14.5, where the swelling behaviour was influenced mainly by the chitosan. The response surface curve similarly shows that the swollen mass was drastically reduced from 364.78 (plain bacterial cellulose film) to 96.16% (bacterial cellulose-1% chitosan film) when the chitosan concentration increased in the blend.

Plain cellulose films and cellulose-chitosan showed similar lower solubility values. Plain cellulose exhibits a capacity of water retention that is five times its dry weight, giving a swelling degree of 364.78%.

The swelling behaviour of the film is affected by polymer network density and hydration capacity. For example, the swelling capacity of plain cellulose is about five times its dry weight higher than that of other hydrocolloid films. Thus, a model developed for forecasting the response of composite polysaccharide-based edible film to variable plasticizer levels shows that the swelling behaviour is influenced mainly by the chitosan (Fig. 14.5). The swollen mass drastically reduced from 364.78% (plain bacterial cellulose films) to 96.16% (bacterial cellulose-chitosan 1% film) when the chitosan concentration increased in the blend (Cazón et al. 2018b). The porous structure of polysaccharide such as bacterial cellulose allows a greater contact surface, easing the physical entrapment of water molecules inside the cellulose matrix (Cazón et al. 2018b).

Chitosan produces a significant decrease in the swelling degree; however, the presence of both PVA and chitosan as reinforcement polymers allowed similar values as plain cellulose to be reached. Thus, the water solubility of edible film increased with addition of PVA to the formulation.

The response surface describing the effect of the concentrations of cellulose, glycerol and PVA on $\%W$ is shown in Fig. 14.6. The value of $\%W$ increases insignificantly with increasing PVA concentration, as a function of the concentration of cellulose and glycerol.

Model Eq. (14.11) predicts the values of water adsorption ($\%W$) as a function of the concentrations of cellulose, glycerol and PVA (Cazón et al. 2018b):

$$(\%W) = 45.12 + 29.95\ \text{Cellulose}(\%) - 3.65\ \text{Glycerol} + 7.37\ \text{PVA}\ (\%) \quad (14.11)$$

Thus, PVA concentration is the main factor affecting $\%W$ value, whereas chitosan and cellulose concentrations had lower and negligible linear effects, respectively, on $\%W$ response value.

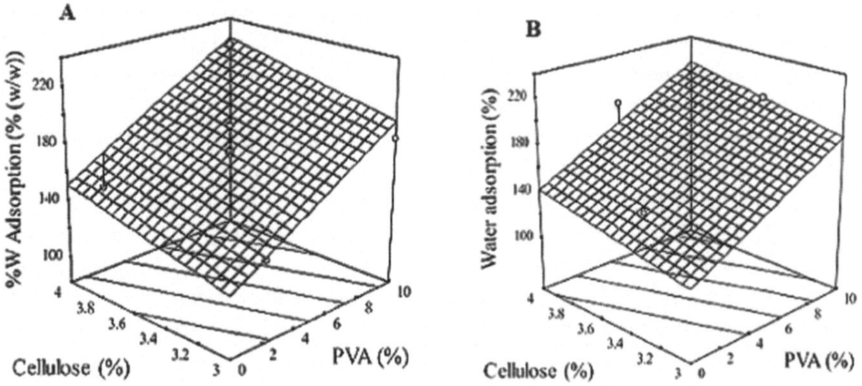

Fig. 14.6 Prediction of the model for the effect of the concentrations of cellulose, glycerol (Gly) and polyvinyl alcohol (PVA) on water adsorption. (**a**) Films with 2.5% (w/w) Gly. (**b**) Films with 5% (w/w) Gly (Cazón et al. 2018b)

14.3.4 Water Vapour Permeability

Water vapour permeability (WVP) of edible films is measured to quantify the amount of water passing through the film per unit of area and time (g/m^{2}sPa) (Cazón et al. 2020). The measurement of water vapour permeability (WVP) is done gravimetrically based on ASTM E96–92 method (Cerqueira et al. 2011). The measured WVP of the films is then obtained as follows:

$$\mathrm{WVP} = \frac{\mathrm{WVTR} \cdot L}{\Delta P} \tag{14.12}$$

where WVTR is the measured water vapour transmission rate through a film, L is the mean film thickness (in metres) and ΔP is the partial water vapour pressure difference (Pa) across the two sides of the film. The values of WVTR of the films are calculated as follows:

$$\mathrm{WVTR} = \frac{\mathrm{SLOPE}}{\mathrm{FILM\ AREA}} \tag{14.13}$$

where the slope of weight loss versus time is obtained by linear regression and the cup test mouth area is the film area (Cazón et al. 2018b). Equation (14.14) can be used to estimates the response of the concentration of glycerol and PVA on the WVP of a cellulose film.

$$\mathrm{WVP} = 2.09 \times 10^{-12} + 2.72 \times 10^{-11}\ \mathrm{Glycerol}\ (\%) + 1.10 \times 10^{-11} \mathrm{PVA}\ (\%) \tag{14.14}$$

Table 14.4 Water vapour permeability data for plain chitosan films at different conditions

Composition	Acid used	Chitosan properties	Conditions	Water vapour permeability (g/msPa)	References
Chitosan 2%	Citric	DD 95%	25 °C	$(4.0–5.1) \times 10^{-10}$	Park et al. (2002)
	Lactic	Mw 370–920 kDa	50% RH	$(4.7–6.9) \times 10^{-10}$	
Chitosan 2%	Acetic	DD ~80%	38 °C 90% RH	(4016–4500) g/m^2 day	Srinivasa et al. (2004)
		Mw 100 kDa			
Chitosan 1%	Acetic	Mw ~150 kDa	25 °C; 50% RH	3.53 g mm/m^2 h kPa	Li et al. (2010)
Chitosan 1%	Acetic	DD 78%	90% RH	0.01322 g m/m^2 day kPa	Srinivasa et al. (2007)
		Mw 200 kDa			
Chitosan 1%	Acetic	High Mw	38 °C, 90% RH	8.07×10^{-13}	Leceta et al. (2013)
Chitosan 1%	Acetic	Low Mw	38 °C; 90% RH	8.07×10^{-13}	Leceta et al. (2013)
Chitosan 2.5%	Acetic	DD 85%	5 °C; 2000 pa	4.14×10^{-10}	Bof et al. (2015)
		Low Mw			
Chitosan 2.5%	Acetic	DD 85%	5 °C; 2000 pa	3.38×10^{-10}	Bof et al. (2015)
		Medium Mw			

DD degree of deacetylation, *RH* relative humidity, *Mw* molecular weight

Previous studies on WVP show that the water vapour barrier of films is significantly ($p < 0.05$) affected by plasticizer concentration, film thickness and relative humidity gradient, polysaccharide source and extraction method (Jridi et al. 2013).

The results from a response surface suggested that the immersion of cellulose-based film in PVA bath imparted a slight decrease in the WVP of the film. Though PVA is a hydrophilic polymer, it decreased the porosity of the film and thus promoted a close packing of the structure (Cazón et al. 2020). Chitosan-based edible films exhibit relatively low water barrier characteristics due to their high hydrophilic nature (Park et al. 2002) as shown in Table 14.4.

In general, hydrophilic biopolymer films show good oxygen barrier property and higher water vapour permeability (Ayranci and Tunc 2003; Siracusa et al. 2008; Aider 2010; Leceta et al. 2013).

14.3.5 Colour, Optical and Microstructural Properties

The opacity of a material is an indication of how much light passes through it and is calculated from reflectance measurements (Casariego et al. 2009). The ultraviolet/visible light and transparency barrier properties of edible film can be determined using a UV-Vis-NIR spectrophotometer within regions 190–2500 nm at 2 nm transmittance interval. The transparency of the films is then calculated from the amount of light (%) transmitted at 600 nm, following the equation below proposed by Han and Floros (1997):

$$\text{Transparency} = (\log\%T600)/x \ (3).$$

where %T600 is the percent transmittance at 600 nm and x is the film thickness (mm).

Cazón et al. (2018b) reported that chitosan reinforced the UV radiation barrier properties of pure cellulose combined with PVA. Also, UV-Vis spectra showed optimal optical barrier properties of regenerated cellulose-based-PVA films against UV radiation and significantly improved the transparency and visual appearance of the films. Studies on pure bacterial cellulose film showed low transmittance values ranging from 0% to 2% in the UV region, like plain regenerated cellulose; however, when chitosan concentration reached 1% (w/w), the transmittance was reduced to 4.80% (Cazón et al. 2018a). A model of cellulose-based edible film obtained by the NaOH/urea/H_2O solution in the ratio (7:14:79, w/w) yielded a transparent film with good UV-Vis light transmission properties, Fig. 14.7. Transmission range of UV light is between 100 and 400 nm. In the range 180–400 nm, pure regenerated cellulose films showed very low transmittance values below 8.5%. The presence of chitosan decreased the transmission of UV light below 4.9%. However, PVA increased the UV light transmittance of the films up to 55%. Again, chitosan

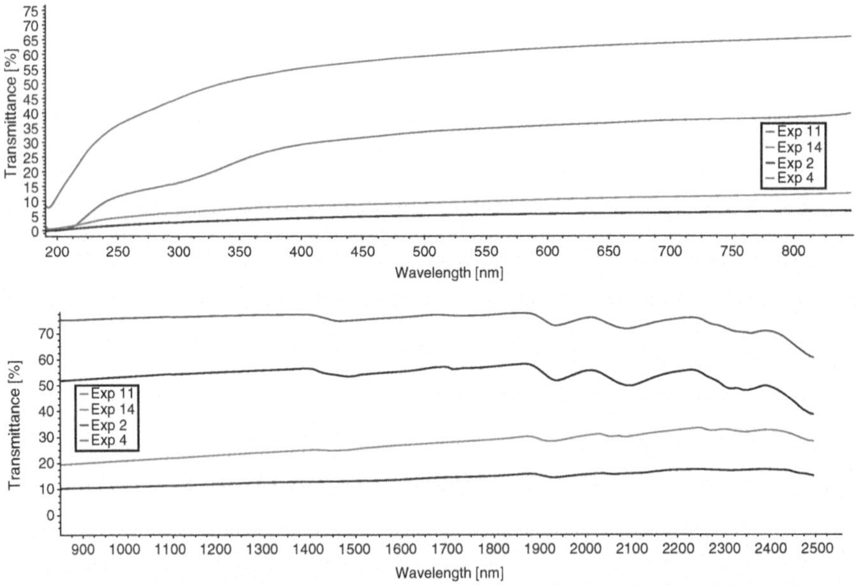

Fig. 14.7 UV-Vis-NIR spectra profile of selected films. Composition: All samples cellulose (4%), Exp 2 (1% chitosan, 0% PVA), Exp 4 (1% chitosan, 4% PVA), Exp 11 (0% chitosan, 4% PVA), Exp 14 (0% chitosan, 0% PVA (Cazón et al. 2018a)

decreases the transmission; films with chitosan and PVA showed a UV light transmittance up to 29%. In the UV region, optimum formulation of edible film showed maximum and minimum transmittance values of 35.1% and 21.6%, respectively. However, film with higher chitosan content and lower PVA reached the minimum transmittance value, 9.2% (Cazón et al. 2018b). Glycerol content imparted a UV-protective effect on the films to prevent oxidative deterioration of food. Furthermore, cellulose-glycerol-PVA films had optimal values of transparency, depicting the values of synthetic polymers. Increasing PVA to 10% showed a very low absorption in the UV region, suggesting a UV-protective effect of the films with glycerol when applied as protective in foods with tendency to undergo oxidative deterioration (Imran et al. 2012). The polymer concentration is a most important factor ($p < 0.05$) affecting the opacity of films, followed by the interaction between polymer and plasticizer and by the presence of lipids (Cazón et al. 2019b).

14.4 Significance

The greatest challenge encountered by the food manufacturer is the loss of quality of food products during storage, which eventually adds to the waste that has become a deadly scourge for the environment. Recent trends in the science of food packaging have shown that biodegradable and edible polysaccharide-based food packages, as food films, can save the environment from waste and their harmful effects and provided improved nutrition as fibres in addition to saving money. Edible packaging is projected as a potential alternative to protecting food quality and improving shelf life by delaying microbial spoilage and providing moisture and gas barrier properties. It is also forecasted as one sustainable, cost-effective solution to synthetic and/or non-biodegradable packaging. Developments in edible packaging and technology have shown promising results in enhancing the quality of food products. The improved mechanical properties allowed the development of polysaccharide-based films resistant to rupture with barrier properties to UV radiation and a high degree of water vapour permeability characteristics.

The presentation in this chapter can help researchers and industries design and formulate a viable, efficient and cost-effective systems for development of edible films for specific food product application. Thus, the developed films could be useful in the food industry as better alternatives to synthetic film, with improved nutritional properties, preventing lipid oxidation in foods, delays senescence in fresh fruits, and acts as antimicrobial pads to inhibit the growth of microorganisms.

14.5 Conclusion

Edible films are pre-formed, thin layer of biopolymers that are applied directly as wrappings on the surface of food product to serve the role of primary packaging. They can be made from one or more members of the polysaccharide family since they (the polysaccharides) are the most abundant, versatile and amenable class of biopolymers. However, since the varied properties desired of an edible film cannot be obtained from a plain or pure polysaccharide, composing a formulation to contain few polymers as composite film could yield film with the desired and appropriate mechanical, structural and physicochemical properties. They are designed to meet specific food product requirement and handling stress.

As shown in the models, composite polysaccharide films with polyvinyl alcohol and glycerol and plasticizer yielded a much edible film than any of the polysaccharides alone. Adding polyvinyl alcohol to polysaccharide-based film formulation improved both the physicochemical and mechanical properties of the edible films including their transparency and visual appearance. Addition of plasticizers such as glycerol and sorbitol significantly improved extensibility, plasticity, toughness and workability properties of the edible films. However, it is suggested product-specific pilot study and evaluation of practical applications and limitations of a model edible packaging should be carried out within the main purpose of keeping food safe with acceptable quality for extended shelf life.

References

Abugoch LE, Tapia C, Villamán MC, Yazdani-Pedram M, Díaz-Dosque M (2011) Characterization of quinoa protein-chitosan blend edible films. Food Hydrocoll 25(5):879–886

Aider M (2010) Chitosan application for active bio-based films production and potential in the food industry. LWT Food Sci Technol 43(6):837–842

Albert S, Mittal GS (2002) Comparative evaluation of edible coatings to reduce fat uptake in a deep-fried cereal product. Food Res Int 35(5):445–458

Alvarado-González JS, Chanona-Pérez JJ, Welti-Chanes JS, Calderón-Domínguez G, Arzate-Vázquez I, Pacheco-Alcalá SU, Gutiérrez-López GF (2012) Optical, microstructural, functional and nanomechanical properties of Aloe vera gel/gellan gum edible films. Rev Mex Ing Quím 11(2):193–210

Antoniou J, Liu F, Majeed H, Qazi HJ, Zhong F (2014) Physicochemical and thermomechanical characterization of tara gum edible films: effect of polyols as plasticizers. Carbohydr Polym 111: 359–365

Aşik E, Candoğan K (2014) Effects of chitosan coatings incorporated with garlic oil on quality characteristics of shrimp. J Food Qual 37(4):237–246

Ayranci E, Tunc S (2003) A method for the measurement of the oxygen permeability and the development of edible films to reduce the rate of oxidative reactions in fresh foods. Food Chem 80(3):423–431

Bangyekan C, Aht-Ong D, Srikulkit K (2006) Preparation and properties evaluation of chitosan-coated cassava starch films. Carbohydr Polym 63:61–71

Barreto PLM, Pires ATN, Soldi V (2003) Thermal degradation of edible films based on milk proteins and gelatin in inert atmosphere. Polym Degrad Stab 79:147–152

Basiak E, Lenart A, Debeaufort F (2017) Effect of starch type on the physico-chemical properties of edible films. Int J Biol Macromol 98:348–356

Benbettaïeb N, Kurek M, Bornaz S, Debeaufort F (2014) Barrier, structural and mechanical properties of bovine gelatin–chitosan blend films related to biopolymer interactions. J Sci Food Agric 94(12):2409–2419

Bof MJ, Bordagaray VC, Locaso DE, García MA (2015) Chitosan molecular weight effect on starch-composite film properties. Food Hydrocoll 51:281–294
Bonilla J, Atarés L, Vargas M, Chiralt A (2013) Properties of wheat starch film-forming dispersions and films as affected by chitosan addition. J Food Eng 114(3):303–312
Bonilla J, Fortunati E, Atarés L, Chiralt A, Kenny JM (2014) Physical, structural and antimicrobial properties of poly vinyl alcohol-chitosan biodegradable films. Food Hydrocoll 35:463–470
Bourtoom T (2008) Edible films and coatings: characteristics and properties. Int Food Res J 15(3): 237–248
Box GEP, Behnken DW (1960) Some new three level designs for the study of quantitative variables. Technometrics 2(4):455–475
Casariego ABWS, Souza BWS, Cerqueira MA, Teixeira JA, Cruz L, Díaz R, Vicente AA (2009) Chitosan/clay films' properties as affected by biopolymer and clay micro/nanoparticles' concentrations. Food Hydrocoll 23(7):1895–1902
Cuq B, Aymard C, CUQ JL, Guilbert S (1995) Edible packaging films based on fish myofibrillar proteins: formulation and functional properties. J Food Sci 60(6):1369–1374
van den Broek LA, Knoop RJ, Kappen FH, Boeriu CG (2015) Chitosan films and blends for packaging material. Carbohydr Polym 116:237–242
Cazón P, Velazquez G, Ramírez JA, Vázquez M (2017) Polysaccharide-based films and coatings for food packaging: a review. Food Hydrocoll 68:136–148
Cazón P, Vázquez M, Velazquez G (2018a) Composite films of regenerate cellulose with chitosan and polyvinyl alcohol: evaluation of water adsorption, mechanical and optical properties. Int J Biol Macromol 117:235–246
Cazón P, Vázquez M, Velazquez G (2018b) Cellulose-glycerol-polyvinyl alcohol composite films for food packaging: evaluation of water adsorption, mechanical properties, light-barrier properties, and transparency. Carbohydr Polym 195:432–443
Cazón P, Velázquez G, Vázquez M (2019a) Characterization of bacterial cellulose films combined with chitosan and polyvinyl alcohol: evaluation of mechanical and barrier properties. Carbohydr Polym 216:72–85
Cazón P, Velazquez G, Vázquez M (2019b) Novel composite films from regenerated cellulose–glycerol–polyvinyl alcohol: mechanical and barrier properties. Food Hydrocoll 89:481–491
Cazón P, Velazquez G, Vázquez M (2020) UV-protecting films based on bacterial cellulose, glycerol and polyvinyl alcohol: effect of water activity on barrier, mechanical and optical properties. Cellulose 27(14):8199–8213
Cerqueira MA, Bourbon AI, Pinheiro AC, Martins JT, Souza BWS, Teixeira JA, Vicente AA (2011) Galactomannans use in the development of edible films/coatings for food applications. Trends Food Sci Technol 22(12):662–671
Cerqueira MA, Souza BW, Teixeira JA, Vicente AA (2013) Utilization of galactomannan from gleditsia triacanthos in polysaccharide-based films: effects of interactions between film constituents on film properties. Food Bioprocess Technol 6(6):1600–1608
Chen Y, Xu L, Wang Y, Chen Z, Zhang M, Chen H (2020) Characterization and functional properties of a pectin/tara gum based edible film with ellagitannins from the unripe fruits of Rubus Chingii Hu. Food Chem 325:126,964
Cheng Y, Zhai X, Wu Y, Li C, Zhang R, Sun C, Wang W, Hou H (2023) Effects of natural wax type on the physicochemical properties of starch/gelatin edible films fabricated by extrusion blowing. Food Chem 401:134081
Elsabee MZ, Abdou ES (2013) Chitosan-based edible films and coatings: A review. Materials Science and Engineering C 33(4):1819–1841
El Seoud OA, Fidale LC, Ruiz N, D'Almeida MLO, Frollini E (2008) Cellulose swelling by protic solvents: Which properties of the biopolymer and the solvent matter? Cellulose 15(3):371–392
Elsabee MZ, Abdou ES (2013) Chitosan based edible films and coatings: a review. Mater Sci Eng C 33(4):1819–1841

Fernandes SCM, Freire CSR, Silvestre AJD, Pascoal Neto C, Gandini A, Berglund LA, Salmén L (2010) Transparent chitosan films reinforced with a high content of nano-fibrillated cellulose. Carbohydr Polym 81(2):394–401

Frone AN, Nicolae CA, Gabor RA, Panaitescu DM (2015) Thermal properties of water-resistant starch—polyvinyl alcohol films modified with cellulose nanofibers. Polym Degrad Stab 121: 385–397

Gómez-Estaca J, De Lacey AL, López-Caballero ME, Gómez-Guillén MC, Montero P (2010) Biodegradable gelatin–chitosan films incorporated with essential oils as antimicrobial agents for fish preservation. Food Microbiol 27(7):889–896

Guo M, Jin TZ, Yang R (2014) Antimicrobial polylactic acid packaging films against Listeria and Salmonella in culture medium and on ready-to-eat meat. Food Bioproc Tech 7:3293–3307

Hajji S, Chaker A, Jridi M, Maalej H, Jellouli K, Boufi S, Nasri M (2016) Structural analysis, and antioxidant and antibacterial properties of chitosan-poly (vinyl alcohol) biodegradable films. Environ Sci Pollut Res 23(15):15310–15320

Han JH (2014) Edible films and coatings: a review. Innovations in food packaging 213–255

Han JH, Floros JD (1997) Casting antimicrobial packaging films and measuring their physical properties and antimicrobial activity. J Plast Film Sheet 13(4):287–298

Imran M, Revol-Junelles AM, René N, Jamshidian M, Akhtar MJ, Arab-Tehrany E, Desobry S (2012) Microstructure and physico-chemical evaluation of nano-emulsion-based antimicrobial peptides embedded in bioactive packaging films. Food Hydrocoll 29(2):407–419

Jridi M, Souissi N, Mbarek A, Chadeyron G, Kammoun M, Nasri M (2013) Comparative study of physico-mechanical and antioxidant properties of edible gelatin films from the skin of cuttlefish. Int J Biol Macromol 61:17–25

Jridi M, Hajji S, Ayed HB, Lassoued I, Mbarek A, Kammoun M et al (2014) Physical, structural, antioxidant and antimicrobial properties of gelatin–chitosan composite edible films. Int J Biol Macromol 67:373–379

Kõrge K, Bajić M, Likozar B, Novak U (2020) Active chitosan–chestnut extract films used for packaging and storage of fresh pasta. Int J Food Sci Technol 55(8):3043–3052

Kechichian V, Ditchfield C, Veiga-Santos P, Tadini CC (2010) Natural antimicrobial ingredients are incorporated in biodegradable films based on cassava starch. LWT - Food Science and Technology 43(7):1088–1094

Khan A, Khan RA, Salmieri S, Le Tien C, Riedl B, Bouchard J, Chauve G, Tan V, Kamal MR, Lacroix M (2012) Mechanical and barrier properties of nanocrystalline cellulose reinforced chitosan based nanocomposite films. Carbohydr Polym 90(4):1601–1608

Leceta I, Guerrero P, De La Caba K (2013) Functional properties of chitosan-based films. Carbohydr Polym 93(1):339–346

Li H, Yu T (2001) Effect of chitosan on incidence of brown rot, quality and physiological attributes of postharvest peach fruit. J Sci Food Agric 81(2):269–274

Li J, Zivanovic S, Davidson PA, Kit K (2010) Characterization and comparison of chitosan/PVP and chitosan/PEO blend films. Carbohydr Polym 79(3):786–791

Liu Z, Ge X, Lu Y, Dong S, Zhao Y, Zeng M (2012) Effects of chitosan molecular weight and degree of deacetylation on the properties of gelatine-based films. Food Hydrocoll 26:311–317

Liu L, Liang H, Zhang J, Zhang P, Xu Q, Lu Q, Zhang C (2018) Poly(vinyl alcohol)/chitosan composites: physically transient materials for sustainable and transient bioelectronics. J Clean Prod 195:786–795

Mohammadi R, Mohammadifar MA, Rouhi M, Kariminejad M, Mortazavian AM, Sadeghi E, Hasanvand S (2018) Physico-mechanical and structural properties of eggshell membrane gelatin-chitosan blend edible films. Int J Biol Macromol 107:406–412

Morgado DL, Frollini E, Castellan A, Rosa DS, Coma V (2011) Biobased films prepared from NaOH/thiourea aqueous solution of chitosan and linter cellulose. Cellulose 18(3):699–712

Myllärinen P, Partanen R, Seppälä J, Forssell P (2002) Effect of glycerol on behaviour of amylose and amylopectin films. Carbohydr Polym 50:355–361

Nešić A, Cabrera-Barjas G, Dimitrijević-Branković S, Davidović S, Radovanović N, Delattre C (2020) Prospect of polysaccharide-based materials as advanced food packaging. Molecules 25(1):135

Nešić A, Cabrera-Barjas G, Dimitrijević-Branković S, Davidović S, Radovanović N, Delattre C (2019) The prospect of polysaccharide-based materials as advanced food packaging. Mol 25 (1):135

Park HJ, Weller CL, Vergano PJ, Testin RF (1993) Permeability and mechanical properties of cellulose-based edible films. J Food Sci 58(6):1361–1364. https://doi.org/10.1111/j.1365-2621.1993.tb06183.x

Park SY, Marsh KS, Rhim JW (2002) Characteristics of different molecular weight chitosan films affected by the type of organic solvents. J Food Sci 67(1):194–197

Pavaloiu RD, Stoica-Guzun A, Stroescu M, Jinga SI, Dobre T (2014) Composite films of poly (vinyl alcohol)-chitosan-bacterial cellulose for drug-controlled release. Int J Biol Macromol 68: 117–124

Pellá MC, Silva OA, Pellá MG, Beneton AG, Caetano J, Simões MR, Dragunski DC (2020) Effect of gelatin and casein additions on starch edible biodegradable films for fruit surface coating. Food Chem 309:125764

Pirozzi A, Pataro G, Donsì F, Ferrari G (2020) Edible coating and pulsed light to increase the shelf life of food products. Food Eng Rev 13:544–569

Razmjoo F, Sadeghi E, Rouhi M, Mohammadi R, Noroozi R, Safajoo S (2021) Polyvinyl alcohol–Zedo gum edible film: physical, mechanical, and thermal properties. J Appl Polym Sci 138(8): 49875

Rhim JW (2004) Physical and mechanical properties of water-resistant sodium alginate films. LWT Food Sci Technol 37(3):323–330

Rojas-Graü MA, Tapia MS, Martín-Belloso O (2008) Using polysaccharide-based edible coatings to maintain quality of fresh-cut Fuji apples. LWT Food Sci Technol 41:139–147

Roohani M, Habibi Y, Belgacem NM, Ebrahim G, Karimi AN, Dufresne A (2008) Cellulose whiskers reinforced polyvinyl alcohol copolymers nanocomposites. Eur Polym J 44(8): 2489–2498

Romero-Bastida CA, Zamudio-Flores PB, Bello-Pérez LA (2011) Antimicrobials in oxidized banana starch films: effect on antibacterial activity, microstructure, mechanical and barrier properties. Rev Mex Ing Quim 10(3):445–453

Rouhi M, Razavi SH, Mousavi SM (2017) Optimization of crosslinked poly (vinyl alcohol) nanocomposite films for mechanical properties. Mater Sci Eng C 71:1052–1063

Sabaghi M, Maghsoudlou Y, Habibi P (2015) Enhancing structural properties and antioxidant activity of Kefiran films by chitosan addition. Food Struct 5:66–71

Sabbah M, Di Pierro P, Cammarota M, Dell'Olmo E, Arciello A, Porta R (2019) Development and properties of new chitosan-based films plasticized with spermidine and/or glycerol. Food Hydrocoll 87:245–252

Saibuatong O, Phisalaphong M (2010) Novo aloe vera-bacterial cellulose composite film from biosynthesis. Carbohydr Polym 79:455–460

Sahraee S, Milani JM, Regenstein JM, Kafil HS (2019) Protection of foods against oxidative deterioration using edible films and coatings: A review. Food Biosci 32:100451

Shivakumara LR, Demappa T (2016) Mechanical and barrier properties of hydroxyl propyl methyl cellulose edible polymer films with plasticizer combinations. J Food Process Preserv 41(4): e13020. https://doi.org/10.1111/jfpp.13020

Silva GGD, Sobral PJA, Carvalho RA, Bergo PVA, Mendieta-Taboada O, Habitante AMQB (2008) Biodegradable films based on blends of gelatin and poly (vinyl alcohol): effect of PVA type or concentration on some physical properties of films. J Polym Environ 16(4): 276–285

Siracusa V, Rocculi P, Romani S, Dalla Rosa M (2008) Biodegradable polymers for food packaging: a review. Trends Food Sci Technol 19(12):634–643

Srinivasa PC, Ramesh MN, Kumar KR, Tharanathan RN (2004) Properties of chitosan films prepared under different drying conditions. J Food Eng 63(1):79–85

Srinivasa PC, Ramesh MN, Tharanathan RN (2007) Effect of plasticizers and fatty acids on mechanical and permeability characteristics of chitosan films. Food Hydrocoll 21(7):1113–1122

Taghizadeh M, Mohammadifar MA, Sadeghi E, Rouhi M, Mohammadi R, Askari F et al (2018) Photosensitizer-induced cross-linking: a novel approach for improvement of physicochemical and structural properties of gelatin edible films. Food Res Int 112:90–97

Tapia-Blácido DR, Maniglia BC, Tosi MM (2018) Transport phenomena in edible films. Polym Food Appl:149–192

Vieira MGA, Da Silva MA, Dos Santos LO, Beppu MM (2011) Natural-based plasticizers and biopolymer films: a review. Eur Polym J 47(3):254–263

Yener FY, Korel F, Yemenicioğlu A (2009) Antimicrobial activity of lactoperoxidase system incorporated into cross-linked alginate films. J Food Sci 74(2):M73–M79

Wahyuningsih K, Iriani ES, Fahma F (2016) Utilization of cellulose from pineapple leaf fibers as nanofiller in polyvinyl alcohol-based film. Indian J Chem 16(2):181–189

Wang S, Lu A, Zhang L (2016) Recent advances in regenerated cellulose materials. Prog Polym Sci 53:169–206. https://doi.org/10.1016/j.progpolymsci

Xu Q, Chen C, Rosswurm K, Yao T, Janaswamy S (2016) A facile route to prepare cellulose-based films. Carbohydr Polym 149:274–281

Zhang Y, Han JH (2008) Sorption isotherm and plasticization effect of moisture and plasticizers in pea starch film. J Food Sci 73(7):E313–E324

Zhang S, Zhang X (2012) Toughness evaluation of hard coatings and thin films. Thin Solid Films 520:2375–2389

Zhang S, Sun D, Fu Y, Du H (2005) Toughness measurement of thin films: a critical review. Surf Coat Technol 198(1–3):74–84

Chapter 15
Polysaccharide-Based Films: Carriers of Active Substances and Controlled Release Characteristics

Kashif Ameer, Mian Anjum Murtaza, Saadia Zainab, Young-Min Kim, Muhammad Umair Arshad, Imran Pasha, Muhammad Abid, and Mi-Kyung Park

15.1 Introduction

Among manufacturing industries, food industry ranks as one of the most significant sector involving variegated products and processes. Along with food processing equipment and quality assurance of food products for intended consumers, food packaging is one of the integral parts in manufacturing and processing flow chain (Raza et al. 2021a). As far as packaging industry is concerned, food packaging is one of the vital and important facets and must meet the following criteria: (1) assurance

K. Ameer (✉) · M. A. Murtaza
Institute of Food Science and Nutrition, University of Sargodha, Sargodha, Pakistan
e-mail: kashif.ameer@uos.edu.pk

S. Zainab · M.-K. Park
School of Food Science and Biotechnology, and Food and Bio-Industry Research Institute, Kyungpook National University, Daegu, Republic of Korea

Y.-M. Kim
Department of Integrative Food, Bioscience and Biotechnology, Chonnam National University, Gwangju, South Korea

M. U. Arshad
Department of Food Science, Government College University Faisalabad, Faisalabad, Pakistan

I. Pasha
National Institute of Food Science and Technology, University of Agriculture Faisalabad, Faisalabad, Pakistan

M. Abid
Institute of Food and Nutritional Sciences, PMAS-Arid Agriculture University Rawalpindi, Rawalpindi, Pakistan

T. Amin et al. (eds.), *Polysaccharide Based Films for Food Packaging: Fundamentals, Properties and Applications*,
https://doi.org/10.1007/978-981-99-4898-7_15

of quality and safety standards up to highest levels and (2) compliance to governmental and international trade regulations (Rashid et al. 2020).

Various factors need to be taken into consideration when selecting food packaging which prominently encompass foodstuffs, packaging materials, transportation, storage, and distribution. Apart from the fundamental functionalities of packaging inkling safety, preservation, and integrity of food products, several other factors play pivotal roles in determining packaging functionality (Wu 2021b). Some of main facets which provide innovatively new solution in food packaging include friendliness to user and environment in compline to modern eco-friendly trends. Use-friendliness usually fulfills the consumers' desires keeping in view the enhanced food quality and consumer convenience with reduced packaging waste disposal (Jiang et al. 2021a, b, c). Moreover, conscience is provided to the consumers by provision of ease of opening, handling, and dispensing during preparation of foodstuffs in the packages. Packaging also serves the main purposes of providing protection from the external atmospheric factors (moisture, temperature, humidity) and high degree of food security from microbial contamination. Other advantages of packaging include aroma permeability and retardation of deteriorative changes in functional properties of packaged foods as well as improving barrier properties owing to prevention of gaseous exchanges of carbon dioxide, nitrogen, and oxygen (Kim et al. 2022; Wu et al. 2021a; Liu et al. 2019).

Traditionally, different materials are employed for packaged food including metals (tin-free steel, laminates, foils, aluminum, and tinplate), papers, glass, paperboards, and plastics. These materials exert significant impact on environment in terms of pollution and lead to increasing tendency in demand of manufacturing sector for nonrenewable resources (Shafique et al. 2022). Food industries still require the usage of traditional food packaging material which significantly contributes to the cumulative packaging waste. However, with the passage of time, increased consumer awareness has been witnessed worldwide regarding adverse effects of conventional packaging especially plastics (although these are useful in many respects); adverse effects include irreversible damage to whole ecosystem including water resources and marine life diversity (Raza et al. 2021b). Hence, it may be implied that apart from ongoing strategies focused at reduction of food packaging waste disposal through recycling, reuse, and source reduction, it is dire need of the time to explore the alternatives of traditional packaging materials with more eco-friendly aspects to fulfill industrial demands (Alipoormazandarani et al. 2015).

Among eco-friendly packaging materials, biodegradable or bio-based innovative packaging stuffs are the most widely envisaged in eco-friendliness context. Biodegradability phenomenon involves the materials' ability to exhibit breakdown and returning to the eco-system after disintegration within a short time period at post-disposal stage (Jiang et al. 2020b). Biodegradable materials usually comprise of natural components and fossil resources that decompose into natural ingredients due to microbial activity in environment. Manufacturers and researchers in food industry are particularly interested in exploitation of two biopolymers named polylactic acid (PLA) and the polyhydroxy butyrate acid (PHB) as biodegradable packaging materials (Jiang et al. 2021d). Bio-based resources are usually employed in manufacturing of biodegradable materials, and such materials include polysaccharides, proteins, and lipids of plant, animal, or microbial origin [microbial polyesters (PHAs) or polysaccharides] (Feng et al. 2022).

In nature, polysaccharides are categorized as the chiefly present material with abundance and affluence as plant and marine biomass provide the largest reservoirs of polysaccharides (Murtaza and Ameer 2022). Engineered bioprocessing involves microbial fermentation that can be tailored to ensure availability of metabolic products with desired functionalities and characteristics.

15.2 Polysaccharides as Carriers of Active Substances

Polymeric carbohydrates comprise of polysaccharides as the principal group with several bio-functionalities and peculiar properties, such as biodegradability, biocompatibility, and nontoxicity for living biomass and intended consumers (Zhao et al. 2021). These characteristics provide the technological advantages for exploitation in food packaging in terms of bio-based films and edible coating. In recent years, several researchers have investigated the polysaccharides for recovering sensor and bioactive materials for usage and implementation of active and intelligent packaging modalities (Jiang et al. 2022a).

The use of active and intelligent packaging is replacing the conventional packaging modalities partially in some instances, and this can cause significant reduction in overall usage of synthetic packaging materials (Wu et al. 2018; Kang et al. 2022). These packaging materials can be further researched to incorporate certain technological characteristics in their functional properties owing to their protective function (inherent or designed) and simplification of packaging structures by incorporating polysaccharide-based materials and bioactive components, with enhanced antioxidant and antimicrobial properties (Ramachandraiah et al. 2022; Ameer et al. 2017b).

15.3 Release Kinetics of Active Substances

Release kinetics of active substances include the multiple film-forming mechanisms of biopolymers. Some of the primary reported mechanisms include ionic, hydrophobic, and electrostatic interactions or intermolecular forces involving disulfide covalent binding and cross-linking (Azeredo and Waldron 2016). In order to achieve suitability for food packaging applications, film preparation must include the processes and operational conditions suitable for food processing flow line, and main operational conditions encompass salt addition, chemical and enzymatic modifications, thermal and pH constraints, drying, reactions with other food-grade chemicals, and use of food-grade solvents (Ameer et al. 2017a; Bashir et al. 2022).

On the other hand, some characteristics of polysaccharide materials (films and coatings) restrict their exploitation in mass utilization for industrial applications

(Azeredo and Waldron 2016). Those particular characteristics include poor water vapor permeability, and mechanical properties also restrict their usage on industrial scale. Furthermore, polysaccharide functionalities can be improved in terms of their natural properties by means of modification and functionalization, which are mandatory to fulfil requirement of achieved desired molecular structures (Jiang et al. 2020a; Zafar et al. 2016; Cazón et al. 2017). Some of the few polysaccharide modification modalities for achieving favorable food packaging properties include synergistic biopolymer combination or interaction with other components. It is precisely well reported in published literature that synergy of two or more packaging materials is mandatory sometimes to provide the best-suited and innovative solution for certain foodstuffs. In such as polysaccharide modification, the fundamental backbone of macromolecules will not subject to any change but will unveil the advanced structural modifications intended for specific packaging applications (Yoon et al. 2020; Ali et al. 2018). Currently, polysaccharides are commonly employed for preparing bio-nanocomposites. Nanoparticles are usually utilized for stabilizing the biopolymers and bio-nanocomposites (Ameer et al. 2020b). Nanoparticles are employed for enhancement of the techno-functional characteristics (barrier properties, mechanical, chemical and thermal stability) and may comprise of nanosilver and nanoclays (Ko et al. 2021; Rafique et al. 2023). Polysaccharides originated from plants, animal, marine, and microbial origin are discussed in further subsections of this chapter.

15.4 Methods Used to Control Active Substance Controlled Release Characteristics

Majority of the polysaccharides employed in food industry got compliance approval from FDA which render their suitability for food packaging sector for safe processing of bio-based materials (Irfan et al. 2023). Owing to great potential of film and gel-forming characteristics, a wide variety of polysaccharides have been innovatively modified to produce thin membrane films and gels for applied purposes in food, pharmacy, and medicine (Cazón et al. 2017). Apart from this, the modern development with regard to polysaccharide-based carrier agents includes the formation of edible films, gels, aerogels, and intelligent packaging films for maintenance and assurance of quality and safety of foodstuffs during transportation and storage and shelf life extension (Zafar et al. 2016; Ali et al. 2018; Boostani and Jafari 2021).

Other supplementary advantages of using bio-based polysaccharide materials are increased protection from external environmental extremes, such as moisture, light, heat and moisture, oxygen, dust particles, enzymes, and gaseous exchanges. Moreover, polysaccharide-based materials usually offer wonderful barrier properties especially when exposed to oxygen and carbon dioxide at elevated and lower relative humidity levels (Azeredo and Waldron 2016; Malik et al. 2023). However, their main demerit features lie in the fact that when polysaccharide-based materials are exposed to water vapors, then permeability might be poor owing to hydrophilic nature and low mechanical stability which can be technically overcome by formation of blended mixtures of polysaccharides or employing lipids as well as systematic incorporation of bioactive components (Benbettaïeb et al. 2016; Jiang et al. 2021a, b, c; Boostani and Jafari 2021).

For intelligent packaging, a wide range of polysaccharide-based bioindicators were investigated by the researchers for propagation of green technology and eco-sustainability (Jiang et al. 2019). As bioindicators in terms of incorporation in polymer matrix, various bioactive components sensitive to pH and temperature have been explored including natural extracts and dyes for their potential application as alternatives of commercially employed toxic chemicals and synthetic counterparts (Raza et al. 2023; Ameer et al. 2020a). Polysaccharide/bioactive components follow the same mechanism of action as those of synthetic counterparts. Usually, incorporation of bioactive component and polysaccharides into polymer matrix is carried out, and final bio-based polymeric material is obtained through lyophilization or casting method (Das et al. 2020; Azeredo and Waldron 2016). Various forms of bioindicators have been reported in published literature, such as in form of edible film indicator, as direct coating on the food product surfaces, and as an insertion pad inside the packaged foodstuff (Ali et al. 2018).

15.4.1 Edible Food Packaging

Films and coatings are included in edible food packaging materials, and these can be utilized in direct manner for foodstuffs or also include usage in form of wrapping for providing several functionalities, such as restricting migration of oxygen, carbon dioxide, lipid, and volatile flavoring compounds (Hong et al. 2022; Zhang et al. 2020). Hence, edible film has a potential role as barrier in extension of shelf life of food products. However, it must be taken into account that packaging materials must not be proof to ethylene, water, and gas in absolute manner because of respiratory activity of fresh produce, fruits, and vegetables. Furthermore, there should be no chance of rendering any mechanical damage, and films and coatings must exhibit good adherence to the food product (Kim et al. 2020; Boostani and Jafari 2021).

Usually, the materials which are used for preparing edible packaging are soluble in water. Usage of organic solvents is not recommended due to possible toxicity and safety issues as well as contamination of foodstuffs. It is imperative to keep in view that such biodegradable materials must be safe for human use and exhibit approval from regulatory and food safety authorities (Ahmad et al. 2022). Other peculiar features lie in the uniqueness of this type of packaging material as compared to conventional counterparts owing to their edible nature and from being the renewable resources. Polysaccharides classified as food-grade chemical constituents are suitable alternatives as raw materials to render formation of edible coatings and films (Tian et al. 2022; Shafique et al. 2023). Moreover, polysaccharide-based materials have the following characteristics: (1) they are colorless, (2) have oil-free appearance, (3) have minor caloric content, and exhibit potential application in elongation of shelf life of food commodities (meat products, vegetables, fruits, and shellfish) owing to significant decline in surface darkening, oxidative rancidity, and dehydration (Chung et al. 2019; Arshad et al. 2020; Cazón et al. 2017). Moreover, along with barrier properties, certain polysaccharides exhibit additional functionality of serving as carriers of bioactive compounds, such as organic acids which are reported to contribute to inhibition of microbial growth (especially *Listeria monocytogenes*, *Salmonella*, and *Escherichia coli*) (Raza et al. 2023). Furthermore, the polysaccharide edible films provide advantage of antioxidants as well as supplementation to prepared product itself. Along with eco-friendly nature of edible coatings and films, these innovative packaging materials may serve as potentially vital source of minerals and vitamins and may help in taste improvement of foodstuffs (Zafar et al. 2016; Cazón et al. 2017).

15.4.2 Intelligent Packaging

Systems involving intelligent packaging are made from those materials which when come into contact with foodstuffs provide monitored information in detailed manner about freshness and quality properties of foodstuffs during their long-term storage or transportation (Puligundla et al. 2012). Intelligent packaging is differentiated from conventional counterparts with respect to the presence of internal or external sensors/indicators on packaging headspace or surfaces and provides qualitative data about the packaged foodstuffs. In this regard, the most important indicators include gas, freshness, and time-temperature-sensitive indicators (Jiang et al. 2022b; Nemes et al. 2020).

15.4.2.1 Freshness Indicators/Sensors

Packaged food products (prominently meat and fish) are susceptible to breakdown of food components and deterioration induced by microbial contamination in headspace of packaged foodstuffs, and freshness indicators play a pivotal role in determining degree of microbial breakdown (Moreira et al. 2015). In general, pH indicator dyes are employed for monitoring sensitive changes in volatile flavoring compounds, closely linked to food spoilage. Usually, polymeric matrix of polysaccharide-based materials comprises of such pH-sensitive dyes which are indicative of visible color changes in the form of response because of possible release of volatile flavoring compounds inside the packaging materials (Nair et al. 2018; Sheikh et al. 2023).

15.4.2.2 Gas Indicators/Sensors

Fresh produce, fruits, and vegetables exhibit respiration phenomenon which exerts significant influence on internal atmosphere of package, and hence, gas indicators are particularly important in monitoring such packaged foodstuffs of high respiratory activity (Nemes et al. 2020). Gas composition inside the package container may subject to variations because of interaction of food components with surrounding environment. The underlying principle of action of carbon dioxide and oxygen indicators is reliant on color changes exhibited by foodstuffs because of chemical or enzymatic modifications (Jiang et al. 2021a, b, c; Jamróz et al. 2019). Principal advantage of gas indicators is that gas indicators are indicative of quality changes of packaged products, and usually these indicators are available in tablet or labeling forms. Gas indicators offer the advantage of not fulfilling requirement of package opening for checking quality of foodstuffs. This also ensures increased food safety and microbial decontamination by means of dragging or installation of quaternary ammonium salt (Jamróz et al. 2019; Zafar et al. 2016; Azeredo and Waldron 2016).

15.4.2.3 Temperature Indicators/Sensors

During supply and distribution of foodstuffs, thermal limits of foodstuffs may vary depending on the food handling, storage, and transportation conditions. Foods are exposed to heat loads and are usually indicated by the temperature indicators/sensors (Nemes et al. 2020). Temperature indictors are indicative of color changes, migration of coloring ingredients, and mechanical deformation (Ali et al. 2018). Two types of indicators are usually employed in food industry: (1) registration indicators (these track changes about extreme conditions, such as product exposure to low or elevated temperatures in comparison with critical temperature) and (2) continuous temperature sensors/indicators (also known as weather temperature indicators which track the weather-induced thermal changes) (Müller and Schmid 2019; Azeredo and Waldron 2016). On commercial scale, time/temperature-type indicators perform operation on basis of several chemical indicators, degree of polymerization reactions, enzymatic hydrolysis reactions, chemical diffusion, etc. Such indicators are preferably employed in various unit operations like packaging, transportation, and frozen foods intended for long-term/short-term storage periods (Ameer et al. 2017c; Kalpana et al. 2019).

15.4.2.4 Aerogels for Food Packaging Applications

Highly solid materials with high degree of porosity and lightweightness are classified as aerogels and are described with the following peculiar characteristics: (1) large surface area, (2) lower thermal conductivity, and (3) high mechanical stiffness (Franco et al. 2018). In general, preparation of aerogels involves replacement of liquid with air from gel by supercritical drying of carbon dioxide.

Particularly, the solvent removal is achieved from gel through compression of CO_2 at operational parameters set above critical pressure and temperature of solvent-CO_2 mixture (de Oliveira et al. 2019). Through this way, homogenous single-phase mixing procedure is employed coupled with avoidance of vapor-liquid interfaces. In gel porous structure, the physical presence of supercritical fluid hinders the emergence of surface tension; thus, pore collapse is inhibited during drying phase (Raza et al. 2021c; Nešić et al. 2018). When the supercritical CO_2 is completely replaced with liquid solvent in gel structure, the slow depressurization is achieved at working conditions above supercritical temperature for release of CO_2 with ease. Hence, it implied that such drying methods ensure retention of completed three-dimensional structure of gel matrix without any degradation or pore collapsing (Aragón-Gutierrez et al. 2020).

Although aerogels have been utilized extensively in medicine, food, and drug delivery systems owing to easier encapsulation of bioactive compounds/nutrients or drugs because of high porosity and surface area, still the keenly focused research on formulation of polysaccharide-based aerogels has been initiated in last years (Nešić et al. 2018). During last decade, the research was aimed at exploration of eco-friendly and technological properties of polysaccharides, such as edibility, biodegradability, and biocompatibility. Furthermore, polysaccharides were investigated for development of multifunctional bio-aerogels as potential replacer of synthetic and toxic chemicals for targeted applications (Yang et al. 2023; de Oliveira et al. 2019). Recent advances in this regard of aerogels include their usage as integral part of active food packaging where insertion of polysaccharide-based aerogels is carried out inside the package and are employed as bioabsorbers of oxygen and humidity. Details of edible packaging materials from different sources are depicted in Fig. 15.1. Additionally, these may serve as the possible carriers of biologically

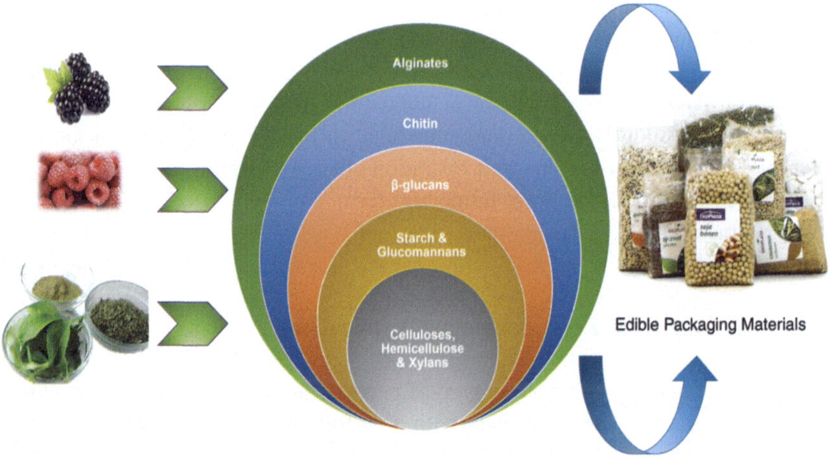

Fig. 15.1 Polysaccharides derived from different sources for exploiting as component of edible packaging films and coatings

active compounds and nutrients of choice through inhibition or hindrance of microbial contamination of foodstuffs (Mikkonen et al. 2013). Generally, polysaccharide-based aerogels exhibit lower thermal conductivity ranging from 16 to 22 mW/mK which is equivalent to those observed in polystyrene foams or silica gels, commonly employed in food and beverage sector for hot and chilled foodstuffs. In bio-based food packaging materials intended for thermosensitive foods, polysaccharide-based aerogels might be utilized to provide mechanical support (de Oliveira et al. 2019; Nešić et al. 2018).

15.5 Role of Polysaccharide-Based Materials as Carriers of Active Substances

15.5.1 Plant-Based Polysaccharides

15.5.1.1 Cellulose

Globally, cellulose is categorized as one of the most widely found polymers from plant sources. All plant cells exhibit the presence of cellulose in their compositions. Some of the fungal and algal species, invertebrates, few gram-negative bacteria, and marine organisms belonging to tunicate family also possess cellulose as constituents of their structural configurations (Novák et al. 2020). Biochemically, cellulose is classified as a linear homopolysaccharide made up of d-glucopyranosyl subunits connected to each other by means of β-(1 $\rightarrow$ 4) glycosidic linkages (Zafar et al. 2016).

In nature, cellulose is most commonly sourced from cotton fibers and wood pulp materials. However, valorization of waste materials of plant origin and byproducts, such as leaves, shoot, stem, shell, husk, peel, and skin as well as sugarcane bagasse, has been utilized as suitable and chief cellulose sources which play an important role on economic and ecological fronts (Aulin et al. 2010). Cellulose has been reported to exhibit appreciable film-forming capability. Moreover, cellulose is easy to handle for synthesis of cellulose derivatives and possesses high chemical stability (De Azeredo et al. 2014). Conversely, poor solubility of cellulose in aqueous medium and organic solvents hampers and restricts the exploitation of cellulose for edible film and coating formations. Derivatization has been suggested as the possible solution to overcome this demerit of cellulose solubility, and it also led to significant improvement in thermoplasticity of cellulose (Ramachandraiah et al. 2022; Wang et al. 2016).

Edible food packaging involves exploitation of various cellulose derivatives, such as cellulose esters, hydroxypropylmethyl cellulose, carboxymethyl cellulose (CMC), methylcellulose, and hydroxypropyl cellulose (Cazón et al. 2017). Hydroxypropyl methylcellulose has been reported to improve barrier properties (especially water vapor permeability) and mechanical strength of edible film with corresponding rises in molecular weight and substation degree. Moreover, in case of deep-fried starch products, reduced fat uptake and moisture loss have been observed from edible coating manufactured by methyl cellulose and hydroxypropyl methyl cellulose (Otoni et al. 2018). Incorporation of antimicrobial components and antioxidant compounds has been employed for functionality enhancement of edible films made by cellulose derivatives. Improvement of biological activity was also achieved by utilization of different additives, such as polysaccharides derived from spent ground coffee, quercetin and rutin-loaded dipalmitoyl lecithin liposomes, candelilla wax, and α-tocopherol. In a report by de Dicastillo et al. (2016a, b), murta fruit extract was mixed with methyl cellulose and led to high antioxidant potential, and methyl cellulose/native Chilean berry extract was also utilized in formulation of edible coating and film. These edible films were ideal substitute for food products with high fat proportions. Carboxymethyl cellulose was used for edible coating with addition of bacteriocin (derived from *Bacillus methylotrophicus* BM47) for increasing shelf life of strawberries (de Dicastillo et al. 2016a, b). Strawberries were subjected to coating with 0.5% of both CMC and bacteriocin-added CMC at storage conditions of 75% RH and 4 °C for storage period of 16 days. It was evident from results that bacteriocin-added CMC caused delayed ripening of fruit up to four days and inhibited fungal growth in comparison with uncoated strawberry samples. Bacteriocin-added CMC coating did not exhibit any adverse effect on strawberry quality, and coated strawberries showed maintenance of antioxidant activity and lower total soluble solid (TSS) content and weight loss during whole storage period (Khan et al. 2020; Jiang et al. 2021d). Moreover, in another instance, acetate films prepared by cellulose and hydroxyethyl cellulose were further subjected to enrichment through introduction of resveratrol inclusion complexes, and this caused significant rise in antioxidant potential against *Campylobacter*, whereas cellulose-based films incorporated with niacin demonstrated higher degree of antimicrobial potential against *Bacillus*, *Listeria*, *Enterococcus*, and *Staphylococcus* spp. (Silva et al. 2016).

Apart from cellulose and its derivatives, cellophane has also emerged as byproduct from regenerated cellulose to exhibit well-defined barrier characteristics, and this rendered it an ideal candidate for food packaging applications. Since last century (from 1930 onward), cellophane usage has been reported for fresh produce, vegetables, and fruits and bakery products (sandwiches and cookies) (Youssef and El-Sayed 2018). Cellulose has been reported to formulate bio-based aerogels for food packages, and such aerogels have high degree of thermal conductivity and mechanical strength as compared to what was found for silicon gels (Peelman et al. 2013). Depending on degree of cross-linking and operational conditions during drying, cellulose-based aerogels could be tailored for modification of textural and morphological properties. With regard to sensing/indicator applications, cellulose-based indictors were employed for monitoring of gaseous exchange and concentration of carbon dioxide and oxygen for quality assurance and control of food products (Ameer et al. 2024; Cazón et al. 2017; Zafar et al. 2016).

15.5.1.2 Hemicellulose

Processing of plants and plant byproducts generates multitude of hemicellulose from wood. Majority of hemicelluloses become entrapped with other components, such as lignin and lignin-carbohydrate complexes due to hydrogen bonding; hence hemicelluloses may subject to variations of structure, composition, localization, and branching patterns. Main peculiar differences of hemicellulose in comparison with cellulose lie with respect to lower molecular mass, shorter chain length, and presence of noncrystalline regions (Cazón et al. 2017; Zafar et al. 2016). Four main groups of hemicellulosic biomass, such as xyloglucans (xylans), aminoglycans (mannans), β-glucans, and xyloglucans, have been reported in nature. Substitution patterns and degree of substitution exert significant influence on the solubility, cell wall components, enzymatic hydrolysis, and gelling characteristics (Wu et al. 2021; Esa et al. 2014). Xylan chain substitution has been recognized as the primary factor influencing water solubility, and oxidation of cereal-derived β-glucan has been reported to improve solubility. Hemicellulose has good film-forming and gelling capacities; however, the major drawback lies in its increased hydrophilic nature which limits their practical applications owing to less resistance to moisture (Zafar et al. 2016). However, hemicellulose consists of various free hydroxyl groups in its side chains and backbone, and these functional groups enable to achieve functionalization and manufacturing of cellulose derivatives with desired properties of crystallinity, hydrophobicity, and solubility (Ali et al. 2018). Other functionality enhancement modalities include hydroxypropylsulfonation, ion-mediated oxidation, and hydroxypropylation. The most profoundly investigated hemicelluloses include glucomannans, xylans, and β-glucans (Khan et al. 2022; Benbettaïeb et al. 2016).

15.5.1.3 Xylans

Most commonly employed hemicelluloses include xylans, and these are ranked as the second most widely found polysaccharides originated from plants (Cazón et al. 2017). β-(1 → 4), (1 → 2), and (1 → 3) linkages have been reported in xylan. Branching type is indicative of xylan subtype. Good sources of xylan include corn cobs, straw, hulls, husk, bran sugarcane, and cereal straw (Zafar et al. 2016). In a research conducted by Mikkonen and Tenkanen (2012), xylan films were developed which exhibited lower grease and oxygen permeability with moderate mechanical stability. Spruce xylan also showed lower oxygen permeability, whereas paper-based xylan laminates were prepared through exploitation of carboxymethyl xylan and a 2-dodecenyl succinic anhydride potential replacer of paper-based coating and polyethylene films (Cazón et al. 2017). As far as aerogels are concentered, very few published data are available until now pertaining to physiochemical characteristics of xylan-based aerogels, and morphological properties of these gels have also been investigated.

15.5.1.4 Glucomannans

Plant hemicelluloses also comprise of glucomannan as member of mannan family. Dicot plant and woody parts of conifers constitute glucomannan in their composition ranging 2–5%. Konjac plant has been recorded as the most profound source of glucomannans (about 60–80% on dry weight basis) (Wu et al. 2019). Moreover, glucomannan is also present as cell wall component of yeast and bacteria. In case of structural configuration, d-mannopyranose and d-glucopyranose units are found in connected manner by β-(1 → 4) bonding pattern. Galactoglucomannans are also found in soft woody parts of different plants (Du et al. 2019).

Edible films were prepared using salep glucomannan, and their efficacy was compared to that of guar gum and locust bean-based films. Salep glucomannan demonstrated better barrier properties for water vapors and oxygen (Kurt and Kahyaoglu 2014). As salep glucomannan exhibits less mechanical strength, therefore glucomannan biopolymers are prepared in synergistic manner with other polysaccharide components like curdlan, ethyl cellulose, and gellan gum (Kurt 2019). Polydopamine addition has been shown to improve mechanical strength and barrier properties and led to enhanced functionalization of konjac glucomannan.

Nanoparticles produced from synergy of gallic acid and chitosan or nisin/chitosan improved antimicrobial activity of konjac glucomannan films against various foodborne pathogens including *Bacillus cereus*, *S. aureus*, *Listeria monocytogenes*, and *E. coli* (Wu et al. 2019). Regarding preservation of mushrooms, konjac glucomannan/carrageenan was evaluated by Zhang et al. (2019) for postharvest storage stability increases. Best physicochemical properties were exhibited by *Agaricus bisporus* mushroom subjected to storage period of 5–12 days with 0.3% of bio-based film prepared by konjac glucomannan/carrageenan.

15.5.1.5 β-Glucans

Naturally, β-glucan is abundantly available as biopolymer and polysaccharide material, and main sources in this regard include cereals, plants, bacteria, algae, yeast, mushroom, and fungi. Cereals are the predominately described sources and hence discussed in detail. Grains consist of internal endosperm cell walls having consistent membrane compartments referred to as aleurone and sub-aleurone layers (Peltzer et al. 2017). β-Glucans are found in highest amounts in barley (3–11%), followed by oats, rye, and wheat in range of 1–7%. β-Glucans mainly comprise of constituent molecular moieties named as cellotriosyl and cellotetraosyl subunits. Generally, they generally exhibit a higher degree of solubility and rising tendency in gelation property (Peltzer et al. 2018).

Edible films and coating are useful as packaging materials for protecting foodstuffs from the oxidation phenomenon owing to physical stability, better barrier properties, and mechanical strength. Main drawbacks of using β-glucans lie on the fact of the higher costs associated with the extraction processes utilized for recoveries and subsequent purification as well as poor water vapor properties. Natural polymer compounds, such as combination of arabinoxylans and β-glucans, have been explored as potential solutions. Different functionality enhancement techniques, such as esterification carboxymethylation and ion-mediated oxidation, positively exert significant influence on structural and physical properties of β-glucan-based composite packaging materials (Razzaq et al. 2016). β-Glucan aerogels have also been explored actively by the researchers for carrying active ingredients and targeted delivery of nutraceuticals, for example, flax lignan and flax oil in food manufacturing industry (Ying et al. 2015).

15.5.1.6 Chitin

Ranked second to cellulose, this is classified as one of the most abundant polysaccharides in nature. Chemically, alkaline N-deacetylation of chitin results in yield of chitosan. Chitin is most commonly found in exoskeleton of majority of insects and crustaceans (Pillai et al. 2009). Moreover, as a byproduct of waste from the shellfish sources, it is volarized in several commercial applications. Chitosan exhibits potent antimicrobial activity, and the most widely accepted hypothesis in this regard includes modification of cell permeability owing to electrostatic interaction between microbial cell membrane and chitosan molecules (Jayakumar et al. 2010). Published literature offers several methods of preparing chitosan-based films. One such method is termed as casting which involves chitosan dissolution in ambient solvents or moderately acidified water followed by addition of plasticizer. Then the solution is poured on flat surface subsequently followed by solvent evaporation (Bouissil et al. 2019). For obtainment of foils, the extrusion is the most widely reported method. During foil or film manufacturing, pressure is progressivity increased inside the barrel followed by ingredient mixing. At end, mixture is allowed to expand after release from the dye; however, chitosan thermo-plasticity is lower, and this is one of the prime disadvantages of chitosan; hence chitosan degrades prior to start the initiation of melting point (Lekjing 2016). Therefore, conversely to the conventionally employed thermoplastic polymer films, chitosan films can be subjected to heat sealing, molding, or extrusion. To overcome this issue, chitosan is usually blended with other thermoplastic polymers, such as poly(butylene succinate), poly(−butylene terephthalate adipate), and poly(butylene succinate adipate) for rendering improved thermal properties (do Amaral et al. 2015). Achieving levels of good film-forming properties in chitosan allows to produce coatings exhibiting good mechanical properties and high degree of selective permeability for O_2 and CO_2. Moreover, chitosan films have great permeability to water vapors, and this restricts their application in food industries because moisture needs to be effectively controlled in order to ensure food preservation especially in case of high-humid scenarios (van den Broek et al. 2015). For improving functional properties of chitosan, several strategies have been devised for chitosan films. This includes addition of plasticizers, modification of solvent type, and pH and degree of deacetylation. Addition of proteins or polysaccharides may also improve functional characteristics (Elsabee and Abdou 2013).

Unfortunately, despite the addition of improvers, the solubility of chitosan remains to be highly sensitive or in some case exhibits water solubility which limits its applications. For prevention of dissolution and swelling of chitosan-based films, chitosan cross-linking has been proposed as viable solution with regents such as genipin, glutaraldehyde, or formaldehyde (Cazón et al. 2017).

15.5.1.7 Starch

As a natural polymer, starch might be the possible alternative material for producing food packaging materials. Starch is one of the chief compositional materials which has been studied extensively owing to its edible nature, cheap and abundant availability, and biodegradability (Cazón et al. 2017). As agricultural biopolymer, starch is found in a variety of plants, such as potatoes, beans, rice, corn, wheat, etc. In compositions of various kernels, it comprises 60% of the total composition, and its separation is relatively easier from other components (Zafar et al. 2016). Two main polysaccharides make up essentially the starch granules: amylose and amylopectin. Most of the starches found in nature are categorized as the semicrystalline materials. On the basis of amylose/amylopectin ratio, the degree of relative crystallinity usually varied from 15% to 45%. In case of amylopectin, short-branched chains usually form the crystalline regions, whereas amylopectin branching points and amylose constitute the formation of amorphous regions (Molavi et al. 2021). In cold water, starch molecules are usually insoluble. However, when starch is subjected to heating, then it causes disruption of crystalline structure and may result in partial solubilization. Starch heating at elevated temperature of 65 °C and 90 °C may lead to formation of hydrogen bonds in starch suspensions, and gelatinization process takes place in irreversible manner. For obtaining homogenous film-forming starch solution, gelatinization of starch granules is necessary in excessive water (>90% w/w) (Cazón et al. 2017). Amylopectin matrix is usually broken during this process and results in releasing of amylose molecules. Owing to water diffusion through starch granules, these changes are brought about followed by increasing tendency in starch crystallite melting (Jimenez et al. 2012). Such complex starch gelatinization is required to prepare film from homogenous solution mixture. For formation of starch-based films, usually two types of techniques are utilized: (1) wet process and (2) dry process. Dry process involves the use of extrusion on the basis of thermoplastic properties of starches. In dry process, starch is subjected to elevated temperatures usually more than the glass transition temperature in the presence of lower moisture content (Cazón et al. 2017). Whereas in case of wet process, fist solubilization of polymer is carried out followed by drying of film-forming solution. Edible-preformed films are preferably manufactured through wet processing method, and sometimes dipping, brushing, or spray-drying of coatings is also performed for application into food products (Cazón et al. 2017). On industrial scale, however, dry method exhibits more suitability as compared to wet drying. Starch-based films exhibit excellent barrier properties especially for oxygen owing to highly developed network of hydrogen linkages in which both crystalline and noncrystalline regions are embedded in alternate fashion. Improvement of barrier properties is usually carried out through addition of high amylopectin content or rising degree of crystallinity (Zafar et al. 2016; Jimenez et al. 2012). As compared to synthetic counterparts, starch-based films have several drawbacks. Mainly starch films have relatively higher tensile strength, but low elongation percentage usually is indicative of poor mechanical characteristics. Starch-based films with high degree of crystallinity are usually less susceptible to moisture and high relative humidity found in surrounding environment (Jimenez et al. 2012). Poor mechanical characteristics of starch-based film also include brittleness which might be attributed to the presence of amorphous regions of amylose molecules. Starch film brittleness issue is usually overcome by addition of plasticizing material and hence leads to achieve improved extensibility and flexibility. Another modality to resolve this issues involves blending of starch molecules with biopolymers, such as polyvinyl alcohol (PVA) which results in enhanced mechanical properties as well as maintenance of biodegradability (Bonilla et al. 2013).

15.5.1.8 Alginates

Usually the brown seaweed belonging to class Phaeophyceae are rich and excellent sources of alginates. Alginic acid is the precursor molecule whereas alginates are produced as results of condensation reactions in the form of salts (Tavassoli-Kafrani et al. 2016). Alginates exist in the form of copolymer linearly arranged and mainly composed of 1–4 b-D-mannuronic acid (M) and a-L-guluronic acid (G). Monomers are usually found in arrangement of alternate fashion in long polymer chains in the form of blocking sequences of GG, MM, or MG blocks (SenturkParreidt et al. 2018). Chemically, one of the prime features of alginate consists of their reaction ability with those of di- or trivalent cations to form alginate-based films. As gelling agents, most appropriately reported are the calcium ions in terms of their efficacy as compared to other mineral ions, such as ferrous, aluminum, manganese, or magnesium (Dehghani et al. 2018). Between M and G blocks of alginates, these ions help to form associations which in turn cause highly ordered three-dimensional network with improved stability. This precise and well-developed network is envisaged as "eggbox model." Such kind of cross-linking between polyvalent cations is usually employed in improvement of aqueous barrier characteristics, cohesiveness, mechanical resistance, and degree of stiffness as well as delayed release of bioactive compounds (Zafar et al. 2016; Carina et al. 2021). Owing to rapid cross-linking with that of calcium ions, gelling areas are created inside the starch films in localized manner which may lead to compromised stability of the films. In this regard, a feasible technique of homogenous matrices by means of slow calcium release has been proposed in an acidic medium (Kontominas 2020). During formation of alginate films, first the solvent is vaporized, and then the added calcium may cause improvement in barrier and mechanical properties through cross-linking. As aforementioned, alginate coatings or films usually exhibit poor barrier properties against moisture; however, their hygroscopic nature helps to retard dehydrations of foodstuffs for which these are applied for practical purpose. Edible films made from the pectin also exhibit similar functional properties to those of alginate films and help to achieve retardation of moisture loss (Zafar et al. 2016; Aziz et al. 2018).

15.6 Limitations and Future Trends

Although polysaccharides are highly promising biomolecules of interest and exhibit great potential to be employed for biocompatible nano-carriers of bioactive compounds and ingredients for nutrient-delivery and nutraceutical applications, however, inherent natural variability of polysaccharides based on size, shape, botanical origin, composition, and molecular weight might be disadvantageous in certain situations when compared to barrier and mechanical properties of their synthetic counterparts prepared through controlled radical polymerization. Therefore, it is need of the time to develop large-scale manufacturing processes for refined polysaccharides. It is also imperative and useful to mention that adoption of advanced chemical separation techniques is required to achieve purification of natural polysaccharides. In this regard, synthesis of biocompatible polymers through engineering techniques is a way forward to explore future possibilities. Another chief concern in this regard includes inherent immunogenicity found in polysaccharide molecules in a similar manner to peptides and proteins. Polysaccharides must be processed to reduce/suppress the immunogenicity to safe levels for their application on industrial scale as carrier agents.

15.7 Conclusions

Applications of polysaccharide-based packaging films as carriers of active substances and controlled release characteristics in food science and nutraceutical industries have opened new era of opportunities owing to the development of novel functional food products with biocompatibility and biodegradability, having good barrier properties as modern alterative to traditionally employed synthetic counterparts. Valorization of carbohydrate polymers and edible starches may also solve the problem of gigantic waste streams as an eco-friendly approach. Mechanical and barrier properties of starch-based packaging films and coatings are still far away from the desired characteristics of their synthetic counterparts in food industry. However, continuous efforts are being made by researchers worldwide for improvement of starch film barrier and mechanical properties by synergistic combination of polysaccharides with those of linkers, cross-linking additives, and improvers. Potentially, commercial applications of polysaccharides in food products are still debatable; however, researchers are keenly interested on global scale to explore suitable combinations of mixing materials to those of food products as the functional characteristics and efficiency of polysaccharide-based coatings and packaging films are dependent on biopolymers and other kinds of additives for improving functional properties and final food products for which edible films are employed. Further research is needed regarding improvement of barrier properties and tensile and mechanical strength of bio-based starch polymers similar to those of synthetically used commercial plastics and fiber coatings. In the future, polysaccharide-based edible films and coatings would be an excellent and wonderful alternative as replacer of synthetic counterparts.

References

Ahmad Z, Ilyas M, Ameer K, Khan MA, Waseem M, Mehmood T et al (2022) The influence of fenugreek seed powder addition on the nutritional, antioxidant, and sensorial properties of value-added noodles. J Food Qual 2022

Ali A, Xie F, Yu L, Liu H, Meng L, Khalid S, Chen L (2018) Preparation and characterization of starch-based composite films reinforced by polysaccharide-based crystals. Compos Part B 133: 122–128

Alipoormazandarani N, Ghazihoseini S, Nafchi AM (2015) Preparation and characterization of novel bionanocomposite based on soluble soybean polysaccharide and halloysite nanoclay. Carbohydr Polym 134:745–751

do Amaral DS, Cardelle-Cobas A, do Nascimento BM, Monteiro MJ, Madruga MS, Pintado MME (2015) Development of a low fat fresh pork sausage based on chitosan with health claims: impact on the quality, functionality and shelf-life. Food Funct 6(8):2768–2778

Ameer K, Bae SW, Jo Y, Lee HG, Ameer A, Kwon JH (2017a) Optimization of microwave-assisted extraction of total extract, stevioside and rebaudioside-a from Stevia rebaudiana (Bertoni) leaves, using response surface methodology (RSM) and artificial neural network (ANN) modelling. Food Chem 229:198–207

Ameer K, Chun BS, Kwon JH (2017b) Optimization of supercritical fluid extraction of steviol glycosides and total phenolic content from Stevia rebaudiana (Bertoni) leaves using response surface methodology and artificial neural network modeling. Ind Crop Prod 109:672–685

Ameer K, Shahbaz HM, Kwon JH (2017c) Green extraction methods for polyphenols from plant matrices and their byproducts: a review. Compr Rev Food Sci Food Saf 16(2):295–315

Ameer K, Das PR, Eun JB (2020a) Advances in enzymatic applications in food industry. In: Microbial fermentation and enzyme technology. CRC Press, pp 85–97

Ameer K, Jo Y, Amir RM, Shahbaz HM, Kwon JH (2020b) Screening and identification of electron-beam irradiated dried spice-mixture products by electronic sensing and standard analytical methods through dose estimation. LWT 125:108957

Ameer K, Kim YM, Murtaza MA, Nasir MA, Khan FA, Khan MA et al (2024) Extraction of oil from oilseeds. In: Extraction processes in the food industry. Woodhead Publishing, pp 149–175

Aragón-Gutierrez A, Arrieta MP, López-González M, Fernández-García M, López D (2020) Hybrid biocomposites based on poly (lactic acid) and silica aerogel for food packaging applications. Materials 13(21):4910

Arshad MS, Kwon JH, Ahmad RS, Ameer K, Ahmad S, Jo Y (2020) Influence of E-beam irradiation on microbiological and physicochemical properties and fatty acid profile of frozen duck meat. Food Sci Nutr 8(2):1020–1029

Aulin C, Gällstedt M, Lindström T (2010) Oxygen and oil barrier properties of microfibrillated cellulose films and coatings. Cellulose 17(3):559–574

Azeredo HM, Waldron KW (2016) Crosslinking in polysaccharide and protein films and coatings for food contact–A review. Trends Food Sci Technol 52:109–122

Aziz MSA, Salama HE, Sabaa MW (2018) Biobased alginate/castor oil edible films for active food packaging. LWT 96:455–460

Bashir O, Hussain SZ, Ameer K, Amin T, Beenish, Ahmed IAM, Nazir N (2022) Influence of anticaking agents and storage conditions on quality characteristics of spray dried apricot powder: shelf life prediction studies using Guggenheim-Anderson-de Boer (GAB) model. Foods 12(1):171

Benbettaïeb N, Gay JP, Karbowiak T, Debeaufort F (2016) Tuning the functional properties of polysaccharide–protein bio-based edible films by chemical, enzymatic, and physical cross-linking. Compr Rev Food Sci Food Saf 15(4):739–752

Bonilla J, Atarés L, Vargas M, Chiralt A (2013) Properties of wheat starch film-forming dispersions and films as affected by chitosan addition. J Food Eng 114(3):303–312

Boostani S, Jafari SM (2021) A comprehensive review on the controlled release of encapsulated food ingredients; fundamental concepts to design and applications. Trends Food Sci Technol 109:303–321

Bouissil S, Pierre G, Alaoui-Talibi ZE, Michaud P, El Modafar C, Delattre C (2019) Applications of algal polysaccharides and derivatives in therapeutic and agricultural fields. Curr Pharm Des 25(11):1187–1199

van den Broek LA, Knoop RJ, Kappen FH, Boeriu CG (2015) Chitosan films and blends for packaging material. Carbohydr Polym 116:237–242

Carina D, Sharma S, Jaiswal AK, Jaiswal S (2021) Seaweeds polysaccharides in active food packaging: a review of recent progress. Trends Food Sci Technol 110:559–572

Cazón P, Velazquez G, Ramírez JA, Vázquez M (2017) Polysaccharide-based films and coatings for food packaging: a review. Food Hydrocoll 68:136–148

Chung N, Ameer K, Jo Y, Kwon JH (2019) Comparison of electronic sensing techniques for screening dried shrimps irradiated using three types of approved radiation with standard analytical methods. Food Chem 286:395–404

Das PR, Ameer K, Eun JB (2020) Role of enzymes in development of functional foods and food products. In: Microbial fermentation and enzyme technology. CRC Press, pp 99–113

De Azeredo HMC, Rosa MF, De Sá M, Souza Filho M, Waldron KW (2014) The use of biomass for packaging films and coatings. In: Waldron K (ed) Advances in biorefineries. Woodhead Publishing, Cambridge, pp 819–874

Dehghani S, Hosseini SV, Regenstein JM (2018) Edible films and coatings in seafood preservation: a review. Food Chem 240:505–513

de Dicastillo CL, Bustos F, Guarda A, Galotto MJ (2016a) Cross-linked methyl cellulose films with murta fruit extract for antioxidant and antimicrobial active food packaging. Food Hydrocoll 60: 335–344

de Dicastillo CL, Rodríguez F, Guarda A, Galotto MJ (2016b) Antioxidant films based on cross-linked methyl cellulose and native Chilean berry for food packaging applications. Carbohydr Polym 136:1052–1060

Du Y, Wang L, Mu R, Wang Y, Li Y, Wu D et al (2019) Fabrication of novel konjac glucomannan/shellac film with advanced functions for food packaging. Int J Biol Macromol 131:36–42

Elsabee MZ, Abdou ES (2013) Chitosan based edible films and coatings: a review. Mater Sci Eng C 33(4):1819–1841

Esa F, Tasirin SM, Abd Rahman N (2014) Overview of bacterial cellulose production and application. Agric Sci Proc 2:113–119

Feng X, Ameer K, Ramachandraiah K, Wu Z, Huo N, Bai X, Nie W, Jiang G (2022) Effects of papaya (Carica papaya L.) seed supplementation on quality attributes, adsorption capacities, and in vitro starch digestibility of wheat bread. J Food Measur Charact 16:3226–3239

Franco P, Aliakbarian B, Perego P, Reverchon E, De Marco I (2018) Supercritical adsorption of quercetin on aerogels for active packaging applications. Ind Eng Chem Res 57(44): 15105–15113

Hong SJ, Park BR, Lee HN, Kang HJ, Ameer K, Kim SJ, Kim YM (2022) Carbohydrate-binding module of cycloisomaltooligosaccharide glucanotransferase from Thermoanaerobacter thermocopriae improves its cyclodextran production. Enzym Microb Technol 157:110023

Irfan S, Murtaza MA, Mueen ud Din G, Hafiz I, Murtaza MS, Rafique S et al (2023) Physico-chemical, microbial, and functional attributes of processed Cheddar cheese fortified with olive oil–whey protein isolate emulsion. Food Sci Nutr 11(3):1247–1256

Jamróz E, Kulawik P, Kopel P (2019) The effect of nanofillers on the functional properties of biopolymer-based films: a review. Polymers 11(4):675

Jayakumar R, Menon D, Manzoor K, Nair SV, Tamura H (2010) Biomedical applications of chitin and chitosan based nanomaterials—a short review. Carbohydr Polym 82(2):227–232

Jiang GH, Ameer K, Eun JB (2019) Effects of carrier agents and relative humidity on the physicochemical and microstructural characteristics of hot air-dried Asian Pear (Pyrus pyrifolia Nakai cv. Niitaka) powder. Food Biophysics 14:235–248

Jiang G, Ameer K, Eun JB (2020a) Encapsulation of hot air-dried Asian pear powders using rice bran dietary fiber. Food Biosci 38:100742

Jiang G, Wu Z, Ameer K, Li S, Ramachandraiah K (2020b) Particle size of ginseng (*Panax ginseng* Meyer) insoluble dietary fiber and its effect on physicochemical properties and antioxidant activities. Appl Biol Chem 63:1-0.

Jiang G, Feng X, Wu Z, Li S, Bai X, Zhao C, Ameer K (2021a) Development of wheat bread added with insoluble dietary fiber from ginseng residue and effects on physiochemical properties, in vitro adsorption capacities and starch digestibility. LWT 149:111855

Jiang G, Ibrahim MS, Ibrahim MK, Zhao C, Butt M, Ameer K et al (2021b) Profiling and characterization of oat cultivars (Avena sativa L.) with respect to bioactive compounds, pesticide residues and mycotoxin. Int J Food Prop 24(1):1187–1201

Jiang G, Ramachandraiah K, Murtaza MA, Wang L, Li S, Ameer K (2021c) Synergistic effects of black ginseng and aged garlic extracts for the amelioration of nonalcoholic fatty liver disease (NAFLD) in mice. Food Sci Nutr 9(6):3091–3099

Jiang G, Wu Z, Ramachandra K, Zhao C, Ameer K (2021d) Changes in structural and chemical composition of insoluble dietary fibers bound phenolic complexes from grape pomace by alkaline hydrolysis treatment. Food Sci Techn 42:e50921

Jiang G, Feng X, Zhao C, Ameer K, Wu Z (2022a) Development of biscuits supplemented with papaya seed and peel: effects on physicochemical properties, bioactive compounds, in vitro absorption capacities and starch digestibility. J Food Sci Technol 59(4):1341–1352

Jiang G, Ramachandraiah K, Wu Z, Ameer K (2022b) The influence of different extraction methods on the structure, rheological, thermal and functional properties of soluble dietary fiber from Sanchi (Panax notoginseng) flower. Foods 11(14):1995

Jimenez A, Fabra MJ, Talens P, Chiralt A (2012) Edible and biodegradable starch films: a review. Food Bioprocess Technol 5(6):2058–2076

Kalpana S, Priyadarshini SR, Leena MM, Moses JA, Anandharamakrishnan C (2019) Intelligent packaging: trends and applications in food systems. Trends Food Sci Technol 93:145–157

Kang HJ, Lee HN, Hong SJ, Park BR, Ameer K, Cho JY, Kim YM (2022) Synthesis and characteristics of a rebaudioside-a like compound as a potential non-caloric natural sweetener by Leuconostoc kimchii dextransucrase. Food Chem 366:130623

Khan FA, Arshad MU, Rizvi SSH, Ameer K, Anjum FM, Imran M, Imran A (2020) Optimization of potato starch gel formulation as green alternative of animal-sourced gelatin. Food Sci Technol 41:626–634

Khan MA, Ameer K, Shakoor S, Ashraf MR, Butt M, Khalid MS et al (2022) Development and characterization of wheat rusks supplemented with Chia (Salvia hispanica L.) flour with respect to physicochemical, rheological and sensory characteristics. Food Sci Technol 42:e53921

Kim SH, Ameer K, Oh JH, Park MK (2020) Characterization of hot water extract from Korean deer velvet antler (Cervus canadensis Erxleben). Korean J Food Preserv 27(6):725–733

Kim SY, Lee HN, Hong SJ, Kang HJ, Cho JY, Kim D, Ameer K, Kim YM (2022) Enhanced biotransformation of the minor ginsenosides in red ginseng extract by Penicillium decumbens β-glucosidase. Enzym Microb Technol 153:109941

Ko JA, Ryu YB, Lee WS, Ameer K, Kim YM (2021) Optimization of microwave-assisted green method for enhanced solubilization of water-soluble curcuminoids prepared using steviol glycosides. Foods 10(11):2803

Kontominas MG (2020) Use of alginates as food packaging materials. Foods 9(10):1440. https://doi.org/10.3390/foods9101440

Kurt A (2019) Rheology of film-forming solutions and physical properties of differently deacetylated salep glucomannan film. Food Health 5(3):175–184

Kurt A, Kahyaoglu T (2014) Characterization of a new biodegradable edible film made from salep glucomannan. Carbohydr Polym 104:50–58

Lekjing S (2016) A chitosan-based coating with or without clove oil extends the shelf life of cooked pork sausages in refrigerated storage. Meat Sci 111:192–197

Liu QR, Wang W, Qi J, Huang Q, Xiao J (2019) Oregano essential oil loaded soybean polysaccharide films: effect of pickering type immobilization on physical and antimicrobial properties. Food Hydrocoll 87:165–172

Malik T, Sharma R, Ameer K, Bashir O, Amin T, Manzoor S, Mohamed Ahmed IA (2023) Potential of high-pressure homogenization (HPH) in the development of functional foods. Int J Food Prop 26(1):2509–2531

Mikkonen KS, Tenkanen M (2012) Sustainable food-packaging materials based on future biorefinery products: xylans and mannans. Trends Food Sci Technol 28(2):90–102

Mikkonen KS, Parikka K, Ghafar A, Tenkanen M (2013) Prospects of polysaccharide aerogels as modern advanced food materials. Trends Food Sci Technol 34(2):124–136

Mohamed SA, El-Sakhawy M, El-Sakhawy MAM (2020) Polysaccharides, protein and lipid-based natural edible films in food packaging: a review. Carbohydr Polym 238:116178

Molavi H, Behfar S, Shariati MA, Kaviani M, Atarod S (2021) A review on biodegradable starch based film. J Microbiol Biotechnol Food Sci 2021:456–461

Moreira MR, Cassani L, Martín-Belloso O, Soliva-Fortuny R (2015) Effects of polysaccharide-based edible coatings enriched with dietary fiber on quality attributes of fresh-cut apples. J Food Sci Technol 52(12):7795–7805

Müller P, Schmid M (2019) Intelligent packaging in the food sector: a brief overview. Foods 8(1):16

Murtaza MA, Ameer K (2022) Food processing industrial byproducts as raw material for the production of plant protein foods. In: Plant protein foods. Springer International Publishing, Cham, pp 109–129

Myrick JM, Vendra VK, Krishnan S (2014) Self-assembled polysaccharide nanostructures for controlled-release applications. Nanotechnol Rev 3(4):319–346

Nair MS, Saxena A, Kaur C (2018) Characterization and antifungal activity of pomegranate peel extract and its use in polysaccharide-based edible coatings to extend the shelf-life of capsicum (Capsicum annuum L.). Food Bioprocess Technol 11(7):1317–1327

Nemes SA, Szabo K, Vodnar DC (2020) Applicability of agro-industrial by-products in intelligent food packaging. Coatings 10(6):550

Nešić A, Gordić M, Davidović S, Radovanović Ž, Nedeljković J, Smirnova I, Gurikov P (2018) Pectin-based nanocomposite aerogels for potential insulated food packaging application. Carbohydr Polym 195:128–135

Novák I, Šauša O, Maťko I, Bilgin ES, Berek D (2020) Microporous carbon fibers prepared by carbonization of cellulose as carriers of particles of active substances. Chem Pap 74(4): 1359–1365

de Oliveira JP, Bruni GP, El Halal SLM, Bertoldi FC, Dias ARG, da Rosa Zavareze E (2019) Cellulose nanocrystals from rice and oat husks and their application in aerogels for food packaging. Int J Biol Macromol 124:175–184

Otoni CG, Lorevice MV, de Moura MR, Mattoso LH (2018) On the effects of hydroxyl substitution degree and molecular weight on mechanical and water barrier properties of hydroxypropyl methylcellulose films. Carbohydr Polym 185:105–111

Peelman N, Ragaert P, De Meulenaer B, Adons D, Peeters R, Cardon L et al (2013) Application of bioplastics for food packaging. Trends Food Sci Technol 32(2):128–141

Peltzer MA, Salvay AG, Delgado JF, Wagner JR (2017) Use of edible films and coatings for functional foods developments: a review. In: Nelson DL (ed) Functional foods sources, health effects and future perspectives. Nova Science Publishers, New York, pp 1–26

Peltzer M, Delgado JF, Salvay AG, Wagner JR (2018) β-Glucan, a promising polysaccharide for bio-based films developments for food contact materials and medical applications. Curr Org Chem 22(12):1249–1254

Pillai CKS, Paul W, Sharma CP (2009) Chitin and chitosan polymers: chemistry, solubility and fiber formation. Prog Polym Sci 34(7):641–678

Puligundla P, Jung J, Ko S (2012) Carbon dioxide sensors for intelligent food packaging applications. Food Control 25(1):328–333

Rafique S, Murtaza MA, Hafiz I, Ameer K, Qayyum MMN, Yaqub S, Mohamed Ahmed IA (2023) Investigation of the antimicrobial, antioxidant, hemolytic, and thrombolytic activities of Camellia sinensis, Thymus vulgaris, and Zanthoxylum armatum ethanolic and methanolic extracts. Food Sci Nutr

Ramachandraiah K, Ameer K, Jiang G, Hong GP (2022) Micro-and nanoplastic contamination in livestock production: entry pathways, potential effects and analytical challenges. Sci Total Environ 844:157234

Rashid F, Ahmed Z, Ameer K, Amir RM, Khattak M (2020) Optimization of polysaccharides-based nanoemulsion using response surface methodology and application to improve postharvest storage of apple (Malus domestica). J Food Measur Charac 14:2676–2688

Raza H, Ameer K, Ma H, Liang Q, Ren X (2021a) Structural and physicochemical characterization of modified starch from arrowhead tuber (Sagittaria sagittifolia L.) using tri-frequency power ultrasound. Ultrason Sonochem 80:105826

Raza H, Ameer K, Ren X, Liang Q, Chen X, Chen H, Ma H (2021b) Physicochemical properties and digestion mechanism of starch-linoleic acid complex induced by multi-frequency power ultrasound. Food Chem 364:130392

Raza H, Ameer K, Zaaboul F, Shoaib M, Zhao CC, Ali B et al (2021c) Physicochemical, rheological, & sensory characteristics of yogurt fortified with ball-milled roasted chickpea powder (Cicer arietinum L.). Food Sci Technol 42

Raza H, Ameer K, Ren X, Liu Y, Kang L, Liang Q et al (2023) Synergistic impact of heat-ultrasound treatment on the properties and digestibility of Sagittaria sagittifolia L. starch-phenolic acid complexes. Int J Biol Macromol 125457

Razzaq HA, Pezzuto M, Santagata G, Silvestre C, Cimmino S, Larsen N, Duraccio D (2016) Barley β-glucan-protein based bioplastic film with enhanced physicochemical properties for packaging. Food Hydrocoll 58:276–283

SenturkParreidt T, Müller K, Schmid M (2018) Alginate-based edible films and coatings for food packaging applications. Foods 7(10):170

Shafique B, Ranjha MM, Murtaza MA, Walayat N, Nawaz A, Khalid W, Mahmood S, Nadeem M, Manzoor MF, Ameer K, Aadil RM (2022) Recent trends and applications of nanoencapsulated bacteriocins against microbes in food quality and safety. Microorganisms 11:85

Shafique B, Murtaza MA, Hafiz I, Ameer K, Basharat S, Mohamed Ahmed IA (2023) Proteolysis and therapeutic potential of bioactive peptides derived from Cheddar cheese. Food Sci Nutr

Sheikh S, Siddique F, Ameer K, Ahmad RS, Hameed A, Ebad A et al (2023) Effects of white mulberry powder fortification on antioxidant activity, physicochemical, microbial and sensorial properties of yogurt produced from buffalo milk. Food Sci Nutr 11(1):204–215

Silva Â, Duarte A, Sousa S, Ramos A, Domingues FC (2016) Characterization and antimicrobial activity of cellulose derivatives films incorporated with a resveratrol inclusion complex. LWT 73:481–489

Tan C, Xie J, Zhang X, Cai J, Xia S (2016) Polysaccharide-based nanoparticles by chitosan and gum Arabic polyelectrolyte complexation as carriers for curcumin. Food Hydrocoll 57:236–245

Tavassoli-Kafrani E, Shekarchizadeh H, Masoudpour-Behabadi M (2016) Development of edible films and coatings from alginates and carrageenans. Carbohydr Polym 137:360–374

Tian Z, Ameer K, Shi Y, Yi J, Zhu J, Kang Q et al (2022) Characterization of physicochemical properties, microbial diversity and volatile compounds of traditional fermented soybean paste in Henan province of China. Food Biosci 50:102045

Wang S, Lu A, Zhang L (2016) Recent advances in regenerated cellulose materials. Prog Polym Sci 53:169–206

Wu S, Wang W, Yan K, Ding F, Shi X, Deng H, Du Y (2018) Electrochemical writing on edible polysaccharide films for intelligent food packaging. Carbohydr Polym 186:236–242

Wu C, Li Y, Du Y, Wang L, Tong C, Hu Y et al (2019) Preparation and characterization of konjac glucomannan-based bionanocomposite film for active food packaging. Food Hydrocoll 89:682–690

Wu Z, Ameer K, Hu C, Bao A, Wang R, Tang W, Chaudhary N, Jiang G (2021a) Particle size of yam flour and its effects on physicochemical properties and bioactive compounds. Food Sci Technol 42:e43921

Wu Z, Ameer K, Jiang G (2021b) Effects of superfine grinding on the physicochemical properties and antioxidant activities of Sanchi (*Panax notoginseng*) flower powders. J Food Sci Technol 58:62–73

Yang HI, Ameer K, Chung YB, Min SG, Eun JB (2023) Optimization of spray drying process for recovery of onion–stevia leaf hot water extract powder using response surface methodology. Food Sci Nutr 11(4):1770–1784

Ying R, Saulnier L, Bouchet B, Barron C, Ji S, Rondeau-Mouro C (2015) Multiscale characterization of arabinoxylan and β-glucan composite films. Carbohydr Polym 122:248–254

Yoon YS, Ameer K, Song BS, Kim JK, Park HY, Lee KC et al (2020) Effects of X-ray irradiation on the postharvest quality characteristics of 'Maehyang'strawberry (Fragaria× ananassa). Food Chem 325:126817

Youssef AM, El-Sayed SM (2018) Bionanocomposites materials for food packaging applications: concepts and future outlook. Carbohydr Polym 193:19–27

Zafar R, Zia KM, Tabasum S, Jabeen F, Noreen A, Zuber M (2016) Polysaccharide based bionanocomposites, properties and applications: a review. Int J Biol Macromol 92:1012–1024

Zhang R, Wang X, Wang J, Cheng M (2019) Synthesis and characterization of konjac glucomannan/carrageenan/nano-silica films for the preservation of postharvest white mushrooms. Polymers 11(1):6

Zhang W, Zhang Y, Cao J, Jiang W (2020) Improving the performance of edible food packaging films by using nanocellulose as an additive. Int J Biol Macromol 166:286–298

Zhao CC, Lu JK, Ameer K (2021) Effects of tofu whey powder on the quality attributes, isoflavones composition and antioxidant activity of wheat flour pan bread. LWT 143:111166

Zheng Y, Monty J, Linhardt RJ (2015) Polysaccharide-based nanocomposites and their applications. Carbohydr Res 405:23–32

Chapter 16
Role of Plasticizers for the Development of Polysaccharide-Based Edible Films

Gulzar Ahmed Rather, Faheem Patwekar, Mohsina Patwekar, Mir Ashiq Hussain, and Huda Nafees

16.1 Introduction

Plasticizers are a type of nonvolatile chemicals with a less molecular weight that are broadly utilized as supplements in the polymer manufactory. The essential job of such a substance is to increase polymer flexibility, extensibility, and processability by bringing down the subsequent request progress temperature and the glass transition temperature (Tg) (Sejidov et al. 2005). They can situate themselves among polymer atoms and slow down polymer chain cooperation to further develop flexibleness and processability. Plasticizers increment the free volume of polymer formation or the atomic portability of polymer particles (Valencia-Chamorro et al. 2011). Plasticizers are for the most part used to work on the habitual action of a polymeric matrix, additionally adjusting the thermic property, aqua retention conduct, and adherent action of the polymeric chain. The mentioned properties influence the fortitude of covering films also and the honesty of end results, which in addition

G. A. Rather (✉)
Department of Biomedical engineering, Sathyabama Institute of Science & technology deemed to be University, Chennai, India

F. Patwekar · M. Patwekar
Luqman College of Pharmacy, Gulbarga, Karnataka, India

M. A. Hussain
Department of Chemistry, Pondicherry University, Puducherry, India

H. Nafees
Faculty of Unani Medicine, Department of Saidla, Aligarh Muslim University, Aligarh, Uttar Pradesh, India

T. Amin et al. (eds.), *Polysaccharide Based Films for Food Packaging: Fundamentals, Properties and Applications*,
https://doi.org/10.1007/978-981-99-4898-7_16

affect drug discharge execution (Erythropel et al. 2014; Somwanshi et al. 2011). The main disadvantage of plasticizers is the increased permeability of gas, water vapor, and solute as a result of diminished film cohesion.

16.2 Plasticizers

Definition: Plasticizers are the components that are added to a synthetic resin to increase plasticity, flexibility, and brittleness (Fig. 16.1). They decrease viscosity and friction while handling in manufacture. In general, plasticizers diminish intermolecular powers long the polymer chains, hence expanding free volume and chain developments (Porter et al. 2009).

The committee of the IUPAC (International Union of Pure and Applied Chemistry) characterized a plasticizer as "a substance or material incorporated in a material (usually a plastic or elastomer) to increase its flexibility, workability or distensibility" (Table 16.1).

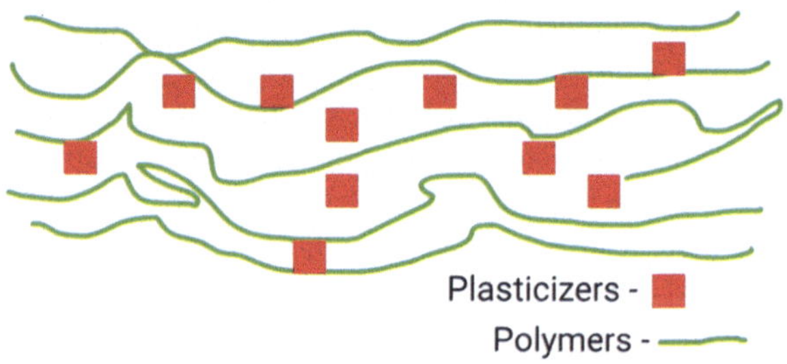

Fig. 16.1 Plasticizers and polymers

Table 16.1 Common plasticizers used in film coating formulation

Class	Examples
Polyhydric alcohols	Polyethylene glycols
	Glycerol
	Propylene glycol
Acetate esters	Triethyl citrate
	Acetyl triethyl citrate
	Glyceryl triacetate
Phthalate esters	Diethyl phthalate
Glycerides	Acetylated monoglycerides
Oils	Castor oil
	Mineral oil

16.3 Polysaccharide-Based Edible Films

Polymeric matrices created using edible polymers make up edible films and coatings. The polymer matrices could be eaten together along with packed foodstuff. Digestible biopolymers are used to create these polymeric matrices. The majority of biopolymers that are now being investigated are either from renewable sources or agro-industrial wastes. Because the majority of biopolymers come from natural sources, they are biodegradable and thus environmentally beneficial. Edible and biodegradable films could be an option in contrast to synthetic packaging materials in a few applications because of their abilities to forestall moisture misfortune, fragrance loss, solute movement, water retention in the food matrix, and on the other hand oxygen penetration (Aider 2010). Polysaccharides are naturally occurring polymers that are frequently employed in the production of edible films and coatings. Food scientist including engineers is working on new edible and biodegradable film materials that could open up new possibilities for developing novel food packaging methods. Cellulose subsidiaries, starches, alginates, and also carrageenans take place as the maximum well-known polysaccharides utilized in the creation of edible films and coatings (Mahajan et al. 2018). In most cases, the expansion of plasticizers is expected to acquire protein-based and polysaccharide-based films. With no plasticizer combined, the films made with multiple polysaccharides are often friable due to collaborations among polymer chains. Plasticizers diminish the intermolecular tensions between adjacent polymer chains, reducing cohesion within the film network. Plasticizers enhance or improve mechanical qualities in this way, lowering deformation tension, hardness, density, and viscosity while increasing polymer chain flexibility and fracture resistance. Plasticizers alter the barrier qualities of hydrophilic materials and in most cases increase water vapor penetration. The inclusion of hydrophobic plasticizers, on the other hand, may result in a reduction in water uptake (Espitia et al. 2014; Vieira et al. 2011).

16.4 Mechanism of the Action of Plasticizers

Considering the superior plasticizing impact, they should be completely blended along with integrated within the PVC polymer network. This is normally acquired through warming and blending as either the resin breaks down in the plasticizer or the plasticizer disintegrates within the resin. The plasticized compound is from there cooled. Various attributes in one and the other the straightforwardness affect the formation of the plasticized compound and furthermore its subsequent mechanical and actual functions of the adaptable manufacturers (Krauskopf and Godwin 2005).

The function of plasticizers is characterized as mediating among each separate filament of polymer, contributing to the disruption of polymer collaborations. It was generally imagined that plasticizers work by installing themselves between the chains of polymers, dispersing them separated (rising free volume) (Maeda and

Paul 1987). The tertiary construction based on the polymer is adjusted toward high permeability and flexibility along with lower strong design. Plasticizers lighten and grow the polymer that helps in defeating their protection from the deformity. The plasticized polymer would distort at a lesser tensile strength when contrasted with no plasticizer. This impact thus improves connection and relies on the glass progress temperature of polymers. Glass transition temperature (Tg) is the degree at which tough shiny polymer is changed over toward a rubbery component. All polymers admit larger glass transition temperature; also the expansion of plasticizers diminishes the glass progress temperature.

The following theories are offered to analyze the mechanism of plasticizers:

1. The lubrication theory
2. The gel theory
3. The free volume theory

During the 1940s, the lubrication theory and gel theory were developed. They are all attempting to clarify a few key aspects of plasticized polymers. Other models were presented in the successive years, during the 1950s. Out of which, free volume theory is unquestionably the specific that provides a major detailed description of plasticization. Maybe this is why, in the year since no new theories have been published, simply certain mathematical portraits established on the free volume theory also link the enlargement of specific features of the plasticized complexes including their plasticizer concentration (Doolittle 1954).

16.4.1 The Lubrication Theory

According to lubrication theory, plasticizers operate under the name of internal lubricants from diminishing frictional power through mixing themselves enclosed by polymer filaments. This hypothesis has been presented in a variety of ways, but the essential premise is that although a plastic item is bent, the polymer molecules must slide up one and all. The plasticizer reduces the internal barrier to sliding by lubricating the movement of the molecule (Doolittle 1965; Casalini et al. 2000).

According to Arnold Kirkpatrick, plasticizers work as follows: some part of the plasticizer particle is joined, pretty much emphatically to some part of the polymer molecule, more or less strongly, causing it to operate as a solvent, while the unattached sections function as a lubricant among the polymer moiety (Kirkpatrick 1940).

As per Clark, good plasticizing includes filling, by certain measure, the huge gaps in the subatomic space grid, thus prompting the development of planes of easy glide. The plasticizer makes up for the shortfalls among the gliding planes and acts as a lubricant, allowing the planes to glide up each and every one whenever strength is enforced. Clark figured out that plasticized plastic is made up of lateral waves of a base of resin along with a plasticizer. These are essential thoughts encouraging the lubricity hypothesis. The polarities attributed to the polymer base including the plasticizer are thought to be most effective in facilitating the selection of potential plasticizers.

Houwink stressed the emphasis on the plasticizers including polymer polarities, as well as the approach of gliding planes, that Clark had popularized years back. Houwink believed that the first step in plasticization is all about dissolving or swelling. The issues of dissolving, swelling, and plasticization are all firmly intertwined. In many polymers, the filament contains a wide range of homopolar connections between the chain molecules; nevertheless, any sort of bond can exist between the chains. All bond connections between the chains totally relaxed. The link among the filaments slightly collapsed and was somewhat undamaged in the event as a concern to restricted swelling (Marcilla and Beltrãn 2017).

16.4.2 The Gel Theory

Subsequently, Aiken concentrated on the outcomes coming out of the tensile creep test against plasticized PVC including various plasticizers. Aiken attempted a few connections between plasticizers' efficaciousness and softening affinity and molecular framework. He responded that certain polar groups within the plasticizer and also polymer aligned themselves on the PVC chains to generate solvating dipoles. Because nonpolar tails are antagonistic to PVC, they are likely to gather together in small bundles, eliminating a huge quantity of unreliable polar polymer chains, which generate powerful polymer-polymer interactions, from a gel complex, and cause viscous flow. The micro-Brownian motion was most likely caused by the plasticizer surrounding the polymer chains (Marturano et al. 2017). The gel hypothesis in food hydrocolloids relates to an inward, three-dimensional honeycomb conformation formed from joining the macromolecules along their length. Plasticizer particles append and frustrate the powers holding polymer chains together and reduce the force holding polymer-polymer attachments. This diminishes the rigid nature of the gel structure, bringing about expanded gel adaptability. Likewise, plasticizer particles that are not joined to polymer structure collected plasticizer spaces that work with the development of polymer particles. This likewise improves gel adaptability (Doolittle 1965; Casalini et al. 2000). Aiken suggested a couple of prospects as follows:

- A powerful harmony inside the solvation-desolvation phenomenon, in this way, plasticizers are able to diffuse straight into the polymer framework, breaking through polymer association for a time, and afterward drift away, permitting the construction to shut behind it in an alternate position.
- A static component site, polymer-plasticizer interactions bear extended mean lifetimes compared to the time scale of segmental motion.

According to Aiken, no absolutely static picture of the polymer-plasticizer structure is adequate to make sense of the mechanical impacts of the plasticizer.

16.4.3 The Free Volume Theory

The free volume hypothesis began a few years after the lubrication theory and gel theory; at the same time, the development of various qualities of polymers as an element of temperature, thermal expansion coefficients, specific volume, or viscosity was sought to be described. The connections among these analogous upon polymer arrangements, for example, the atomic weight or terminal gathering composition, the existence of additional monomer, and moreover obviously the presence of plasticizers, were also explained (22). Free volume is brought about and supported by the movement of polymer end or side groups. Unplasticized PVC at room temperature (under its glass transition point) acts restricted movements and also impacts the exceptionally tough and inflexible component. When plasticizers are combined with a polymer, free volume is expanded; as a consequence, different developments in the polymer filaments occurred more quickly, bringing down the glass transition (Ramos-Devalle and Gilbert 1990).

The free volume of a crystal, glass, and also fluid might be characterized by the contrast between the volume saw at outright nil temperature and the volume estimated for the genuine crystal, glass, or fluid at a particular temperature (Sears and Darby 1982).

The following formula is used to calculate free volume

$$V_f = V_t - V_o$$

where V_f is the free volume, V_t is the specific volume at temperature T, V_o is the specific volume at reference temperature (e.g., absolute zero).

Increasing the free volume of a polymer system could be done in a variety of ways, like the following (Gracenea et al. 2004; Duda and Zielinski 1996):

- Raising the number of end groups (decreasing the molecular weight)
- Raising the number or length of side chains
- Expanding the opportunity for primary chain development by consideration of sections of less steric obstacle and less intermolecular fascination
- Incorporation of a viable compound with lower atomic weight that can do all abovementioned three ways
- Increasing temperature

16.5 Different Types of Plasticizers

Several varieties of plasticizers are used; some commonly used plasticizers are discussed below (Fig. 16.2) (Ghosh 2015):

1. Phthalates
2. Adipates

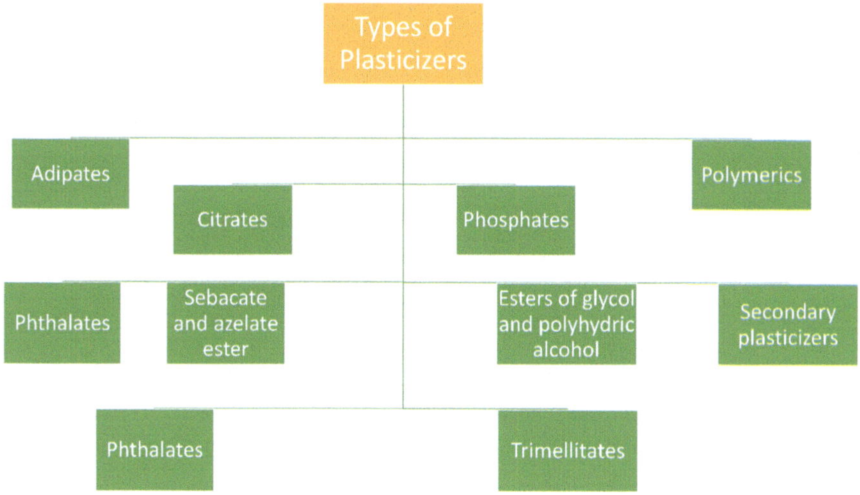

Fig. 16.2 Types of plasticizers

3. Citrates
4. Phosphate
5. Polymerics
6. Esters of glycol and polyhydric alcohol
7. Sebacate and azelate ester
8. Secondary plasticizers
9. Trimellitates

16.5.1 Phthalates

Phthalates are formulated by esterification of phthalic anhydride or phthalic acid acquired through oxidation of ortho-xylene or naphthalene. Phthalates are widely used plasticizers. They conquer around 92% of all plasticizers that are inert, colorless, and oily liquid with high boiling points and are stable for a long period of time. Because of its dissolving nature in organic solvents, such as hexane and oil, phthalates are tragically/acceptably dissolved in various bodily fluids like saliva and blood plasma (Wypych 2017):

(a) Bis(2-ethylhexyl) phthalate (DEHP): Low molecular weight ortho-phthalate
(b) Diisononyl phthalate (DINP) and diisodecyl phthalate (DIDP): High molecular weight ortho-phthalate

Diethylhexyl phthalate and diisononyl phthalate are two of the most regularly used PVC plasticizers. By the chemical structure (Fig. 16.3), it is easy to see that DEHP, which is still the very commonly applied plasticizer in pharmaceutical

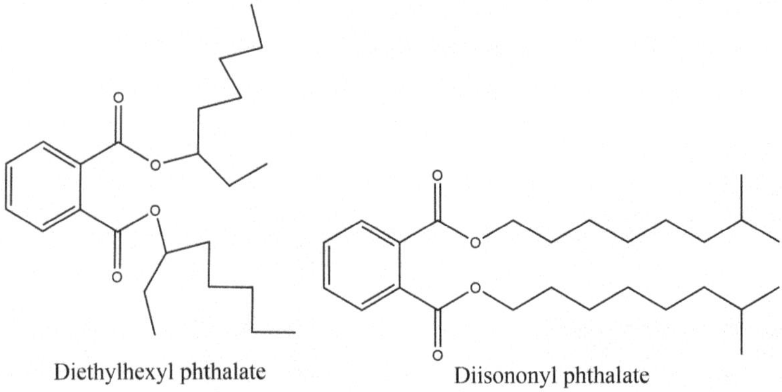

Fig. 16.3 Structure of diethylhexyl phthalate and diisononyl phthalate

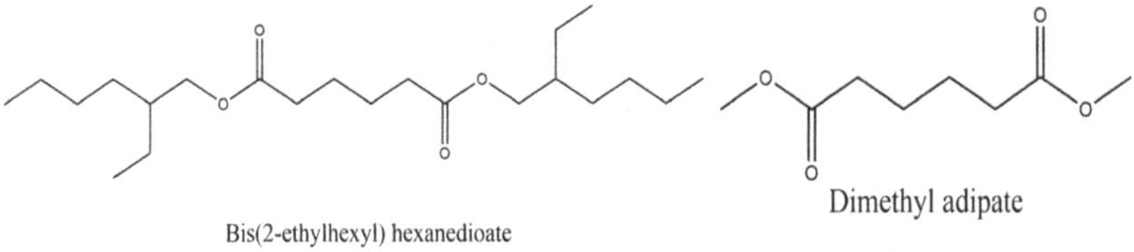

Fig. 16.4 Structure of bis(2-ethylhexyl) hexanedioate and dimethyl adipate

employments, has been replaced by DINP which is made up of longer hydrocarbon chains and has a larger molecular weight; hence it has a lower solubility and migration rate (Peijnenburg 2008; Giamh et al. 1978).

16.5.2 Adipates

Adipates are produced from alcohol; they have high volatility and are effective at a low viscosity. Adipate-based plasticizers are utilized as a primary plasticizer but are typically adopted for insignificant temperature or protection from ultraviolet rays (Fig. 16.4)

Examples are bis(2-ethylhexyl)adipate (DEHA), dioctyl adipate, and dimethyl adipate.

16.5.3 Citrates

Citrates are tetra esters, formed when one mole of citric acid reacts with three moles of alcohol. The lone hydroxyl group in citric acid gets acetylated. They come in the

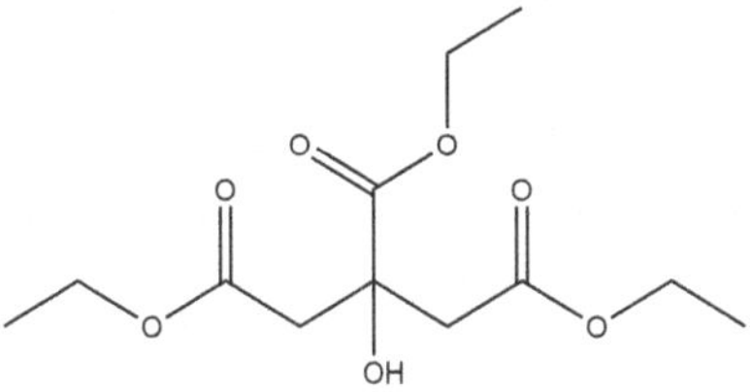

Triethyl citrate

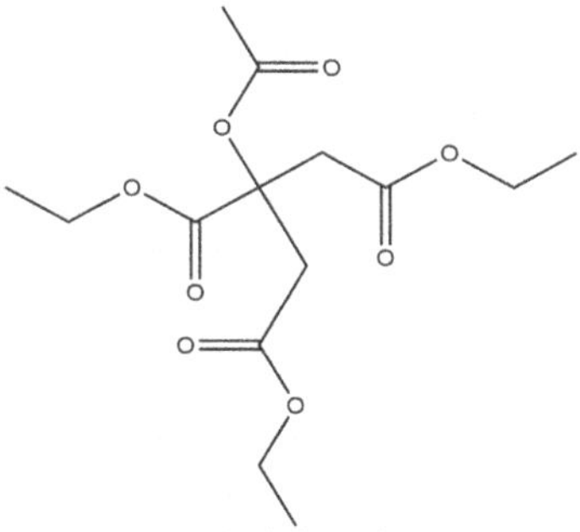

Acetyl triethyl citrate

Fig. 16.5 Structure of triethyl citrate and acetyl triethyle citrate

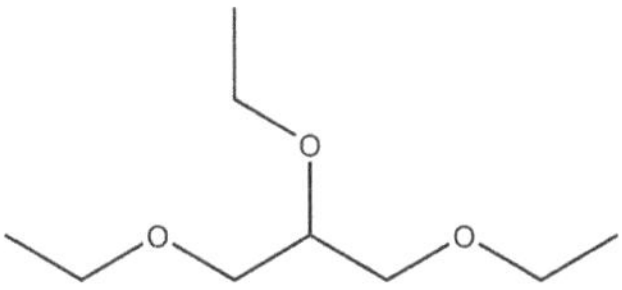

Triethyl glycerin

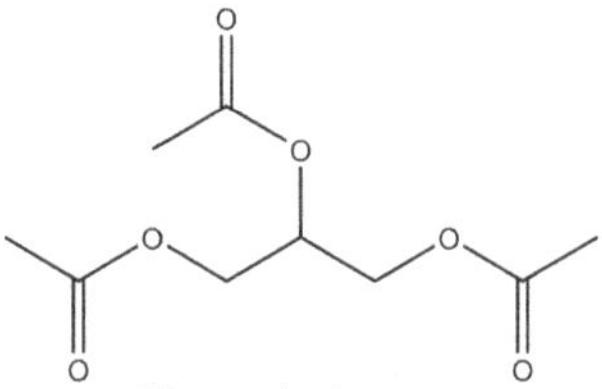

Glycerol triacetate

Fig. 16.6 Structure of triethyl glycerin and glycerol triacetate

form of an oily liquid that is virtually colorless and odorless. They have primarily been employed as plasticizers in pharmaceutical plastics and food association packaging with cellulosic, PVC, polyvinyl acetate, and other polymers (Fig. 16.5). Blood bags and food wraps are two applications for citrate esters. Plasticizers based on citric acid are one of the leading candidates to replace phthalates (Labrecque et al. 1997; Park et al. 2004).

Examples are triethyl citrate, acetyl triethyl citrate, tributyl citrate, acetyl tributyl citrate, etc.

16.5.4 Phosphate

Phosphates show great viability with polyvinyl chloride (PVC). The main advantage of phosphate esters is that they have great low-temperature execution and further developed fire retardancy contrasted relative with phthalate (Fig. 16.6) (Chen et al. 2016).

Examples are triaryl phosphate, alkyl diary phosphate, triethyl glycerin, glycerol triacetate, etc.

Triaryl and alkyl diary phosphates are the main class of fire-resistant phosphate plasticizers utilized alongside PVC, explicitly to accomplish fire retardancy or potentially little smoke production. Phosphates are essential plasticizers concerning PVC and able to be used alone or in a cost-effective combination. Triaryl phosphate has good flame retardancy and low volatility. Alkyl dairy phosphate has great low-temperature adaptability, yet it is highly unpredictable and attempts to lower fire retardancy than the triaryl phosphate. A few phosphates have specified clearance in food and pharmaceutical equipment standards, despite being confined to applications needing increased flame and smoke performance (Rahman and Brazel 2004).

16.5.5 Polymerics

Polymerics are prepared by a reaction between dibasic carboxylic acids and one or more glycols, hence one being with a two-phase framework, with chemically equivalent phases. A two-phase system is formed when a "compatible" plasticizer is added, with one phase being pure crystalline polymer and the other being a compatible amorphous blend. The fact that the molten mix is a single homogenous phase at temperatures above the crystalline melting point justifies calling such blends "compatible." These are produced in various varieties of viscosities. Handling becomes difficult with rising in viscosity (Hammer 1978).

Examples are dmex, 334F polymeric plasticizer, and admex, 523 polymeric plasticizers.

16.5.6 Esters of Glycol and Polyhydric Alcohol

Esters of glycols are glyceryl triacetate and triethyl citrate. Compared to epoxidized polyols aside from glycerol, epoxidized esters of polyols other than glycerol have received very little attention. Because unsaturated acids and a variety of polyols are both relatively affordable, these raw materials might be mingled to make esters suited to the contemplation of making epoxy plasticizers (Hensch and Wilbur 1958).

Polyhydric alcohols are propylene glycol, glycerol, and polyethylene glycol. A composite of polyhydric alcohol and polyethylene (PU) has been successfully developed for use as fabric coating material. The outcome of several polyhydric alcohols close to the characteristics of polyethylene composite was studied. They are soluble in water and utilized in aqueous film coating (Wang et al. 2006) (Fig. 16.7).

16.5.7 Sebacate

Sebacate plasticizers contain sebacate which are esters of sebacic acid. Castor oil, a bio-based raw material, is commonly used to make sebacic acid. They stick to other materials easily; thus they can be combined with steel, glass, metals, and other solid

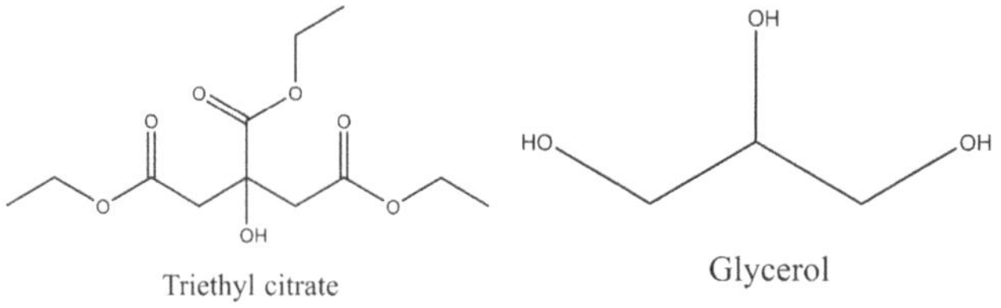

Fig. 16.7 Structure of triethyl citrate and glycerol

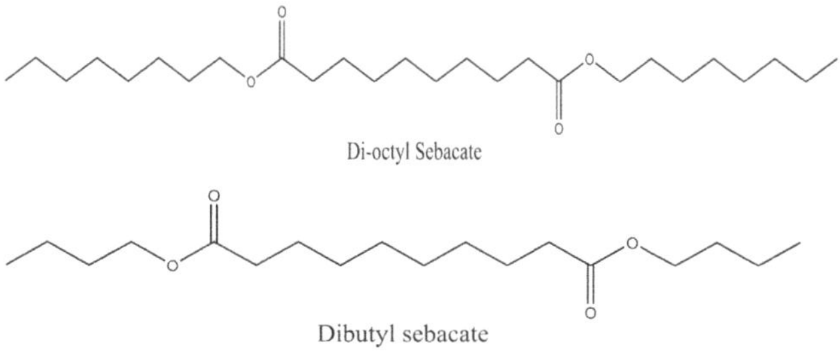

Fig. 16.8 Structure of dibutyl sebacate

materials to create inflexible plastic materials. Sebacate-based plasticizers furnish great similarity along with a scope of plastic components and engineered rubbers (explicitly nitrile rubber and neoprene). They have a great level of fire resistance. When a liquid is added to a polymer, it produced a non-tacky product that has a lower processing temperature and is softened and more flexible than the polymer alone. Even at low temperatures, they are pliable. Adipate and phthalates have similar sorption characteristics (Wypych 2013).

Examples are dibutyl sebacate (DBS), di(2-ethylhexyl)sebacate, and di-octyl Sebacate (DOS)

Dibutyl sebacate (DBS) is an organic material that is the dibutyl ester of sebacic acid. It has excellent compatibility with a variety of plastic materials, excellent low-temperature qualities, and strong oil resistance (Fig. 16.8). Di-octyl sebacate (DOS) is a plasticizer of the ester type. It has been utilized as a standard in PVC compounds that require a high-quality cold-resistant PVC plasticizer with high plasticizing efficiency and minimal color distortion (Müller 2017).

16.5.8 Secondary Plasticizers

Secondary plasticizers are commonly recognized as extenders. Most of these plasticizers contain chlorinated paraffin that are hydrocarbons chlorinated to a degree of 30–70%. The rise in the content of chlorine rises the viscosity and fire retardancy. Due to the presence of epoxy content, they perform as lubricants (Ali et al. 2020). A

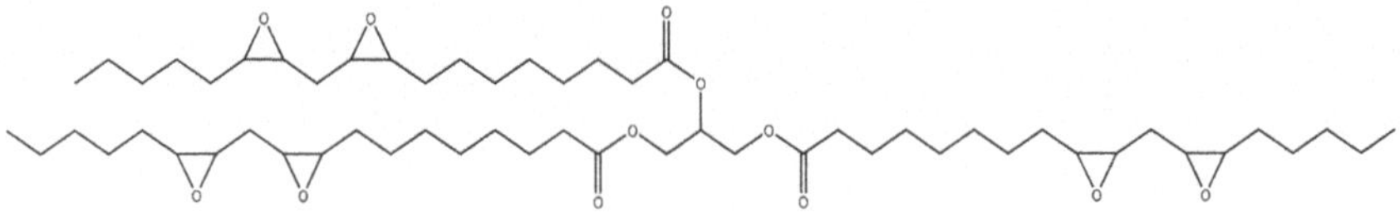

Epoxidized soybean oil (ESBO)

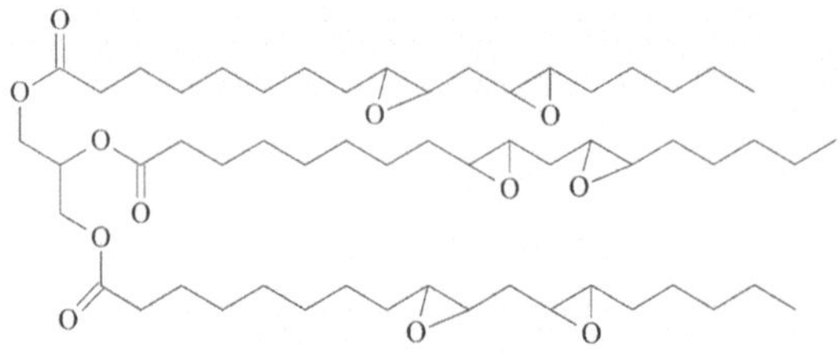

Epoxidized linseed oil (ELO)

Fig. 16.9 Structure of secondary plasticizers

secondary plasticizer ordinarily cannot be utilized as the sole plasticizer in a plasticized polymer. Secondary plasticizers are permitted to have a low polymer affinity or potentially great volatility. They might possibly have functional groups that permit them to solvate the polymer at handling temperatures. Secondary plasticizers come in a variety of forms (Fig. 16.9).

Examples are epoxidized soybean oil (ESBO) and epoxidized linseed oil (ELO).

Epoxidized soybean oil (ESBO) and epoxidized linseed oil (ELO) are utilized to improve the achievement of primary plasticizers (Sastri 2010).

16.5.9 Trimellitates

Trimellitate plasticizers carry reduced volatility, superior extraction resistance, and admirable processability at the same time in comparison to phthalates. Trimellitates are not able to classify as phthalate-free plasticizers because they contain small fragments of phthalates. Trimellitates are utilized in automobile interiors and different operations, where protection from strong temperatures is mandatory. They admit incredibly flat instability (Kambia et al. 2001) (Fig. 16.10).

Examples are tris-(2-ethylhexyl) trimellitate (TOTM), tri-(2-ethylhexyl) trimellitate (TEHTM), tri-(isononyl)trimellitate (TINTM), and tri-(isodecyl)trimellitate (TIDTM).

Plasticizer tris-(2-ethylhexyl) timellitate (TOTM) is an excellent PVC plasticizer for applications that require moderate volatility. Tri-(2-ethylhexyl) trimellitate (TEHTM) is a PVC plasticizer that is handled as a substitute for di-(2-ethylhexyl) phthalate (DEHP) in medical devices (Eckert et al. 2016; Iwase et al. 2014).

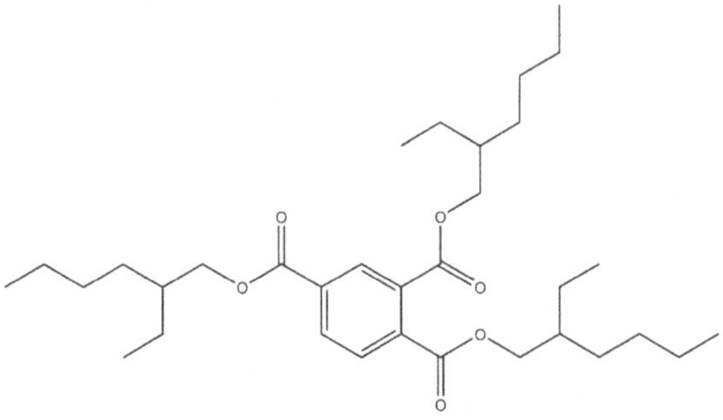

Tris-(2-ethylhexyl)trimellitate

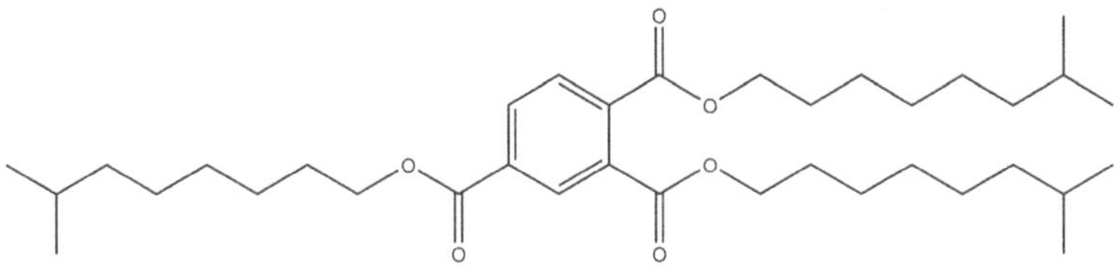

Tri-(isononyl)trimellitate

Fig. 16.10 Structure of trimellitates

16.6 Future Prospects

The food business is perhaps the main business area with extremely different procedures and products. Food packaging is an unavoidable aspect of the food industry and furthermore the quality of raw materials and processing equipment. Food packaging is an important commercial sector in the packaging industry, and like food, it must fulfill high-quality and safety standards, as well as government rules and policies, in a plan to be fortunate in the merchandise (Kim et al. 2014). The food itself, packing materials, stockpiling, transportation, and distribution are all factors to consider when it comes to food wrapping, which is to protect the integrity, preservation, and security of foodstuffs; numerous different perspectives influence the usefulness of the packaging. A few vital viewpoints which offer various open doors for advancement and new arrangements in food packaging are user-friendly and eco-accommodating patterns (Marsh and Bugusu 2007). The developing food industry that actually applies conventional food packaging straightforwardly adds to the increment of packaging waste. However, customers around the world are becoming more conscious that traditional packaging, particularly plastic products, was an established immense harm to the climate, water assets, and the whole environment. As a result, alternative packaging materials that are eco-friendly are needed like never before, in addition to other strategies for decreasing food packaging waste, such as source reduction, recycling, and reusing (Ljungberg 2007).

Polysaccharides are the most common raw materials since they come from the biomass of plants and marine organisms. Microbial fermentation, on the other hand, is a bioprocess that can produce a metabolic product with specific features and functioning. Polysaccharide-based materials may simplify the overall packaging structure due to their protective qualities, which are either inherent or developed with the inclusion of antibacterial, antioxidant plasticizers or other biologically active components. Plasticizers are required to improve the capability of digestible polymers to design useable films and coatings, despite the fact that they make up a minor percentage of edible film formulations. Permeability and mechanical action of films to generate them extra appropriate for food coatings, the number, and variety of plasticizers used are all significant factors to consider.

16.6.1 Application in Food System

16.6.1.1 Edible Food Packaging

Thin film and coatings that can be directly applied to food products or used as food wraps are included in edible food packaging, and they serve a variety of tasks. The materials used to make edible packaging are mostly water-soluble. Hydrophilic plasticizers are used in these types of packaging systems. Furthermore, because it is edible and made from renewable resources, this sort of packaging represents a distinct category of materials from other types of packaging (Ghaani et al. 2016).

16.6.1.2 Intelligent Packaging

The goal of using edible films and coatings is to increase food excellence and serviceable time by decreasing the usage of nonrecyclable synthetic polymer films, which are expensive. The edible coating can preserve food from microorganisms and mechanical corruption, in conjugation with offering an additional exciting presentation for consumers, in addition to preserving it against humidity modification, deterioration, and scent vanish. Food stability can be improved by selecting the right plasticizers for each application. To employ edible films for food pouches, you must first determine the heat-sealing temperature that can provide a reliable heat seal. The initiation temperatures of thermal transitions differed between plasticizers. This is also a significant plasticizer issue, as plasticizers are required for continuous extrusion film production. Finally, as previously stated, sensory characteristics of digestible films and coatings are conditionally researched in order to assure the end user's pleasure (Krochta and De Mulder-Johnston 1997).

16.7 Conclusion

The vast majority of polysaccharide and protein-based films and coatings are fragile. Plasticizers are needed to improve the film's compliance and endurance in order to alleviate this problem. However, as the extent of plasticizer in the coat grows, the permeability of the coat hikes. When a plasticizer is chosen correctly for a certain polymer, the mechanical characteristics of the film can be improved while the permeability of the film is reduced to a minimum. Plasticizer affinity, adaptability, durability, and budget must all be considered when choosing a plasticizer.

Plasticization takes place in the amorphous (increased molecular mobility) area; the capacity of plasticizers to disrupt hydrogen bonding and protein chains is dependent on the number and category of plasticizers used. In general, adding more plasticizers decreases the film glass transition temperature (Tg). Reducing polymer molecular weight, which increases end groups of the polymer chain, reduces intermolecular pressures accompanying polymer chains; all so increasing polymer free volume is another way of incrementation of film flexibility. Different types of plasticizers can be utilized to get the desired action. Combing plasticizer types can make use of each plasticizer's benefits while minimizing its drawbacks.

References

Aider M (2010) Chitosan application for active bio-based film production and potential in the food industry: review. LWT Food Sci Technol 43(6):837–842. https://doi.org/10.1016/j.lwt.2010.01.02

Ali M, Lu Y, Ahmed S, Khanal S, Xu S (2020) Effect of modified cardanol as secondary plasticizer on thermal and mechanical properties of soft polyvinyl chloride. ACS Omega 5(28):17111–17117. https://doi.org/10.1021/acsomega.0c00826

Casalini R, Ngai KL, Robertson CG, Roland CM (2000) α- and β-relaxations in neat and antiplasticized polybutadiene. J Polym Sci 38:1841–1847

Chen J, Liu Z, Li X, Liu P, Jiang J, Nie X (2016) Thermal behavior of epoxidized cardanol diethyl phosphate as novel renewable plasticizer for poly(vinyl chloride). Polym Degrad Stabil 126:58–64. https://doi.org/10.1016/j.polymdegradstab.2016.01.018

Doolittle AK (1954) The technology of solvents and plasticizers. John Wiley & Sons, New York

Doolittle AK (1965) The effects of plasticizers on crystallinity permeability, and mechanical properties of methylcellulose films. In: Bruins PF (ed) Plasticizer technology. Chapman and Hall, London, pp 1–20

Duda JL, Zielinski JM (1996) Free-volume theory. In: Neogi P (ed) Diffusion in polymers. Marcel Dekker, New York, pp 143–171

Eckert E, Münch F, Göen T, Purbojo A, Müller J, Cesnjevar R (2016) Comparative study on the migration of di-2-ethylhexyl phthalate (DEHP) and tri-2-ethylhexyl trimellitate (TOTM) into blood from PVC tubing material of a heart-lung machine. Chemosphere 145:10–16. https://doi.org/10.1016/j.chemosphere.2015.11.067

Erythropel HC, Maric M, Nicell JA et al (2014) Leaching of the plasticizer di(2-ethylhexyl) phthalate (DEHP) from plastic containers and the question of human exposure. Appl Microbiol Biotechnol 98(24):9967–9981

Espitia PJP, Du W-X, de Jesús Avena-Bustillos R, de Fátima Ferreira Soares N, McHugh TH (2014) Edible films from pectin: physical-mechanical and antimicrobial properties—a review. Food Hydrocoll 35:287–296. https://doi.org/10.1016/j.foodhyd.2013.06.005

Ghaani M, Cozzolino CA, Castelli G, Farris S (2016) An overview of the intelligent packaging technologies in the food sector. Trends Food Sci Technol 51:1–11. https://doi.org/10.1016/j.tifs.2016.02.008

Ghosh DP (2015) Plasticizers in cosmetic technology. Plasticizer class. https://www.slideshare.net/vedg007/plasticizer-class-ppt

Giamh CS, Chan S, Atlas EL (1978) Phthalate ester plasticizers: a new class of marine pollutant. Science 199(4327):419–421. https://doi.org/10.1126/science.413194

Gracenea JJ, Saura J, Suay J (2004) The influence of the critical pigment volume concentration (CPVC) on the properties of an epoxy coating: Part II. Anticorrosion and economic properties. Prog Org Coat 50:68–74. https://doi.org/10.1016/j.porgcoat.2003.10.014

Hammer CF (1978) Polymeric plasticizers. In: Paul DR, Newman S (eds) Polymer blends. Academic Press, London, pp 219–241. https://doi.org/10.1016/B978-0-12-546802-2.50013-0

Hensch EJ, Wilbur AG (1958) Epoxidized ester of glycols and pentaerythritol: application as plasticizers for poly(vinyl chloride). Ind Eng Chem 50(6):871–872. https://doi.org/10.1021/ie50582a027

Iwase H, Oiso S, Kariyazono H, Nakamura K (2014) Biological effects of the plasticizer Tris (2-Ethylhexyl) Trimellitate. Clin Pharmacol Biopharm S2:004. https://doi.org/10.4172/2167-065X.S2-004

Kambia K, Dine T, Azar R, Gressier B, Luyckx M, Brunet C (2001) Comparative study of the leachability of di(2-ethylhexyl) phthalate and tri(2-ethylhexyl) trimellitate from haemodialysis tubing. Int J Pharm 229(1–2):139–146. https://doi.org/10.1016/S0378-5173(01)00840-7

Kim Y, Min B, Kim KW (2014) General characteristics of packaging materials for food system. In: Han JH (ed) Innovations in food packaging. Academic Press, London, pp 13–35. https://doi.org/10.1016/B978-0-12-394601-0.00002-3

Kirkpatrick A (1940) Some relations between molecular structure and plasticizing effect. J Appl Phys 11(4):255–261. https://doi.org/10.1063/1.1712768

Krauskopf LG, Godwin A (2005) Plasticizers. In: Wilkes CE, Daniels CA, Summers JW (eds) PVC handbook. Hanser, Munich, pp 173–193

Krochta JM, De Mulder-Johnston C (1997) Edible and biodegradable polymer films: challenges and opportunities. Food Technol 51:61–74

Labrecque LV et al (1997) Citrate esters as plasticizers for poly (lactic acid). J Appl Polym Sci 66(8):1507–1513. https://doi.org/10.1002/(SICI)1097-4628(19971121)66:8%3C1507::AID-APP11%3E3.0.CO;2-0

Ljungberg LY (2007) Materials selection and design for development of sustainable products. Mater Des 28(2):466–479. https://doi.org/10.1016/j.matdes.2005.09.006

Maeda Y, Paul DRJ (1987) Effect of antiplasticization on gas sorption and transport. III. Free volume interpretation. Polym Sci B Polym Phys 25:957–980. https://doi.org/10.1002/polb.1987.090250503

Mahajan BVC, Tandon R, Kapoor S, Sidhu MK (2018) Natural coatings for shelf-life enhancement and quality maintenance of fresh fruits and vegetables—a review. J Postharvest Technol 6:12–26

Marcilla A, Beltrãn M (2017) Mechanisms of plasticizers action. In: Wypych G (ed) Handbook of plasticizers, 3rd edn. ChemTec Publishing, Alicante, pp 119–134. https://doi.org/10.1016/B978-1-895198-97-3.50007-X

Marsh K, Bugusu B (2007) Food packaging—roles, materials, and environmental issues. J Food Sci 72(3):R39–R55. https://doi.org/10.1111/j.1750-3841.2007.00301.x

Marturano V, Cerruti P, Ambrogi V (2017) Polymer additives. Phys Sci Rev 2(6):20160130. https://doi.org/10.1515/psr-2016-0130

Müller C (2017) A viable and sustainable alternative to dioctyl sebacate plasticizer. https://www.pcimag.com/articles/103867-a-viable-and-sustainable-alternative-to-dioctyl-sebacate-plasticizer

Park H-M, Misra M, Drzal LT, Mohanty AK (2004) "Green" nanocomposites from cellulose acetate bioplastic and clay: effect of eco-friendly triethyl citrate plasticizer. Biomacromolecules 5(6):2281–2288. https://doi.org/10.1021/bm049690f

Peijnenburg WJGM (2008) Phthalates. In: Jargensen SE, Fath BD (eds) Encyclopedia of ecology. Academic Press, London, pp 2733–2738. https://doi.org/10.1016/B978-008045405-4.00419-5

Porter S, Sackett G, Liu L (2009) Development, optimization, and scale-up of process parameters: pan coating. In: Qiu Y, Chen Y, Zhang GGZ, Liu L, Porter WR (eds) Developing solid oral dosage forms. Academic Press, London, pp 761–805. https://doi.org/10.1016/B978-0-444-53242-8.00033-3

Rahman M, Brazel CS (2004) The plasticizer market: an assessment of traditional plasticizers and research trends to meet new challenges. Prog Polym Sci 29(12):1223–1248. https://doi.org/10.1016/j.progpolymsci.2004.10.001

Ramos-Devalle L, Gilbert M (1990) PVC/plasticizer compatibility: evaluation and its relation to processing. J Vinyl Technol 12(4):222–225. https://doi.org/10.1002/vnl.730120409

Sastri VR (2010) Polymer additives used to enhance material properties for medical device applications. In: Plastics in medical devices, Plastics design library. William Andrew Publishing, London, pp 55–72. https://doi.org/10.1016/B978-0-8155-2027-6.10005-4

Sears JK, Darby JR (1982) Mechanism of plasticizer action. In: Sears JK, Darby JR (eds) The technology of plasticizers. Wiley-Interscience, New York, pp 35–77

Sejidov FT, Mansoori Y, Goodarzi N (2005) Esterification reaction using solid heterogeneous acid catalysts under solvent-less condition. J Mol Catal A Chem 240(1–2):186–190. https://doi.org/10.1016/j.molcata.2005.06.048

Somwanshi SB et al (2011) Effects of drug-polymer ratio and plasticizer concentration on the release of metoprolol polymeric microspheres. Int J Pharm Res Dev 3(3):139–146

Valencia-Chamorro SA et al (2011) Antimicrobial edible films and coatings for fresh and minimally processed fruits and vegetables: a review. Crit Rev Food Sci Nutr 51(9):872–900. https://doi.org/10.1080/10408398.2010.485705

Vieira MGA, da Silva MA, dos Santos LO, Beppu MM (2011) Natural-based plasticizers and biopolymer films: a review. Eur Polym J 47(3):254e263. https://doi.org/10.1016/j.eurpolymj.2010.12.011

Wang S, Sue H-J, Jane J (2006) Effects of polyhydric alcohols on the mechanical properties of soy protein plastics. J Macromol Sci A 33:557–569. https://doi.org/10.1080/10601329608010878

Wypych A (2013) Sebacates. In: Wypych A (ed) Plasticizers databook. Elsevier, London, pp 557–585. https://doi.org/10.1016/B978-1-895198-58-4.50030-4

Wypych G (2017) Health and safety issues with plasticizers and plasticized materials. In: Handbook of plasticizers, 3rd edn. ChemTec Publishing, Alicante, pp 681–743. https://doi.org/10.1016/B978-1-895198-97-3.50019-6

Chapter 17
Challenges in Commercialization of Polysaccharide-Based Films

Aasima Rafiq, Abida Jabeen, Mariya Nayeem, Bushra Manzoor, and Tawheed Amin

17.1 Introduction to the Challenges in Commercialization of Polysaccharide-Based Films

Polysaccharides are biodegradable natural polymers that are commonly utilized to make packaging material due to their extensive renewability, availability in nature, film-forming ability, biocompatibility and cost-effectiveness. They typically form hydrogen-bonded network and thus prevent oxygen penetration. Plants provide starch and cellulose materials, but starch has a significant disadvantage due to its water loving nature, which results in bad water barrier and mechanical capabilities (Souza et al. 2012). Chitosan, an amino-polysaccharide compound, is yet another important polysaccharide substance with intrinsic antibacterial properties making it ideal for use in disposable packing films. Films made out of a solo polymer fail to meet the standards for food packaging. As a result, much research has been conducted to improve the poor qualities of a single polymer film employing other variables like cross-linking agents, nano-composites, etc.

A. Rafiq (✉)
Sher-e-Kashmir University of Agricultural Sciences and Technology of Kashmir, Shalimar, Srinagar, Jammu and Kashmir, India

Krishi Vigyan Kendra Srinagar SKUAST-K, Srinagar, India

A. Jabeen · M. Nayeem · B. Manzoor · T. Amin
Sher-e-Kashmir University of Agricultural Sciences and Technology of Kashmir, Shalimar, Srinagar, Jammu and Kashmir, India

T. Amin et al. (eds.), *Polysaccharide Based Films for Food Packaging: Fundamentals, Properties and Applications*,
https://doi.org/10.1007/978-981-99-4898-7_17

17.1.1 Starch

Native starches have restricted application in industries due to a variety of drawbacks, such as restricted digestion, low shear resistance, poor solubility, excessive viscosity, excessive syneresis and retrogradation proneness. As a result, starches are commonly treated to various physical, chemical and enzymatic treatments in order to provide required qualities (Santana and Angela 2014). Yet, starch films are very hydrophilic, with excessive water vapour solubility and permeability and low mechanical strength. Plasticizers, nanomaterials and biopolymer additives can all be used to enhance the barrier and mechanical characteristics of starch films. On this basis, starch bio-nanocomposite films are said to have improved properties, but there is still a lot of room for extensive research in this direction.

17.1.2 Cellulose

Cellulose, existing in the cell walls of microorganisms and plants, is the most prevalent organic and polymer source that is renewable. Hydroxypropyl methylcellulose (HPMC), carboxymethyl cellulose (CMC), and methyl cellulose (MC) are cellulose ether derivatives that are water soluble and therefore extensively utilized in the packaging industry. Despite their outstanding film-forming and gas barrier capabilities, cellulose ether derivatives' applicability in the sector of food packaging has been constrained because of their high water solubility, water barrier property, swelling ability and weak mechanical properties. This drawback can again be potentially sorted with the addition of cross-linking and nanomaterials.

17.1.3 Chitosan

Chitin is a nitrogenous linear polymer found in crustacean exoskeletons, insect cell walls and fungal cell walls. Chitosan is made by deacetylating chitin in an alkaline environment. Degree of deacetylation of chitosan and molecular weight determines its solubility (Cazón et al. 2017). It is nontoxic, biodegradable, antifungal and antibacterial polysaccharide. It has low oxygen permeability and excellent film-forming properties and acts as a barrier to water vapour (Aguirre-Loredo et al. 2016). It cannot, however, be extruded or moulded like other thermoplastic polymers. Furthermore, the films exhibit weak water barrier qualities and are resistant to heat sealing. All of these flaws render chitosan film unsuitable for marketable food packaging. Therefore, water barrier properties of chitosan should be improved.

17.2 Improving Performance and Lowering Costs

One of the most recurring difficulties in the commercialization of polysaccharide-based films is the production cost. Also, the performance of various synthetic films as packaging material is still considered to be better than the polysaccharide ones in terms of some functional properties like the CO_2/O_2 permeability, water vapour absorption, etc. Therefore, polysaccharide films composed of single polymer cannot be used, and an addition of cross-linking agents, nano-composites and blends of other polymeric compounds is required to improve performance. As a result, a hike in the cost of production is inevitable for the commercialization of these films. However, a technological advancement is applicable only when the cost of production is minimized or optimized to the maximum. This is why constraints can be observed in the application and development of polysaccharide-based films in the food packaging industry.

The leap from laboratory to industry for these films on a large scale is only possible through cost optimization. Extensive research in this regard is being prioritized by the researchers' world over. A range of inexpensive additives required for the manipulation of various necessary properties has become the modus operandi in recent years. For instance, citric acid may be utilized as both a plasticizer and as a cross-linker in many polysaccharide films. Because of the creation of an ester bond among the hydroxyl groups of the polysaccharide and carboxyl groups of citric acid, the oxygen barrier and mechanical capabilities of the films improve drastically (Shi et al. 2008); the inclusion of sorbic acid or potassium sorbate to films can improve their gas barrier properties.

Further research, however, is required not only in the improvement of the film compositions and operating conditions but also in the production cost optimization to increase commercialization of polysaccharide-based films in the food packaging industry.

17.3 Opportunities in Commercialization of Polysaccharide-Based Films

As there is a growing interest of consumer in biodegradable packaging materials made from biopolymers, thus industry and researchers are always trying to improve the commercialization of polysaccharide-based films through various means:

1. The most advantageous features of polysaccharide-based edible films are their eatability and inherent biodegradability (Guilbert et al. 1996; Krochta 2002). The biodegradability of these films makes them environmentally friendly. This is one of the important points for commercialization of edible films as trend of green plastics is growing and companies demand packaging material which is eco-friendly.

2. Polysaccharide-based food coatings and films are potential enhancement systems of food quality, shelf life, safety and functionality. They can be used as individual packaging materials, food coating materials and active ingredient carriers and to separate the compartments of heterogeneous ingredients within foods. The active ingredient carrier function is very useful for the addition of quality preserving agents as well as nutrients and nutraceuticals, resulting in an upgraded quality level of products such as coloured/flavoured confectionary, glazed bakery, flavoured nuts and vitamin-enriched rice.
3. Edible coatings and films offer numerous convenience advantages. Augmented surface strength of fragile products makes handling easier. Coated fruits and vegetables have much higher resistance against bruising and tissue damage caused by physical impact and vibration. Besides this protective function for foods, edible films and coatings are utilized in the food and pharmaceutical industries to develop single-dose pre-measured pouches of food ingredients and drugs, as well as to mask the eventual undesirable taste of medicines (Gennadios and Weller 1990).
4. Important roles of edible coatings and films are quality enhancement and maintenance. They can retard surface dehydration, moisture absorption, oxidation of ingredients, aroma loss, frying oil absorption, ripening, ageing and microbial deterioration of food products. In addition to the physical and chemical quality enhancement, edible films and coatings contribute to visual quality, surface smoothness, flavour carriage, edible colour print and other marketing-related quality factors. Edible coatings maintain the quality of foods even after the package is opened (Krochta 1997).
5. Polysaccharide-based edible films like other edible films are discrete forms, created and applied for specific packaging uses. They can be used to maintain separation of various components within a single food product to inhibit migration of moisture, oxygen, CO_2, aromas, lipids, etc. The use of edible films and coatings for separating different components in composite foods can improve the standard of a food commodity (Krochta and De Mulder-Johnston 1997) and offers a great opportunity for commercialization.
6. Polysaccharide-based commercialized edible coatings and films are rare as preformed self-supporting materials but as pre-mix ingredients. This makes it very difficult to estimate the demand in the market, especially when this is not significant compared to demand for other specialty ingredients and additives.
7. Distributors of gums, cellulose and derivatives and other materials like food-texturing agents, food-grade surfactants, waxes and lipid derivatives are common in this sector. Recently, lipid materials specifically designed to be used as edible water barriers have been proposed for commercial use. Most of them are composed of specifically designed acetoglycerides, either alone or mixed with waxes.
8. The market for biodegradable plastic materials is currently expanding, but it remains a niche market. Eighty-seven percent of biodegradable materials will result from renewable resources. This market still has great diversity in producers and materials, but attempts are being made to decrease this variety.

9. Interdependent relationship among different polysaccharide/gum mixtures have gotten a lot of interest in the food sector as they can strengthen mechanical, barrier and rheological properties of products (Bak and Yoo 2018).
10. Availability of feasible technologies to improve the mechanical or physico-chemical properties of films by application of extrusion or promotion of reaction between film components is crucial. Physical (heating, irradiation, pressure, ultrasound), chemical (chemical cross-linking, acids, alkali) or enzymatic treatments can improve functional properties of films. Thus to satisfy the new production systems, the composition of film-forming materials and the film-forming mechanisms can be modified as per need of the company or product.
11. The polysaccharide-based edible films are more reliable than protein films because they are generally brittle and susceptible to cracking due to the strong cohesive energy density of the polymers (Lim et al. 2002). Moreover, the highly hydrophobic nature of protein-based film limits their ability to provide desired edible film functions.
12. Plasticizers could enhance flexibility, lower brittleness and have a significant impact on the film's physicochemical qualities (Yang and Paulson 2000). Water is the "natural" plasticizer for most carbohydrates; however, it is unstable (Ashikin et al. 2010). Furthermore, stable plasticizers such as glycerol were utilized in edible films to improve the films' physicochemical qualities (Paolicelli et al. 2018), sorbitol (Gao et al. 2017), polyethylene glycol (PEG) (Pongjanyakul and Puttipipatkhachorn 2007), xylitol, mannitol (Santana and Kieckbusch 2013) or low molecular weight sugar (Olivas and Barbosa-Cánovas 2008). Plasticizers with a variety of water-loving hydroxyl groups are particularly suitable with the polysaccharides that make up the film network, improving the films' mechanical and barrier properties (García et al. 2000).
13. While biopolymers have the potential to partially replace the existing plastics, their employment in the packaging industry is limited because of their high water vapour permeability, fragility and water solubility or gas permeability, among other factors (Mu et al. 2012). Addition of nanoparticles of different hydrocolloids to the edible films improves their mechanical characteristics and water vapour barrier emphasizing the significance of sub-micron size of particles, which tends to increase the components' surface contact that engages with the film matrix to change parameters like tensile strength and elongation at breaking, trying to make the films tougher and enhance their ability to expand without tearing.
14. The potential paybacks of polysaccharide-based edible films or any other polymer films as carriers of flavours, antimicrobial agents, probiotics, colouring agents, antioxidants, vitamins and nutraceutical and novel techniques to increase edible film delivery qualities can further increase the opportunities for market potential of edible films.
15. Availability of modern analytical equipments provides opportunities for checking the safety concerns that are related to the allergenicity and potential toxicity of the polysaccharide-based edible films incorporated with some new material and microflora alteration of the packaged food with coating. Thus, it

provides edible packaging companies to follow the regulatory issues and enhance the commercialization of films for a variety of food products.

16. Specific directives needed for unfastening the packages, eating the packaged foods that are coated and discarding the old packaging substances should be mentioned on package material. Thus, these help the consumer to make use of edible films in more reliable manner.
17. Novel technologies like ohmic heating, pulsed electric field, etc. can be used for enhancement of edible film for wide commercialization. As the mass transfer of gas is primarily a function of the amorphous phase, because the crystal line phase is usually assumed to be impermeable, also a high degree of crystallinity is directly related to good mechanical properties. Application of novel techniques enhance the crystallinity of the polymers that include the structural regularity of the polymer chains, polymer chain mobility (which allows variable conformation), the repeated presence of side chains (which engage in intermolecular bonding) and the absence of bulky side chains (which interfere with the crystal lattice formation) (Souza et al. 2010).

In general, a continued growth of processing and functionality is needed in order to overcome the commercial breakthrough, which is currently being faced by the edible film manufacturers.

17.4 Conclusion

Biopolymer-based eco-friendly packaging materials are becoming increasingly popular. Consumer urge for high-quality meals, health issues and environmental worries about the dumping of nonrenewable food containers all contribute to this interest. Food goods benefit from biodegradable films and coatings because they protect them against biological, chemical and physical deterioration. Biodegradable coatings and films can easily enhance food product physical strength, eliminate particle clumping and improve aesthetic and tactile aspects on product surfaces. As per current studies on the implementation of biodegradable biofilms in the food packaging sector, polysaccharide-based films are the most viable. Needs for various biofilm applications in the food packaging sector have prompted researchers to study several origins of polysaccharides because of their distinct features. To accommodate the polysaccharide, particularly starch, in order to make the required film, several materials can be used, with the goal of producing films with acceptable qualities that are free of dangerous substances. Using different polymers in films to boost antibacterial impact is another way to adjust mechanical characteristics and permeability of polysaccharides. Current research is focused on improving polysaccharide films by employing unique approaches that can improve the characteristics of these films.

References

Aguirre-Loredo RY, Rodríguez-Hernández AI, Morales-Sánchez E, Gómez-Aldapa CA, Velazquez G (2016) Effect of equilibrium moisture content on barrier, mechanical and thermal properties of chitosan films. Food Chem 196:560–566

Ashikin WHNS, Wong TW, Law CL (2010) Plasticity of hot air-dried mannuronate- and guluronate-rich alginate films. Carbohydr Polym 81:104–113

Bak JH, Yoo B (2018) Intrinsic viscosity of binary gum mixtures with xanthan gum and guar gum: effect of NaCl, sucrose, and Ph. Int J Biol Macromol 111:77–81

Cazón P, Velazquez G, Ramírez JA, Vázquez M (2017) Polysaccharide-based films and coatings for food packaging: a review. Food Hydrocoll 68:136–148

Gao C, Pollet E, Avérous L (2017) Innovative plasticized alginate obtained by thermo-mechanical mixing: effect of different biobased polyols systems. Carbohydr Polym 157:669–676

García MA, Martino MN, Zaritzky NE (2000) Microstructural characterization of plasticized starch-based films. Starch Stärke 52:118–124

Gennadios A, Weller CL (1990) Edible films and coatings from wheat and corn proteins. Food Technol 44(10):63–69

Guilbert S, Gontard N, Gorris LGM (1996) Prolongation of the shelf life of perishable food products using biodegradable films and coatings. LWT Food Sci Technol 29:10–17

Krochta JM (1997) Edible protein films and coatings. In: Damodaran S, Paraf A (eds) Food proteins and their applications. Marcel Dekker, New York, pp 529–549

Krochta JM (2002) Proteins as raw materials for films and coatings: definitions, current status, and opportunities. In: Gennadios A (ed) Protein-based films and coatings. CRC Press, Boca Raton, pp 1–41

Krochta JM, De Mulder-Johnston C (1997) Edible and biodegradable polymer films: challenges and opportunities. Food Technol 51(2):61–74

Lim LT, Mine Y, Britt IJ, Tung MA (2002) Formation and properties of egg white protein films and coatings. In: Gennadios A (ed) Proteins-based films and coatings. CRC, Boca Raton, pp 233–252

Mu CD, Guo JM, Li XY, Lin W, Li DF (2012) Preparation and properties of dialdehyde carboxymethyl cellulose crosslinked gelatin edible films. Food Hydrocoll 27:22–29

Olivas GI, Barbosa-Cánovas GV (2008) Alginate–calcium films: water vapor permeability and mechanical properties as affected by plasticizer and relative humidity. LWT Food Sci Technol 41:359–366

Paolicelli P, Petralito S, Varani G, Nardoni M, Pacelli S, Di Muzio L, Tirillo J, Bartuli C, Cesa S, Casadei MA (2018) Effect of glycerol on the physical and mechanical properties of thin gellan gum films for oral drug delivery. Int J Pharm 547:226–234

Pongjanyakul T, Puttipipatkhachorn S (2007) Alginate-magnesium aluminum silicate films: effect of plasticizers on film properties, drug permeation and drug release from coated tablets. Int J Pharm 333:34–44

Santana A, Angela M (2014) New starches are the trend for industry applications: a review. Food Public Health 4:229–241

Santana AA, Kieckbusch TG (2013) Physical evaluation of biodegradable films of calcium alginate plasticized with polyols. Braz J Chem Eng 30:835–845

Shi R, Bi J, Zhang Z, Zhu A, Chen D, Zhou X, Zhang L, Tian W (2008) The effect of citric acid on the structural properties and cytotoxicity of the polyvinyl alcohol/starch films when molding at high temperature. Carbohydr Polym 74(4):763–770

Souza BWS, Cerqueira MA, Teixeira JA, Vicente AA (2010) The use of electric fields for edible coatings and films development and production: a review. Food Eng Rev 2:244–255

Souza AC, Benze RFES, Ferrão ES, Ditchfield C, Coelho ACV, Tadini CC (2012) Cassava starch biodegradable films: influence of glycerol and clay nanoparticles content on tensile and barrier properties and glass transition temperature. LWT Food Sci Technol 46(1):110–117

Yang L, Paulson AT (2000) Mechanical and water vapour barrier properties of edible gellan films. Food Res Int 33:563–570

SPRINGER NATURE

GPSR Compliance

The European Union's (EU) General Product Safety Regulation (GPSR) is a set of rules that requires consumer products to be safe and our obligations to ensure this.

If you have any concerns about our products, you can contact us on ProductSafety@springernature.com

In case Publisher is established outside the EU, the EU authorized representative is:

Springer Nature Customer Service Center GmbH
Europaplatz 3
69115 Heidelberg, Germany

The manufacturer's authorised representative in the EU is Springer Nature Customer Service Centre GmbH, Europaplatz 3, 69115 Heidelberg, Germany. If you have any concerns regarding our products, please contact ProductSafety@springernature.com

Printed and bound by CPI Group (UK) Ltd, Croydon, CR0 4YY
01/10/2026
02225863-0013